OUR ENVIRONMENT
AN INTRODUCTION TO PHYSICAL GEOGRAPHY

THIRD EDITION

OUR ENVIRONMENT

AN INTRODUCTION TO PHYSICAL GEOGRAPHY

DONALD KEITH FELLOWS
Los Angeles Mission College

JOHN WILEY & SONS
New York • Chichester • Brisbane • Toronto • Singapore

Cover: Watercolor by **Albert Drogin**

Copyright © 1975, 1980, 1985 by John Wiley & Sons, Inc.

All rights reserved. Published simultaneously in Canada.

Reproduction or translation of any part of this work beyond that permitted by Sections 107 and 108 of the 1976 United States Copyright Act without the permission of the copyright owner is unlawful. Requests for permission or further information should be addressed to the Permissions Department, John Wiley & Sons.

Library of Congress Cataloging in Publication Data:
Fellows, Donald Keith.
   Our environment.

   Includes bibliographies and index.
   1. Physical geography. I. Title.
GB55.F36 1985   551   84-22165
ISBN 0-471-88193-7

Printed in the United States of America

10 9 8 7 6 5 4

ABOUT THE AUTHOR

Donald K. Fellows received his bachelor's degree at the University of California at Los Angeles and his master's degree from California State University, Northridge. He is widely traveled, having covered much of the Pacific Ocean area during and after World War II; the Soviet Union, from European Russia to Siberia, Central Asia, and the region south of the Caucasus; Great Britain and Ireland; Western Europe and Scandinavia; parts of Mexico and Canada; Alaska; Peoples Republic of China; South Africa; Kenya and Tanzania; Greece, Egypt, and Israel; Fiji, New Zealand, Australia, Tahiti, India and Nepal; and most of the United States. His systematic interests in geography are in the human-cultural fields and their intimate relationship with the physical environment. He has writen *A Mosaic of America's Ethnic Minorities* (published by John Wiley), which is a study of America's most visible minorities. He is a member of the Association of American Geographers, American Geographical Society, National Council for Geographic Education, Association of Pacific Coast Geographers, National Geographic Society, California Geographical Society, and the Los Angeles Geographical Society.

PREFACE

This approach to the study of our physical environment is not intended to be the final, analytical, theoretical, and all-inclusive answer. No attempt is made to boggle the student's mind with intricate or overly technical terminology, nor is the assumption made that the student is incapable of understanding scientific matters. Instead, the reality and the excitement of our planet are explained in a geographic language that any nonprofessional can understand and enjoy.

I ask the following questions of those of you who have not planned careers in the physical sciences, who are aiming for other professional or vocational fields, and who may be wondering how this course will be of value to you. Are you concerned with the quality of your physical and cultural environment? Do you have a clear understanding of the role played by our earth in the great system we call our universe? Do you believe that the energy "crisis" now facing all of us concerns you? Can you as an individual do anything to help alleviate the problems of overuse, misuse, and a general lack of conservation of our natural energy resources—coal, petroleum, and natural gas? Would you like to learn about alternative energy resources that can replace those that are declining to the point of no return?

To make you a part of the answer, this course will teach you about the intimate relationship of the sun to the earth, and how all the elements within this system—climates, vegetation, water, soils, and the sometimes awe-inspiring surface features—affect one another.

We begin with an explanation of the earth system as an integral part of the universe, and then describe the true shape of the earth, how it is portrayed on globes and maps, and its role as a "spaceship" orbiting through the universe, powered by energy from the sun. Two case studies, *Thinking Metric* and *Solar Energy: Friend or Foe?*, are included. Then we focus on the importance of the atmosphere, the differences in air temperatures and pressures across the earth, local winds and wind systems, moisture, clouds, and storms, the earth's water supply, and the differing climates of the world and their influence on vegetation, soils, and human endeavors. Three case studies, *Geothermal Energy, Energy from the Oceans,* and *Biofuel—Energy from Vegetation,* are included.

Next we begin to consider the earth's structure by examining the earth's crust and its deep interior, and the powerful forces that uplift mountains, cause the continents to move, and make the earth's land surface and sea bottom so interesting. One case study, *Nuclear Energy—Fission or Fusion?*, is included. Finally, we take a close look at the earth's sculptured face and explain how the dynamic moving forces within the earth are countered on the surface by the downgrading, leveling processes of gravity, rain, and the different forms of running water, glaciers, oceans, and winds.

The epilogue, "The Human Environment—Tomorrow?," takes a second look at the system, of which we are an important part, and offers some suggestions about how the information learned can be put to positive use in the effort to conserve energy and preserve our natural physical environment. There are four appendices: Appendix A, "Time Zones"; Appendix B, "Topographic Maps"; Appendix C, "Ground Referencing Systems"; and Appendix D, "The Metric System—How to Use It."

To further the learning process, there are several questions at the end of each chapter, based on the material covered. Students should answer these questions before proceeding to the next section. There are selected citations and suggested additional readings throughout the book; they are intended for the students who desire to expand their knowledge beyond these pages.

I would like to express my appreciation to those who have aided and encouraged me during the revision of this book: Patricia Chapla, Butte Community College, California; Richard Dorf and Janice Hamrin, Energy Extension Service, University of California, Davis; Susan Scott, California Energy Commission; Arnold Court, California State University, Northridge; Mark Powell, Los Angeles Pierce College; Richard Raskoff, Los Angeles Valley College; William Russell, Los Angeles City College; and my students.

In addition, I would like to thank various people associated with John Wiley & Sons: Katie Vignery (Geography Editor), David Smith (Production Supervisor), Linda Gutierrez (Photographic Editor), and the Illustration Department.

Donald Keith Fellows

CONTENTS

1 ON THE THRESHOLD 1
EARTH AS A SYSTEM 2
EARLY THEORIES OF EARTH'S SYSTEMS 5
HUMANITY AND NATURE 7
CASE STUDY: THINKING METRIC 9
REVIEW AND DISCUSSION 10
NOTES 10
ADDITIONAL READINGS 10

2 PORTRAYING THE EARTH 13
EARTH'S SHAPE: ANCIENT CONCEPTS 14
EARTH AS A SPHERE 15
MAPS 27
REVIEW AND DISCUSSION 48
NOTES 49
ADDITIONAL READINGS 49

3 SPACESHIP EARTH 51
EARTH AND THE SOLAR SYSTEM 52
EARTH'S POWER PLANT: SOLAR ENERGY 64
CASE STUDY: SOLAR ENERGY—FRIEND OR FOE? 81
REVIEW AND DISCUSSION 85
NOTE 86
ADDITIONAL READING 86

4 EARTH'S DYNAMIC ATMOSPHERE 89
COMPOSITION OF EARTH'S ATMOSPHERE 90
AIR TEMPERATURES 96
AIR PRESSURES AND CIRCULATION 101
REVIEW AND DISCUSSION 119
ADDITIONAL READING 119

5 MOISTURE: CLOUDS AND STORMS 123
THE HYDROLOGIC CYCLE 124
EARTH'S WATER BALANCE 124
MOISTURE AND PRECIPITATION 125
STORMS 137
REVIEW AND DISCUSSION 145
NOTES 146
ADDITIONAL READING 146

6 THE EARTH'S WATERS 149
THE ATMOSPHERE 150
SOIL MOISTURE 150
GROUNDWATER 152
CASE STUDY: GEOTHERMAL ENERGY 159
STREAMS, RIVERS, AND LAKES 162
ICECAPS AND GLACIERS 167
THE OCEANS 169
EARTH'S WATERS: A SECOND LOOK 183
CASE STUDY: ENERGY FROM THE OCEANS 184
REVIEW AND DISCUSSION 185
NOTES 185
ADDITIONAL READING 186

7 EARTH'S CLIMATES 189
KÖPPEN'S CLASSIFICATION OF CLIMATES 190
TROPICAL CLIMATES 194
SUBTROPICAL CLIMATES 201
MIDLATITUDE CLIMATES 206
HIGH-LATITUDE CLIMATES 213
HIGHLAND CLIMATES 217
CLIMATE AND HUMANITY 217
REVIEW AND DISCUSSION 220
NOTE 221
ADDITIONAL READING 221

8 EARTH'S VEGETATION 223
THE VEGETATIVE CYCLE 224
THE STRUCTURE OF PLANTS 231
DISTRIBUTION OF NATURAL VEGETATION 237
VEGETATION AND CLIMATE 245
CASE STUDY: BIOFUEL—ENERGY FROM VEGETATION 246
REVIEW AND DISCUSSION 247
NOTES 247
ADDITIONAL READING 247

9 THE ROLE OF SOILS 249
FORMATION OF SOILS 250
SOIL CLASSIFICATION 259

REVIEW AND DISCUSSION 264
ADDITIONAL READING 264

10 COMPOSITION OF THE EARTH 267
GEOLOGIC TIME 268
EARTH'S INTERIOR 271
EARTH'S CRUST 273
EARTH MATERIALS 274
CASE STUDY: NUCLEAR ENERGY—FISSION OR FUSION? 283
REVIEW AND DISCUSSION 285
NOTES 286
ADDITIONAL READING 286

11 LAND SURFACE AND SEA BOTTOM 289
TECTONIC FORCES 290
LANDFORM CLASSIFICATION 314
REVIEW AND DISCUSSION 318
NOTES 319
ADDITIONAL READING 319

12 HOW THE SCULPTURING BEGINS 321
WEATHERING 322
MASS WASTING 328
REVIEW AND DISCUSSION 332
ADDITIONAL READING 332

13 EMBOSSED BY RUNNING WATER 335
SCULPTURING PROCESS OF RUNNING WATER 336
DEVELOPMENT OF STREAM VALLEYS 345
GROUNDWATER ETCHINGS 349
REVIEW AND DISCUSSION 353
NOTE 353
ADDITIONAL READING 353

14 CARVED BY OCEANS 355
CONTINENTAL SHELVES 356
WAVE ACTION 360
LANDFORMS OF WAVE EROSION 364
LANDFORMS OF WAVE DEPOSITION 366
SHORELINES 370
REVIEW AND DISCUSSION 373
NOTES 374
ADDITIONAL READING 374

15 ENGRAVED BY GLACIERS 377
THE FORMATION OF GLACIERS 378
CLASSIFICATION OF GLACIERS 382
REVIEW AND DISCUSSION 400
NOTES 400
ADDITIONAL READING 400

16 SCOURED BY THE WINDS 403
WIND EROSION 406
WIND TRANSPORTATION 407
WIND DEPOSITION 408
REVIEW AND DISCUSSION 415
NOTES 415
ADDITIONAL READING 415

EPILOGUE   THE HUMAN ENVIRONMENT—TOMORROW? 417

APPENDIX A   TIME ZONES 421

APPENDIX B   TOPOGRAPHIC MAPS 426

APPENDIX C   GROUND REFERENCING SYSTEMS 441

APPENDIX D   THE METRIC SYSTEM—HOW TO USE IT 445

CREDITS 447

INDEX AND GLOSSARY 449

OUR ENVIRONMENT

AN INTRODUCTION TO PHYSICAL GEOGRAPHY

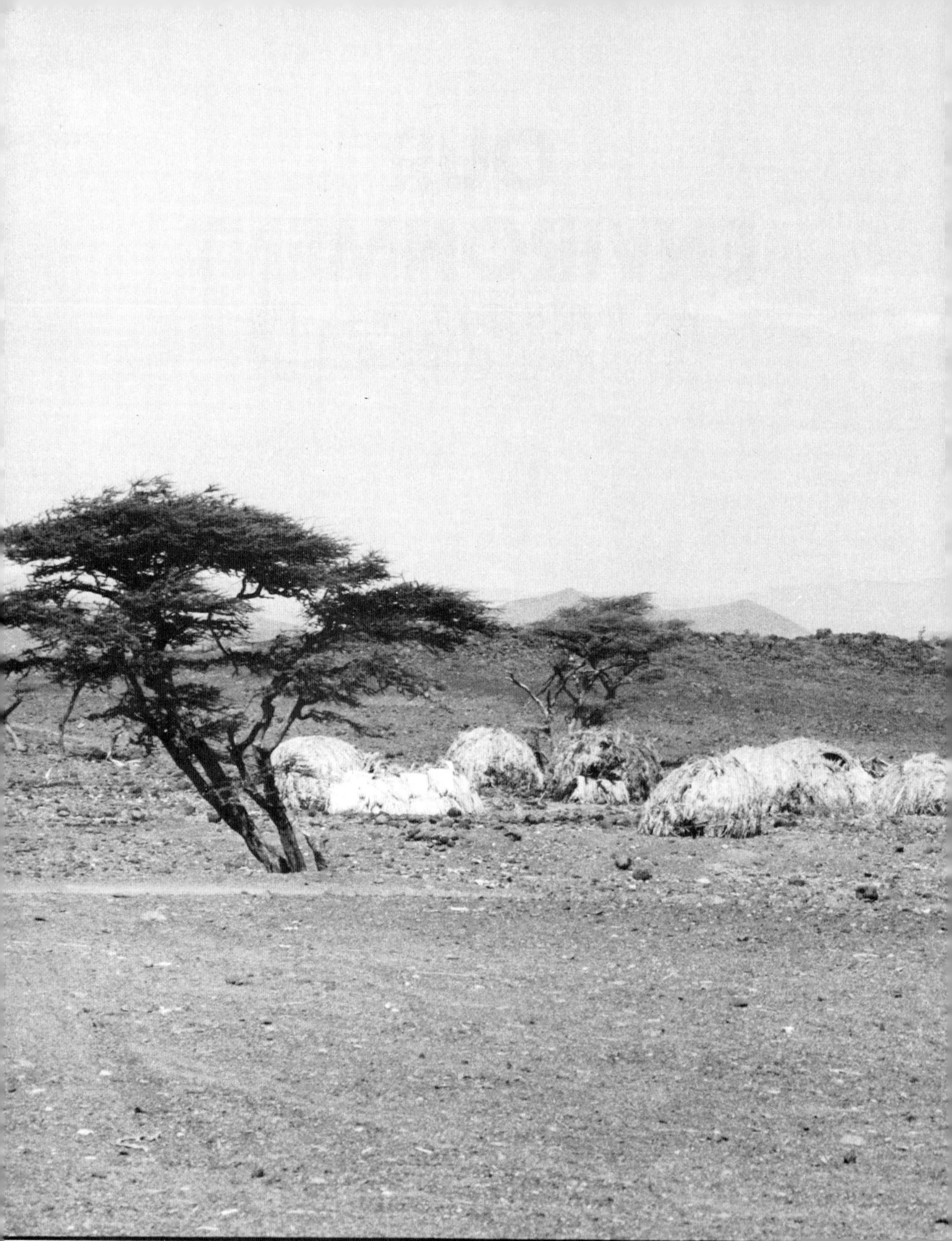

ONE

ON THE THRESHOLD

He who does not rouse himself when it is time to rise, who, though young and strong is full of sloth, whose will and thought are weak, that lazy and idle man never finds the way to knowledge.
THE BUDDHIST DHAMMAPADA

There are three classes of people in the world. The first learn from their own experience—these are wise; the second learn from the experience of others—these are the happy; the third neither learn from their own experience nor the experience of others—these are fools.
LORD CHESTERFIELD

The more extensive a man's knowledge of what has been done, the greater will be his power of knowing what to do.
BENJAMIN DISRAELI

The object of this book is to help you become more aware of life here on earth, gain a greater understanding of yourself and your relationship to earth, and comprehend the seriousness of the problems of pollution, overpopulation, and energy shortages. This study of our physical environment stresses that what we know as "the earth" is actually one vast system, with all the components interconnected. This system is divided into four areas, or spheres: the *atmosphere* (the envelope of air), the *hydrosphere* (earth's waters), the *biosphere* (vegetation and other life forms), and the *lithosphere* (the rocky crust of the earth itself). These spheres exist together, and a change in any one of them affects all the others. This knowledge, although too often disregarded in the past, is not new. In the second century A.D., one of the more compassionate of the Roman Caesars, Marcus Aurelius, wrote:

> *Regarding the universe often as one living being, having one substance and one soul; and observe how all things act with one movement; and how all things cooperate as the causes of all that exists; observe too the continuous spinning of the thread and the single texture of the web.*[1]

Perhaps now, if you heed these words and those spoken by Disraeli and try to acquire the knowledge of what humans have done to the earth, you will be able to help correct the wrongs.

You are now on the threshold of a voyage of discovery, a voyage that will take you all over the earth, an exciting journey that will enable you to form opinions based on fact, not emotion. You will stand in a much better position than those who devise theories out of dreams and "visions" instead of from experience.

EARTH AS A SYSTEM

To understand truly our physical environment, we need to know more about the elements, the component parts, that make up the complex structure we call earth. Geographers are trained to look at the earth's system in a highly comprehensive way. They use the work of many specialists (Table 1-1) as they conduct their research. They consult pedologists to learn about soils; they study the work of botanists, geologists, zoologists, and agronomists; and they collate the compiled material. To understand the effect of the sun and moon on our planet and its waters, they study astronomy. To understand the movements and the work of oceans, they consult oceanographers.

TABLE 1-1 **The Earth Sciences.**

THE EARTH SCIENCES	THEIR OBJECTIVES
Geodesy	Investigations to determine the true shape of the earth
Physical oceanography	Study of the oceans, waves, currents, and the ocean floor.
Astronomy	Study of the relationship of the earth to the sun, moon, and stars.
Climatology	Examination of the effects, over long periods of time, of solar radiation—sun energy—on the earth and the physical environment.
Meteorology	Study of weather—that is, short-term atmospheric conditions, day-to-day variations in climate.
Hydrology	Study of earth's freshwater supply: streams, rivers, lakes, groundwater.
Plant geography	Investigation of the vegetative cover of the earth: desert, shrubs, grasslands, forests.

TABLE 1-1 (Continued)

THE EARTH SCIENCES	THEIR OBJECTIVES
Pedology	Study of the soils of the earth.
Petrology	Study of the rocks of the earth.
Minerology	Study of the minerals of the earth.
Geology	Examination of the structure of the earth.
Geomorphology	Examination of the face of the earth: its landforms.
Cartography	Presentation, by means of charts and maps and graphs, of information gleaned from any source.
Geography	Analysis of the features of the earth: their location and distribution and how they are related to each other.

To understand the precarious balance between all of these portions of the earth's system, geographers do a great deal of research, so their answers to the earth's problems will have objectivity and validity. The physical geographer must try to understand how all the different elements of our earth fit together, how they mesh, relate, and affect each other.

Geographers want to know why people, places, and things differ from place to place. They are interested in the *location* of things—not merely that they exist in whatever form, but *where* they are and *why* they exist where they do. They want to explain the *distribution* of people, places, and things. Physical geographers seek the reasons for the distribution of volcanoes, fjords, tornadoes and hurricanes, lakes and streams, forests and grasslands. Human geographers consider all this information and try to understand the relationship of the various elements of earth's system to the distribution of human population (Fig. 1-1) across the earth. As a beginning, compare the population map with the maps that portray the world's vegetation (Figs. 8-2, 8-3, and 8-4) and climates (Fig. 7-2).

Some geographers prefer to approach their investigation and study by dividing the earth into sections, or regions, then making a thorough study of a particular region's climate, soils, vegetation, landforms, religions, political preferences, industrial activities, and so on. This is termed the **regional** approach. Others, however, prefer to consider the physical or cultural elements system by system all around the earth, which is termed the **systematic,** or topical, approach. In this manner the systematic geographer comes to understand the spatial interaction of all the elements. For example, geographers using the systematic approach to investigate the wind systems of the earth would study the influence of the sun in causing high- and low-pressure zones, the movement of air from one place to another, and the influence of the rotating earth on wind direction. They would also be concerned with how the wind in an area affects the overall climate, rainfall, fog conditions, and vegetation. Are there any similarities between California and Illinois, or between Moscow, U.S.S.R., and Moscow, Idaho? Some geographers want to step back and take the broad world view, and some want to zero in on small-area complexities.

Here we will look at the earth system by system, feature by feature, physical element by physical element; we will show how all the different physical elements fit together and what effect they have on different peoples' selections of places of habitation. In order, we will consider the earth itself: its shape; how it is portrayed on globes and maps; its weather and climate; its waters, vegetation, and soils; its

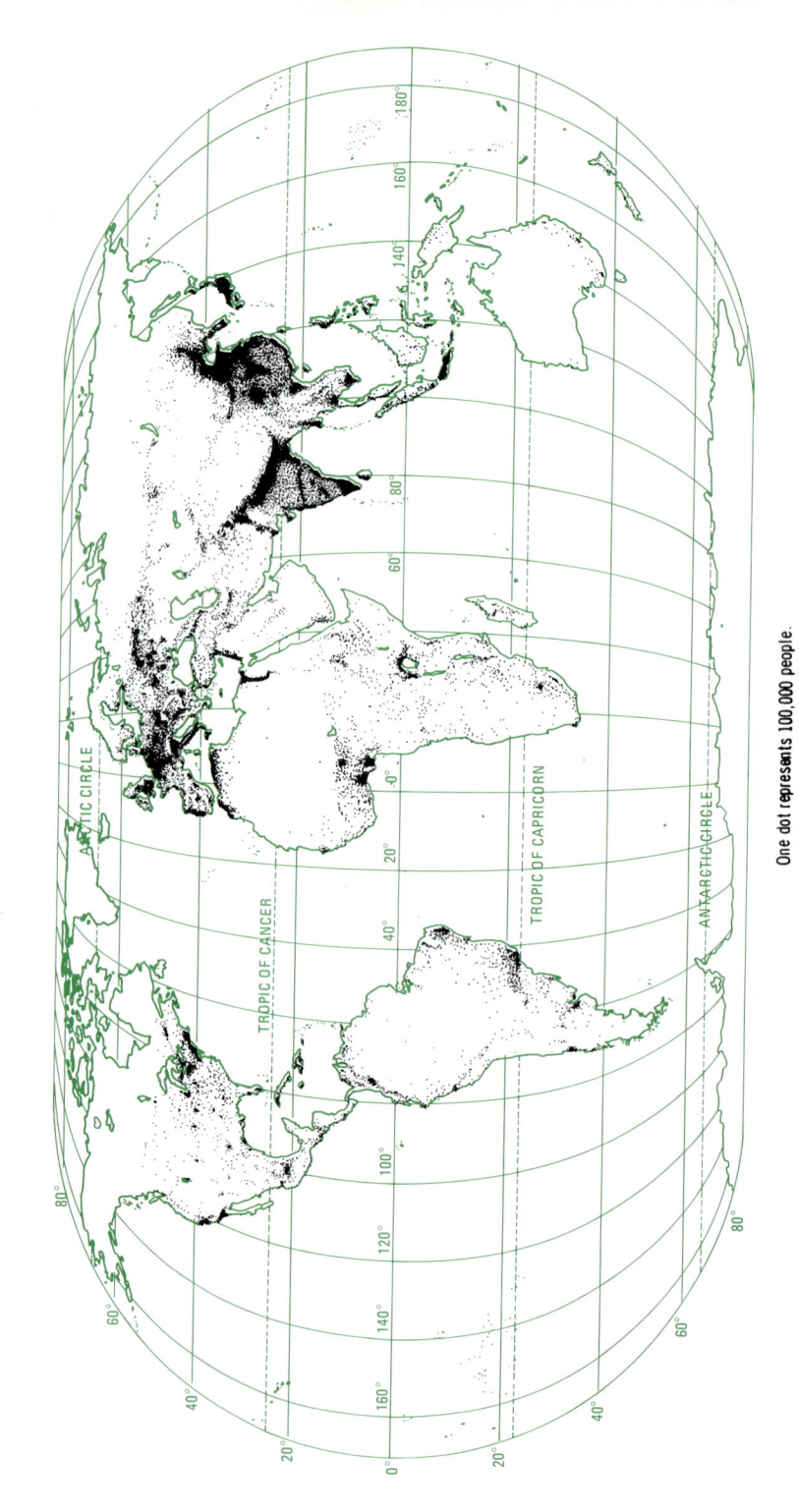

FIGURE 1-1 World distribution of human population

One dot represents 100,000 people.

structure and composition; and the dynamic forces at work above and below the surface that have given our earth its fantastic and beautiful sculptured face.

When we finish, you will have a better understanding of your earth, your home.

EARLY THEORIES OF EARTH'S SYSTEMS

It has been a relatively short span of time—perhaps a few thousand years—during which human life has been somewhat secure, that people could sit and ponder the complexities of their existence, and consider the earth under and about them and the great universe beyond. They evolved a number of theories and ideas to attempt to explain their origin and their place in the scheme of things. Many of these people were profound thinkers, and they came up with answers that were satsifying for a time. One of these, the Chinese sage Kung Futze (Confucius), said that the four seasons pursue their course without the aid of man, and all things are constantly being produced. He thought that the spirits of heat and cold, sun, wind, rain, stars, forests, streams, mountains, and valleys could be made happy and won over to man's side by offering sacrifices.[2] Another Chinese, Chuang Tze, also pondered the mysteries of the sky and wrote:

> *The sky turns round; the earth stands still; sun and moon pursue one another. Who causes this? Who directs this? Who has leisure enough to see that such movements occur? Some think there is a mechanical arrangement which makes these bodies move as they do. Others think that they revolve without being able to stop. The clouds cause rain; rain causes clouds.*[3]

And one of the great Jewish prophets, the second Isaiah, wrote:

> *Who hath measured the waters in the hollow of his hand, and ruled off the heavens with a span, and inclosed the dust of the earth in a measure, and weighed the mountains with a balance and the hills in scales?*[4]

In those days there was insufficient knowledge to discover the reasons behind the changing of the seasons, the transition from day into night and night into day, the formation of stream valleys, or the beneficent or detrimental effect of the environment on humanity. Instead, all these phenomena were regarded as the work of nature spirits or gods, supernatural beings such as Zeus of the Greeks, Shang-ti of the Chinese, Dyaus Pitar of the horsemen of the steppes of Central Asia, Ahura Mazda of the Persians, Yahweh of the Hebrews, or Odin of the Scandinavians.

From Metaphysical to Physical Theories

Change is inevitable, however, and in later times new ideas arose as humans tried to come up with better answers. Consider the ancient Greek philosophers, such as Homer (ninth century B.C.), who thought that the earth was a flat disk surrounded by the river "Ocean"; Anaximander (sixth century B.C.), who thought the earth was a giant cylinder hanging suspended in space; and Pythagoras (sixth century B.C.), who advanced the purely philosophical theory that the earth was neither flat nor cylindrical, but was a sphere, a ball-shaped bit of matter floating in space.

None of these men concerned themselves with the question of the actual origin or creation of the earth; they were interested only in explaining the various phenomena they witnessed, using the best logic they could muster. Their answers were satisfying to many, for those were times when men traveled far but understood little of the true vastness of the earth. They wrote extensively of their travels, but most of these early writers merely described what they saw, and their observations were colored by the basic knowledge of the time.*

It might be said that today we do the same thing, except that we have more data, more statistical information, better scientific equipment, and considerably more knowledge and experience to compare or contrast the reasons for the differences and similarities in the earth's phenomena. The thinkers before our time should not be laughed at or blamed for coming up with erroneous conclusions, any more than we should be today.

Environmental Determinism

Another of the eminent Greeks, the noted metaphysician and philosopher Aristotle (fourth century B.C.), attempted to explain why the weather and climate varied from one place to another in the world he knew. He divided the earth into three regions of climate, or *klimata:* (1) the torrid zone, which included the Great Desert (Sahara) and the grasslands of the south; (2) the frigid zone to the north, a land of forests and snow; and (3), in between, the temperate zone, the region surrounding the Mediterranean Sea. This was a considerable step forward for the time. Aristotle's thinking reflected the extensive explorations the Greeks were making into Africa, Asia, and Europe. His position was similar to ours today in relation to our explorations into outer space. We do not scoff at Aristotle's *klimata,* but we do shake our heads at his interpretation of the influence of climate on the affairs of people, particularly as it might be applied to modern society's technical or cultural achievements.

Aristotle rationalized that the "white-skinned barbarians" who lived in the cold, snowy frigid zone would never amount to much because the climate in northern Europe was too severe. As for the torrid zone, he believed that the "black-skinned savages" should not be expected to be very progressive because life would ever be too easy for them. They would be doomed for all time to wallow in idleness. Aristotle concluded that Greece, because of the ideal climate of the Mediterranean Sea, was the center of the universe. He believed that the "olive-skinned peoples" of the Mediterranean region were obviously destined to rule the world or, at least, to show the way to all other races.

One might well smile over Aristotle's simplistic analysis, but his fourth-century B.C. reasoning persisted through the nineteenth century. Indeed, the concept of **environmental determinism** grew through time as other respected scholars, Charles Darwin and Sir Charles Lyell (England), Friedrich Ratzel (Germany), Ellsworth Huntington and Ellen Churchill Semple (United States), among others,* reinforced his theory. Aristotle's theory was shaken by the growing awareness of the existence of great civilizations and outstanding cultures that evolved in both the frigid and torrid zones: the Vikings, Germans, Russians, and English in the frigid zone, and the Inca of Peru, the Maya of Yucatan, and the empires of Ghana, Mali, and Songhai of Africa in the so-called torrid zone.

The twentieth century, however, brought the theories of environmental determinism to an end. Most geographers of today, along with other theoreticians in the earth sciences (Table

---

*For an interesting accounting of these courageous explorers, see M. Cary and E. H. Warmington (1963), *The Ancient Explorers* (Baltimore, Md.: Penguin Books). See also E. H. Bunbury (1959), *A History of Ancient Geography,* Vols. 1 and 2 (New York: Dover Publications).

*For an interesting discussion of the various concepts of environmental determinism with reference to the influence or direct control of the physical environment on humankind, see A. H. Meyer and J. H. Streitelmeir (1963), *Geography in World Society: A Conceptual Approach* (Philadelphia, Pa., and New York: Lippincott), Chap. 1, pp. 2–25.

1-1), accept the idea that the physical environment does exert significant influence, but they do not believe that it controls our destiny or direction. Other mammals may be forced to adapt to their environment, but humans have free choice. If a particular society does not like the weather or climate where it resides, it can move, or it can alter its climate by modifying its structures. Today, we use insulation, put trees around our homes, and install heating or air-conditioning units. In other words, humans can—within limits—adapt to their environment, leave it, or make it conform to their own desires.

HUMANITY AND NATURE

Perhaps the very freedom to control their environment has brought modern societies to the predicament in which they find themselves—a world becoming dangerously polluted. In general, we might point the finger at the technological advances of "Western culture," since not all culture groups believe that humans rate above nature. Consider, for example, the nomadic Masai of Kenya and Tanzania in East Africa. They still believe that humans and animals—and the soils and vegetation—are one. Humans are not to do as they please with other living forms or the earth itself. In their opinions, humans are, at best, caretakers and protectors.

And, far to the northeast, the Hindus of 3000 or 4000 years ago came up with this hymn to the oneness of life.

> *All hail to heaven! All hail to earth!*
> *All hail to air! All hail to air!*
> *All hail to heaven! All hail to earth!*
> *Mine eye is sun and my breath is wind, air*
> *is my soul and earth my body.*
> *I verily who never have been conquered*
> *give up my life to heaven and earth for*
> *keeping.*[5]

It would seem that the Eastern religions have placed humans not on pedestals but directly *in* nature, an integral part of nature. Traditionally, ancient thinkers of eastern Asia, such as Gautama the Buddha and Confucius: held a strong ecological approach to humans and their relationship to their environment. It is quite possible, however, that the tremendous increase in the population of Asia during the past century will shake this tradition and bring

Geodesy

Physical Oceanography

Astronomy

8 ON THE THRESHOLD

Cimatology

Meteorology

Hydrology

Plant Geography

Pedology

Petrology

Mineralogy

Geology

Geomorphology

about a greater despoilation of their once sacred earth.

One must also be wary of placing too much blame on Western peoples. Perhaps some have taken the words of the Bible's Genesis 1:26 too literally.

Let us make man in our image, after our likeness, and let him have dominion over the fish of the sea, the birds of the air, the domestic animals, the wild beasts, and all the land reptiles.

Perhaps, also, the violators of our earth should have read further, to Genesis 2:15, and heeded these words.

Then the Lord God took the man and put him in the garden of Eden to till it and look after it.

If the latter statement had been stressed as strongly as the first, humanity might not have decimated the forests, the birds, and the beasts of the fields and seas.

To reverse this trend, to bring about a greater ecological awareness, we must know more about the earth on which we reside. We must realize that we are not *above* nature, but are an integral part of it. It is to be hoped that we can learn and understand what might be termed a *land ethic*—a realization that the earth is not just a toy to be used and thrown away. The earth—its waters, soils, vegetation, landforms, and atmosphere—is our home, our place of residence. In reality, it is not the environment so much as the manner in which we use it that counts.

CASE STUDY   **THINKING METRIC**

On December 23, 1975, the United States officially adopted the metric system—the International System of Units (SI)—as national policy with passage of the Metric Conversion Act. The reason for making this move was obvious: The United States was the only major nation in the world that had not adopted metrification, and if we were to continue to be involved in international trade, we had no alternative but to go metric. Our system, the British System, was inherited from the English. Because most Americans are so used to the customary system of inches and pounds, the move to centimeters, meters, liters, and kilos has been very slow.

Why are we so resistant? Consider these timeworn phrases: A miss is as good as a kilometer; 28 grams of prevention are worth 450 grams of cure; or how about a 25.4-millimeter worm millimetering its way across the ground? We must now face the inevitable and start "thinking metric."

After you have carefully examined Appendix D, which gives you the common metric/English conversion units, try to think of yourself as you relate to metrification. As an example, consider your body. If you are 5 feet 6 inches tall and weigh 150 pounds, you are a 1.7-meter person who weighs 68 kilograms. When you think about air temperature and you think metric, you will respond to 10°C (50°F) as being cold, 20°C (68°F) as being relatively nice and warm, and 30°C (86°F) as being hot. The doorknob in your house is probably about 1 meter from the floor. A paper clip weighs about 1 gram, a dime weighs 2 grams, and a golf ball weighs 46 grams. A quart of milk is a little smaller than a liter, a meter is a little longer than a yard, and a kilogram is slightly heavier than 2 pounds.

REVIEW AND DISCUSSION

1. From your personal experience and background, how would you compare the approaches of the ancient geographers to those of today?
2. Discuss Aristotle's theory of the *klimata*. Is there any validity to his theories on the division of the earth into three climatic zones?
3. After doing some outside reading of the works of Ellen Churchill Semple and Ellsworth Huntington, discuss their concept of environmental determinism.
4. What do you think of the idea that "people are an integral part of nature?"
5. Explain the importance of the geographic concept of location.
6. Explain the difference between the regional and systematic approaches to the geographic study of the earth.
7. Compare and contrast the various earth sciences: geodesy, oceanography, astronomy, climatology, meteorology, hydrology, pedology, geology, and geomorphology.
8. What do you think of the concept that the earth is one great interrelated system?
9. What do you think about the United States adopting the metric system?
10. If the temperature was 40°C, would you feel extremely cold, comfortable, or devastated by the heat?

NOTES

[1] Irwin Edman, compiler (1945), *Marcus Aurelius and His Times* (Roslyn, N.Y.: Walter J. Black), p. 41. With permission.

[2] Robert O. Ballou, ed. (1977), *The Portable World Bible* (New York: Viking Press), pp. 486–490.

[3] Ballou, p. 561.

[4] Isaiah 40:12–13.

[5] Ballou, p. 37.

ADDITIONAL READING

On the history of geography, its character, its essence, and its evolution from ancient times to the present, see:

Bunbury, E. H. (1979). *A History of Ancient Geography.* Vols. 1 and 2. New York: Humanities.

Broek, Jan O. M. (1966). *Compass of Geography.* Columbus, Ohio: Charles E. Merrill.

Carter, George F. (1975). *Man and the Land.* 3rd ed. New York: Holt, Rinehart and Winston. See, especially, the introduction.

Cary, M., and E. H. Warmington (1963). *The Ancient Explorers.* Baltimore, Md.: Penguin Books.

Meyer, A. H., and J. H. Streitelmeier (1963). *Geography in World Society: A Conceptual Approach,* Chap. 2. Philadelphia, Pa., and New York: Lippincott.

Semple, Ellen Churchill (1911). *Influences of Geographic Environment.* New York: Henry Holt.

Warntz, William, and Peter Wolff (1971). *Breakthroughs in Geography.* New York: New American Library. Contains "the key writings of the great historical figures of geography."

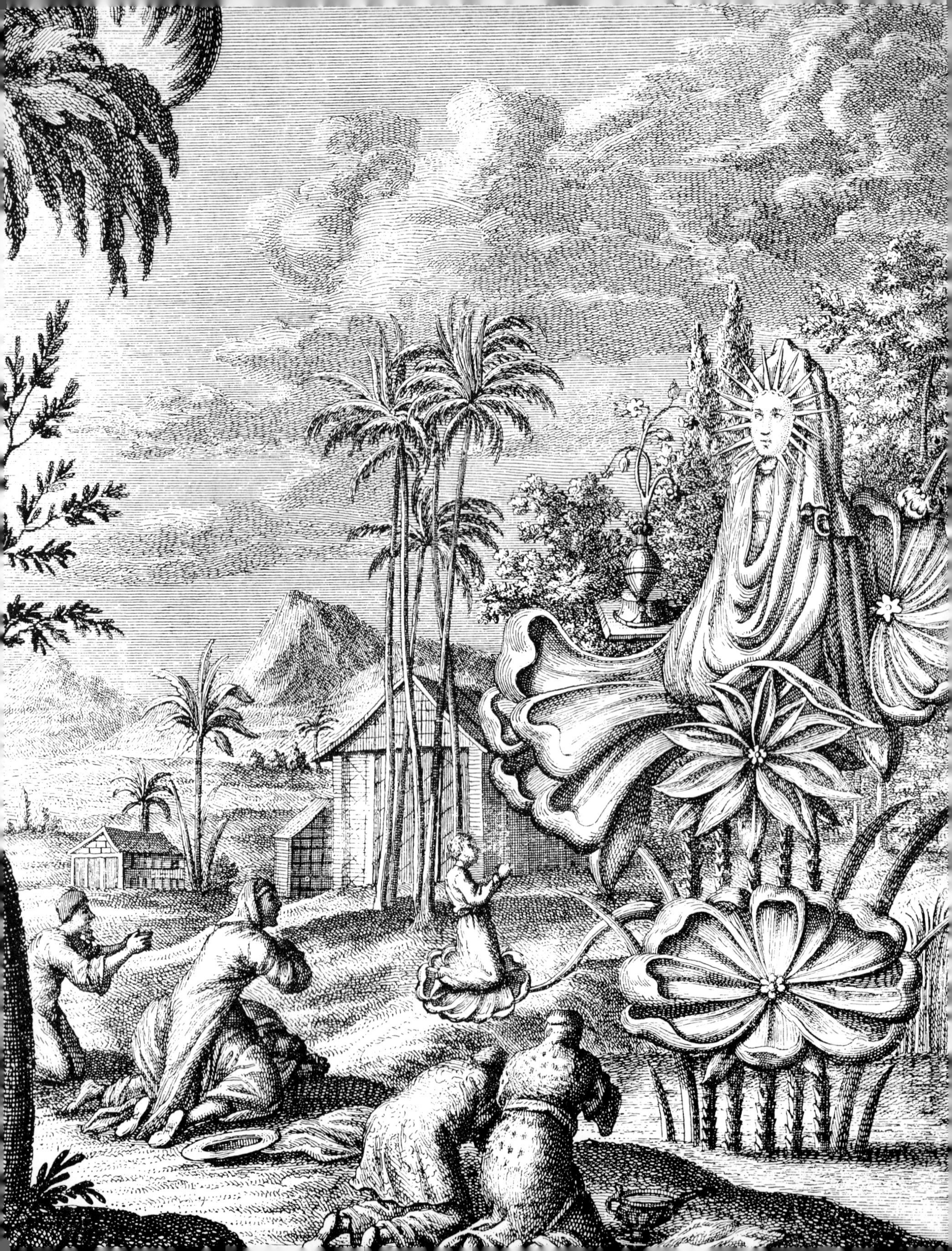

TWO
PORTRAYING THE EARTH

Since the beginning of human existence on the planet Earth, individuals have attempted to portray the land about them. Some have done it verbally, with songs, chants, and poems, as they described the directions, distances, and landmarks of their important routes. Others tied sticks together to form a layout of the region about them, or they inscribed lines in a mass of wet clay. These methods worked fairly well for those who knew only that bit of earth space about them, but think of the problems that arose when knowledge of the actual size and shape of the earth began to enter their understanding.

It was a relatively simple matter for our ancestors to construct maps of the vicinity of their homes or hunting grounds, but it became quite a different story when people became concerned with extensive areas and vast distances. Before maps of a continent or the earth itself could be constructed, a series of questions had to be answered, such as the size and shape of the earth, and how to make a flat representation of a round globe without distortion. The first mapmakers had to devise ways to pinpoint the location of places, to figure true sailing courses across large oceans, and to make the map readable and understandable to any who might want to use it.

Today we feel quite satisfied with our solutions, but it took a long time to find them. For thousands of years humans tried to unravel the puzzles.

EARTH'S SHAPE: ANCIENT CONCEPTS

Some 2000 years before the beginning of the Christian era, the Hindus compiled a book of Indian scriptures from which we learn:

Ancient Egyptian concept of heaven and earth The goddess Nut's starry body forms the celestial "vault of heaven," and over her form sails the sun-god Ra once each day. Beneath Nut lies the earth-god Geb, whose body forms the earth's crust.

In the beginning, this was nonexistent. It became existent, it grew. It turned into an egg. The egg lay for the time of a year. The egg broke open. The two halves were one of silver, the other of gold. The silver one became the earth; the golden one, the sky; the thick membrane of the yoke, the mist with the clouds; the small veins, the rivers; the fluid, the sea.[1]

Centuries later, halfway around the world, an Apache of the American Southwest believed that in the beginning there was no earth, no world; there was nothing but darkness, water, and cyclone, but:

there were the gods who had always existed and who had the material out of which everything was created. They made the world first, the earth, the underworld, then they made the sky. They made the earth in the form of a living woman and called her Mother. They made the sky in the form of a man and called him Father. He faces downward and the woman faces up. He is our Father and the woman is our Mother.[2]

Several thousand years ago the peoples of the island of Crete and the nomadic horsemen of the steppes of Central Asia held similar views, although the latter emphasized the masculine and tended to downgrade the importance of the earth as a "mother." The Egyptians thought differently. They conceived of the earth as masculine, a recumbent earth-god, Geb, from whose body sprouted all the world's vegetation. Above him was the sky-goddess Nut,

whose long, curved body touches the earth only with the tips of her toes and fingers. It was the starry belly of the goddess which men saw shining in the night above them.[3]

The number of theories that try to explain the origin of earth and its physical shape are as varied as the peoples who devised them. Homer, the illustrious Greek writer of the ninth century B.C., thought of the earth as a kind of a raft floating in space, surrounded by a mighty river he called *ocean*. Tahitians of the South Pacific thought of the earth as being a giant tortoise swimming leisurely through space. Babylonians of the Middle East believed the earth was a hollow mountain around which an ocean swirled, and the sky was like a curved ceiling through which peeked the stars and sun. Almost all of these ancient theories considered the earth as fixed in a permanent position in space, while the sun, moon, planets, and stars marched overhead in a regular procession.

EARTH AS A SPHERE

With the passage of time, people began to believe that the earth was not merely flat like a plate but, instead, had curvature. Imagine a group of people standing atop a hill and peering seaward toward an offshore island 50 miles (80 kilometers) away. The people, deciding that the island might be easy to defend against attack by hostile neighbors, descend the hill and begin to swim out to the island. But when they look up to get their bearing, they discover to their dismay that the island has disappeared. Had some capricious god or spirit of the sea snatched the island away? Returning to the hilltop, they are aghast. There is the island. The confusion may have then given way to a new thought: The earth's surface must curve. They could see it only when they were high enough for their line of sight to encompass the island as it rose above the curvature. Figure 2-1 offers an explanation as to how such an event could occur. Of course, the earth's curvature is understood today by anyone who has seen photographs of the earth taken from satellites in orbit.

As the centuries passed, new theories evolved. Pythagoras (sixth century B.C.) philosophized that the only perfect shape in the universe was a sphere. Since the earth (in his mind) was the center of the universe, and therefore perfect, it *had* to be a sphere. This concept became common knowledge among

16 PORTRAYING THE EARTH

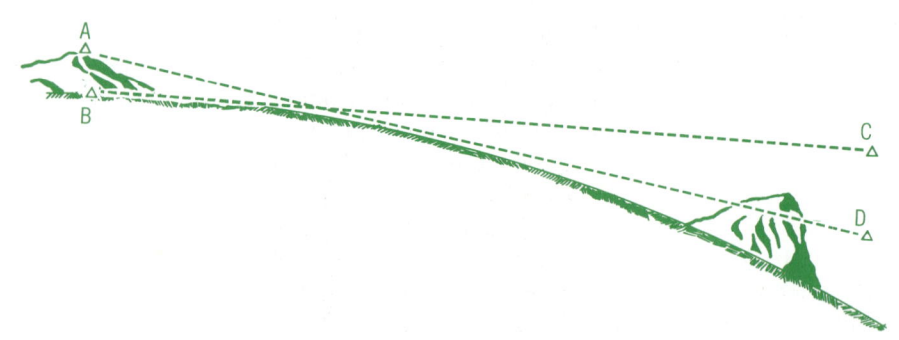

FIGURE 2-1 Curvature of the earth Line A-D represents the line of sight of a person standing on a hilltop looking toward an island. Line B-C is the line of sight of a swimmer. If the island were 50 miles (80 kilometers) distant, with an elevation of 1320 feet (400 meters), it would be clearly visible to the person at point A. The swimmer at point B, however, cannot see the island because the earth makes such a curve at that distance. In fact, the island would have to have an elevation of at least 1650 feet (500 meters) to reach the swimmer's line of sight.

the early Greeks and Romans. However, the first person of record actually to postulate some earth measurements was Eratosthenes of Alexandria, Egypt, during the third century B.C. Determining angles of noonday shadows cast by the midsummer sun at two different places along the Nile River and measuring the land distance between the two places, he estimated that the arc, or angle, between these two places from the center of the earth was about one-fiftieth of a complete circle. Assuming that the earth was a sphere, he multiplied the distance by fifty and came up with a figure of the earth's circumference that was a little over 25,000 miles (40,000 kilometers). Although he did not have accurate measurements with which to work, his errors counteracted each other, and he arrived at this figure, which was within 100 miles (160 kilometers) of today's accurately figured circumference of 24,902 miles (39,843 kilometers).*

By the second century of the Christian era the scientific study of the shape of the earth was advanced by two important men of history: Crates and Ptolemy. The philosopher Crates constructed a globe 10 feet (3 meters) in diameter that, because of the lack of adequate and accurate data, did not portray the landmasses and oceans of the world as we know them to be today. He was soon followed by a geographer turned mapmaker named Ptolemy, who put his specialties of astronomy and geography to work to prove that the earth was a sphere. He, too, constructed a globe and from it developed a map projection in an attempt to create a flat representation of his theoretical round earth. He used lines representing parallels of latitude and meridians of longitude (developed seven centuries earlier by Herodotus*) to locate places on his map of the known world.

---

*The following two treatments of the work of Eratosthenes are recommended. David Greenhood (1964), *Mapping* (Chicago, Ill.: University of Chicago Press), pp. 39–40; and Desmond King-Hele (October 1967), "The Shape of the Earth," *Scientific American*, 217 (4), p. 4 (offprint 873).

*Herodotus (fifth century B.C.) believed the earth was wider in its east-west dimensions than it was from north to south. He developed terms such as *longitude* (for the longer east-west directions) and *latitude* (for the lateral, or lesser, north-south directions). His thinking affected mapmaking for centuries.

EARTH AS A SPHERE 17

The curvature of the Earth This photograph taken from the Apollo 11 spacecraft, which was 98,000 nautical miles from the Earth, clearly shows the curvature of the Earth. Because of the heavy cloud layer only Saudi Arabia can be clearly seen.

Although Ptolemy's map carried forward the errors long held by geographers and navigators that the earth was considerably wider from east to west than it was from north to south, it was used for more than 1000 years.† Had it not been for his error in assuming that Spain and Asia were closer than they really were, it is possible that Christopher Columbus might never have set out on his epochal journey to "far Cathay" (China). Although Columbus did not reach the Indies and the people he met were not "Indians," he did add further proof to the theory that the earth was round.

Old beliefs, ideas and traditions die hard, however, and there are still some people who will not accept the "round earth" doctrine. In England, for example, a small group of supposedly well-informed people formed the Flat Earth Society of Dover and, despite a dwindling membership,* these staunch believers in the "flat earth theory" scoff at our so-called proofs. They verbally attack America's space accomplishments as merely fraudulent attempts to deceive the people and to promote

†For an interesting look at the actual words used by Ptolemy to explain his theories, see William Warntz and Peter Wolff (1971), "Ptolemy: The Classical Synthesis," *Breakthroughs in Geography* (New York: New American Library), chap. 1, pp. 15–60.

*Today, the International Flat Earth Research Society is headquartered in Lancaster, California.

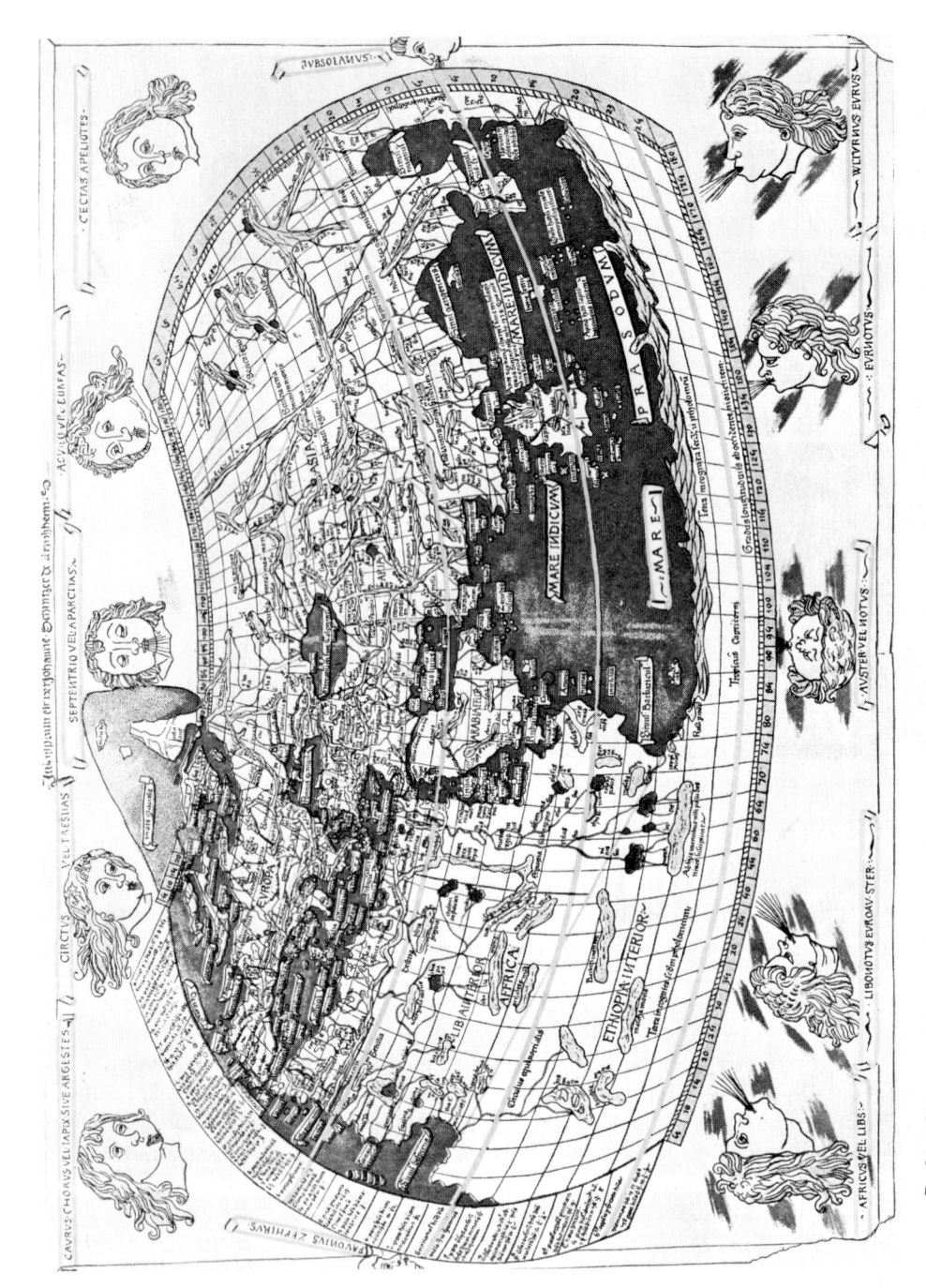

Ptolemy's map of the world—second century A.D. (Norman J. W. Thrower, *Maps & Man: An Examination of Cartography in Relation to Culture and Civilization*, © 1972, p. 45. Reprinted by permission of Prentice-Hall, Inc., Englewood Cliffs, New Jersey.)

the vested interests of manufacturers of maps and globes. As they cling to the ancient belief that the earth is the center of the universe, they dismiss television broadcasts and satellite photographs from outer space as "cinematic trick photography made in Hollywood."[4]

With all due respect for the rights of people to believe as they will, there is no need for deception. The earth is round—that is, it has a roundish shape, slightly elliptical (oval) in form (Fig. 2-2). The distance from one side of the earth to the other, measured through the equator, is about 7927 miles (12,757 kilometers). The distance from the North Pole to the South Pole, measured through the center of the earth, is about 27 miles (43 kilometers)

Earth from space December 1972; and **Apollo 17** flashes toward the moon. The members of the crew look back to see the dazzling interaction of the atmosphere with the rays of the sun; swirling clouds spiraling out from Antarctica, and Africa, a not-so-dark continent.

20 PORTRAYING THE EARTH

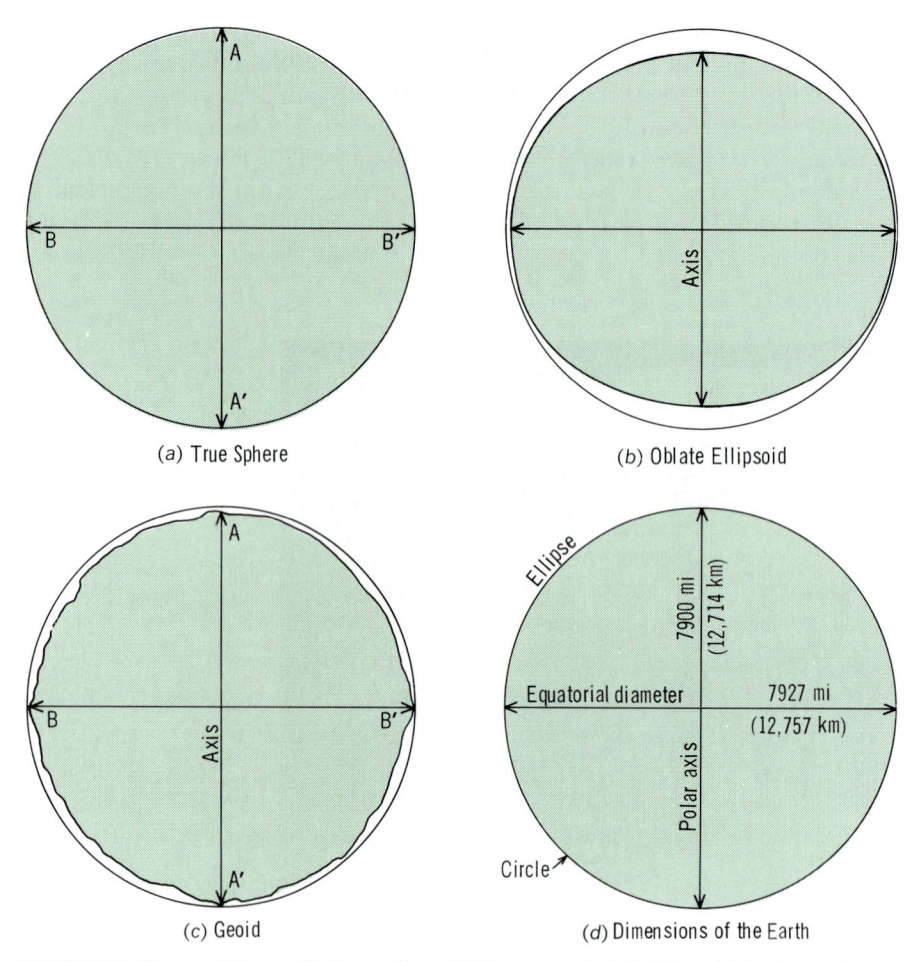

FIGURE 2-2 Shape of the earth Among these highly exaggerated sketches, (*a*) is a true sphere, and the distance from A to A' is the same as that from B to B'. (*b*) shows the distortion from the true sphere caused by the earth's rotation: The earth bulges slightly around the middle; thus, the distance from B to B' is greater than that from A to A'. (*c*) is a geoid. Although not to true scale, it is representative of the earth's elliptical shape and the mountainous highs and oceanic lows of the earth's surface. (*d*) gives the dimensions from pole to pole (A to A') and through the earth at the equator (B to B').

shorter. Its form is that of an ellipse, but one that is squashed in a little at the poles and a little fatter about its equatorial middle. The reason for the bulge around the equator is the earth's rotation. The spinning about the polar axis causes a displacement of the earth's bulk and a subsequent downpressing of the polar regions. The earth's shape, therefore, is that of an **oblate ellipsoid.**

A closer look at the earth also tells us that it does not have the smooth shape of an elliptical bowling ball. It has, instead, an uneven surface. The elevations of the landmasses are much higher than those of the ocean floors, In addition, there are great heights, such as Mt. Everest, towering to an elevation of 29,028 feet (8708 meters) above sea level, and great depths in the oceans, such as the Mariana Trench, which is 35,800 feet (10,740 meters) below sea level. Obviously, therefore, the earth has an uneven surface. Because of this, the **geodesists** (scientists who try to determine the exact shape of the earth) call the earth's shape a **geoid.** This implies that our earth has a unique shape, apparently unmatched anywhere else in our solar system. Certainly we could not describe the shape of the planet Mars as that of a geoid, because Mars has its own distinctive shape and form.

Although some earth scientists may require precise mathematical measurements in trying to access the size and shape of the earth accurately, it is not necessary for us to concern ourselves with such miniscule differences. For our purposes, we can assume that the earth is a sphere.

The Geographic Grid

One of the things a geographer needs to do is to define location. Although we can rely on names or descriptions—street addresses, zip codes, names of cities, or verbal directions—a more accurate way is to use the **geographic grid.** This is simply a system of locating places on the globe or a map by using lines running east-west and north-south. These lines are parallels of latitude and meridians of longitude and, in order to understand these devices, we must first comprehend the differences between great circles and small circles.

Great Circles and Small Circles Great circles always divide a sphere into two equal parts; small circles do not. If you were to take an orange, which is a sphere, and cut it in half, the resulting two pieces would be equal in size. The circumference of the exposed face of either half would be a complete circle, called a **great circle.** If you were to take another orange and cut it into two pieces, but your knife did not slice through the center of the orange,

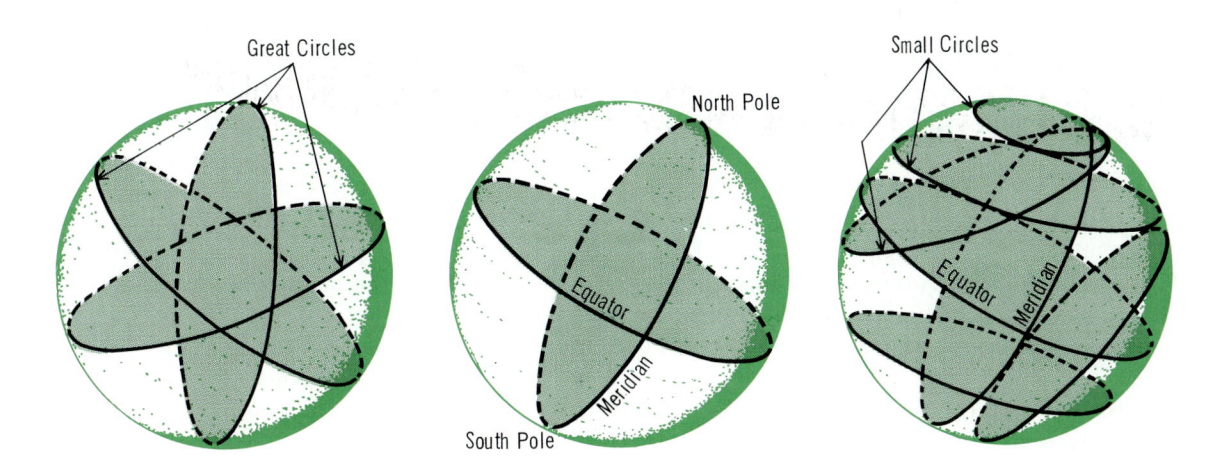

FIGURE 2-3 Great circles and small circles

you would have two pieces, but they would be of different sizes. The circle representing the cut face would be a complete circle, but it would not be a great circle; it would be a **small circle.** All circles drawn on the face of a globe (which represents the earth) that have as their center the exact center of the earth, or those passing through both of the poles, are great circles. Consider Fig. 2-3. Figure 2-3a displays a number of great circles, drawn on a globe. All of these lines, or circles, on the globe have as their center the center of the earth. In Fig. 2-3b only two great circles are shown, but these are the two most important ones. The north-south great circle is composed of two meridians (a meridian is half of a great circle), one being the **prime meridian** (or zero meridian). The other meridian would be 180° away, on the other side of the earth. The east-west great circle in this sketch is the equator. Figure 2-3c has a number of circles, but only two are great circles. The others are small circles, and it is obvious that their centers are not at the center of the earth.

This concept of great and small circles is most important in navigation for both ships and planes, because the shortest distance between two different locations on the earth's surface is always along a great circle connecting the two points. Ship's courses are generally referred to as "great-circle routes," and today's jets, flying from Los Angeles to London, usually fly a great-circle route that takes them over the southern tip of Greenland. If a jet flew over New York en route to London, the distance would be much greater, because such a path would be along a small circle.

Meridians of Longitude and Parallels of Latitude **Meridians of longitude** are halves of great circles. As the circumference of any circle is equal to 360°, a meridian is 180° in length, stretching from pole to pole (Fig. 2-4). If you look at a globe, you will note that all the meridians run in a true north-south direction. Spaces between meridians along the equator are greater than those along parallels of latitude away from the equator, for the simple reason that meridians converge at the poles. Any number of meridians can be drawn on a globe, and every place on the earth's surface has its own meridian. On most globes, however, the meridians are spaced about 10° or 15° apart, and the first meridian, the prime meridian, which runs through the Royal Observatory at Greenwich near London, England, is the starting place. Figure 2-4b shows how meridians of longitude are determined, and you can see that longitude is the angular distance measured east or west of the prime meridian. Meridians to the west of Greenwich are referred to as meridians of west longitude; those to the east are east longitude. When the angular count reaches 180° east or west, the count stops. This is the 180° meridian, also known as the **International Date Line.**[*]

Parallels of latitude are the other lines used in determining the geographic grid. Knowing an area's meridian of longitude is only half the solution. We need to know just where along a meridian the area is located. We do this by finding which parallel of latitude crosses the meridian at that place. Parallels of latitude (see Fig. 2-4) are small circles, except for the equator, and they are always parallel to the equator and to each other. They run in a true east-west direction, and they always intersect meridians at right angles, except at the poles, where there are no parallels. Any number of parallels can be drawn on a globe, but, again, most globes have them spaced at 10° or 15° intervals, starting at the equator, and

---

[*]For clarification of the role of the 180° meridian as the International Date Line, see Appendix A, on time zones.

EARTH AS A SPHERE 23

(a)

(b)

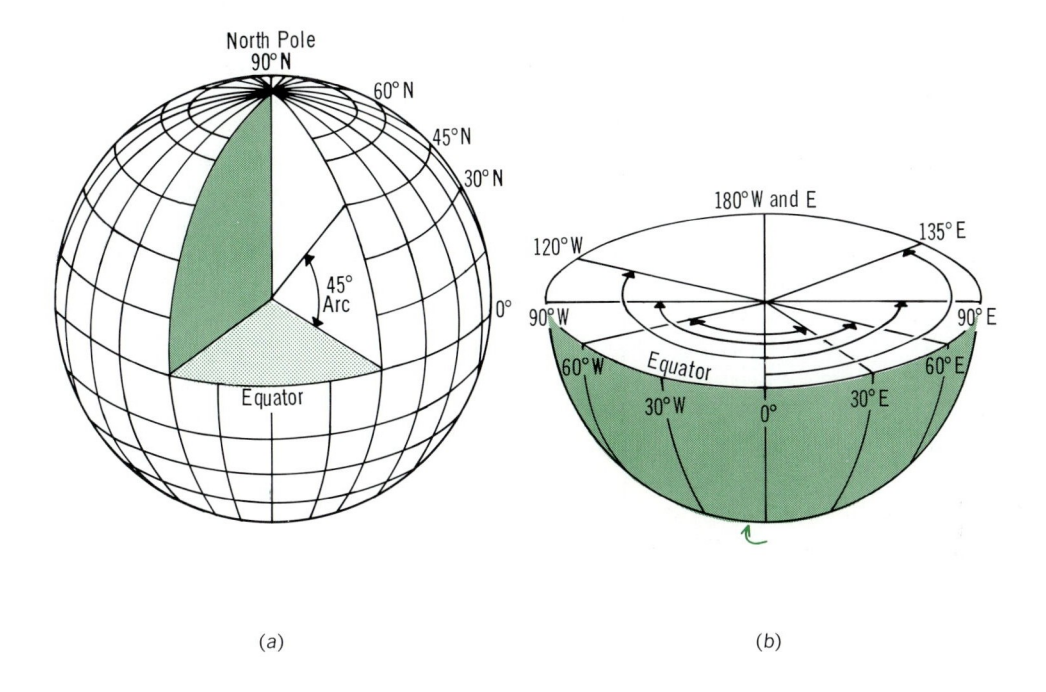

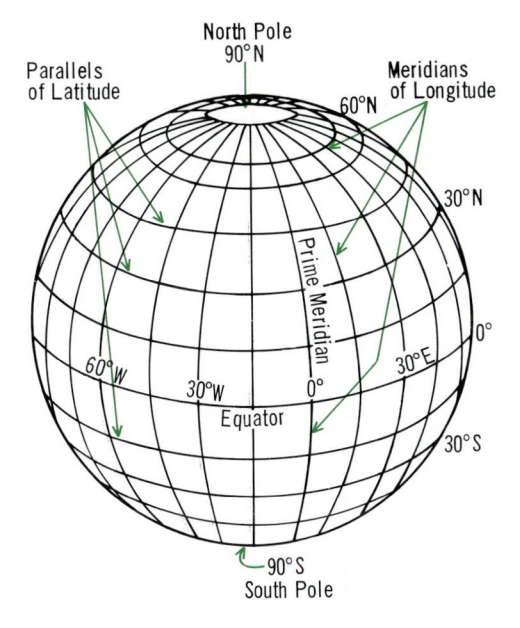

(c)

FIGURE 2-4 Meridians of longitude and parallels of latitude

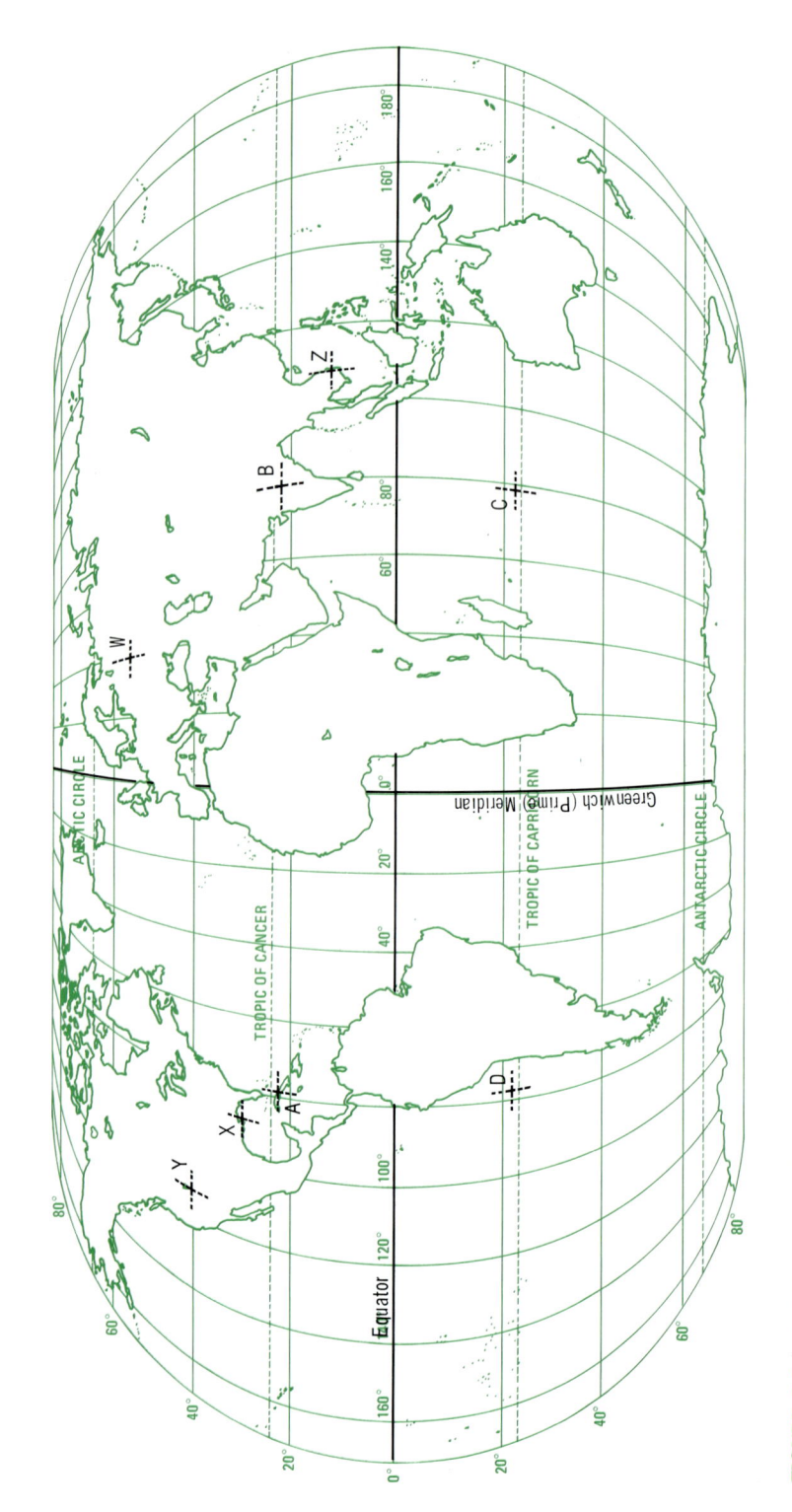

FIGURE 2-5 Using the geographic grid to locate places Place A is at latitude 22° north, longitude 79° west and is in Cuba. Place B is at latitude 22° north, longitude 79° east and is in India. Place C is at latitude 22° south, longitude 79° east and is between Africa and Australia in the Indian Ocean. Place D is at latitude 22° south, longitude 79° west and is off the coast of northern Chile. What are the geographic grid positions of places W (Moscow), X (New Orleans), Y (Great Salt Lake), and Z (Saigon)? It might help if you were to consult your world atlas.

0° parallel, and ending up to 90° north or south (the poles).

Therefore, to find an area's geographic grid position, we must determine both its latitude and its longitude. Where the lines intersect is the area's geographic grid position on the earth. Another important point to remember is this: Merely knowing the degree of latitude and longtitude is not enough; we must also know whether the latitude is north or south and whether the longitude is east or west. Consider, for example, the geographic grid for Puerto Rico. It will not suffice to say "latitude 18°, longitude 67°," because there are four places on earth that have such a position. We must go further and complete the geographic grid this way: "Puerto Rico's geographic grid position is latitude 18° north, longitude 67° west." Without the north or west directions you would not know where Puerto Rico is located. Consider the example in Fig. 2-5, which locates four positions that all have the same latitude and longitude (latitude 22°, longitude 79°).

Determining Latitudinal and Longitudinal Position

It is easy to pick up a map and point to your latitude—if you know where you are. But what if you are lost? You would then need an instrument called a **sextant.** This device is used to measure the angle between the horizon and a celestial body, such as the sun or the North Star (Polaris). For example, if you are in the Northern Hemisphere, taking a position facing north, you sight through the eyepiece of the sextant and measure the angle between the horizon and Polaris. That horizon angle is your latitude. Polaris can be used as a sighting point only in the Northern Hemisphere, however, because it cannot be seen south of the equator. If you are on the equator, Polaris seems to be resting on the horizon. As you move northward, Polaris rises 1° in angle for every degree north you move. For example, if you stopped along the way and found that your horizon angle was 60°, your latitude would be 60° north. At the North Pole the angle would be 90°, since Polaris would be directly overhead.

Determining your longitude is somewhat less complicated. You do not need a sextant, but you do require two precise clocks, called **chronometers,** for telling the time. One chronometer tells you what time it is in Greenwich, England, through which the prime meridian passes, and the other tells you the local time. Because the earth rotates around its polar axis once every 24 hours, it moves from west to east 15° every hour (360° divided by 24 hours equals 15°). If you know the time at Greenwich and your local time, all you need do to determine your longitudinal position is to calculate the number of hours difference (Fig. 2-6) and multiply that number by fifteen. If your time is earlier in the day, you will be west of Greenwich; if your time is later than it is on the prime meridian, your longitude will be east. Furthermore, if you keep in mind that international time is counted from 0000 (midnight) to 1200 (noon) to 2400 (midnight again), it is easier to figure the difference. Consider this: If your local time is 0900 (9:00 A.M.) and it is 1200 hours at Greenwich, your position is 3 hours earlier in the day, or 45° west of Greenwich, and your longitudinal position is 45° west. One-half hour represents 7° 30′. Therefore, if your time was 1430 hours, you would be 2½ hours east of Greenwich, or 37° 30′; thus, your longitudinal position would be 37°30′ east.*

Earth's Shape and Determining Distances

Until now we have been operating on the premise that the earth is a sphere in employing the parallels of latitude and meridians of long-

---

*For additional information on longitude and time, see the section titled "Earth's Revolution" on p. 54 and Appendix A, "Time Zones" (p. 421).

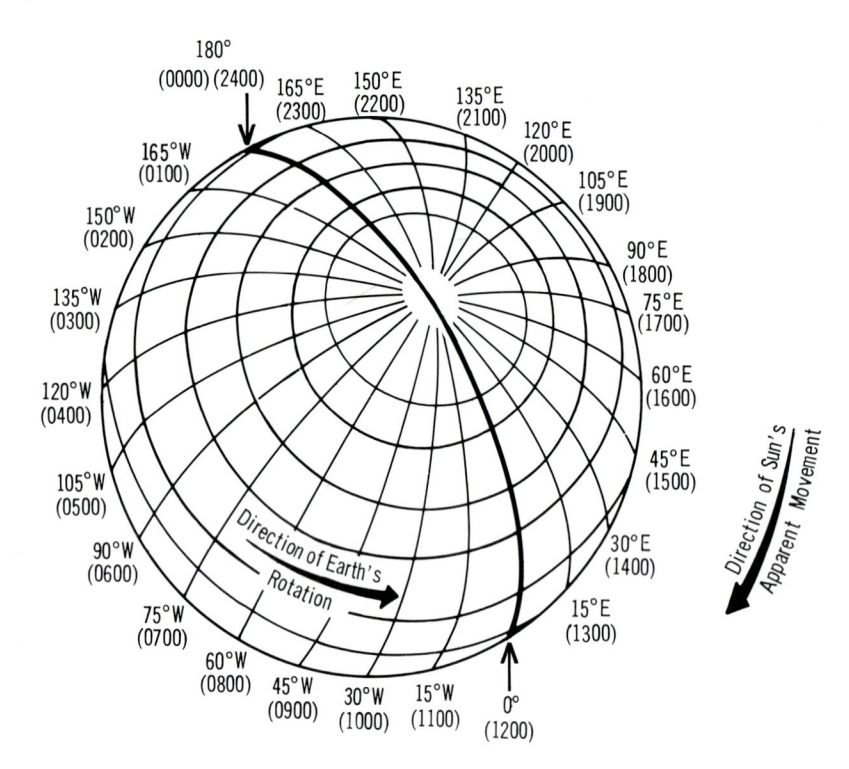

FIGURE 2-6 Longitude and Time The earth rotates about its polar axis 360° every 24 hours, a 15° distance each hour. Time difference between the 0 meridian and the 180° meridian is 12 hours. When it is noon along the 0 meridian, it is midnight along the 180° meridian. In this illustration the times shown relate to a noon (1200 hours) time at Greenwich, England, along the 0 (or prime) meridian.

itude to locate positions on the earth. Now it is time to reconsider the shape of the earth and its effect on the geographic grid and the determination of distances on the earth's surface. If the earth were a true sphere and its circumference were exactly 25,000 miles (40,000 kilometers), each of the 360 degrees would measure 69 miles, or 60 nautical miles (111 kilometers). By further subdividing each degree into minutes and each minute into seconds (and remember, we are talking about distance, not time) the results would show that 1 minute would equal 1.15 miles or 1 nautical mile (1.84 kilometer), and 1 second would represent 0.019 miles, or about 100 feet (0.03 kilometers). If we had a geographical grid position of latitude 20°30′10″ north, longitude 70°20′20″ west, we would know that the area represented would be 1415 miles (2264 kilometers) north of the equator and 4853 miles (7764 kilometers) west of Greenwich.

Unfortunately, however, we cannot ignore the elliptical shape of the earth. The length of 1° of latitude or longitude does not remain constant over the face of the earth (Fig. 2-7, Table 2-1). The only place where 1° equals 69 miles (111 kilometers) is along the equator.* Along the 60° parallel, we discover that the length of 1° of longitude is only 34.67 miles (55.8 kilometers). The distance between the equator and 10° north is about 687 miles (1100 kilometers), but the distance between 60° and 70° north is only about 6 miles (9.6 kilometers)

---

*Although the precise measurement of 1° along the equator is 110.57 kilometers for latitude and 111.32 for longitude, geographers normally use 111 kilometers when determining distances where absolute accuracy is not required.

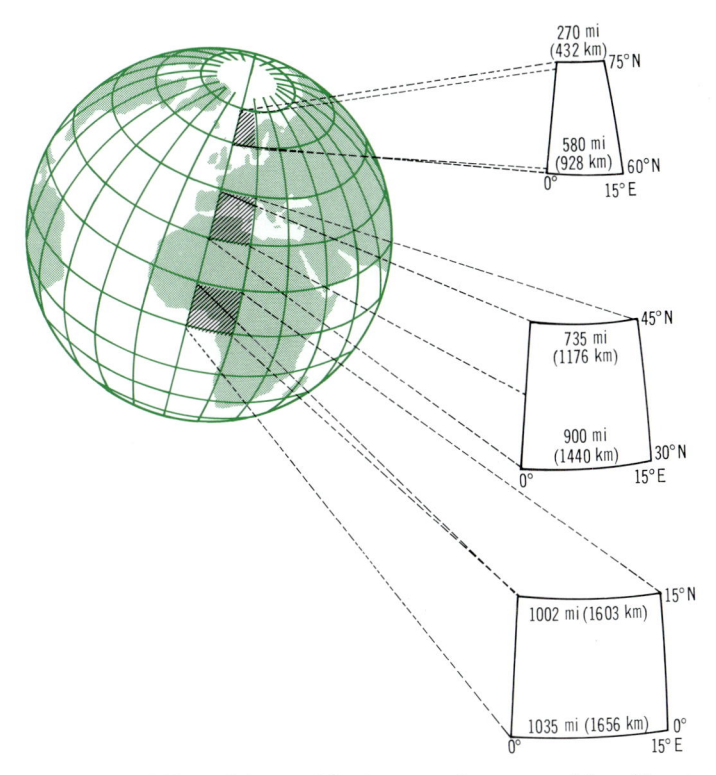

FIGURE 2-7 Effect of the earth's shape on the geographic grid

longer, showing us that distances measured along meridians are not greatly distorted. Consider the difference between meridians, as measured along a parallel. Distance between the prime meridian and 15° east longitude is 1035 miles (1656 kilometers) along the equator, but along the 60° north parallel the distance shortens to 580 miles (928 kilometers)—a shortening of 455 miles (728 kilometers). If you take another look at Figure 2-4, the reason for these variances becomes obvious: The meridians converge as they approach the poles.

MAPS

It was mentioned earlier that cartography is considered to be one of the earth sciences.

Most people assume that cartography—the making of maps, graphs, and charts—falls directly within the province of geography. This is not the case, however; cartography stands on its own feet as a science and as an art form that requires great skill. Cartography can be of assistance to any of the earth sciences, and it can also serve history, anthropology, botany, sociology, and numerous other disciplines. Yet geographers usually are cartographers, too—not always highly skilled, but capable of evaluating maps and charts. Cartography, then, is a *tool* for geographers. Although it may be an overstatement, there is some justification in the saying "If it cannot be explained on a map, it isn't geography." Geographers use maps to explain and show the interaction between places,

TABLE 2-1 Effect of Earth's Shape on the Geographic Grid.

Latitude degrees	LENGTH OF 1° OF LATITUDE Miles	Kilometers	LENGTH OF 1° OF LONGITUDE Miles	Kilometers	Latitude degrees	LENGTH OF 1° OF LATITUDE Miles	Kilometers	LENGTH OF 1° OF LONGITUDE Miles	Kilometers
0	68.704[a]	110.57	69.172[a]	111.32	50	69.115	111.23	44.552	71.70
5	68.710	110.58	68.911	110.90	55	69.175	111.33	39.766	64.00
10	68.725	110.60	68.129	109.64	60	69.230	111.42	34.674	55.80
15	68.751	110.64	66.830	107.55	65	69.281	111.50	29.315	47.18
20	68.786	110.70	65.026	104.65	70	69.324	111.57	23.729	38.19
25	68.829	110.77	62.729	100.95	75	69.360	111.63	17.960	28.90
30	68.879	110.85	59.956	96.49	80	69.386	111.67	12.051	19.39
35	68.935	110.94	56.725	91.29	85	69.402	111.69	6.049	9.74
40	68.993	111.03	53.063	85.40	90	69.407	111.70	0.000	0.00
45	69.054	111.13	48.995	78.85					

[a] A statute mile is shorter than a nautical mile by 796 feet. When traveling about the landmasses, we use statute miles, but aircraft and ships at sea use nautical miles. We are driving at 50 miles per hour, and we mean 50 statute miles per hour. An airplane traveling at 600 *knots* is traveling 600 nautical miles per hour. There are 69 statute miles for every 60 nautical miles.

Source: Adapted from U.S. Geological Survey *Bulletin No. 650,* "Geographic Tables and Formulas," S.S. Gannett (1916), pp. 36–37.

people, and the various elements of nature, as well as their relationships and their distribution.

A map, however, is not the earth. It is not even the small piece of earth that it purports to represent. It is *symbolic,* a representation of something that is too big to be observed and analyzed in a small place. A map is used to portray either the entire earth or a small portion of it. If you were to thumb through the pages of an atlas, you might be awed by the variety of maps presented: maps showing political divisions; climatic, soil, or vegetation regions; population distribution; location of rivers; wheat-growing areas; or oceans and their currents.

Actually, one of the most important things for any of us to know is our location on this planet. If you should become lost in a strange city, you could consult a map of the city and pinpoint your position. If you plan to go hiking or camping in some out-of-the-way wilderness area, you can obtain a map of the region and find the specific locations of the places you want to visit, thus saving yourself the uncertainty of random wandering.

Today, one can go into a department store, stationery shop, or bookstore and discover an endless variety of maps. Perhaps it is natural for us to take this for granted, but it has not always been this way.

The Evolution of Mapmaking

For thousands, perhaps millions, of years humans knew little about the earth about them. They lived in a frightening world, eking out a precarious existence, living from day to day, wandering from place to place in the everlasting search for food. Although they had no maps, did they know where they were? To be sure, they had to be intimately aware of their terrain, their bit of earth space. Using the knowledge gained in their travels, they were able to fix an area's location by observing large trees, huge rocks, mountains, or streams—and maybe the stars and the sun. An individual was compelled to know where he was, else how would he dare go out on a hunt that might last for days? His only means was to observe carefully the physical features of the earth about him, so he could find his way back to his cave, where his mate and children awaited him.

The first maps probably were drawn when people gathered into tribal groups. A hunter, wishing to share with other hunters his knowledge of a valley where herds of wild game abounded, may have constructed a crude map by scratching in the dirt in front of his cave. With the passage of time, new methods evolved, such as using a piece of charcoal to mark out the location of a distant water hole on a piece of tree bark. Still later, men used fire-hardened sticks to press lines onto flat pieces of wet clay. The clay slab would then be baked over a flame, and the result was a symbolic, easily portable representation of a portion of the earth.

Archaeologists have found some of these old maps in the ruins of cities that were once the pivotal points of ancient empires, such as Akkad, Sumer, and Babylon, in the broad valley of what was once known as Mesopotamia, the Land of the Two Rivers (Tigris and Euphrates). The rulers needed maps in order to maintain their control over the people and lands. Phoenicians, earliest of the Mediterranean peoples to sail the seas, made charts that were usable only within sight of land. The Greeks, too, made maps as they increased their knowledge and expanded their empire. Africa's Arabs, on the other hand, were not proficient in mapmaking, although they did manage to map the positions of important oases and trade routes across the deserts.

Polynesians, who long ago colonized islands in the South Pacific, came from somewhere out of the west, perhaps from the vicinity of Southeast Asia, and sailed eastward across the mysterious ocean to the islands of the Pacific. Although they did not have maps as we know them, their knowledge of navigation approached the level of a true science. They were able to sail across the vast reaches of the Pacific by using their "stick-and-shell" maps, which were based on the positions of stars and the movement of ocean currents and waves. However, it was not until the Chinese invented the **magnetic compass** that ships could confidently venture forth into unknown regions. Navigators had long used the North Star and the sun to determine their positions at sea, but imagine the difficulties they faced when the sky was heavily overcast or when storm clouds obscured the rocky headlands and dangerous shoals. The introduction of the compass made it possible for seafarers to travel anywhere, any time. They could sail beyond the horizon and discover new worlds to conquer, to exploit, to settle. The compass and the map were and still are indispensible tools.

When the Romans began to ride the seas, they became excellent mapmakers; it is interesting to note that our term *map* comes to us from the Latin *mappa*, which means "cloth" or "napkin." Is it possible that Roman seamen drew sketches of their travels on the tablecloths of their favorite seafront grog shops? Were those tablecloths later used as maps?

Round Globe to Flat Map

The only valid representation of the earth is the globe. It is spherical, as is the earth, and it can help you to gain an understanding of the earth's shape, the location of the various continents, and the geographic grid, and to visualize the reasons behind the changing of the seasons. However, the globe has one great drawback: It is difficult to cart around. Can you imagine taking a globe with you on a round-the-world flight or trying to fit it into the glove compartment of your car? This is, of course, why we use flat maps. Maps may not be as accurate as a globe with regard to distance and direction or in the size of areas in different parts of the world, but they are convenient to insert into textbooks, newspapers, and pockets. Their drawbacks are, especially, in their inherent distortions. David Greenhood, in his book *Mapping*, explains it this way: "Here, then, is an instance where the unscrupulous propagandist can get in his dirty work. He can give a picture of one country being impressive or insignificant as compared to another by using a map in which the countries are not shown in their truly equivalent sizes."[5] And this warns us that if we are really to understand a map, we must know what properties it must have.

How can you flatten a round object without changing its characteristics? To understand the problem, let us pretend that an orange, the kind with a thick peel, is a globe. With a marking pen, draw the parallels, meridians, and continents on the orange peel. Then cut down each line of meridian almost to the equator. Carefully peel the orange (Fig. 2-8) and flatten out the peeling. What does it look like? Does it still resemble the globe? What about the large empty spaces?

This is the problem that confronted mapmakers after they came to believe the earth was round: how to turn a round globe into a flat map and not have a lot of voids. If the meridians are supposed to run in a true north-south direction and converge at the poles, and the parallels are supposed to run in a true east-west direction, be equidistant, and remain parallel to each other and the equator, how does the mapmaker keep them that way on a flat map? No one has as yet found the perfect answer, because all flat maps must have some distortion (Fig. 2-9). From this we can infer that the maps with which we are acquainted have been manipulated, altered, and modified in order to be of use to us. This means that car-

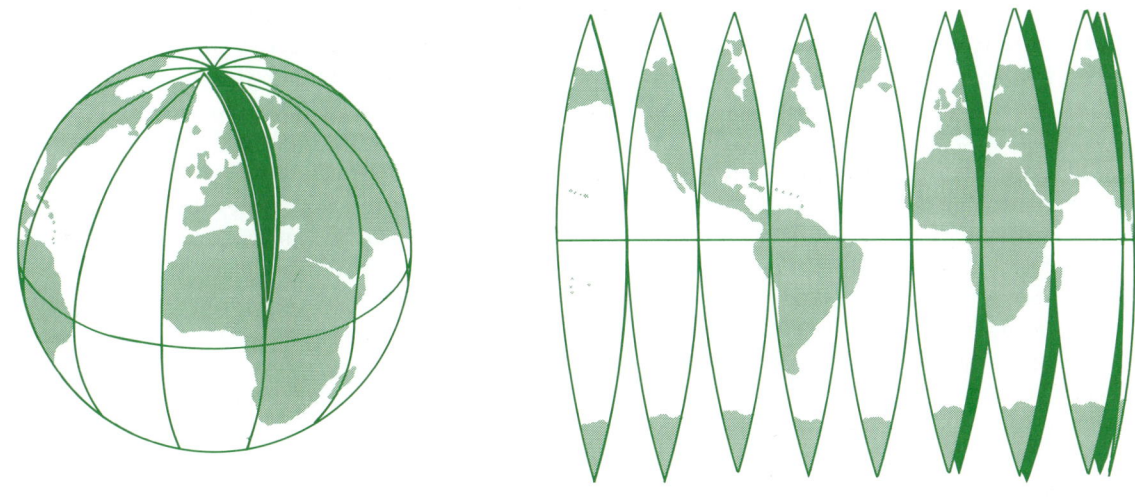

FIGURE 2-8 Round globe to flat map: peeling an orange

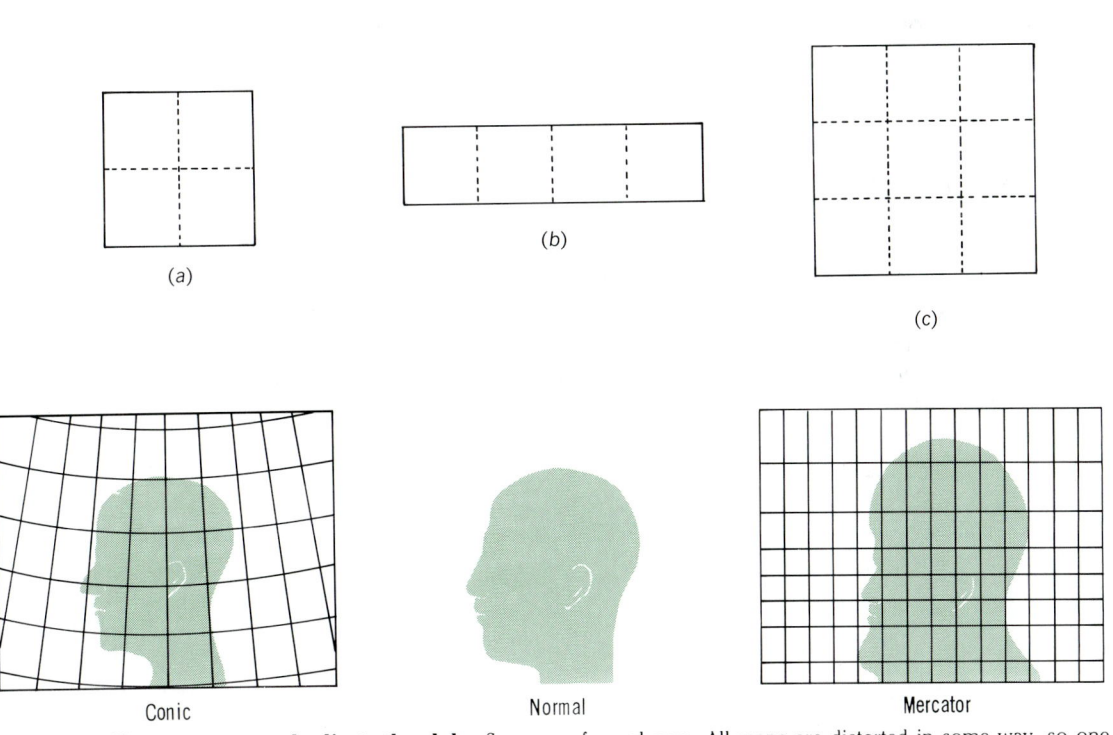

FIGURE 2-9 Flat maps cannot duplicate the globe Some maps are conformal, which means that areas on the map have the same *shape* as they do on the earth or the globe. Look at (*a*) and (*c*). (*a*) and (*c*) are not the same size. They differ in the amount of *area* enclosed. (*a*) and (*b*), however, enclose the same amount of area (earth space) even if they do not have the same shape; thus, (*a*) and (*b*) are considered to be of equal area. All maps are distorted in some way, so one uses a particular projection that is designed for a particular purpose. For example, the Mercator projection is used for ship navigation; azimuthal projections for air travel; and conic projections for maps that portray midlatitude regions such as the United States or Europe.

tographers must consider which of the qualities that a globe possesses are most important for the job at hand. They must first ask which of the various properties must be emphasized. Should the areas on the map be equal to each other in area, just as they are on a globe? But what if the areas will not look just as they do on a globe? Should the map be constructed so that an airline navigator can take true direction from his map? These questions must be answered before cartographers can proceed to construct map projections.

Properties of Map Projections

Properties of map projections include equal area (equivalence), conformality (conformal), equal distance (equidistant), and azimuthal (zenithal). Some map projections are limited to just one of these properties, some have two or more, but cartographers try to soften the deviations from a global representation. If the shapes of islands or coastlines are to be shown "correctly" (to look as they do on a globe), the property to be attained is **conformal.** If areas conform to their actual shape, South America and Greenland will look as they do on a globe. They will not be shown in the proportion to other areas or to each other—as you can see by examining a conformal projection, such as the one developed by Mercator in the sixteenth century (Fig. 2-10). Compare these two landmasses as to shape and size, then examine them on a globe. Is Greenland really larger than South America? It might be noted that there is less distortion of small areas on the map and areas along or near the equator.

Map projections that are considered **equal area** portray areas in the same relation to each other as they are on a globe. This means that all areas on the map are in the proper proportion to each other. On an equal-area world map, 1 square inch measured over North

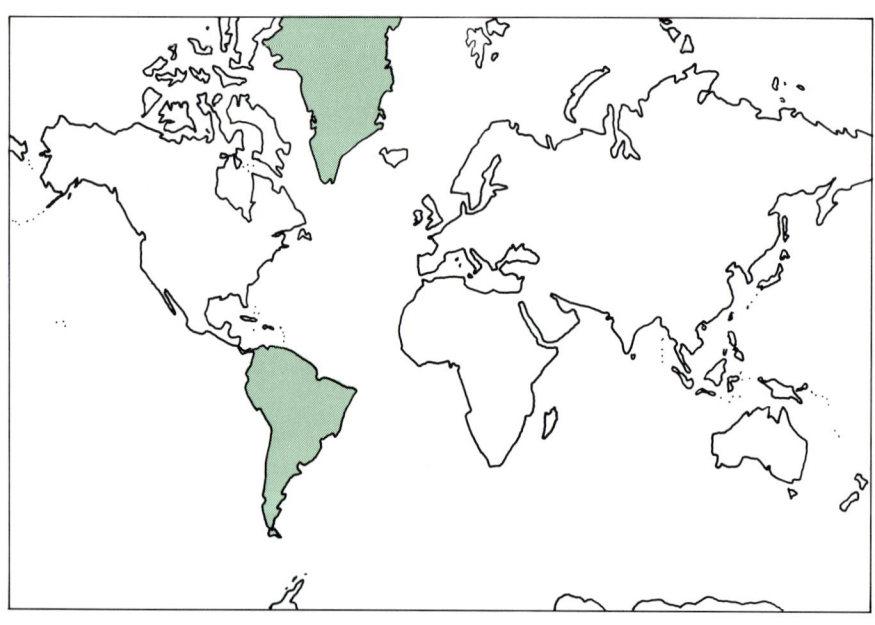

FIGURE 2-10 Mercator's projection: conformal emphasized.

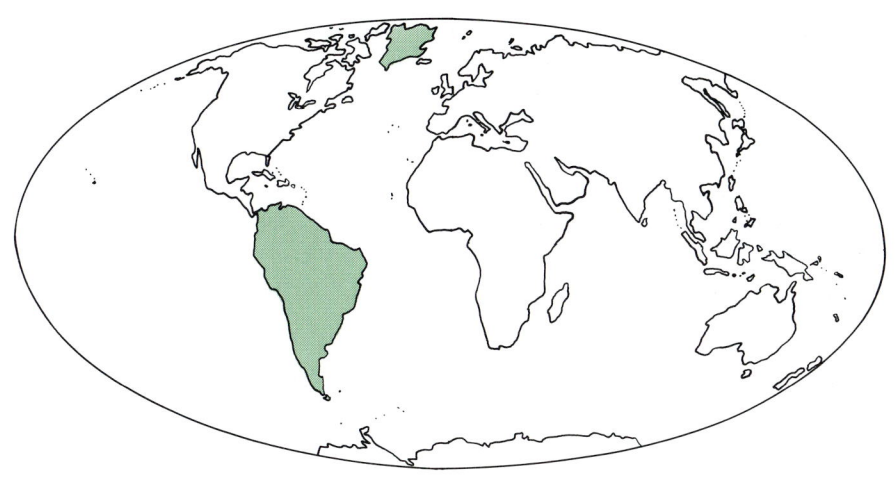

FIGURE 2-11 Mollweide's projection: equal area emphasized

America will enclose the same amount of area as 1 square inch measured over Africa or Siberia. In other words, areas on the map will enclose the same amount of space they would on a globe (Fig. 2-11).

If accuracy in measuring distance is important for the projection, the property of **equal distance** is emphasized, which means that distances can be measured with accuracy from the focal point of the map (Fig. 2-12). This focal (or central) point can be set at one of the poles, or it can be placed over any location on the earth. Many aerial navigation charts (maps) are constructed with this emphasis. Some, however, are more concerned with direction, and the property of distance will be

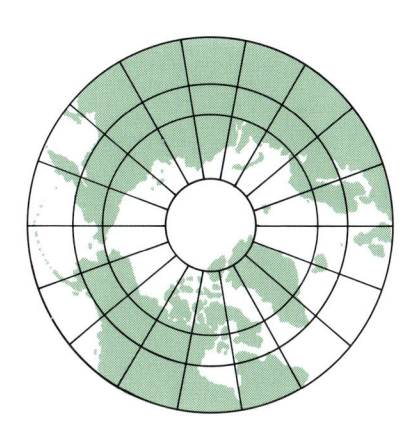

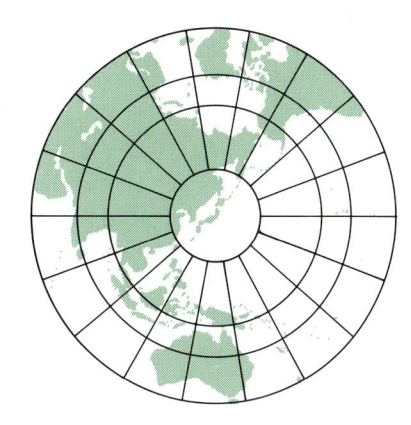

FIGURE 2-12 Azimuthal map projections (*a*) is a zenithal projection with the central, focal point at the geographic North Pole; (*b*) has its focal point at Tokyo, Japan. Both are equal-distance map projections.

sacrificed. In this instance, the projection is **azimuthal,** and all directions taken from the center are true compass directions. If the center were placed on New York's Kennedy International Airport, the navigator could lay out the flight path with ease.

A map, then, is a concession; it is not the whole truth. Each of the map projections we will consider later in this chapter have been formulated to handle a specific job—to be used for ship or air navigation, to portray the entire earth, to focus on certain midlatitude regions, to be used in newspapers, or to be used in textbooks. A projection engineered to portray the United States does not have to be the same kind needed to portray the world.

It might be well for us to examine some of the more important elements in the construction of map projections.

Map Essentials

Once a cartographer has settled on the kind of projection needed, a series of questions must be answered. How much of an area am I going to show? A city or a portion of a city? A state or a larger region encompassing several states? An entire continent—or a hemisphere? Or a whole earth? What scale shall I use—large or small? How shall I indicate directions? What symbols are needed? What will be my title? Figure 2-13 indicates one cartographer's answer. Let us consider these essentials.

Map Scale Maps are not merely pictures drawn to a person's whim; they are representations of the earth or a portion of it. They must be realistic; they must also be accurate in as many respects as possible. One way the cartographer attempts to make a map meaningful is to employ a scale that indicates the ratio between distances on the map and distances on the earth space being represented. There are three types of map scales in use today; some maps indicate only one, some have all three (Table 2-2). Any of these scales can be used for very small areas (such as your college campus) or for extremely vast areas (such as the Western Hemisphere or the entire earth).

Figure 2-14 illustrates the difference between large- and small-scale maps. If the size of the area to be portrayed on a map is small and a considerable amount of detail must be shown

TABLE 2-2 Map Scales.

A map scale is the ratio of the distance on a map as it corresponds to the actual distance on the earth's surface. Without an accurate scale, a map is useless and unreadable.

VERBAL SCALE Also known as the *stated scale*. The ratio is given in words: "1 inch (2.5 centimeters) equals 100 miles (160 kilometers)," or "1 inch (2.5 centimeters) on a map represents, or is equal to, 100 miles (160 kilometers) on the earth's surface."

GRAPHIC SCALE Also known as the *bar, line,* or *linear scale*. The ratio is shown as a line marked off in units (specified intervals or increments):

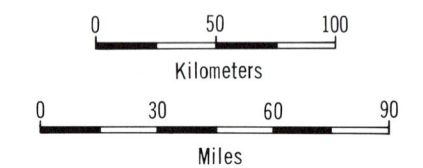

```
0           50          100
|—————|—————|
        Kilometers
0      30      60      90
|———|———|———|
          Miles
```

REPRESENTATIVE FRACTION SCALE Also known as the *R.F. scale*. The ratio is given as a fraction without specifying a definite unit (such as inches, centimeters, yards, meters, miles, or kilometers). The ratio states (1:63,360 or 1/63,360) that one unit on the map represents a specified number of the same kind on the earth's surface. For example, 1:63,360 could mean that 1 inch on the map equals 63,360 inches on the earth.

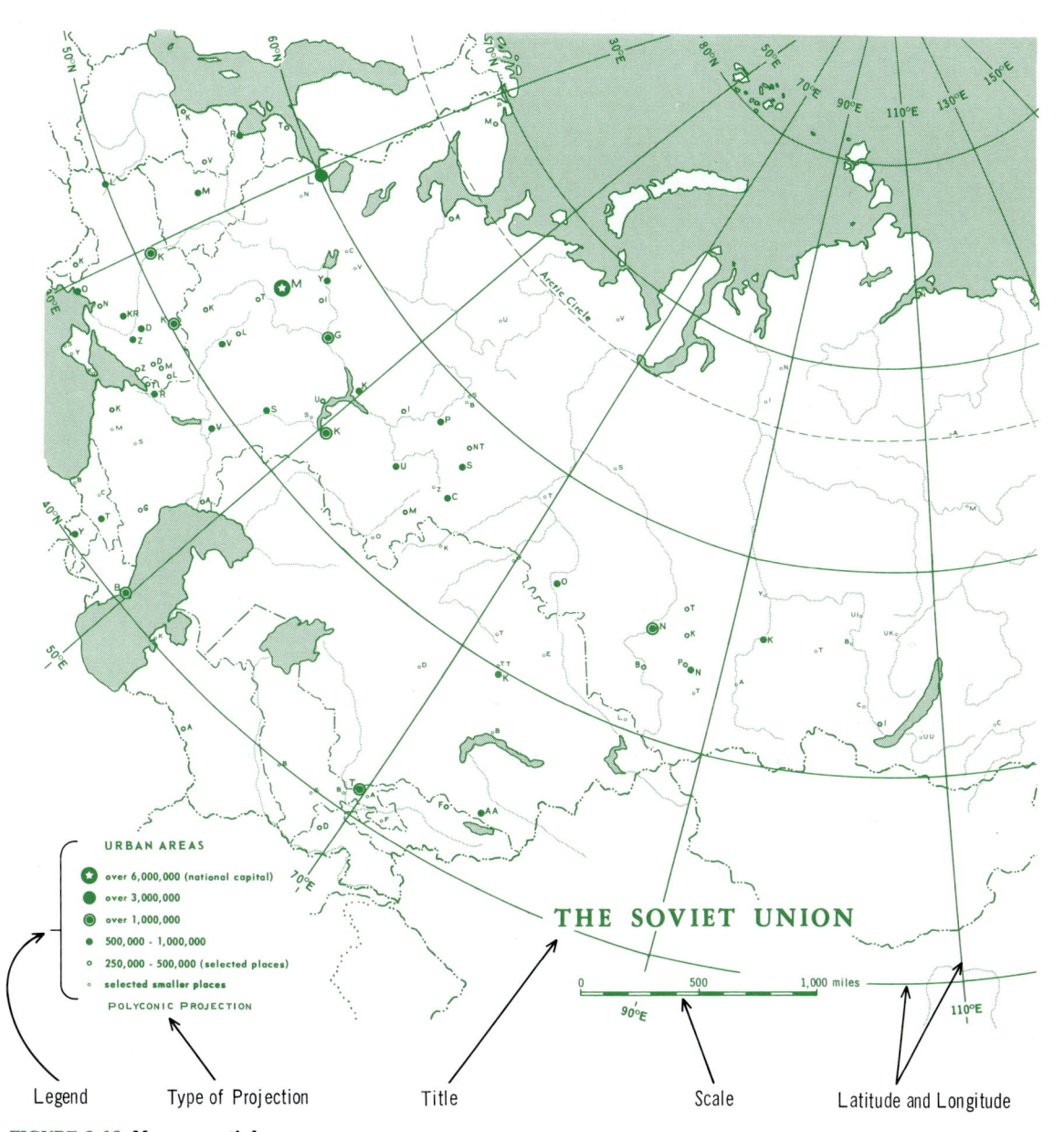

FIGURE 2-13 Map essentials

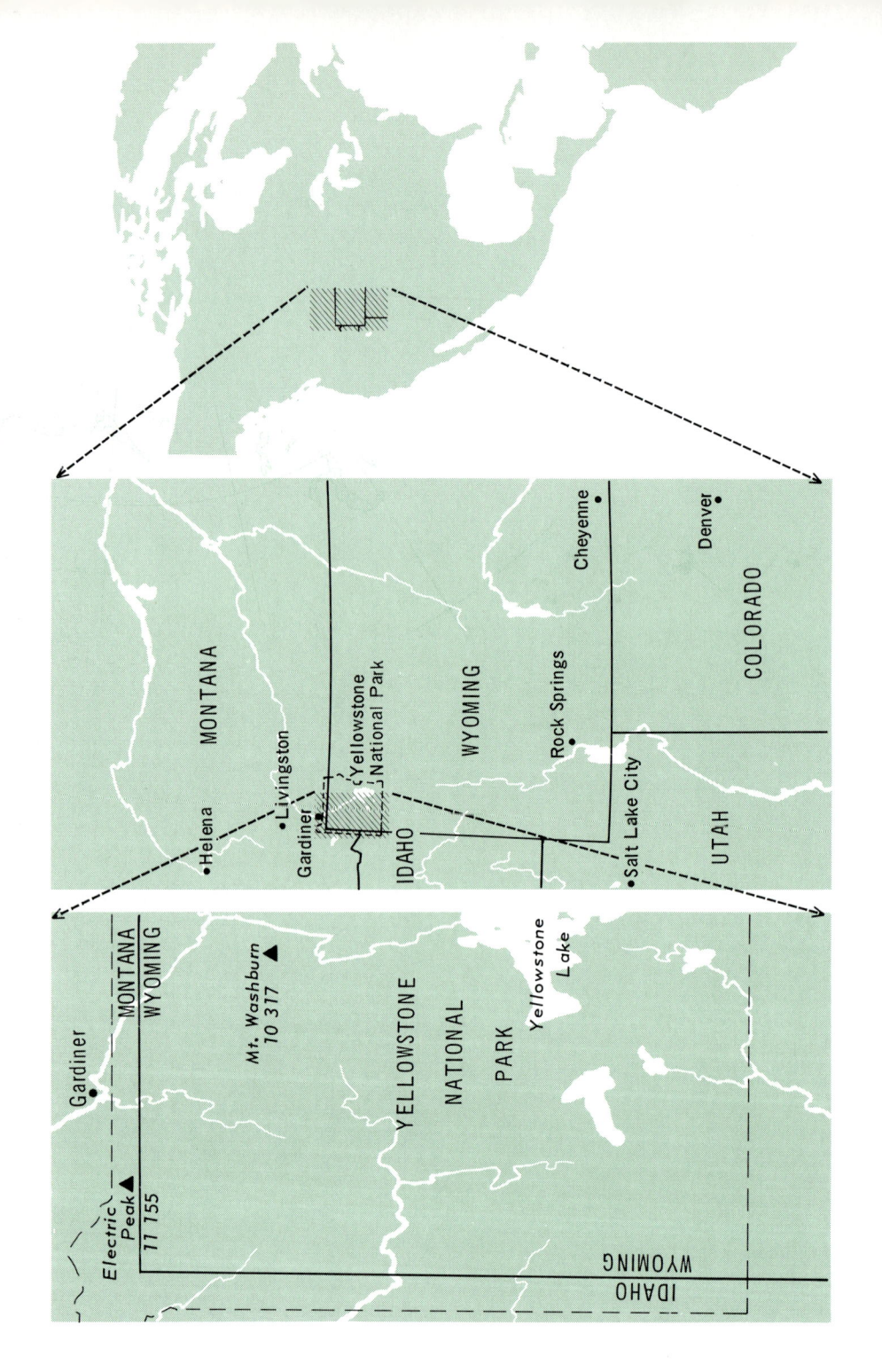

FIGURE 2-14 Large-scale to small-scale maps

(such as houses, factories, roads, paths, farms, railroads, power lines, churches, schools, and so forth), a **large scale** will be used. As an example, on some large-scale maps 1 inch (2.5 centimeters) equals 24,000 inches (60,000 centimeters), or 0.26 miles (0.42 kilometers). If, however, a much larger portion of the earth's surface is to be portrayed (such as a province or state, nation, hemisphere, or the entire world), the amount of detail will be severely reduced, since the cartographer must compress a vast area onto one small page. On some **small-scale** maps 1 inch (2.5 centimeters) equal 100,000,000 inches (250,000,000 centimeters), or 1600 miles (2560 kilometers).

Although the verbal and graphic scales are self-explanatory, the representative fraction (R.F.) scale is sometimes confusing. To clarify it, remember that the bottom number (the denominator) holds the key. The larger the denominator, the smaller the scale. An R.F. scale of 1/15,000 is a much larger scale than one calling for 1/15,000,000. To further clarify, we can say that on a map of 1/15,000, 1 inch represents 15,000 inches (about 0.23 miles or 0.36 kilometers). On a map with an R.F. scale of 1/15,000,000, 1 inch represents 15 million inches, or about 236 miles (377 kilometers). Obviously, the R.F. with the smaller denominator is a much larger scale. Just as with any fraction, the smaller the denominator, the larger the fraction. Thus, 1/10 is larger than 1/10,000,000.

Map Legends and Symbols Any map that has something to say, some information to get across, has a title. For example, thumb through the pages of this book and examine the various maps. Each has a title, a caption that tells you what the particular map is attempting to explain. A title alone, however, usually is not sufficient. There must be additional explanation of just what all the symbols—colors, dots, cross-hatching, or shadings—mean. This is the job of the map's **legend** (Fig. 2-15). The purpose of the legend is to explain those symbols. (Fig. 2-15c uses dots to show the distribution of human population on the face of the earth. Figure 2-15a uses graduated circles to show the relationship of one area or place to another, regarding percentages or the proportion of a segment of the population of a city to the total. Figure 2-15b uses colors and numbers to explain the distribution of soils.

On many of the maps that use color, certain colors seem to be standardized. For example, blue usually denotes water features, such as streams, rivers, lakes, seas, or oceans. On certain maps (such as the topographic maps made by the U.S. Geological Survey) green is used for vegetation, black is used for buildings and cities, and red double lines are used for major highways. Some maps have special symbols, such as ✝ for churches, ⌐ for schools, and ⌒ for intermittent streams. For additional examples, see Fig. B-2 in Appendix B.

Later in this text you will find another type of symbolism: lines sweeping across the map, swerving this way and that—boundary lines that connect points of equal value and separate areas of different values. Some geographers call these lines *isolines;* others refer to them as *isopleths* or **isarithms.** The root *iso* comes to us from the Greek *isos,* meaning equal. Isarithms refer to lines that have an exact value base; isopleths have a relative or non-absolute value. **Isotherms** (see Figs. 4-4 and 4-5) are lines on a map that connect points that have (at a specific point in time) the same air temperature. **Isobars** (see Figs. 4-8 and 4-9) are lines that connect places with equal atmospheric (barometric) pressure. These various isolines will be discussed frequently in later chapters.

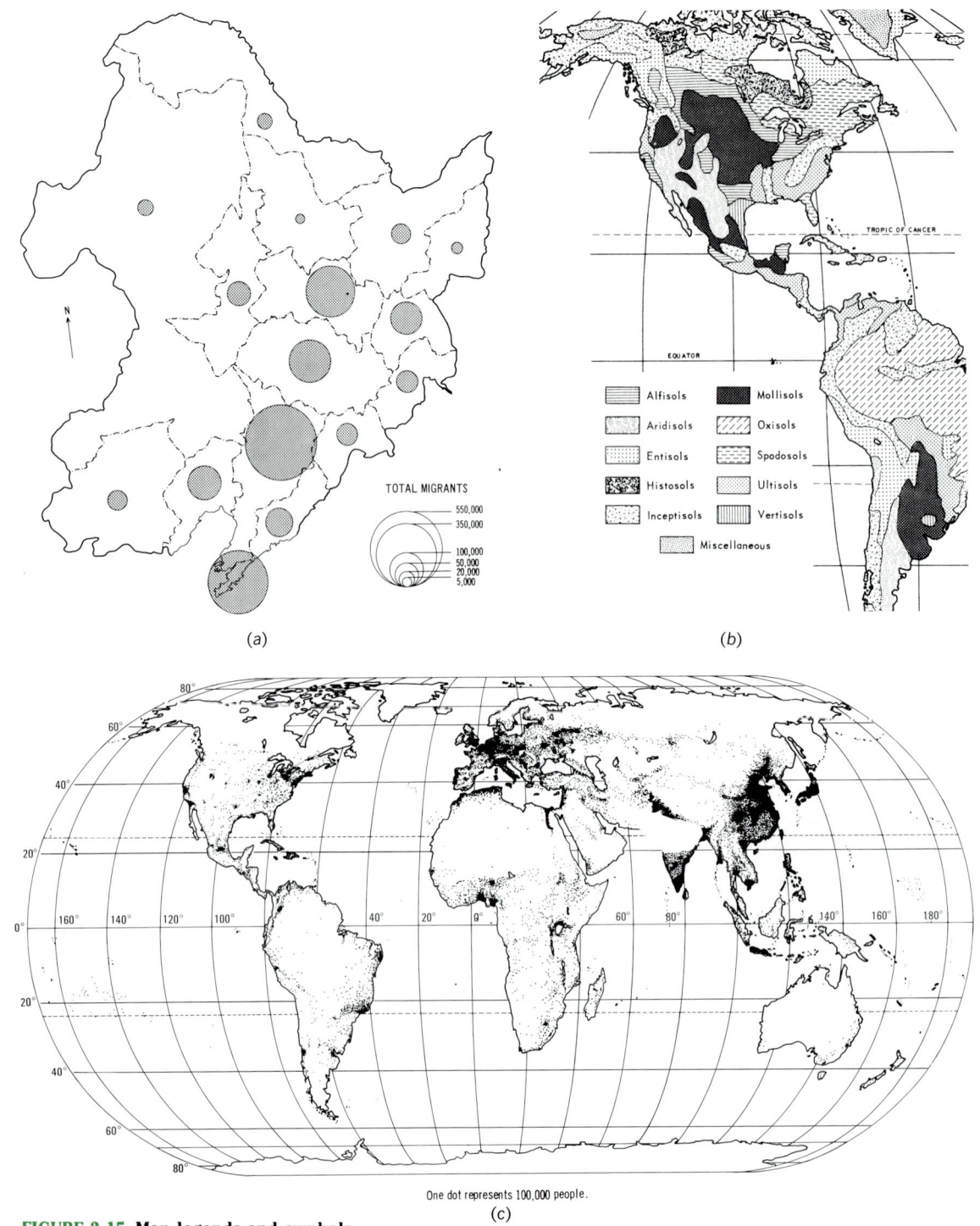

FIGURE 2-15 Map legends and symbols

Map Direction When you pick up a map, do you immediately know where north is? Do you realize that there is not just one north, but three? Is north one particular spot on the earth? If so, how could you pinpoint north on a projection such as that shown in Fig. 2-16a? If every meridian runs in a true north-south direction, how can there be an infinite number of norths? Obviously, there cannot be. If you check Fig. 2-16b, the Northern Hemisphere, you can see "true north," which is more properly called the **geographic north.** Now, what happens when you hold a magnetic compass in your hand, as in Fig. 2-16c? Does the magne-

(a)

(b)

(c)

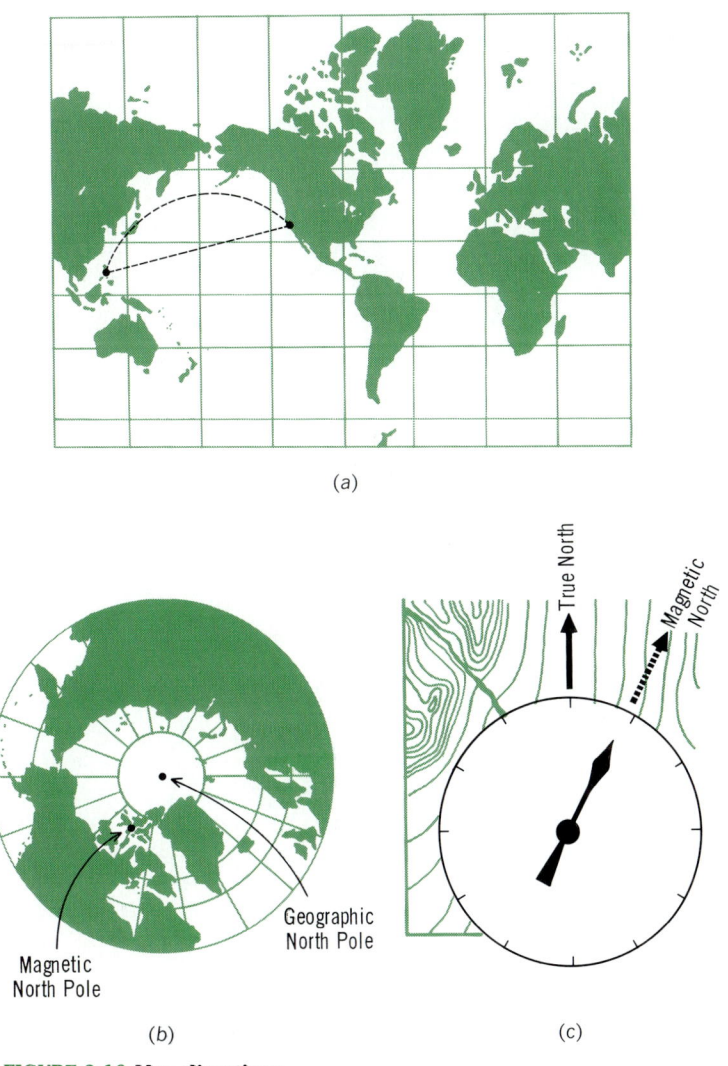

FIGURE 2-16 Map directions

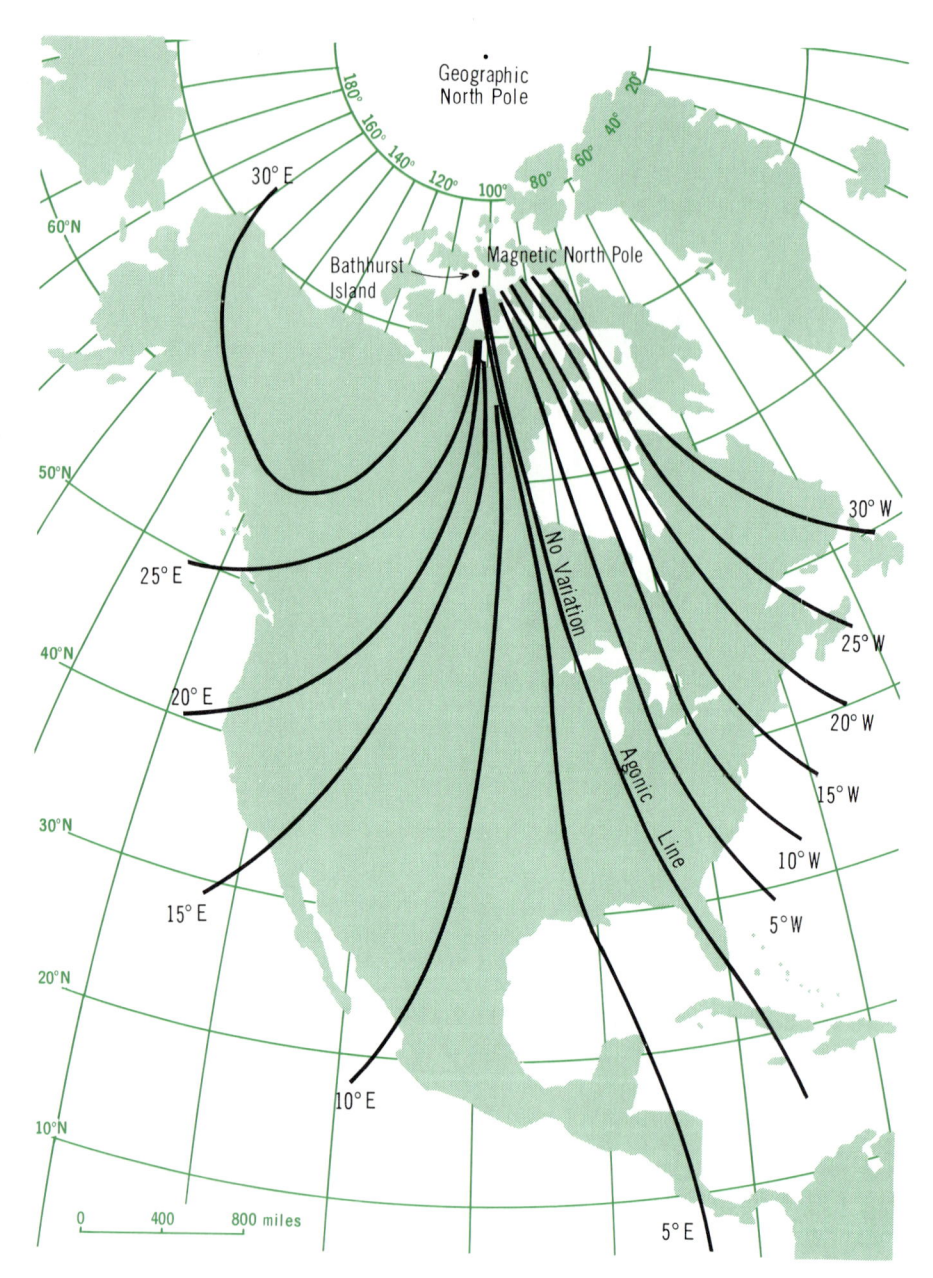

FIGURE 2-17 Isogonic map of North America If your position is to the west of the agonic line, your compass will point to the right (to the east) of true north. It will point toward the magnetic North Pole.

tized needle point toward the geographic north? No, because it aims toward the magnetic North Pole, which, in the mid-1980s, is located about latitude 76° north, longitude 102° west, near the Canadian island of Bathurst. In the Southern Hemisphere, the magnetic compass needle points towards today's magnetic South Pole, located about latitude 68° south, longitude 145° east, on the edge of Antarctica, directly south of Tasmania. These magnetic poles are not in a permanent line because they are constantly changing position, sometimes reversing after more than 200,000 years. This happens because the earth holds a mass of liquid, rather fluid mass of iron around its solid inner core, and this generates the powerful magnetic field within the earth that is also emitted out into space. The difference (which we will cover in more detail in Appendix B) is called the **magnetic declination** (or **variation**). This variation does not remain constant, because the magnetic north is always changing slightly. We must make allowances for the variation, and many maps indicate the number of degrees your compass needle will diverge from geographic north. If you were to crisscross North America, checking carefully to determine the exact amounts that your needle was "off," you would come up with a map (Fig. 2-17) covered with **isogons** (lines connecting points of equal magnetic variation away from the geographic north). You would then be able to use your magnetic compass accurately with your maps. You will note that there is one isogon marked 0°, where there is no magnetic variation (or declination from the geographic north). This is called the **agonic line,** and it is the only isogon where all points on it line up with both the magnetic and the geographic North Pole.

The last of the norths is called **grid north,** and this refers to the "bending" of the meridians while laying out the U.S. Land Survey System (see Appendix C, "Ground Referencing Systems"). Ownership of parcels of land was determined by establishing an initial, or principal, meridian in the area being surveyed (based on geographic north) and a baseline (a parallel of latitude). The lines bordering properties should have been aimed at the geographic north, but the local need for rectangular coordinates caused the surveyors to ignore (slightly, maybe as much as 3°) the true north-south meridianal lines. You might say this is a warning not to trust the borders of a map, because they may not point to the geographic north.

Types of Map Projections

Our purpose here is not to compete with the many excellent texts on cartography,* but merely to acquaint you with some of the ways cartographers have tried to portray our earth. There are two basic ways to develop a map projection and in both methods the same result is sought: to have the parallels and meridians intersect exactly as they do on a globe. One way is to project the round face of the earth onto a flat plane by using a lighted (from the inside) globe. The parallels, meridians, continents, islands, and oceans are cast like shadows onto the plane. The other way is to use mathematics (geometry and trigonometry) to produce a special projection suited for a special purpose.

Cylindrical Projections
To construct **cylindrical projections,** an internally lighted, transparent globe is placed inside a cylinder composed of something like clear plastic (Fig.

---

*Those interested might consult David Greenhood (1964), *Mapping* (Chicago, Ill.: University of Chicago Press); Erwin Raisz (1948), *General Cartography* (New York: McGraw-Hill); A. H. Robinson and R. D. Sale (1969), *Elements of Cartography,* 3rd ed. (New York: Wiley); Judith Tyner (1973), *The World of Maps and Mapping* (New York: McGraw-Hill).

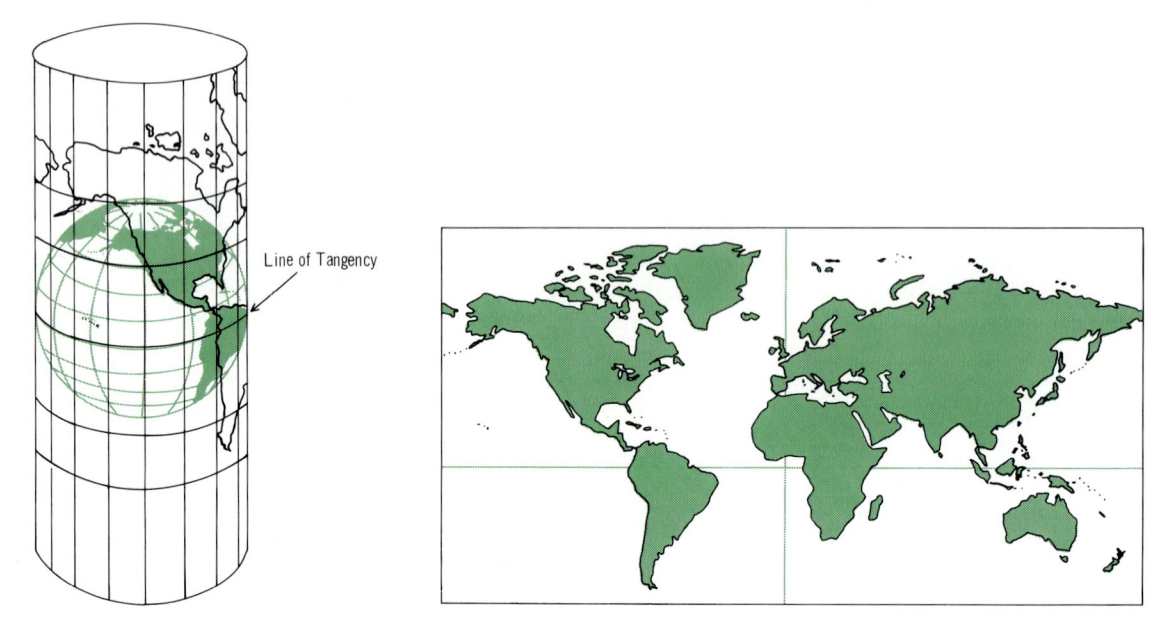

FIGURE 2-18 Construction of a cylindrical projection

2-18). When the light is turned on, the lines representing the parallels and meridians, and the continents and other landforms that have been printed on the globe, are projected onto the covering cylinder (just as movies are projected onto a screen). The line where the cylinder touches the globe is called the **line of tangency,** and only along this line, or a few degrees on either side, is there accuracy. Away from the line of tangency, the parallels are spaced farther and farther apart, and different scales must be used for measuring distance. As for the meridians, they no longer converge as they approach the poles. Instead, they are parallel to each other and are equally spaced from the equator to the poles.

Of all the cylindrical projections constructed, only the Mercator is conformal, and it is used primarily for ship navigation. This is because navigators can ignore the tremendous distortions, since a straight line drawn along the route from one port to another will cut every meridian at the same angle. Thus, the ship has only to steer a compass course along that line, with only slight direction changes, and it will travel the shortest distance between the two ports.*

Conical Projections To construct **conical projections,** a clear plastic cone is placed over an internally lighted globe (Fig. 2-19) with the apex of the cone placed atop the globe. For a simple conic projection, the cone touches the globe along one line of tangency (standard parallel). When the light is turned on, the geographic grid and landforms printed on the globe are transferred to the plastic cone. The meridians appear just as they do on the globe: They converge toward the pole. Parallels, how-

---

*The line that intersects all meridians at the same angle is called a *rhumb line,* or *loxodrome.*

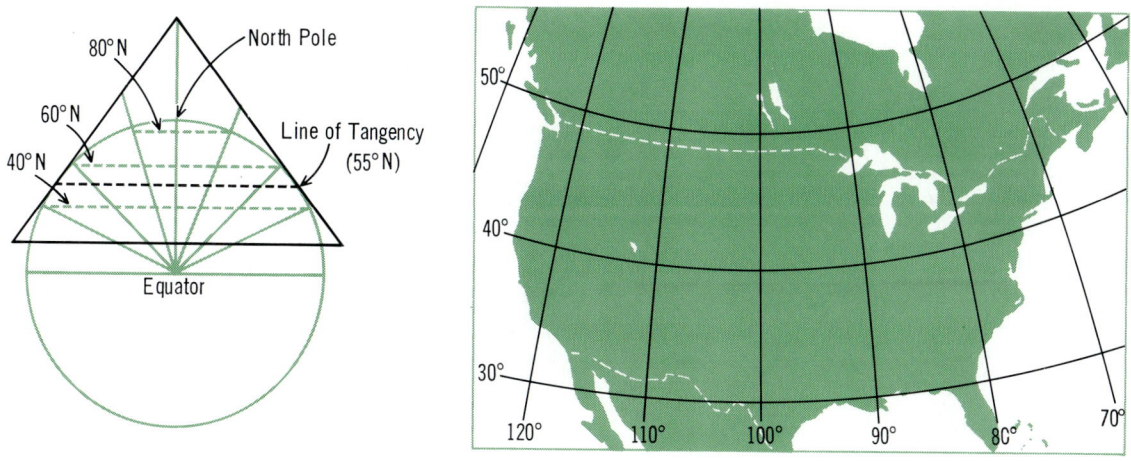

FIGURE 2-19 Construction of a conical projection

ever, are not equally spaced, unless the cone intersects the globe at more than one parallel, in which case a *modified conic* projection, one that has less distortion of areas, results.

Another form of conical projection, the Lambert conic conformal (Fig. 2-20), was developed by a cartographer named Lambert in the eighteenth century. Lambert laid out the parallels mathematically, so that the shapes of the continents were conformal. This projection is widely used to portray midlatitude regions, such as the United States, Europe, or the Soviet Union.

Azimuthal Projections For **azimuthal projections** a flat transparent sheet is placed over a lighted globe (Fig. 2-21), and here the point of tangency is the center of the projection. In the

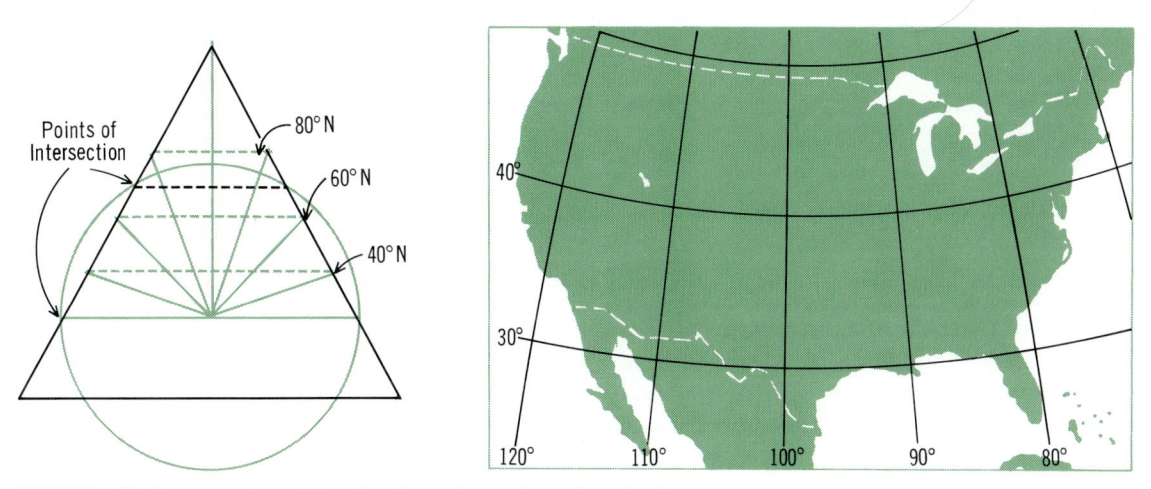

FIGURE 2-20 Construction of Lambert's conic conformal projection

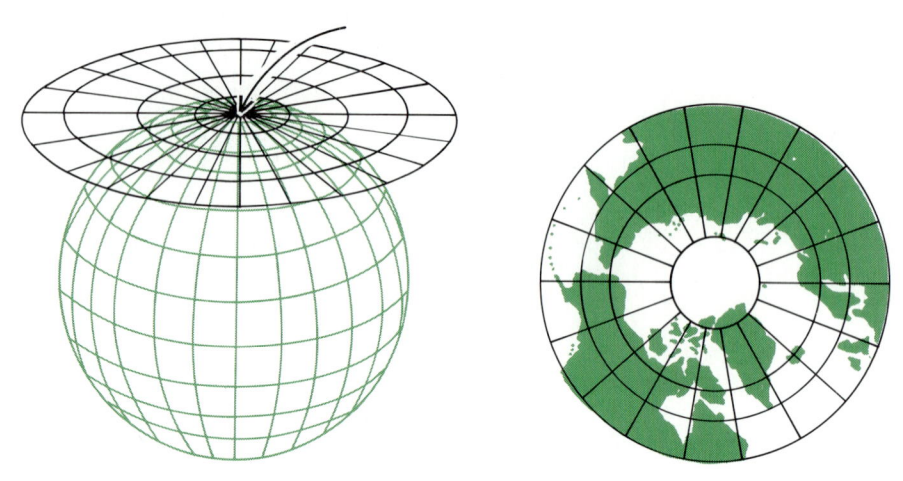

FIGURE 2-21 **Construction of an azimuthal projection**

case of a polar-azimuthal projection, the meridians radiate out from the central point (the pole) like spokes on a bicycle wheel. Lines drawn from that point to any other location on the map are straight lines, part of a great circle. There is also a true compass direction from that point. Azimuthal projections can have any point on earth for their central, or focal, point, and then we use different names: *oblique, azimuthal equal area, orthographic, polar stereographic, equatorial stereographic,* and so on. All of these designations are useful, and each projection has a specific use, whether it is to portray entire hemispheres and continents or relatively small areas, or whether it is for newspaper reproduction, airline navigation charts, or ocean sailing charts.

Miscellaneous Projections Included in the category of miscellaneous projections are maps that have a combination of properties and that have been formulated through mathematical equations. In some cases, the meridians and parallels have been manipulated to suit the mapmaker's purpose. A good example of this would be the homolographic projection, in which the meridians have been turned into ellipses and the globe has been stretched out from side to side. The distance between the poles is about half the distance along the equator, yet the projection remains equal area. One of the most popular of these projections among today's geographers was developed by Dr. Paul Goode. It is an interrupted homolosine (Fig. 2-22) projection, which means that it is a combination of two different mathematical projections, homolographic and sinusoidal. In this projection each continent is centered on its own meridian, and the distortion of the landmasses is minimized in the higher latitudes by splitting the oceans.

"Eckert IV" (Fig. 2-23), the map projection for this text, is an equal-area projection devised by Max Eckert to minimize distortion and crowding of the landmasses and oceans in the higher latitudes. Despite some shearing of the continents and islands near the polar regions, this projection is an excellent one for showing world distributions. The parallels are straight lines that are mathematically designed

MAPS 45

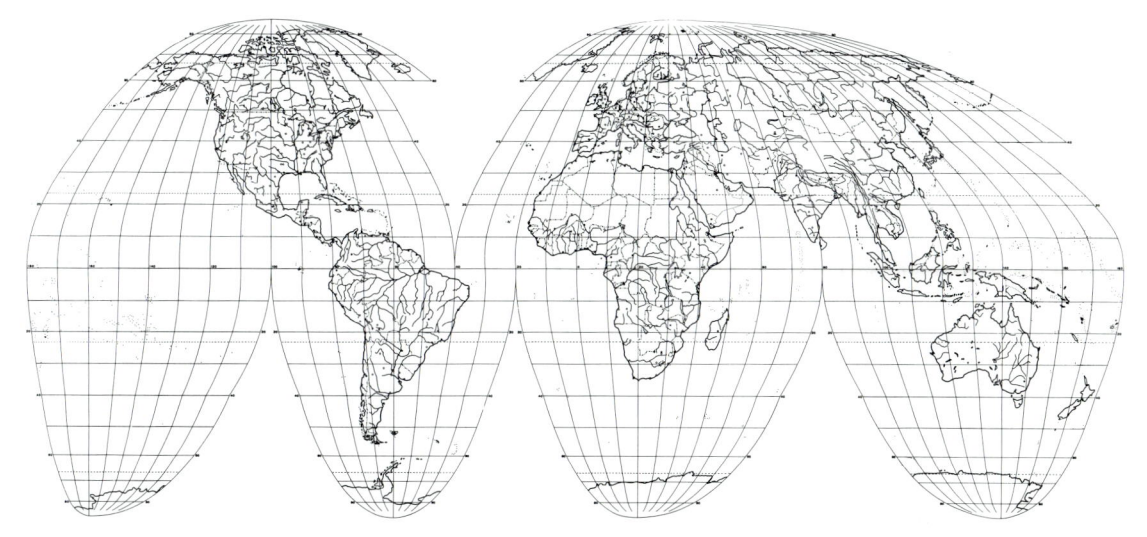

FIGURE 2-22 Goode's interrupted homolosine projection (Copyright by The University of Chicago, Department of Geography.)

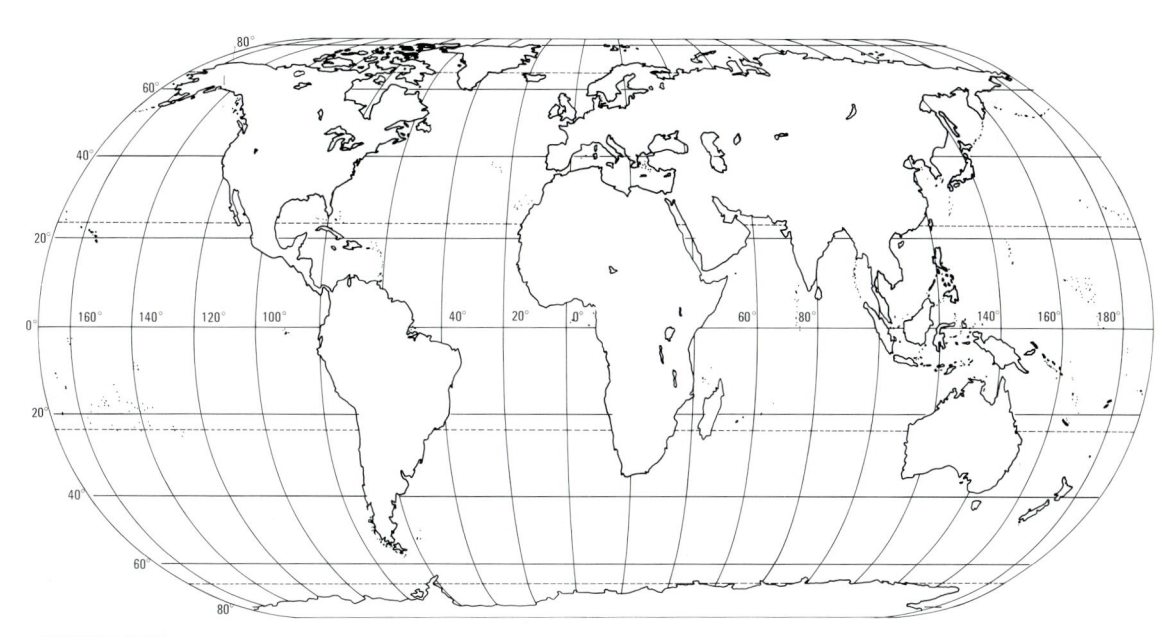

FIGURE 2-23 Eckert IV projection

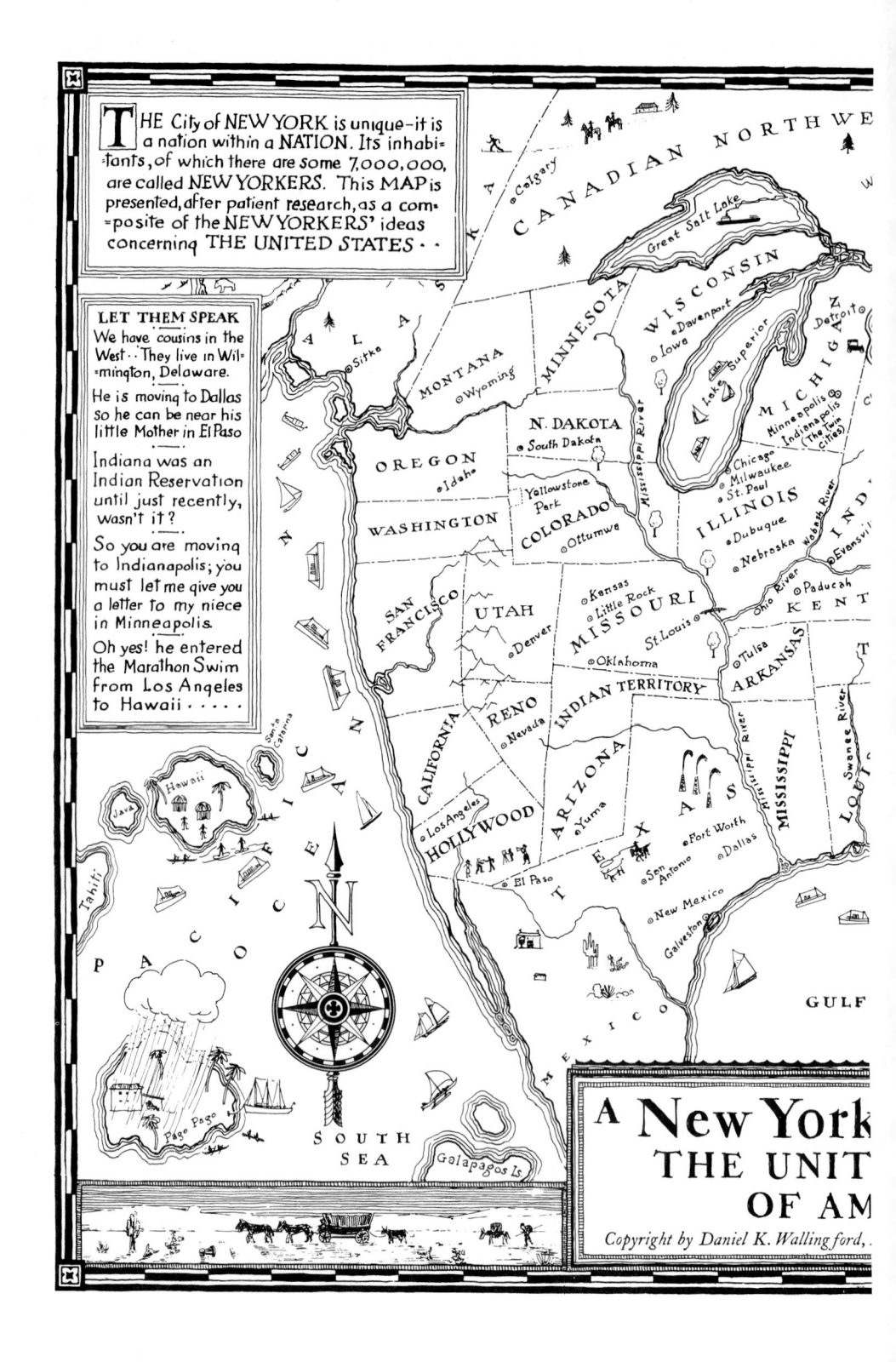

THE City of NEW YORK is unique – it is a nation within a NATION. Its inhabitants, of which there are some 7,000,000, are called NEW YORKERS. This MAP is presented, after patient research, as a composite of the NEW YORKERS' ideas concerning THE UNITED STATES ··

LET THEM SPEAK

We have cousins in the West·· They live in Wilmington, Delaware.

He is moving to Dallas so he can be near his little Mother in El Paso

Indiana was an Indian Reservation until just recently, wasn't it?

So you are moving to Indianapolis; you must let me give you a letter to my niece in Minneapolis.

Oh yes! he entered the Marathon Swim from Los Angeles to Hawaii·····

A New York
THE UNIT
OF AM

Copyright by Daniel K. Wallingford,

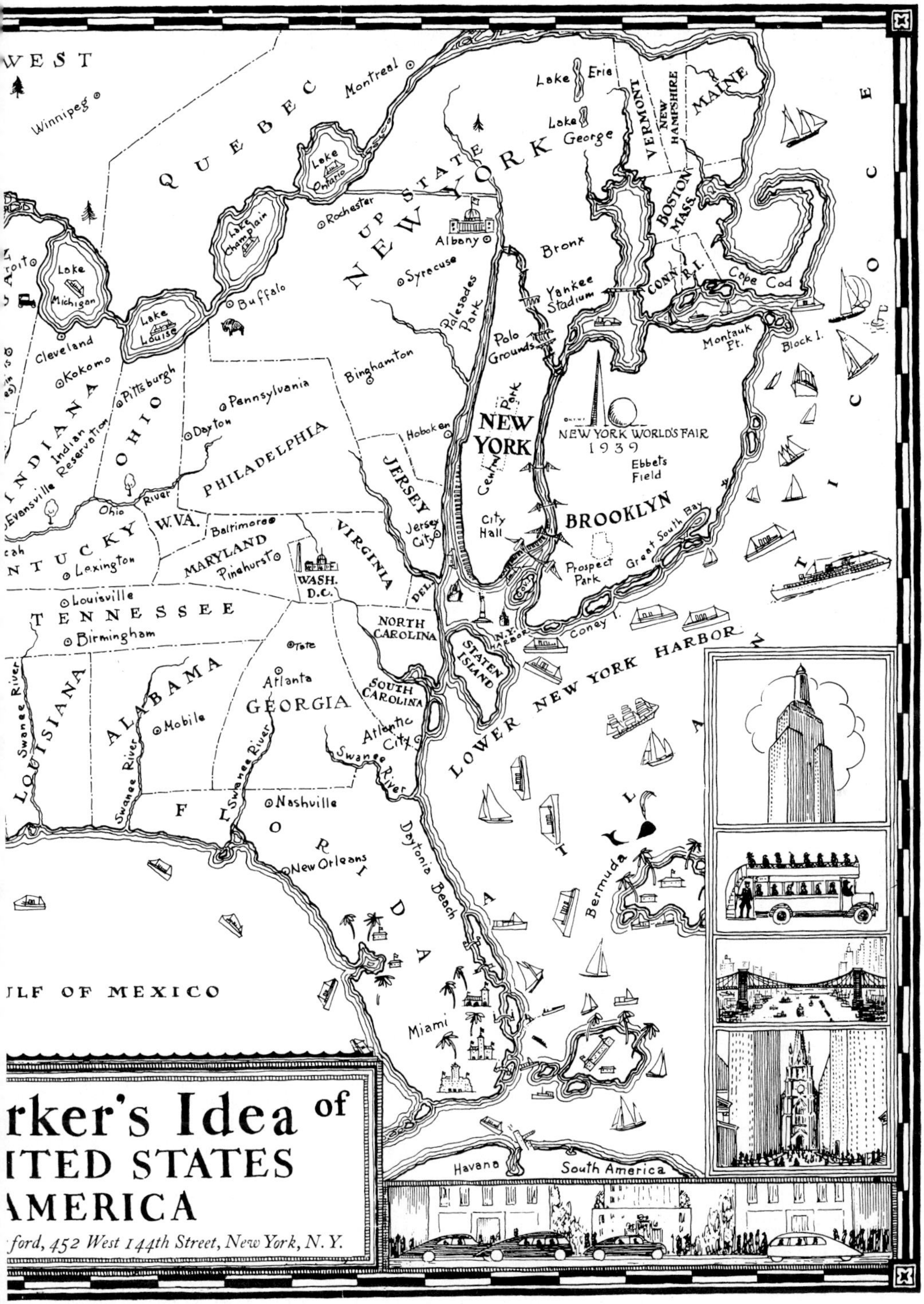

rker's Idea of
ITED STATES
AMERICA
ford, 452 West 144th Street, New York, N.Y.
QUEBEC
Montreal
Winnipeg
Lake Erie
Lake George
VERMONT
NEW HAMPSHIRE
MAINE
BOSTON MASS.
CONN.
R.I.
Cape Cod
Montauk Pt.
Block I.
Lake Ontario
Lake Champlain
Lake Michigan
Lake Louise
UP STATE NEW YORK
Rochester
Albany
Syrecuse
Buffalo
Bronx
Yankee Stadium
Palisades Park
Polo Grounds
Cleveland
Kokomo
Pittsburgh
Pennsylvania
Dayton
Binghamton
INDIANA
Indian Reservation
OHIO
Evansville
Ohio River
PHILADELPHIA
Hoboken
NEW YORK
Central Park
NEW YORK WORLD'S FAIR 1939
Ebbets Field
JERSEY
Jersey City
City Hall
BROOKLYN
Great South Bay
Prospect Park
W.VA.
Baltimore
MARYLAND
Pinehurst
VIRGINIA
WASH. D.C.
DEL.
KENTUCKY
Lexington
Louisville
TENNESSEE
Birmingham
Coney I.
N.Y. HARBOR
STATEN ISLAND
NORTH CAROLINA
SOUTH CAROLINA
LOWER NEW YORK HARBOR
Tate
Atlanta
GEORGIA
ALABAMA
Mobile
LOUISIANA
Swanee River
Atlantic City
Nashville
FLORIDA
New Orleans
Daytona Beach
Bermuda
ATLANTIC OCEAN
Miami
GULF OF MEXICO
Havana
South America

to be diminished in their spacing toward the poles. The meridians are elliptical and do not converge to a point at the poles; therefore, the poles are represented as straight lines that are one-half the length of the equator.

Cartograms Another category of map projections is the cartogram, where distortion is deliberate. In this type of projection no attempt is made to make the maps conformal, equal area, equidistant, or azimuthal. Some are intended to be humorous. For example, think what the results would be if someone asked a resident of Brooklyn—someone who had never left the neighborhood—to make a map of the United States. The result would show the influence of movies, television, and magazines on the mind of the untraveled person.

Sometimes a map is made to illustrate some special point or feature, such as the population distribution of the world, country by country. A map like this would be quite unlike the representation in Figure 1-1. There would be no dots or circles to show population density, because each country is drawn to a size that is in proportion to its population. Meridians and parallels in this projection are figments of the artist's imagination, added to make the cartogram more "realistic." It is an interesting way to point out the disparities between land area and population.

Physiographic Diagram Another way to portray the earth in a graphic way is through physiographic diagrams (see Fig. 11-3). Here, the cartographer attempts to picture the land as it might look from space. Pioneers in this field were A. K. Lobeck and Erwin Raisz.*

With improvements in cameras used in outer space, it may not be long before our maps will be actual photographs of the earth's surface—with the cultural features superimposed for clarification. We have only begun to use these magnificent photographs, and already geologists have discovered fracture zones and surface conditions never before suspected. In addition, the new field of remote sensing (which we will discuss in Appendix B) offers tremendous possibilities for mapping as well as for exploration of our planet. The future of aerial mapping promises to be exciting.

*Among some of the fine works produced by Erwin Raisz, see (1948), *General Cartography* (New York: McGraw-Hill); (1956), *Mapping the World* (New York: Abelard-Schuman); and (1962), *Principles of Cartography* (New York: McGraw-Hill).

REVIEW AND DISCUSSION

1. Explain the term *line of sight*.
2. After doing some additional research on the experiments conducted by Eratosthenes, explain his method for determining the circumference of the earth.
3. Why do you suppose it took so long for humans to accept the theory that the earth was a sphere?
4. What error did Ptolemy commit in making the map projection that caused Columbus to reach the New World instead of China?
5. Who was the first person of record to actually construct a globe?
6. Explain the terms *sphere, spheroid, ellipse, ellipsoid, oblate, oblate ellipsoid,* and *geoid*. Which do you prefer?
7. Briefly define the term *geographic grid*.
8. Explain the difference between great circles and small circles.
9. What is a meridian? Explain the term *meridian of longitude*.
10. What is the prime meridian?
11. What is a parallel of latitude?
12. How many parallels are great circles? How many meridians are great circles?

13. What is a sextant? How does it aid in determining the latitude of a place?
14. How can you use the North Star (Polaris) to determine your latitude?
15. What do meridians of longitude have to do with time?
16. What times of day are the following: 1800; 2300; 0800; 1200; 2400?
17. What effect does the earth's ellipsoidal shape have on the distance between meridians of longitude?
18. Discuss the evolution of mapmaking.
19. Why is the magnetic compass important?
20. Why is an understanding of map distortions important?
21. Explain these map-projection properties: equal area, conformality, equal distance.
22. What is meant by the term *azimuthal*?
23. Explain the difference between large-scale and small-scale maps.
24. Explain these map scales: verbal, graphic, representative fraction.
25. Why is the map legend important?
26. What is an isoline?
27. What is the relationship of the isogonic lines to magnetic variation?
28. What is the difference between grid north and geographic north?
29. What are the main differences between cylindrical and conical projections?
30. What is a rhumb line?

NOTES

[1] Robert O. Ballou, ed. (1977), *The Portable World Bible* (New York: Viking Press), p. 53.

[2] Jacquetta Hawkes (1962), *Man and the Sun* (New York: Random House), p. 61. Copyright © 1962 by Jacquetta Hawkes. With permission.

[3] Richard Aldington and Delano Ames, translators (1959), *Larousse Encyclopedia of Mythology* (London: Paul Hamlyn), p. 11. Reproduced by permission of the Hamlyn Publishing Group Limited from *Larousse Encyclopedia of Mythology*.

[4] Robert C. Toth (July 25, 1969), "Flat Earthist Leader Refuses to Give Up" (UPI), *Los Angeles Times*.

[5] David Greenhood (1964), *Mapping* (Chicago, Ill.: University of Chicago Press), p. 117.

ADDITIONAL READING

Brown, Lloyd A (1979). *The Story of Maps*. Boston, Mass.: Dover.

Bunbury, E. H. (1979). *A History of Ancient Geography*, Vols. 1 and 2. New York: Humanities.

Larson, Edwin E., and Birkeland, Peter (1982). *Putnam's Geology*. 4th ed. New York: Oxford University Press.

Raisz, Erwin (1962). *Principles of Cartography*. New York: McGraw-Hill.

Robinson, A. H., and R. D. Sale (1978). *Elements of Cartography*. 4th ed. New York: Wiley.

Thrower, Norman J. W. (1972). *Maps and Man. An Examination of Cartography in Relation to Culture and Civilization*. Englewood Cliffs, N.J.: Prentice-Hall.

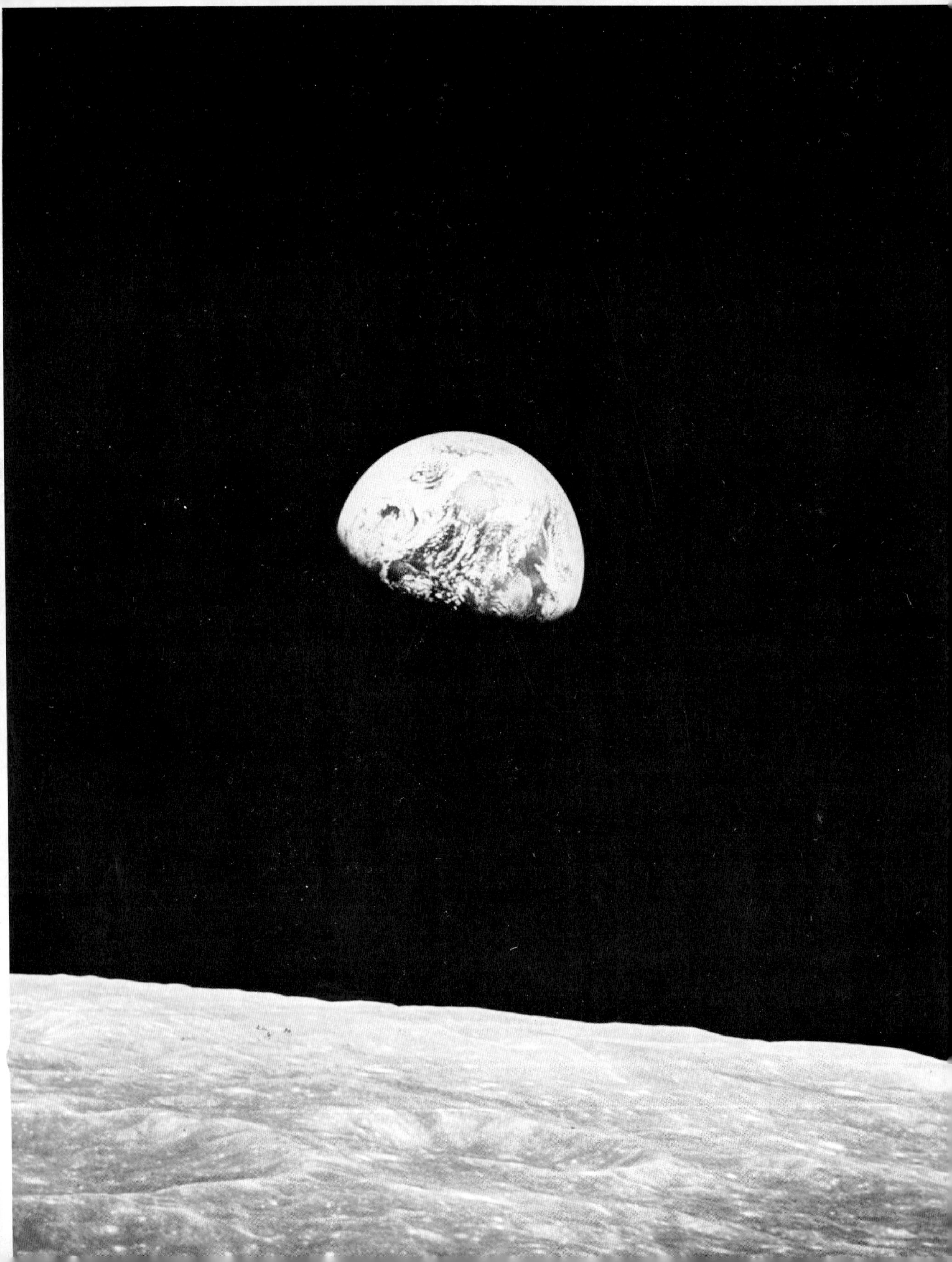

THREE
SPACESHIP EARTH

The waters deluge man with rain, oppress him with hail, and drown him with inundations; the air rushes in storms, prepares the tempest, or lights up the volcano; but the earth, gentle and indulgent, ever subservient to the wants of man, spreads his walks with flowers, and his table with plenty; returns, with interest, every good committed to her care; and though she produces the poison, she still supplies the antidote; though constantly teased more to furnish the luxuries of man than his necessities, yet even to the last she continues her kind indulgence and, when life is over, she piously covers his remains in her bosom.

PLINY THE ELDER

Many of the ancient peoples thought of the earth as a "supermother" who responded to the loving attentions of a "sky father"; these two, of course, looked after human needs. Later civilizations thought of the earth as a motionless body, hanging suspended in space, surrounded by the moving stars, planets, moon, and sun, all of which made regular orbits around the earth.

Today, however, we think of the planet Earth more as a spaceship hurtling through the vast, unknown reaches of the universe. Astronomers and earth scientists realize that every body in space is moving, and each has its own path. The earth is to the universe as a drop of water is to the world ocean.

Our task now is to examine the role that the earth plays in the immense scheme of things, and the reasons behind earth's movements. Some of the questions we will attempt to answer apply directly to the earth's power plant, the engine that makes it function so smoothly—the sun. We will examine the relationship between the sun and the earth, the controls over the heating and cooling processes, and the effect of the changing seasons on earth and its life forms.

EARTH AND THE SOLAR SYSTEM

Despite the wishful thinking of the ancient philosophers, our earth is not the center of the universe. It is but one tiny part of one solar system. With the sun at the center, planets extend billions of miles into space (Fig. 3-1). Diminutive Mercury is the closest to the sun, followed in order by Venus, Earth, Mars, Jupiter, Saturn, Uranus, Neptune, and Pluto. Our present position and size in relation to the sun, and our present stage in planet evolution, provide us with surface and atmospheric conditions that allow plant and animal life, in the forms we know, to survive and evolve. It is quite possible, as some scientists suggest, that there are millions of "earths" like ours revolving around stars or suns somewhere out there in space, but whether they have life forms with which we are familiar is a debatable question. If our space program continues, perhaps someday we will know.

Our position in the solar system is a delicate one. We are, evidently, in exactly the right place—neither too close nor too far from the sun. Mercury, on the other hand, is much too close. The *Mariner* flight passed this small planet in March 1974 and sent us photographs of what seems to be a dead and sterile planet, barren of what we might consider life forms. It is about one-third of earth's distance from the sun, and daytime temperatures soar into the hundreds of degrees Celsius*; at night the temperature falls to hundreds of degrees below freezing. Almost the same situation exists on the planet Venus. Although this planet is about the same size as the earth, it apparently makes only one rotation on its axis each year, and its surface temperature was measured by *Explorer II* at over 600°F (316°C). We know little of Venus's surface because of its dense cloud cover.

Mars, however, has surface temperatures somewhat akin to those of earth. We have assumed Mars to be a lifeless planet, yet our space probes have indicated that there is some atmosphere present. Only time and continued exploration will reveal whether there is enough oxygen and nitrogen in the atmo-

---

*Worldwide, the most commonly used system for measuring temperature is called *Celsius* (also *centigrade*), based on the freezing point for fresh water at sea level at 0°C. The boiling point is 100°C. This system is quite different from the one that has long been used in the United States—the *Fahrenheit* system, which was based on a "zero point," the temperature created when equal amounts of snow and salt are mixed. Fresh water would freeze at 32°F and boil at 212°F from the zero point. For a listing of conversion factors and a closer look at the *Metric system*, see Appendix D.

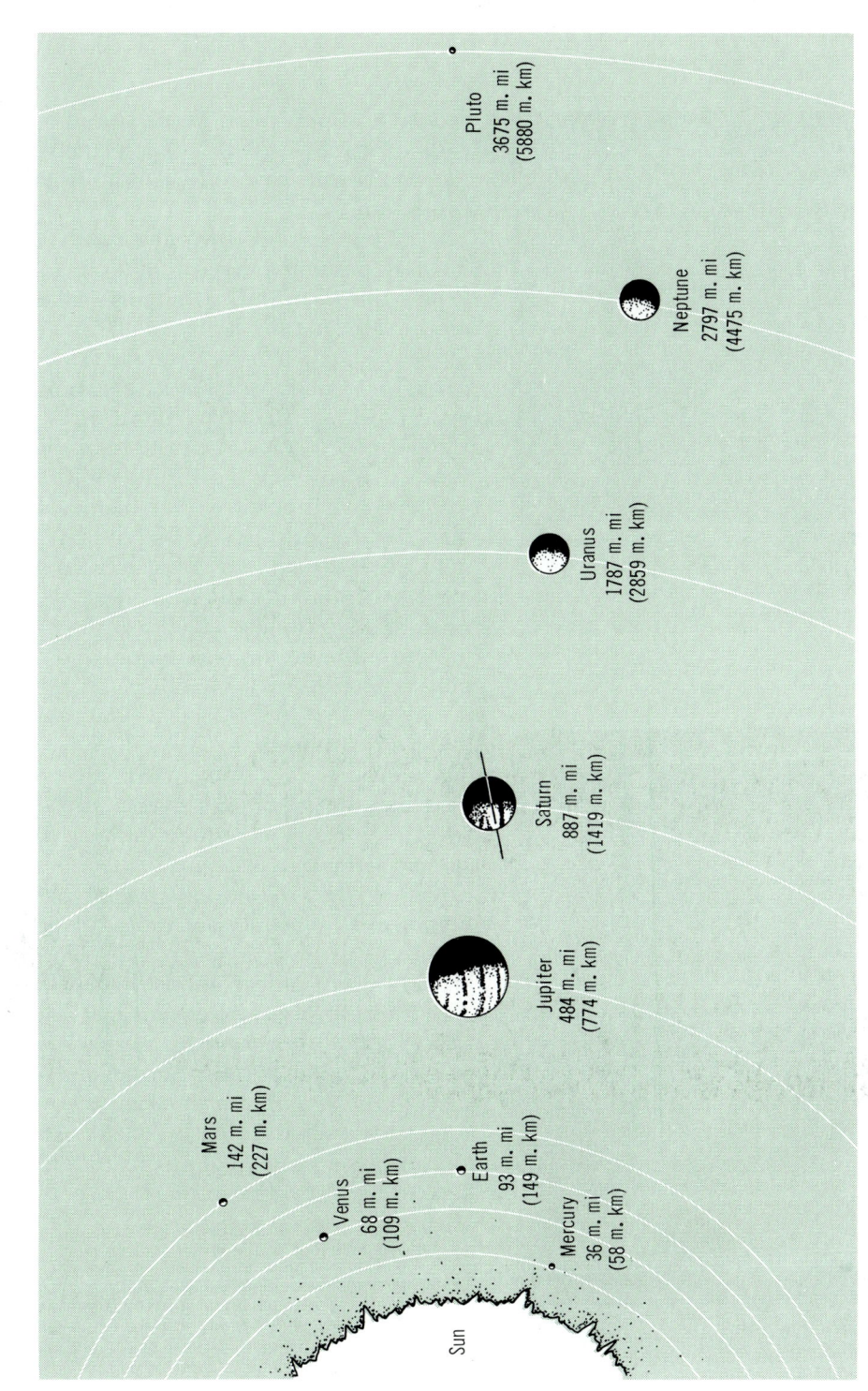

FIGURE 3-1 Our solar system Distances of the planets from the sun are given in millions of miles. (Distances in kilometers are given in parentheses.) (Not drawn to scale.)

The planet Mercury This view of Mercury is a photomosaic that was onstructed on 18 photos taken at 42-second intervals by Mariner 10 on March 29, 1974.

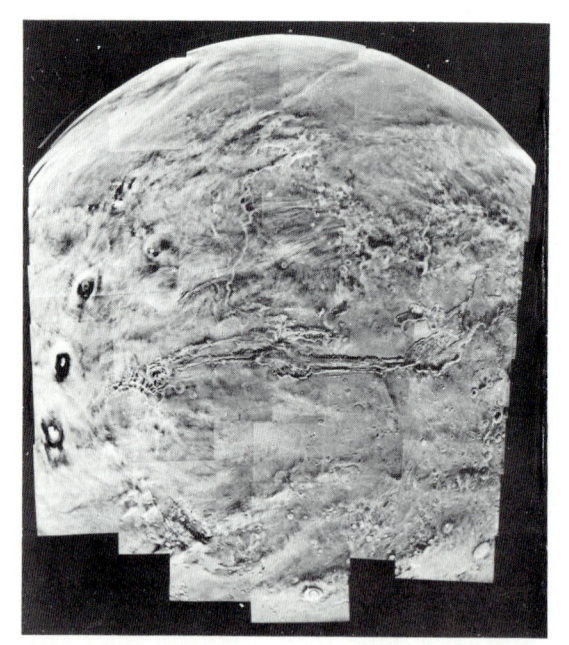

The red planet Mars Three photos of the Northern Hemisphere make up this photomosaic taken by the Mariner 9 spacecraft in August 1972.

sphere to support life, or whether it is mostly carbon dioxide. As for other planets, we do not, at present, have sufficient information to project answers.

To understand why our earth has such a favored position over the other planets, we look to the science of astronomy. The astronomers tell us that the source of all our energy is the sun. The sun is a spherical body that is constantly emitting radiant, electromagnetic energy, with surface temperatures averaging 11,000°F (6000°C). It operates like a gigantic furnace, furnishing the earth with the energy that makes life possible. One might ask, How does the earth remain in its position, neither falling into the sun nor flying away from it out into space? For the answers, we will take a closer look at the movements of the earth, its orbit around the sun (revolution) and its spinning around its axis (rotation).

Earth's Revolution

The earth revolves (orbits) around the sun in an elliptical (oval) path, which causes it to swing outward to its farthest distance from the sun (**aphelion**) on or about July 4 and then to close in on the sun to its closest approach (**perihelion**) on or about January 3.* Figure 3-2 portrays this great looping path. The earth's orbital speed varies (although its rotational speed around its axis remains constant), and the earth slows down as it approaches aphelion and speeds up as it nears perihelion. With respect to the sun, the earth takes 365¼ days to complete a full orbit. This is called the **tropical** (or solar) **year** and requires us to add a day once every 4 years (leap year) on February 29

*The terms *aphelion* and *perihelion* correspond to the terms *apogee*, the point at which an orbiting body is at its farthest distance from the body being orbited, and *perigee*, the point of closest approach. *Helion* refers to the sun and comes from the Greek word *helios*, "the sun."

EARTH AND THE SOLAR SYSTEM 55

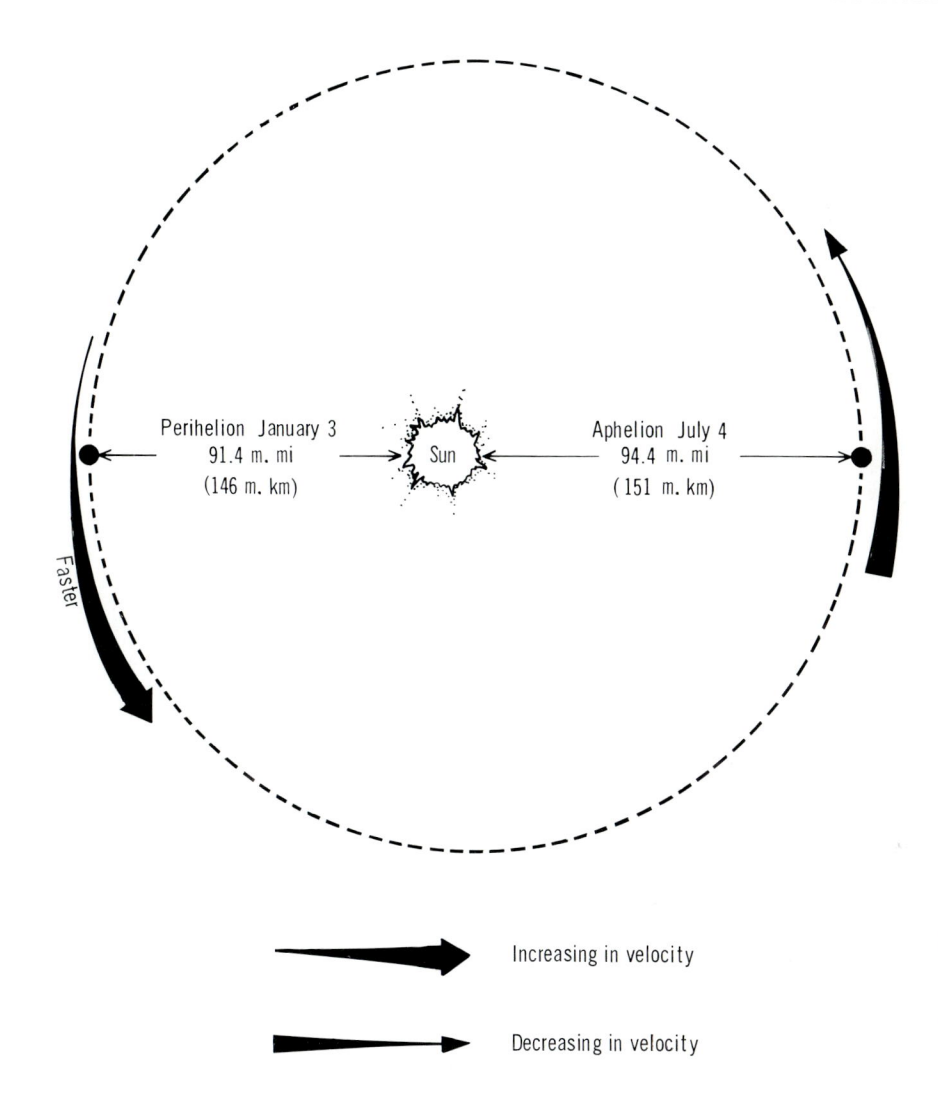

FIGURE 3-2 Earth's orbit around the sun The shape of the arrows indicates both direction and velocity of the earth in its orbit.

to make up for our use of 365 24-hour days. If we calculate the duration of the orbit by relating to a distant star, the orbit would take approximately 365 days, 6 hours, and 9 minutes. This is called the **sidereal year** (*sidereal* means with respect to the stars).

Astronomers are very precise. They make periodic checks of the earth's actual position sidereally and, on New Year's Eve 1972 they delayed the start of the new year by 1 second in order to bring earth time into proper alignment. The sidereal day is exactly 23 hours, 56 minutes, 4.09 seconds long. The 24-hour day to which we are accustomed came to us from ancient Sumeria in the Middle East. The Sumerians used the numbers 12 and 60 as their basic measurement, allowing 12 hours for day and 12 hours for night. The idea of 12 came from their observation of lunar months.

If our earth were to swing too far away, we

would not receive the necessary amount of solar radiation to sustain our forms of life. The earth would slowly freeze, and all things would die. On the other hand, if we came too close to the sun, our atmosphere and our waters would boil away. The earth would become a lifeless, inert mass, floating through space to become eventually, perhaps, similar to the planet Mercury. How, then, does the earth manage to avoid these catastrophes and remain just the right distance from the sun? The answer lies in two opposing, counteracting, balancing forces called *inertia* and *gravity*.

In gravity, masses of matter, large and small, are attracted toward one another. This is called the *universal law of gravitation*. The

**THE FRIDAY NIGHT BLIND DATE
or, How To Make Use of the Law of Inertia!**

If you have a beautiful girl as your blind date and you want her to end up in your lap, all you have to do is make a sudden, right turn! She cannot help it. She is obeying Newton's law of inertia: Any body acted on by a force (the moving car) will continue to move in the direction supplied by that force unless acted on by another force (the turning of the car).

○ Man   ● Woman

↑ Direction of moving car

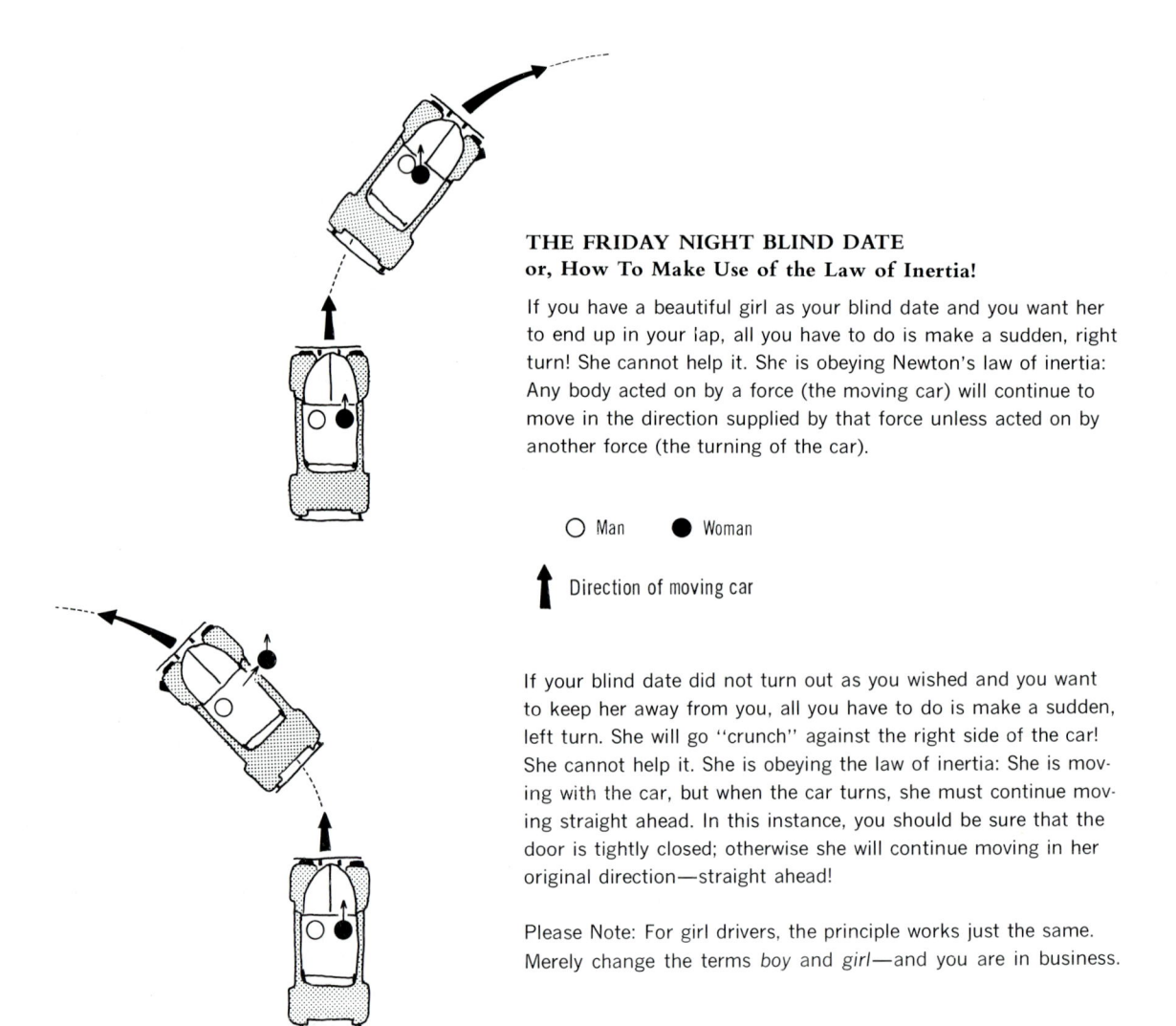

If your blind date did not turn out as you wished and you want to keep her away from you, all you have to do is make a sudden, left turn. She will go "crunch" against the right side of the car! She cannot help it. She is obeying the law of inertia: She is moving with the car, but when the car turns, she must continue moving straight ahead. In this instance, you should be sure that the door is tightly closed; otherwise she will continue moving in her original direction—straight ahead!

Please Note: For girl drivers, the principle works just the same. Merely change the terms *boy* and *girl*—and you are in business.

FIGURE 3-3 Newton's first law of motion: inertia

power and the strength of this pull of gravity depends on the *mass* (size and weight or bulk) of the object. A smaller or lighter mass will tend to be pulled toward—or gravitate to—a larger or heavier mass (and vice versa, to a lesser degree). Thus it is with the moon in its relation to the earth, and the earth in its relation to the sun. If it were not for the balancing force called inertia, the moon would crash onto the earth, and the earth would swing down into the fiery maw of the sun.

To explain this force, in 1687 the English mathematician Isaac Newton formulated his first law of motion. In this law Newton explained that a moving body will try to resist any outside or external force that might try to move it from its path (Fig. 3-3). This means that the earth is trying to move through space in a straight line, but the sun is exerting a gravitational pull that the earth cannot resist. The earth, therefore, moves in a looping, elliptical path around the sun. It is a constant, never-ending struggle—one force against the other—and, fortunately for us, neither force so far has gained the upper hand. The same struggle occurs here on earth. Inertia tries to fling humans, animals, buildings, and even our atmosphere off and away from the earth's surface and out into space, but the gravitational attraction of the earth overcomes that force, and here we stay.

Earth's Rotation

While the earth is revolving around the sun in its orbit, it is at the same time moving in another way. It is rotating (spinning) eastwardly around its axis (an imaginary line that pierces the earth through the poles, north to south). The speed of rotation at the equator is about 1030 miles (1648 kilometers) per hour, slowing down to about 850 miles (1360 kilometers) per hour at Los Angeles, 500 miles (800 kilometers) per hour at Anchorage, Alaska, and zero at the North Pole. This is very fast, but many scientists believe that our present rotational speed is considerably slower than it was several billion years ago. They say that it is highly probable that the earth rotated around its axis once every 5 hours. Imagine a 2½-hour day, followed by a 2½-hour night. Think of what the earth must have looked like then: a really oblate spheroid—much wider at the equator than it is now.

Do we feel this speed? No, because everything about us—the air, clouds, trees, firmly held to the earth by gravity—is moving *with* the earth. Yet we know that there is movement, even if we cannot feel it, because it causes night and day. The earth rotates around its axis once every 24 hours, and this continuous rhythm has a tremendous influence, not only on life conditions, but also on the direction and speed of the winds and the oceans, the length of the day, and the apparent sunrise and sunset.* The relative uniformity of time during which areas of the earth receive solar radiation might be likened to the even cooking of a chicken revolving on a barbecue spit.

Humans of ancient times did not understand the movements of the earth. They thought the sun orbited the earth, rising in the east and setting in the west. It was not until the mid-nineteenth century that proof of the earth's rotation was found. A Frenchman named Foucault affixed a heavy iron ball to a long steel wire that was, in turn, fastened to a ball joint in the ceiling of a high-domed cathedral in Paris. He pushed the ball and started it swinging to and fro. He had theorized that a pendulum would not change its course unless it were acted on by some external force. He set up little wooden pegs on the floor around the pendulum. The swinging iron ball began to knock them down, each in turn. At first he

---

*The influence of the earth's rotation on the direction of winds and ocean currents, called the *Coriolis effect*, is covered in detail in Chapter 4.

Foucault's pendulum

thought the pendulum had altered its direction; then he realized that the pendulum had not swerved from its path, but the earth had rotated beneath the pendulum. The iron ball seemed to rotate clockwise in its arc, which actually meant that the earth was doing just the opposite. The earth was rotating from west to east, a counterclockwise direction for the Northern Hemisphere.* Further experiments showed that the nearer one was to the poles, the better the swinging pendulum worked at proving the earth's rotation. The reason is simple. At the poles the pendulum swings back and forth over a rotating surface, while on the equator the surface of the earth rushes underneath in a straight line.

Today, we can observe the path of weather and communication satellites as they orbit the earth. Their paths seem to wander over the earth's face (Fig. 3-4), but this meandering movement is illusory. The satellites follow a constant great-circle path around the earth (that is, their direction is steady and undeviating), as the earth is rotating, moving, beneath them, so the satellite does not pass over the same places each time it circles the earth. For example, a satellite crosses over the city of Honolulu, Hawaii, at noon, but 45 minutes later, as it again sweeps across the Pacific Ocean, it passes west of Honolulu. Why? Because the earth—and Honolulu—have moved eastward. Remember that looking down over the North Pole would enable you to see that the earth rotates in a counterclockwise direction (Fig. 3-5). But if you were looking at the equator, the west-to-east rotational direction would be obvious.

Earth Movements and the Seasons

Let us once again consider the earth's relationship to the sun. We know that the earth revolves around the sun and, at the same time, rotates around its own axis. This movement—revolution—and the inclined axis, ac-

---

*For further information regarding Foucault and his pendulum, consult W. E. Johnson (1907), *Mathematical Geography* (New York: American Book Company), pp. 54–57. Also, if your college does not have such a pendulum, visit your nearest observatory, such as the one in Griffith Park, Los Angeles, or the one in Golden Gate Park, San Francisco.

EARTH AND THE SOLAR SYSTEM 59

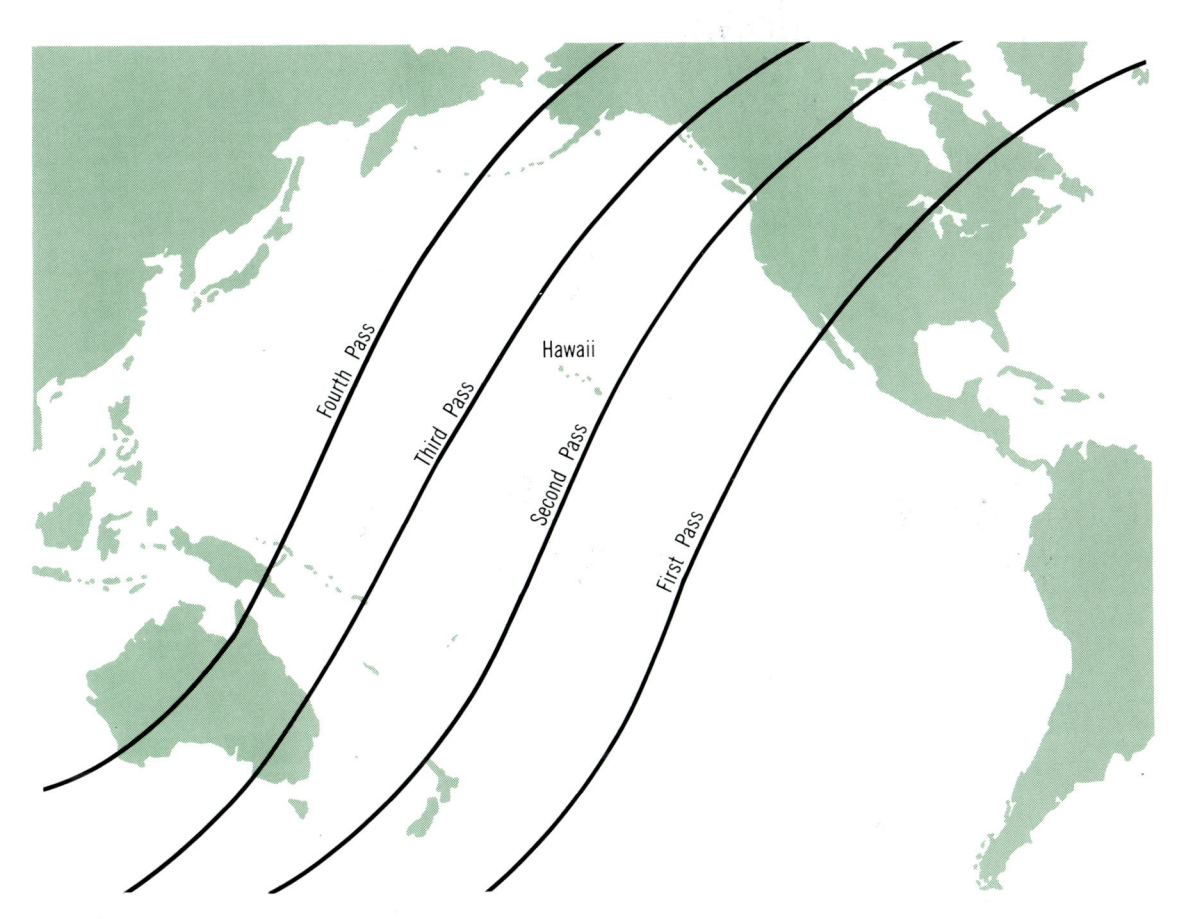

FIGURE 3-4 Path of a satellite This diagram represents the movement of the earth to the east and the apparent "westward" movement of a satellite each time it passes around the earth.

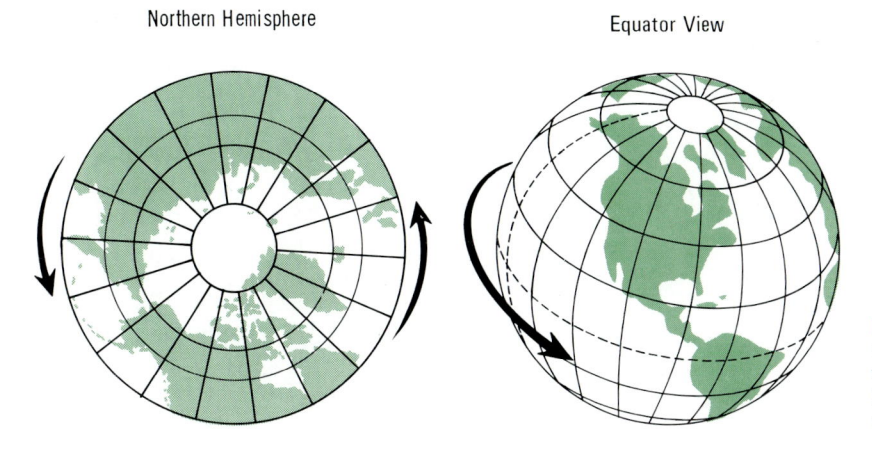

FIGURE 3-5 The earth's rotation about its axis as viewed from above the North Pole and an oblique view of the equator

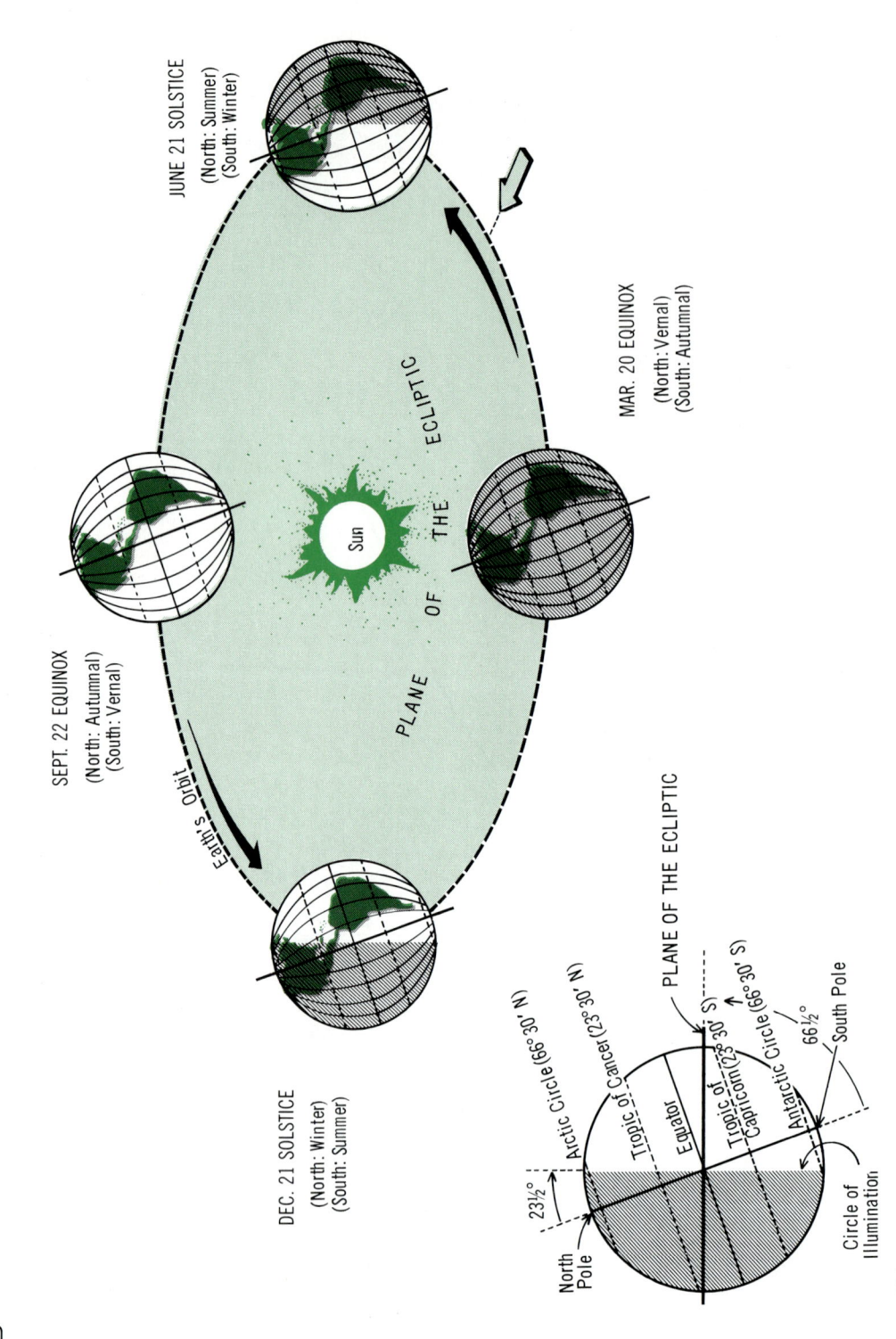

FIGURE 3-6 The earth's orbit and the plane of the ecliptic

count for our changing seasons. As for the inclined axis, have you ever noticed that globes never stand straight up? Figure 3-6, which details the earth's orbit and the different seasons, shows that the earth's axis is tilted. Think, for example, of the outer edge of the earth's orbit as something like the edge of a large dinner plate, with the sun right in the middle of the plate. That plate, that path of the earth's orbit, is termed the **plane of the ecliptic.** The earth's axis is inclined, or tilted, away from that plane by an arc of 66°30′. Some prefer to think of the tilt as the inclination away from a line perpendicular (at right angles) to the plane of ecliptic, to the amount of 23°30′. This inclination causes different regions of the earth to receive different amounts of solar energy, depending on the time of the year and the latitudes of the various regions.

From around March 20 until September 22, the Northern Hemisphere is tilted toward the sun; consequently, that hemisphere receives more of the sun's direct radiant energy than does the Southern Hemisphere. During the Northern Hemisphere's summer solstice (June 21), the region north of the Arctic Circle has daylight for 24 hours, and the days are longer all over the Northern Hemisphere (Fig. 3-7). Exactly the opposite is true for the Southern Hemisphere; nights are longer, and the area south of the Antarctic Circle has 24 hours of darkness. On December 21 the seasons are reversed. The Southern Hemisphere is then tilted toward the sun, and it is summer south of the equator and wintertime in the north.

During the periods of the **solstices,** the vertical, direct rays of the sun are over either the Tropic of Cancer (latitude 23°30′ north) on June 21 or the Tropic of Capricorn (latitude 23°30′ south) on December 21. Remember, the two hemispheres are exactly opposite during the solstices; the Northern Hemisphere's summer is the Southern Hemisphere's winter, and vice versa. Also, the days are longer and the nights shorter during the summer solstice, and the opposite is true during the winter solstice. Only the equator has days and nights of equal length throughout the year.*

Between the times of the two solstices, at the midway points in the earth's journey around the sun (about March 20 and September 22), neither hemisphere receives more direct solar energy than the other. The direct, vertical rays of the sun strike only at the equator. The earth is still tilted, but the inclination is at right angles to the plane of the ecliptic, so both hemispheres have 12-hour days and 12-hour nights. This is the time of the **equinoxes;** March 20 is called the *vernal* (or spring) equinox in the Northern Hemisphere and the *autumnal* (or fall) equinox in the Southern Hemisphere. On September 22 it becomes the autumnal equinox for the north, the vernal for the south.

Between the solstices and the equinoxes, the length of day increases or decreases, depending on the hemisphere, and the nights decrease or increase in length in corresponding opposition. If you were an astronaut out in space, looking down at the earth, you would see that one-half of the earth is always in shadow while the other half is always in daylight. Of course, the rotation of the earth prevents any one side of the earth from getting too much solar radiation. The line between day and night, the line that divides the earth into dark and light, is called the **circle of illumination,** or the *terminator* (Fig. 3-8). The line passes through the poles during the equinoxes, paralleling the meridians as the earth rotates. During the solstices, the line is inclined 23°30′

---

*The term *solstice* comes to us from the Latin *solstitium: sol* ("sun"), *stitium* ("stands"); therefore: "a time when the sun seems to remain motionless in the heavens." *Equinox* comes from the Latin *aeques* ("equal") and *nox* ("night"); thus: "a time when days and nights are of equal length."

	PERIHELION			VERNAL EQUINOX				SUMMER SOLSTICE				APHELION			AUTUMNAL EQUINOX				WINTER SOLSTICE		
	Jan			Mar				June				July			Sept				Dec		
Year	d	h		d	h	m		d	h	m		d	h		d	h	m		d	h	m
1984	03	14		20	02	24		20	21	02		02	23		22	12	33		21	08	23
1985	03	12		20	08	14		21	02	44		05	10		22	18	07		21	14	08
1986	01	21		20	14	03		21	08	30		05	02		22	23	59		21	20	02
1987	04	15		20	19	52		21	14	11		03	17		23	05	45		22	01	46
1988	03	16		20	01	39		20	19	57		05	16		22	11	29		20	07	28
1989	01	14		20	07	28		21	01	53		04	04		22	17	20		21	13	22
1990	04	09		20	13	19		21	07	33		03	21		22	22	55		21	19	07
1991	02	19		20	19	02		21	13	19		06	07		23	04	48		22	00	54
1992	03	07		20	00	48		20	17	14		03	04		22	10	43		21	06	43
1993	03	19		20	06	41		21	01	00		04	14		22	16	22		21	12	26
1994	01	22		20	12	28		21	06	48		05	11		22	22	19		21	18	23
1995	04	03		20	18	14		21	12	34		03	18		23	04	13		22	00	17
1996	03	23		20	00	03		20	18	24		05	10		22	10	00		21	06	06
1997	01	16		20	05	55		21	00	20		04	11		22	15	56		21	12	07
1998	04	13		20	11	55		21	06	03		03	16		22	21	37		21	17	56
1999	03	05		20	17	46		21	11	49		06	14		23	03	31		21	23	44
2000	02	21		19	23	35		20	17	48		03	16		22	09	27		21	05	37
1984–1999 GMT	03	16		20	18	00		21	12	41		04	23		23	03	51		21	23	56
PST	03	08		20	10	00		21	02	41		04	15		22	19	51		21	15	56

FIGURE 3-7 Dates and times (Pacific Standard Time) of points in earth's orbit for the northern Hemisphere (day = day; h = hour; m — minute; GMT = Greenwich Mean Time; PST = Pacific Standard Time.) (From information supplied by the U.S. Naval Observatory and Dr. Arnold Court.)

Summer solstice, Bodo, Norway This series of photos taken before, during, and after midnight at Bodø, Norway (67° north latitude), demonstrates that the land of the midnight sun is not a figment of the nordic imagination.

from the polar axis, making an angle of 66°30′ with the parallels of latitude.

How do the changing seasons affect humans? Those who live near the equatorial regions scarcely know what changing seasons are, but think of the people who live in Bodø, Norway, or Pt. Barrow, Alaska—lands where summertime means daylight for almost 24 continuous hours and wintertime means night for nearly 24 hours. In June, they get up in the morning with bright daylight all around, and when they go to bed at night, there is still daylight. If people are psychologically attuned to going to bed when it is dark, what do they do when there is no night? And consider what happens to their nervous systems 6 months later, when there is little or no daylight? Some people get depressed during these long winters, and suicides are common during this period.

From the high northern latitudes, the sun appears far away to the south and relatively low on the horizon. During the summer, twilight seems to linger because of the low horizon

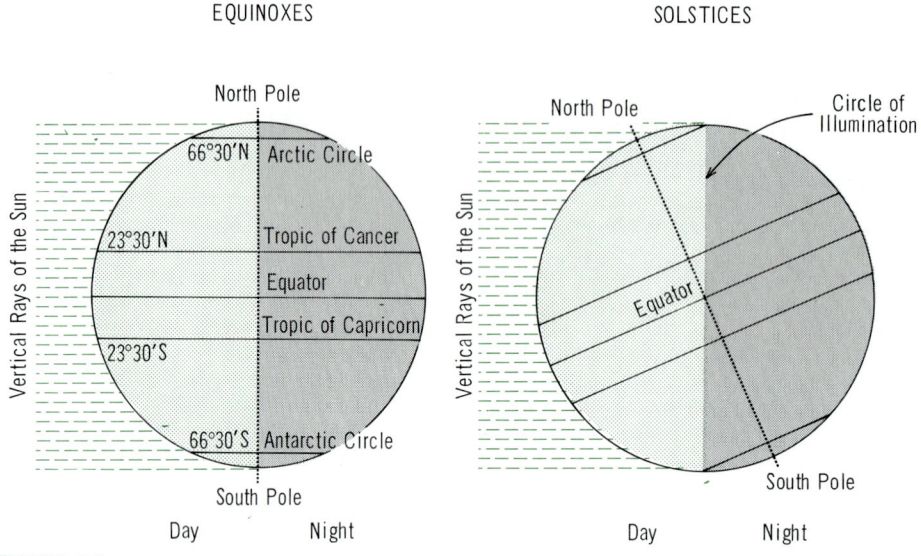

FIGURE 3-8 The circle of illumination

angle of the sun, but this is not the case for areas close to the Tropic of Cancer, such as the Hawaiian islands. In Honolulu the sun appears to be almost overhead throughout the year, and when sunset comes, it comes with great abruptness. If you wish to photograph the brilliant colors of a Hawaiian sunset, you should be ready and waiting, because it all happens, seemingly, in a matter of seconds.

EARTH'S POWER PLANT: SOLAR ENERGY

The sun is not a gigantic ball of flaming gases hanging out in the center of our solar system, but it can be likened to an enormous furnace that is constantly emitting massive amounts of electromagnetic energy in every possible direction (Fig. 3-9) in the form of visible light rays (41 percent), infrared and heat rays (50 percent), and ultraviolet, gamma, and X rays (9 percent). Fortunately for us, however, our earth is at the correct distance from the sun for its size to receive the energy we need. In addition, we have our atmosphere to protect us from excessive doses of solar energy.

Solar Energy and the Earth's Atmosphere

Consider what might happen if our atmosphere did not exist and incoming solar radiation could strike the earth's surface without interference. Although it is unlikely this will ever occur, we had advance warning of what could happen when humans learned how to manufacture their own form of radiant energy—the atom bomb. When a nuclear device is exploded, the nearby earth is showered with deadly gamma rays, X rays, ultraviolet rays, and visible light and heat rays. If people are too close to the scene of the blast, they might be blinded by the intense light, burned by the heat, and scarred by the gamma rays. If they are not in the near vicinity, but happen to be downwind, they might not be instantly killed but will experience scalding and burns, as did

EARTH'S POWER PLANT: SOLAR ENERGY 65

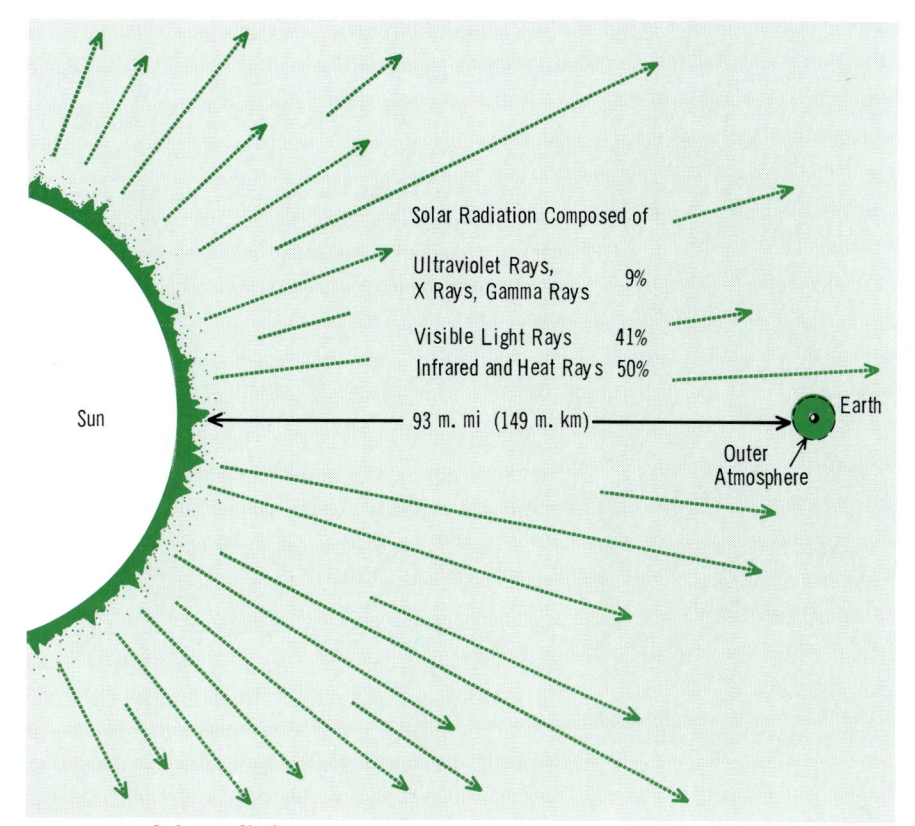

FIGURE 3-9 Solar radiation

Solar Radiation Composed of
Ultraviolet Rays, X Rays, Gamma Rays   9%
Visible Light Rays   41%
Infrared and Heat Rays   50%

Sun — 93 m. mi (149 m. km) — Earth
Outer Atmosphere

the Japanese fishermen who were too close to Bikini Atoll when the great test bomb was exploded.

Surrounding the earth and extending far out into space to a distance of approximately 50,000 miles (80,000 kilometers) is the earth's magnetic field. Within this vast region, often referred to as the *magnetosphere,* there is another form of solar energy, **corpuscular energy.** This energy issues from the sun as clouds of electrified particles that seem to have a direct effect on television and radio communication and on the earth's magnetic field. Also known as magnetic storms, they occur about once a month, gushing forth from the sun to spray our entire solar system with tiny molecules of helium, hydrogen, oxygen, and nitrogen. Our earth, being a magnet, catches a certain percentage of these solar particles at each pole. The particles cause friction with the nitrogen and oxygen in our atmosphere, creating vivid displays of green and red luminescence that skip across the polar skies, giving us the curtains of light we call the *aurora borealis* at the North Pole and *aurora australis* at the South Pole.

It is possible that these solar particles could cause severe problems for us, but there are two belts composed of high-energy-radiation particles (protons and electrons) that en-

Aurora Borealis

circle the earth and protect us by absorbing much of the incoming corpuscular energy. The belts—called the *Van Allen radiation belts*, after the scientist who discovered them—are positioned at two elevations around the earth: 2500 miles (4000 kilometers) and 10,000 miles (16,000 kilometers). The radiation contained within these belts (Fig. 3-10) could be dangerous to astronauts attempting to go into space; it could interfere with their electronic gear, radar equipment, radios, telemetric equipment, and compasses. However, knowledge of the possible dangers enabled NASA to move ahead with confidence, and our astronauts have been able to navigate out through the "holes" in the doughnut-shaped rings above the poles as they journeyed to the moon.

We have examined the part played by the sun in controlling earth's movements, and we have seen how the earth's atmosphere is affected by the incoming solar energy. Let us now investigate what happens to the incoming solar radiation (usually abbreviated to one word, **insolation**) after it reaches the earth's surface.

Solar Energy and the Heat Balance

As we have seen, the earth and its atmosphere receive only a fraction of the total emission by the sun. We have learned also that much of the original solar radiation, particularly the ultraviolet, gamma, and X rays, has either been reflected back into outer space or trapped at the outer levels of the atmosphere. Only those visible light and infrared rays that have escaped such reflection or entrapment are left to continue the journey toward the earth. By the time the surface of the earth is reached (Fig. 3-11), about 35 percent of the original incoming solar radiation is reflected into space, 19 percent has been or will be absorbed by the lower atmosphere, and the remaining 46 percent is or will be absorbed by the earth itself.

Does this mean that the 65 percent of incoming solar radiation that reaches our atmosphere or the earth's surface is retained permanently? Obviously, the answer is no; if this were so, the temperature of the lower atmosphere would rise to a level that might make life on earth impossible. If, on the other hand, the opposite were to occur and the heat loss became excessive, the temperature would drop below freezing and life as we know it might disappear. Although there have been a number of times during earth's history when each of these extremes has occurred, a state of equilibrium has been established—a kind of

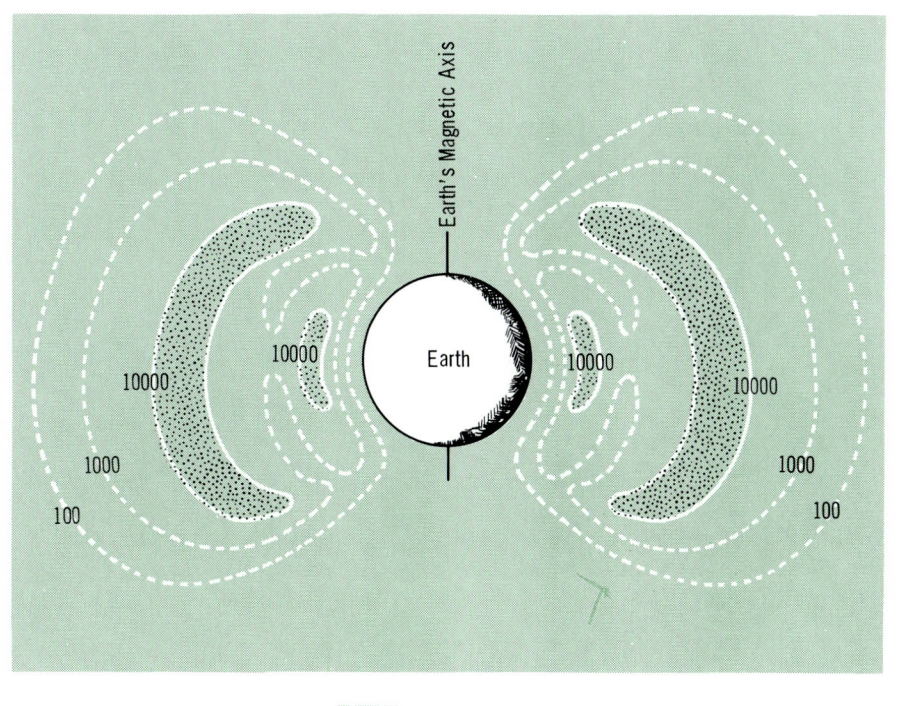

FIGURE 3-10 Van Allen radiation belts High-energy protons and lower-energy electrons that create a penetrating type of radiation make up Van Allen radiation belts (and their accompanying force fields). The intensity of radioactivity in the various belts and force fields is indicated by numbers (measured by a Geiger radiation counter in "counts per second"). (Adapted from U.S. Navy, "The Upper Atmosphere," U.S. Navy Weather Research Facility Chart NWRF 26-1161-051, Washington, D.C.: U.S. Government Printing Office.)

heat balance in which the earth sends back into outer space an amount of radiant energy equal to the amount received. This *energy balance* is vital to the earth system, and many factors go into its maintenance, such as radiation and re-radiation (popularly illustrated as the "greenhouse effect"), compression and expansion, condensation and evaporation, and methods of energy transfer (for example, conduction, convection, and advection). In addition, there are a number of controls that exert influence on the heating and cooling processes.

Controls over Heating and Cooling

Insolation is not received equally by the various levels of the earth's atmosphere, or by different locations on the earth. Figure 3-12 illustrates some of the obstacles that insolation must overcome before it can get on with the task of supplying the earth and the lower atmosphere with heat and other forms of energy. By the time the lowest level of the atmosphere has been reached, much of the insolation has either been absorbed or transformed into other forms of energy as it races

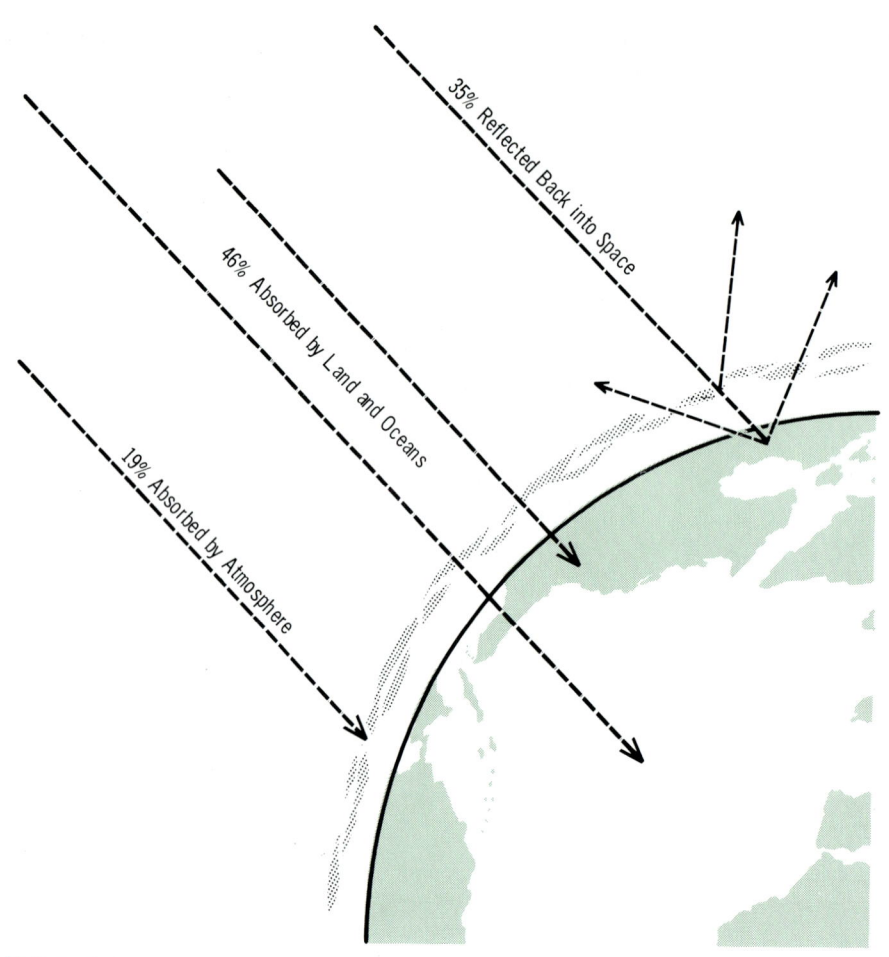

FIGURE 3-11 Incoming solar radiation (insolation)

through the outer atmosphere. As you can see, insolation is controlled by reflection or the scattering of solar particles as they encounter other particles of matter. Within the lower atmosphere another term for reflection is often used—*albedo.*

Albedo Basically, albedo means reflection of radiant energy by an object, expressed as a percentage. The percentage for heat or visible light rays is much higher for a snowfield than it is for a dark pine forest. A dark item such as a black suit would have a lower albedo than a white suit, which you can easily prove on a warm summer day. Insolation would be readily absorbed by the dark clothing, thereby heating you considerably, but the white or light-colored clothing would reflect the insolation and you would feel cool. Meteorologists have calculated the albedo for different substances and find that clouds and regions of ice and snow consistently have the highest albedo (up to about 80 percent) grasslands are intermediate (about 25 percent), and oceans and other water bodies have the lowest (about 5 percent). The earth, as a whole, averages an albedo of around 30 percent. Note that if the energy arrives at a low glancing angle, the albedo increases. One can easily see why in places near the equator the albedo would be higher during mornings and afternoons than at

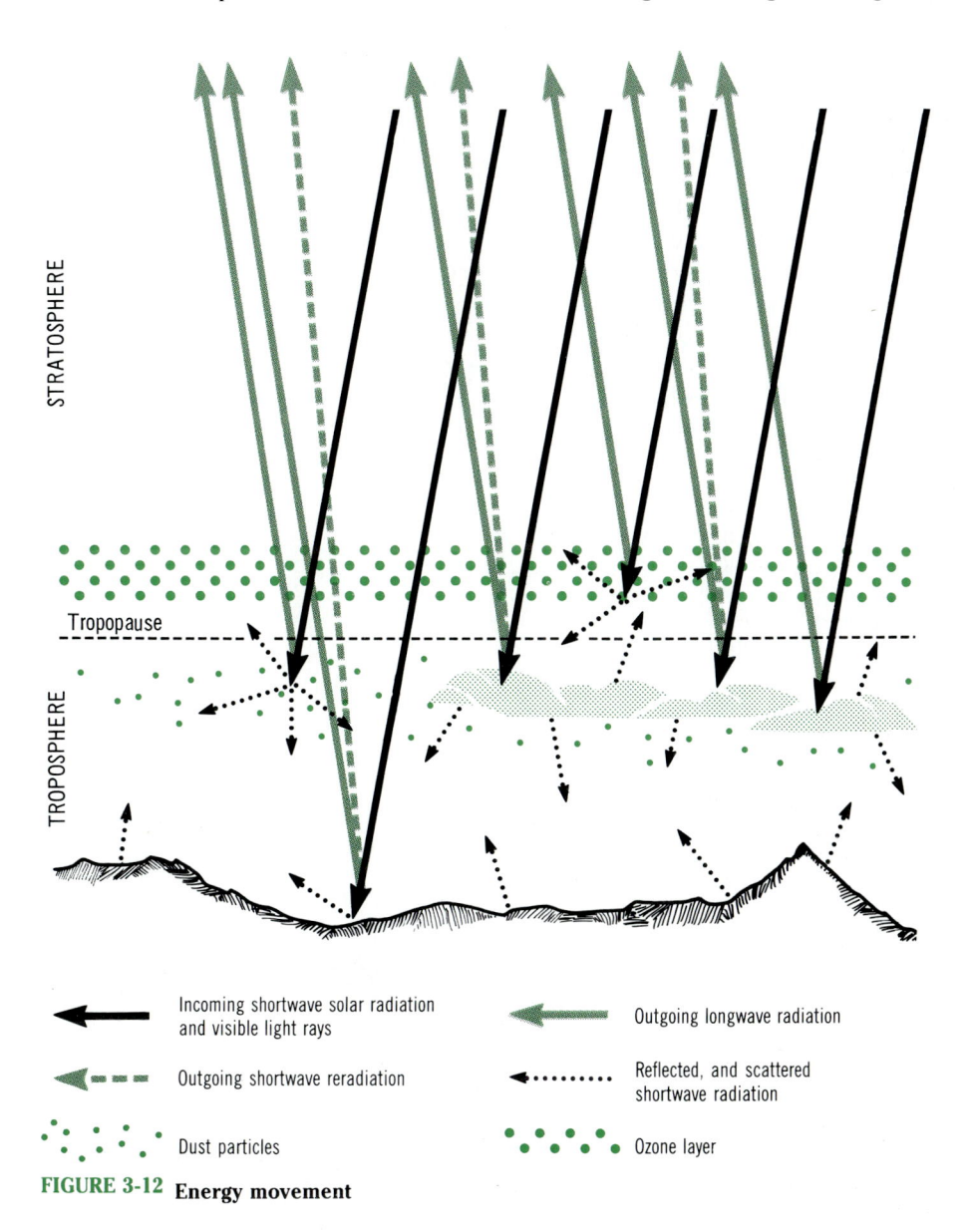

FIGURE 3-12 **Energy movement**

noon, and why the albedo would be higher on polar snowfields than on equatorial Mt. Kenya.

Land and Water Differences The differences in the absorptive qualities of land and water constitute important controls in the heating and cooling of the earth, but land and water do not absorb insolation equally. Oceans heat up very slowly, much more slowly than land does, but the heating goes much deeper. Whereas insolation will be absorbed only a few inches down into soil, it will penetrate more than 6 feet (2 meters) in tropical waters. Consider the difference in the warmth of sand alongside an inland lake or near the ocean and the water itself. If it were in midsummer, at noon or early afternoon, when the insolation was at its highest, the sand would be hot to touch, but if you burrowed with your toes an inch or two into the sand, you would find cold sand. If the temperature of the sand were, say, 90°F (32°C) and the temperature of the body of water were near 70°F (21°C), you would have no difficulty noticing the difference: The water would seem very cold to you. But what happens when the sun has set? The sand loses its heat rapidly and feels cold. The lake or ocean does not lose heat as rapidly, and its temperature falls perhaps 2° or 3°. Compared to the sand, the lake or ocean would seem warm.

There may be differences in surface temperatures in winter and summer, but, generally, the oceans and other large bodies of water are so stable that they are a moderating influence on nearby land. Contributing factors to this, in addition to the reason just discussed, include the movement of warm and cool ocean currents, the upwelling (convection currents)

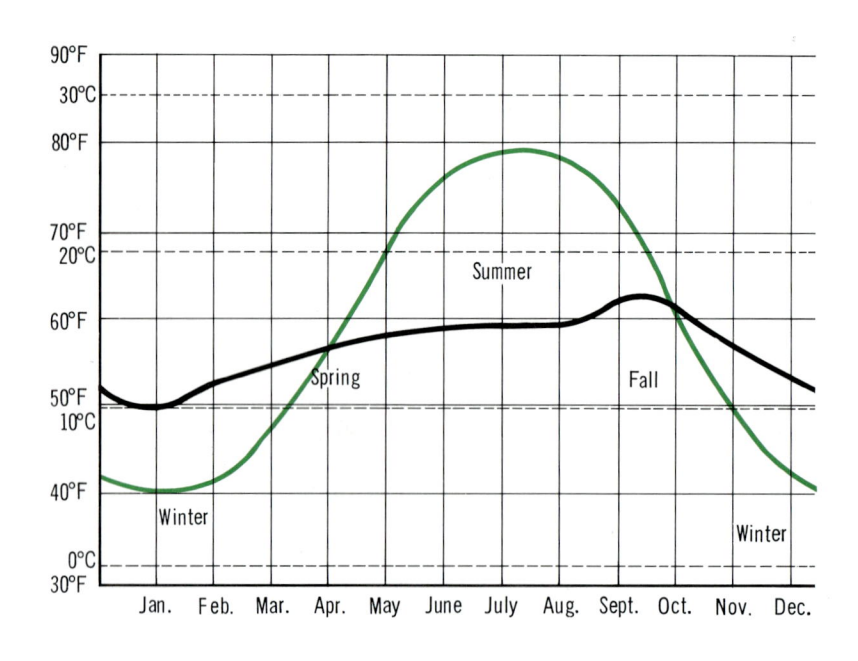

FIGURE 3-13 Influence of land and water on air temperature of two cities at about the same latitude

San Francisco, Calif. lat. 37°47′N (coastal)

Nashville, Tenn. lat. 36°10′N (inland)

of cool water near hot lands, evaporation that continues all year and is especially appreciated during the hot summer months, the great size and depth of the oceans, and the immense volume of water near the equator that warms and retains its heat for long periods.

As an example, compare the differences in monthly temperatures between two cities that are positioned at approximately the same latitude (Fig. 3-13), but one is located next to the ocean and the other is inland. What if you were overly sensitive to cold weather and someone told you that Nashville would be a better place for you to live because its average temperature for the year is higher than that of San Francisco? Would you be getting the entire story? Nashville's average temperature is 59.5°F (15°C) and San Francisco's is about 56.5°F (13.6°C). If you consult the monthly average temperatures, you will note that Nashville's winter temperatures range from 40°F (4.4°C) to about 45°F (7.2°C), while winter in San Francisco averages about 55°F (12.8°C). Consider also the annual range, from the coldest to the warmest months, and you will find that Nashville has a variation of 40°F (4.4°C) and San Francisco has a range of only about 12°F (−11°C). Thus, you might decide that San Francisco's temperature stays fairly even all through the year, while Nashville is subject to extremes.

Latitude Another important control over an area's receipt of insolation is latitude. When we discussed the yearly orbit of the earth around the sun and the resulting changing of the seasons, it was pointed out that the earth's axis is tilted to the plane of the orbit. Thus it is obvious that not all areas on the earth's surface receive the same amount of insolation. Even during the equinoxes, when neither the Northern or Southern Hemisphere is tilted toward or away from the sun, insolation is not the same for all regions.

Consider Fig. 3-14 and contrast the possible receipt of insolation for points X, Y, and Z during the solstices and equinoxes. We can see that the Northern Hemisphere is in its summer solstice (Fig. 3-14a) and point X would have a longer day; thus its receipt of insolation would be considerably greater than at point Z.

The opposite occurs during the winter solstice (Fig. 3-14c). During the equinoxes (Fig. 3-14b) both X and Z receive the same amount of insolation. But how does point Y compare? As it is only 10° south of the equator, its receipt of insolation will be almost constant throughout the year. An important factor is the angle of impact of insolation. For point Y the angle is nearly vertical, while for points X and Z the solar rays will strike at an oblique angle, with resulting high albedo. In addition, insolation coming in at an angle must travel through a greater amount of atmosphere, and when it strikes the earth's surface, it must cover a greater area (Fig. 3-15), thus losing much of its intensity.

Albedo, the time of day, the season of the year, and latitude play important roles in determining the amount of insolation any area will receive, but there are other controls, such as the presence in some areas of great mountain ranges, broad deserts, heavy forests, icecaps and snowfields, the presence or absence of cloud cover, and powerful wind systems.

Heating Processes
Earlier in this chapter we learned that incoming solar radiation warms the atmosphere very little as it passes through it. The main heating processes are performed on or near the earth's surface, and they include radiation (and reradiation), conduction, compression, and condensation. It will become evident, as we proceed, that some of these heating processes also perform the task of cooling the earth and modifying the extremes of temperature that otherwise might result.

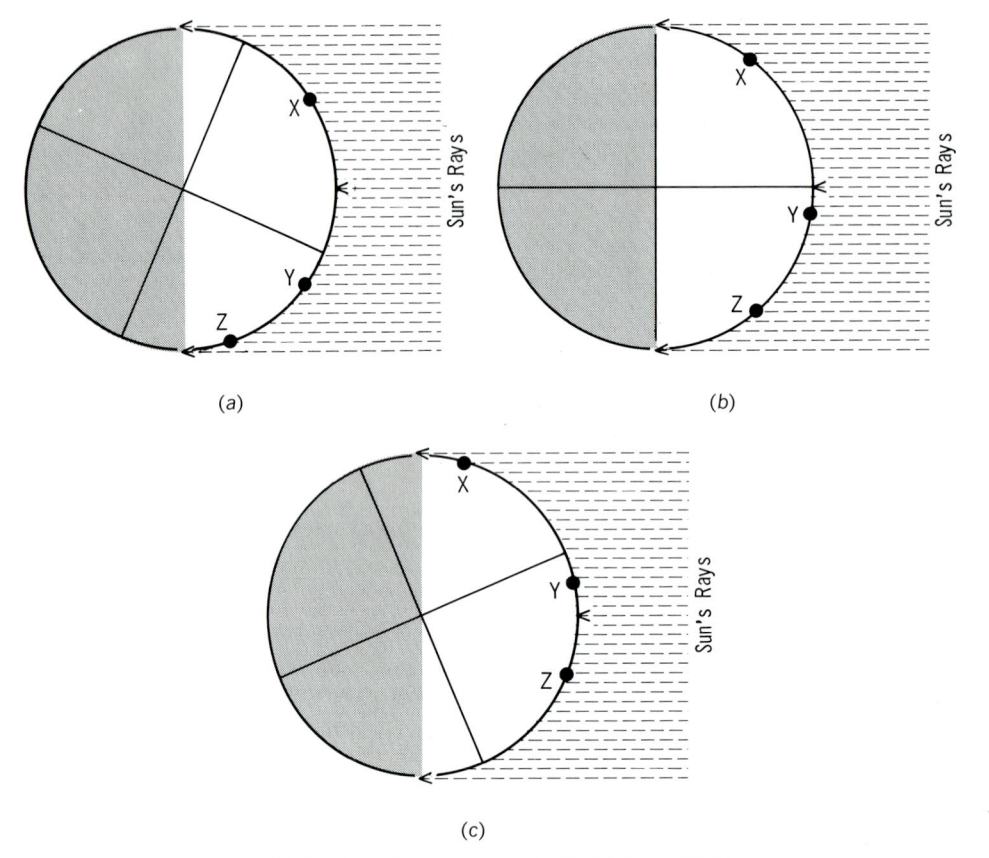

FIGURE 3-14 Influences of latitude and the seasons (*a, b*) Solstices, (*b*) Equinoxes.

Radiation and Reradiation The primary heating processes are **radiation** and **reradiation**, the result of the absorption of incoming shortwave solar energy by the earth and the subsequent reradiation of longwave energy into the atmosphere.* Because the production of shortwaves requires a very hot radiating body, the earth can only produce forms of longer-length energy waves. This longwave radiation is sometimes referred to as *infrared*,

*Both shortwave and longwave radiant energy travel at the speed of light—186,000 miles (298,000 kilometers) per second.

ground, or *terrestrial* radiation. When the ground is warmed under the impact of insolation, the molecules become excited and begin to vibrate rapidly, causing the release of the infrared radiation. These waves rise rapidly into the lower atmosphere, where about 90 percent of them are trapped and absorbed by particles of carbon dioxide, ozone, and water vapor. Eventually, the atmosphere radiates longwave radiant energy back toward the earth, thus cooling the atmosphere while warming the earth.

This process by which our earth and atmosphere are heated has been likened to what

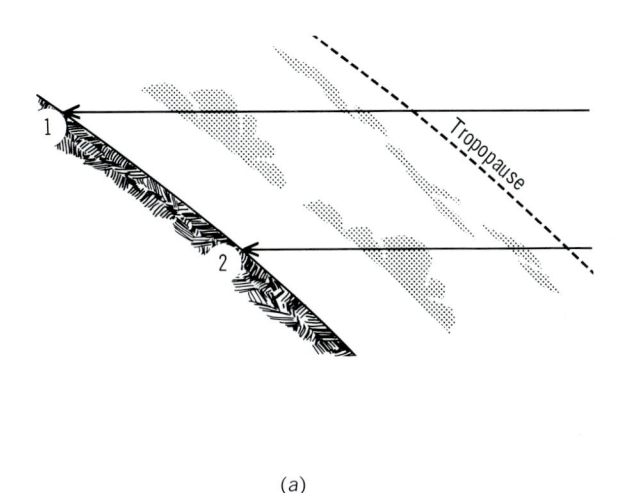

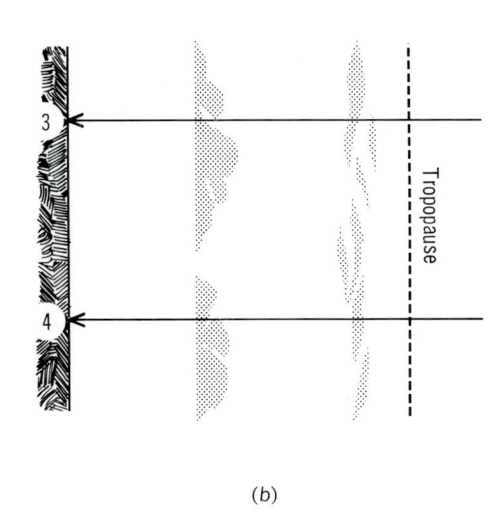

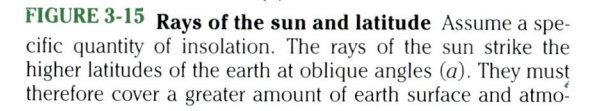

FIGURE 3-15 Rays of the sun and latitude Assume a specific quantity of insolation. The rays of the sun strike the higher latitudes of the earth at oblique angles (*a*). They must therefore cover a greater amount of earth surface and atmosphere than do rays that strike the surface at or near the equator (*b*). Vertical rays are concentrated on a smaller amount of earth space and travel through less atmosphere.

happens when you leave your car out in the sunlight with the windows closed. It gets hot inside. Why? Because shortwave solar energy can easily penetrate the car's windows, and once inside the car, it is absorbed by the upholstery and reradiated as long waves, which cannot escape the car's interior. Obviously, the car heats up.

Another analogy would be a greenhouse, a glass-enclosed building where even in cold climates one can grow tropical plants. Here, again, the powerful shortwave energy passes through the panes of glass and is absorbed by the plants, clay pots, soils, wooden benches, and the earth and is reradiated into the enclosed space. The longwave radiation cannot escape through the glass and is deflected back and forth within the greenhouse. As this process (Fig. 3-16) continues, the temperature inside the greenhouse builds up to, perhaps, a nice 70°F (21°C), even though the outside temperature may be a very cold 40°F (4°C). Meteorologists have long referred to this process as the **greenhouse effect** because the panes of glass actually absorb the longwave radiation and then reradiate it back down. It is not truly a reflection process, and not exactly like that which occurs within our atmosphere. It is becoming more common today for meteorologists to refer to the warming of our earth and atmosphere by radiation as the *atmosphere effect* instead of as the greenhouse effect.*

What would happen if the atmosphere became too dense and did not allow 10 percent of the longwave radiation to escape to outer space? Temperatures within the atmosphere and on the earth would increase to a point of high humidity and extremely high temperatures. Earth would have a tropical age. Water from the oceans, lakes, and streams would evaporate excessively and form clouds

---

*For an interesting explanation of the actual workings of a greenhouse and our atmosphere, see Robert C. Fleagle and Joost A. Businger (1980), *An Introduction to Atmospheric Physics*, 2nd. ed. (New York: Academic Press).

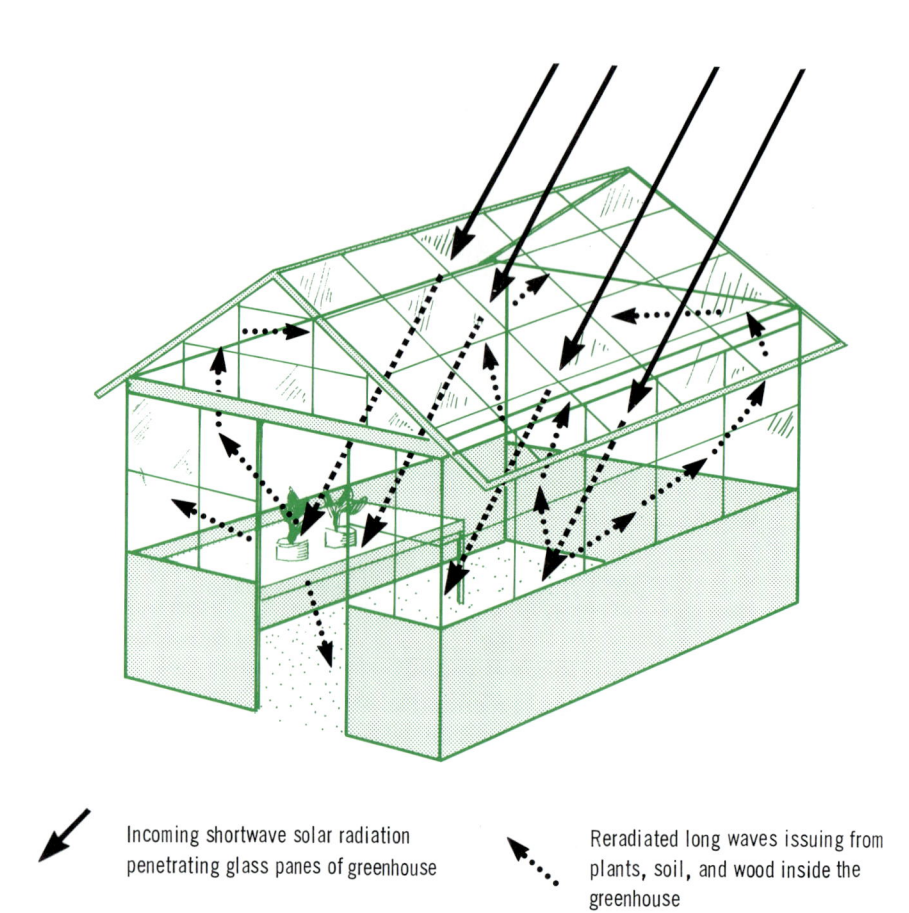

FIGURE 3-16 The greenhouse effect The length of an energy wave depends on the heat of its source. Solar energy waves are strong, short, and move with great intensity and speed. Heat that is reradiated from the earth or from objects such as pottery, vegetation, and people is warm, not hot, and the resulting heat waves are weaker, longer, and slower. Infrared and X-ray photography, in which the heat emanating from a body (or the bones within) is photographed, are partly based on this principle.

Incoming shortwave solar radiation penetrating glass panes of greenhouse

Reradiated long waves issuing from plants, soil, and wood inside the greenhouse

at the tropopause. This newly formed cloud layer would prevent incoming solar radiation from penetrating our atmosphere, and the air temperature would drop, bringing on an ice age. Earth scientists have assured us that there have been many ice ages, alternating with interglacial periods, during the last million years. Can we predict when the next might occur? The last ice age ended about 10,000 years ago.

Conduction Another process of heating is **conduction,** in which heat will move by direct contact from a hot or warm object to a colder one. Some substances are good conductors, others are not. Wood is a bad conductor, but an iron bar or a soup spoon is a good one. Air is a poor conductor and, as a result, cold air settling over warm earth will heat up only a few centimeters from the surface. A lake or ocean warmed by a summer sun will heat up to great depths and will consequently be able to warm the bottom layer of a settling cooler air mass for a relatively long period of time by the simple passage of warmth from the water to the air. Land, on the other hand, does not heat up to any great depth. Furthermore, land loses most of its heat by radiation, and what is left to warm cooler air by conduction is minimal.

EARTH'S POWER PLANT: SOLAR ENERGY **75**

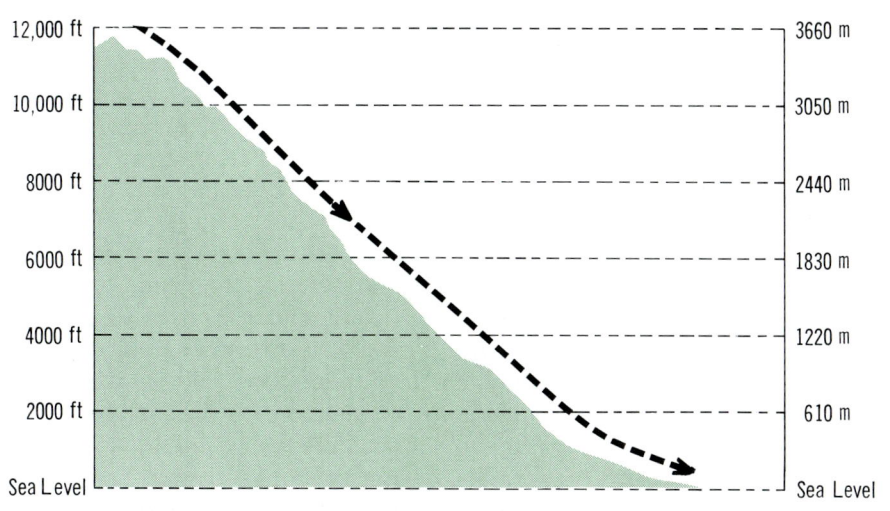

FIGURE 3-17 Heating of air by compression The temperature of an air mass changes with differences in altitude. If the air is still, nonmoving, the normal lapse rate (or change) is 3.5°F per 1000 feet (0.65°C per 100 meters). If the air mass is moving down the side of a mountain, or from a higher elevation to a lower one, the change increases to about 5.5°F per 1000 feet (1°C per 100 meters) as it descends. In the illustration, an air mass descending from 12,000 feet (3660 meters) increases its temperature 66°F (36.6°C) by the time it reaches sea level. On nights when the skies are clear and the air pressure is low, cold air will form close to the ground, and if there is a slope, the heavier, colder, denser air will flow beneath any warmer, lighter air present, producing a cooling wind that is called *katabatic* (or mountain) wind. This is a form of air drainage.

Compression The heating process of **compression** occurs when an air mass descends from a higher to a lower elevation and gains in temperature as it moves downward (Fig. 3-17). The warming rate averages about 5.5°F for every 1000 feet of descent (1°C per 100 meters). Inhabitants of Samarkand (Soviet Uzbekistan) are constantly being reminded of this process by the air masses blowing down from the Pamir Mountains of Afghanistan. These hot, dry winds, called the **Afghanets**, carrying great clouds of fine dust particles, blot out the southeastern horizon, and make life uncomfortable for the Uzbeks.

Residents of Santa Ana, California, also feel the warming effects of descending air masses, especially during winter months, when these **Santa Ana winds** move down from the Great Basin region of Nevada and Utah.[†] These winds are the result of the settling of air that moves poleward from the equator at high altitudes. As the warm air cools, becoming heavier and denser, it begins to sink, spiraling outward from its central area in a clockwise motion. The moving air cannot remain in this plateau region. It must move out, gravitating down from the plateau toward the Pacific Ocean, blowing southwesterly down through mountain passes and across the Mojave Desert,

[†]*Santa Ana* is the correct term for this local winter heating process. It was given its name in the eighteenth century by Spanish explorers who were puzzled by 70°F (21°C) temperatures in midwinter in the region south of present-day Los Angeles. Similar warming winds are to be found in many places: *chinook* for the region east of the Rockies; *foehn* in Switzerland and Europe; *samoon* in Iran.

descending from an average elevation of about 6000 feet (1800 meters) to the near-sea-level elevation of the Los Angeles region. The descending air mass gains about 33°F (18°C) in its downward passage; and if, for example, it began in Nevada at a cool 55°F (12.8°C), its temperature would be increased to about 88°F (31°C) on arrival in Los Angeles. In addition, the funneling process (being squeezed between mountains) causes these drying winds to be quite gusty. Brush and forest fires are particularly frequent in Santa Ana or Afghanets areas. This warming of moving air is called **adiabatic heating,** which means that the air mass neither gains nor loses any of its energy. The opposite of this adiabatic process—when a warm air mass rises and begins to cool as it looses heat—is called *adiabatic cooling.*

Condensation In the heating process of **condensation,** water in a gaseous state (as vapor) is converted to a liquid (water) or a solid (ice or hail) state, and heat (called *latent heat*) is released into the air mass. This occurs when a warm air mass is lifted to a cooler and higher elevation or moves over a cold landmass at a low elevation, and then cools until it can no longer retain the water it holds in its gaseous state. This level, called the **dew point,** is the temperature at which condensation results, when moisture-laden air reaches complete saturation. The point at which this occurs varies with the temperature of the air mass. The warmer the air, the more water vapor it can retain; the cooler the temperature, the less it can hold. Any excess must be released. Note, however, that not all condensation results in rain, snow, or hail. When the dew point is reached, moisture might become fog, dew, or frost (if the air temperature is at or below the freezing point).

Cooling Processes

The temperature of the atmosphere would rise to uncomfortable levels in most regions were it not for a number of cooling processes. Some of the previously discussed heating actions also produce cooling. For example, when heat is radiated from the land or waters, the temperature of the radiating object or substance is reduced in equal measure. The same applies to the process of conduction when, by direct contact, heat is transferred from one object to another. For example, when cold air overlies a warm body of water, heat will move from the water to the air and the water will lose temperature. If fog or rainfall results from condensation, the land on which this form of moisture falls will be cooled as heat is transferred to the moist air mass. Some of the more significant cooling processes are the result of evaporation, expansion, and advection.

Evaporation In the process of **evaporation,** which is the opposite of condensation, water in a liquid or solid state is transformed into a gas (water vapor). Instead of releasing heat (like the latent heat of condensation), the surrounding air mass loses heat as the molecules race wildly about, colliding with each other and finally breaking free to take some of the water's energy with them. If you have ever exercised on a hot, dry day, you know that the perspiration emitted by your body is immediately taken up by the air and you feel cool. But if the air around you is almost to the saturation point and thus cannot accept more moisture, conditions are called *humid.* You would feel hot and sticky and the beads of perspiration would stay on your body. You would not feel cooled. If, however, a breeze were blowing, the ability of the air to accept more moisture would be increased and you would feel the cooling effect of the wind. We might say that evaporation depends on a number of factors: the presence or absence of wind, the aridity (dryness) of the air, and the temperature. On a windless midsummer day, people in humid Jackson, Mississippi, might be more uncomfortable than people in hot and dry Phoenix, Arizona.

Expansion The process of **expansion** is the opposite of compression. As an air mass will warm when the molecules are pressed more tightly together, so the air mass will cool if the molecules are no longer confined in a small amount of space and can move about more freely. This has nothing to do with outside temperatures. There is no transfer of heat by processes such as conduction or radiation to the surrounding air. It is purely automatic and is made possible because of differences in air pressure over an area.

This is how expansion occurs. Insolation warms the earth, causing the overlying air to be heated. As the air mass warms, it expands and is lighter in density. It begins to rise and, as it rises, the pressures surrounding it become weaker, thus allowing the molecules within the rising, warm air mass to separate and move farther and farther apart. Meteorologists call this form of cooling **adiabatic cooling.** The upward movement of the expanding air mass is caused by an energy transfer process called **convection** (vertical movement or uplift of an air mass), and the rate (amount) of cooling is referred to as the **lapse rate** (see Fig. 3-18).

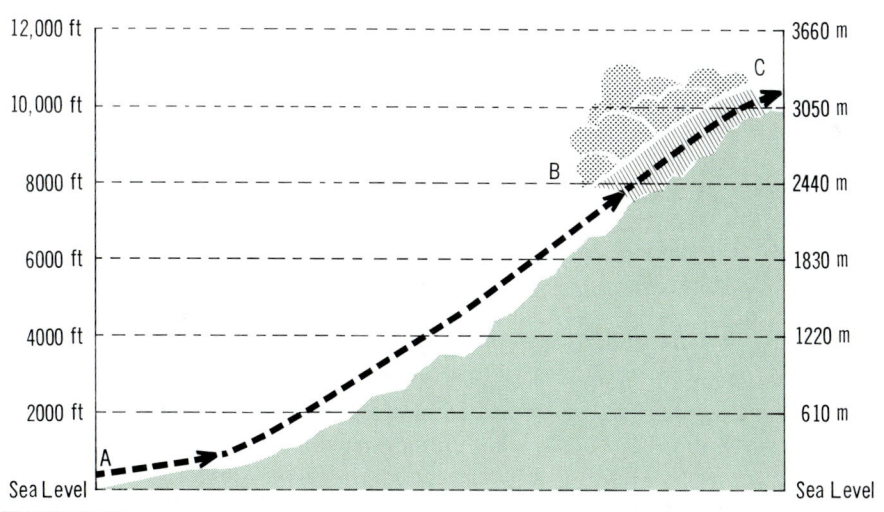

FIGURE 3-18 Adiabatic lapse rates In this diagram a warm air mass moves against a mountain and is forced to rise to higher elevations. If its temperature at sea level, point A, was 80°F (27°C), and if the dry adiabatic lapse rate was in effect (and not interfered with by snow or rain), the temperature loss of the rising air mass would be at the rate of 5.5°F per 1000 feet of rise. Thus the air mass's temperature at point B, the 8000-foot (2440-meter) elevation, would decline to 36°F (2.2°C). If condensation were to occur at point B and continue to the top of the mountain, point C, elevation 10,000 feet (3050 meters), the temperature of the air mass would come under the influence of the wet-adiabatic lapse rate (3.2°F per 1000 feet) and decline to 29.6°F (−1.3°C).

There are three types of lapse rates—normal, dry adiabatic, and wet adiabatic—but only one has anything to do with quiet, unmoving air. This is the *normal* lapse rate. In this case, the change in temperature is strictly regulated by differences in altitude. As discussed earlier, the atmosphere is densest near the earth's surface. Obviously, this means that it is less dense at higher elevations. The rate of decline is about 3.5°F for every 1000 feet of altitude increase (0.65°C per 100 meters). Suppose you wanted to test this by taking a thermometer with you on a balloon ride up to 10,000 feet (3000 meters) above the earth. If the temperature of the air at sea level were, say, 90°F (32.2°C), it would be about 17.5°F (9.75°C) cooler at the 5000-foot (1500-meter) mark, and 35°F (19.5°C) cooler at the top of your ascent, where the air temperature would be 55°F (12.7°C).

When an air mass is moving, the lapse rate increases because the act of movement causes the air mass to lose additional heat. The heat loss now becomes 5.5°F per 1000 feet of altitude increase (1°C per 100 meters). This is termed the *dry adiabatic* lapse rate. If the warm air mass is also heavily laden with moisture and the convectional uplift moves it to a height where the loss of temperature decreases its moisture-carrying capacity, dew point is reached and condensation may occur. If this happens (remember our discussion on page 76), latent heat is produced and the heat loss is diminished to a lapse rate of only about 3.2°F per 1000 feet (0.6°C per 100 meters). This is termed the *wet adiabatic* lapse rate. For an example of these lapse rates in action, see Figure 3-18.

Advection The horizontal movement of air, **advection,** also is an important cooling process. Cool or cold air is heavier and denser than warm air, and as heated air rises from its heating surface, the heavy air slides into what has become an area of low air pressure.* Residents of the American Midwest can testify to the welcome intrusions of cool air from Canada that occasionally temper their long, hot summers. This can also happen along the shoreline of a lake or ocean, when cool winds blow inland to replace warm air rising over towns and cities. Note, however, that advection can work the other way, too, as when a warm air mass glides north from the Caribbean or the Gulf of Mexico to plague an already overheated southeastern United States with additional humidity and, sometimes hurricanes.

Temperature Inversions The usual situation is for temperatures to be cooler the farther away from the earth's surface one gets. However, what happens in the heating and cooling processes and to the lapse rates of air temperatures when the reverse situation occurs—temperatures increase rather than decrease with altitude? Meteorologists call this a **temperature inversion.**

Static and Dynamic Inversions There are two kinds of temperature inversions: static and dynamic. **Static** inversions usually are brought on during long winter nights that are cool, clear, calm, and dry. They occur generally during high-pressure conditions that permit rapid radiation of heat from the earth. If there is little air movement (either vertically or horizontally), rapid ground radiation is given off and the earth becomes cooler than normal. The air in contact with the ground becomes

---

*A column of air has weight, and if the air is cooling and settling, the pressure near the earth's surface will increase, thus creating a high-pressure area. If, on the other hand, the air above the ground is warmed and the column of air above it begins to rise, the pressure is diminished and a low-pressure area results. This will be covered in further detail on page 101.

chilled by conduction as it touches the cold ground and, therefore, loses heat. The resulting cold layer of air acts as an effective barrier against any intrusion from above by warm air. *Dew* will result with a lowering of the temperature. If the temperature drops below freezing, *frost* will occur. If the air mass is damp, a *radiation fog* will occur in some valley areas, such as in California's San Joaquin Valley.

Dynamic inversions are created when a mass of heavy, cold air settles over an area, sliding down the sides of mountains and hills and forming great pools of cold air on the valley bottom, effectively shutting off the people and the crops on the ground from any warm air aloft. Farmers often try to take advantage of this movement of cool air down a slope (a process called *air drainage*) by planting their crops on the hillsides where the moving air does not linger long enough to chill the crops.* Farmers who do use the bottom lands often try to protect their crops by using smudge pots to raise air temperatures, by implementing nighttime irrigation, or by installing wind fans. In seriously plagued areas farmers often plant cotton, sugar beets, or feed crops such as alfalfa—crops that can thrive in cold air.

Temperature Inversions and the Urban Environment What, if any, are the adverse effects of inversions? If an inversion does exist, what happens when a city erupts into its frenzied daytime activity, when people awake to go about their daily tasks? At first the sky is clear and the air is cool and easy to breathe; then the automobiles, trucks, buses, trains, jets, and factories begin their production of various gaseous mixtures such as carbon monoxide and sulphur dioxide, which unite with ozone

*These downslope winds are called *katabatic winds*.

particles in the atmosphere. The resulting pollution begins to build up, and the noxious **smog** is born.

As the city heats up from its receipt of insolation, the gaseous mixtures rise, but they cannot break through the upper levels of the inversion layer, so they thicken and spread back to the earth's surface. The inhabitants of the city—whether it is Los Angeles, New York, or Moscow or Irkutsk—begin to have a number of problems. They may experience stuffy noses, watering eyes, and aching lungs, and they may find that distant mountains can no longer be seen. Citizens of Los Angeles often joke about the "good old days" when "on a clear day you could see Catalina Island."

Why are smog-producing temperatures inversions relatively common in urban areas?* One meteorologist explained that cities develop inversions because they tend to be warmer than the surrounding countryside. There are a number of reasons for this. First, there is a difference in surface materials. Concrete buildings, for example, heat to a greater depth and retain their heat longer than soil does. Second, buildings and other city structures absorb more heat than do features of the natural landscape, because of their shapes and closeness to each other. Third, factories, automobiles, and residential heating systems generate a consid-

*For a thorough accounting of the effect of cities on the environment, see William P. Lowry (August 1967), "The Climate of Cities," *Scientific American*, 217(2) (offprint no. 1215); R. A. Bryson and J. E. Ross (1972), "The Climate of the City," in T. R. Detwyler and M. G. Marcus (eds.), *Urbanization and Environment* (Belmont, Calif.: Duxbury Press), pp. 51–68; J. E. Peterson (October 1969), *The Climate of Cities: A Survey of Recent Literature*, Publication AP-59 (Washington, D.C.: U.S. Department of Health, Education, and Welfare, National Air Pollution Control Administration); J. R. Norwine (July 1973), "Heat Island Properties of an Enclosed Multi-Level Suburban Shopping Center," *Bulletin of the American Meteorological Society*, 54(7), pp. 637ff.

erable amount of heat. Fourth, cities retain little fallen rain or snow because most of such precipitation runs off through drainage systems, leaving little to evaporate and function as a coolant.[1]

Basically, when warm air, for whatever reason, becomes separated from the ground by an intruding mass of cold air or, because of radiation, by a rapid loss of heat at ground level, a temperature inversion is created.

Smog formation of Los Angeles, California, Birmingham, Alabama, and Martigny, Switzerland. Believe it or not.

CASE STUDY SOLAR ENERGY—FRIEND OR FOE?

Now that we have considered some of the results of the sun's electromagnetic radiation, do you consider the sun your friend or your foe? If you have been badly sunburned after a too-long exposure to the invisible ultraviolet rays, or if you find yourself sagging limply during a long, hot, humid day, it is probable that you think of the sun as an adversary. But what about those cold winter days when the sun's warmth envelops you when you are forced to leave the artificially heated home or classroom?

The sun's rays carry a life-sustaining energy vital to everything on earth; it is no wonder that the sun has been revered by humanity from primeval days and was worshipped as a god in early civilizations. But today we know the sun in a different way. We understand the physical processes of solar energy and know how to harness its bounty. We use it creatively, and it serves us well.

Solar energy is here, now. We have experienced it, although we may not have recognized it for what it is. Solar heating and cooling is as common as:

The hot interior of a car parked for hours in the sun.

A cat curled up on a sunny windowsill on a crisp autumn afternoon.

A refreshing breeze cooling the house on a summer evening.

Despite all the new technologies that promise us new sources of energy, we are choosing solar energy because it is abundant, nonpolluting, and free. It is not affected by changes in international politics or trade or by inflation at home. It is available when needed. It is dependable, even in winter. Solar energy will heat your water and your pool, and it will cool and heat your home.

A SOLAR HISTORY*

Solar energy is today. It was also yesterday. Perhaps the earliest and most constant use of the sun's energy has been heating and cooling living space. The basic principles of solar heating and cooling rely on a simple fact. During summer the sun appears high in the sky, and its rays fall almost straight down on us. In winter, as the earth tilts away from the sun, the sun appears lower on the horizon, its path further to the south.

> *One of the easiest and most effective ways to use solar energy is to take advantage of the natural course of the sun as it travels the sky from north to south and back again as the seasons change. If a building faces south, the sun's warming rays will shine directly into it during the winter, providing natural heat. During the summer, the structure will stay cooler if the windows are shielded from the sun by overhangs. A critical factor is the direction the building faces.*

The early Greeks understood this principle of natural solar heating and cooling in the 5th century B.C., entire cities were laid out on an east-west axis, so one long side of each building faced south. Aristotle and Socrates both spoke at length of the benefits of siting a building to take advantage of the sun. By the first century A.D., the Romans had glass and built their famous baths with large south-facing windows, to warm the rooms with sunlight.

A classic example of natural solar design is Mesa Verde, Colorado, where the Indians built their adobe homes among the cliffs of the mesa. During the winter the sun would shine directly into the cliff dwellings and warm them naturally. The thick adobe walls stored the heat of the day, and at night the warmth would radiate into the living space. During the summer, when the sun is high, the overhanging cliffs would shade the structures to keep them cool. The Indians of Mesa Verde built their homes about 1200 A.D. Today, almost 800 years later, their natural heating and cooling system is still working.

In the early 1900s favorable economics prompted the birth of the solar water-heating industry in Southern California. In 1914 soft coal sold for $13 a ton in California, twice the national average. Natural gas cost more than ten times what it costs

---

*This material furnished by the California Energy Commission, 1111 Howe Avenue, Sacramento, California 95825.

The Cliff Palace, Mesa Verde, Colorado—1200 A.D.

today, and electricity cost more than gas. Whatever fuel was used, heating water for the home was an expensive proposition.

Sunlight, on the other hand, was plentiful and free. Solar water heating systems were introduced in 1891. The early models were simply shallow water tanks mounted on the roof of the house with plumbing leading to the kitchen and bathroom. In 1909 Southern Californians were introduced to the Day and Night Solar Water Heater. By adding a separate, insulated hot water storage tank, the solar-heated water stayed heated, day or night.

In 1930 rooftops in Los Angeles and such nearby towns as Pasadena and Monrovia were dotted with solar collectors. The solar industry was also booming in Florida and parts of the southwest. But as electricity, natural gas, and other fuels became more readily available and dropped in price, the use of solar energy declined.

SOLAR CALENDAR

5th Century B.C.   The Greeks use solar orientation extensively. Entire Greek cities are built to take advantage of the sun's heating and cooling properties.

1st Century A.D.   With the advent of glass, the Romans carry the solar principles learned from the Greeks still further. Southfacing glass windows trap the sun's heat to warm public baths, villas, and expensive homes.
Sabinius records the first "sun rights" law.

5th Century A.D.   The Romans make violation of a property owner's solar access a civil offense.

EARTH'S POWER PLANT: SOLAR ENERGY

1200	The Indians of Mesa Verde use the principles of passive solar design to heat and cool their dwellings.
1891	"The Climax," the first commercial solar water heater, is patented. The simple black box rooftop heater cuts natural gas consumption 40 percent.
1895	"The Climax" Solar Water Heater comes to California.
1897	30 percent of the homes in Pasadena, California have solar water heaters.
1900	More than 1600 "Climax" Solar Water Heaters are operating in Southern California.
1909	The "Day and Night" Solar Water Heater is introduced. A major design advance separates the solar heat collector and hot-water storage unit, to cut natural gas consumption for heating 75 percent.
1913	Record freeze hits Southern California. William Bailey designs nonfreezing solar water-heating system.
1914	"Day and Night" Solar Water Heater expands marketing territory to Northern California, the Hawaiian Islands, and Arizona.
1920	This is a peak sales year for "Day and Night": 1000 solar systems sold. Natural gas is discovered in the Los Angeles Basin.
1923	Bailey sells patent rights on "Day and Night" Solar Water System to Florida entrepreneur for $8000 and an Oldsmobile touring car. The solar water heater will see its greatest success in Florida.
1941	At least 60,000 solar water heaters installed in Florida.
1950–1970	United States energy consumption increases 5 percent annually.
1972	United States uses 35 percent of the world's energy.

THE MECHANICS OF SOLAR

To understand how sunshine works to heat your home or office, imagine a garden hose that has been laying in the sun for several hours. When you turn on the faucet, out comes hot water—not the cool water you would expect. The sun's energy is absorbed by the hose and is transmitted as heat to

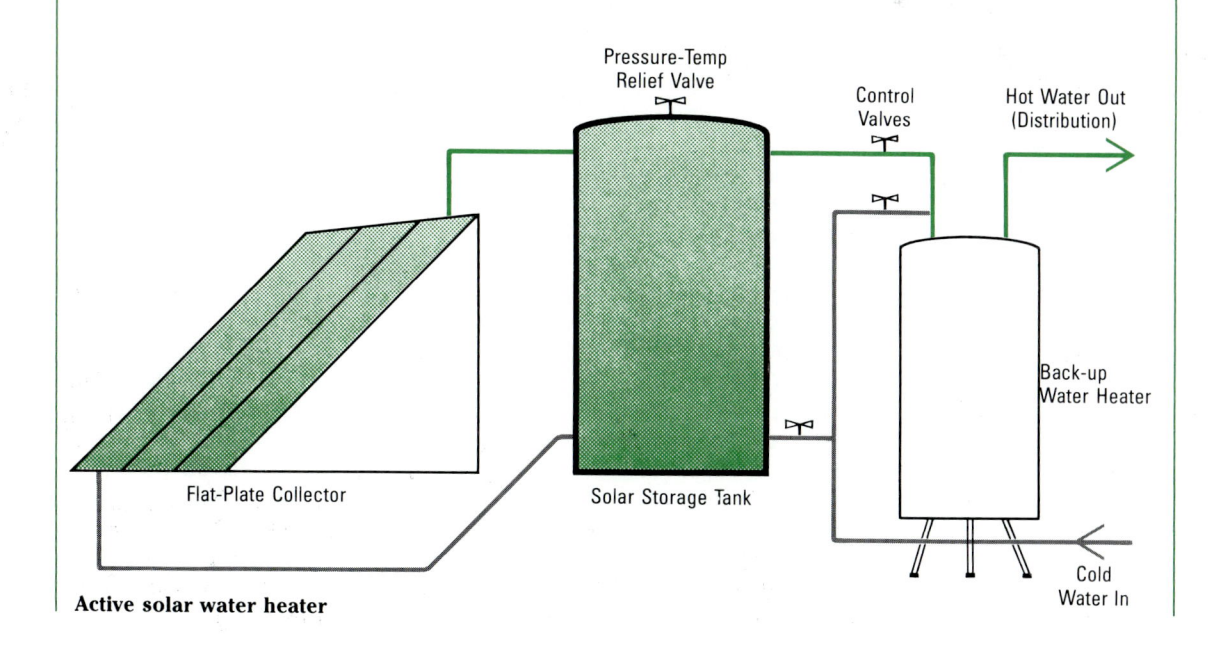

Active solar water heater

Home in Pomona Valley, California, 1911, with Day and Night water heating solar collectors on the roof.

the water inside it. The hose acts as a solar collector.

The *collector* is the first functional element of a solar heating system. Sunlight falling on the collector heats it, just as it heats the garden hose, and the collector in turn raises the temperature of the water or air that carries the heat to where it is used. The other components of a solar heating system are *storage,* where the heated water or air is stored for later use at night or on cloudy days; the *distribution system,* which delivers the warmed liquid or air from the collector to storage, and from storage to where it is needed in the building; and the *controls* which regulate the flow of heat.

An active solar system depends on collectors separate from the living and working space of the structure. Active systems usually tie into an existing heating system or water heater, which are active systems themselves. In fact, mechanically there is not much difference between an active solar system and a conventional heating system. The difference lies in the source of the heat. A solar system gets its heat from the sun; a conventional system, from electricity or natural gas.

EARTH'S POWER PLANT: SOLAR ENERGY 85

The first operating solar collector in New York City: the first step in the revitalization of a portion of Manhattan's tenement district.

Collectors should be installed on a south-facing roof at an angle to receive the most available sunlight, whatever the season. Since energy is collected only when the sun is shining, there must be some means of storing the excess heat collected during sunny hours. Usually the storage unit is a tank of water or a bed of rocks which hold the heat until it is needed.

At this point it becomes obvious that solar energy is our friend, not our foe. As we look to the future, knowing that the earth's energy resources are declining at an alarming rate, we can be grateful for the wide variety of applications of solar energy that are available to us and that will be in use before the end of this century. As you delve further into the possibilities, you will learn that we will be using active solar systems for air conditioning, generation of electricity through a process known as photovoltaics, and solar furnaces that can provide heat up to 6000°F (3342°C) for metallurgical work.

For additional information on this growing field of solar energy, contact your state's Energy Commission, or the Department of Energy (DOE) in Washington, D.C. The following books are also recommended.

Beckman, William A., Sanford Klien, and John Duffie (1977). *Solar Heating Design.* New York: Wiley.
Cook, Early (1976). *Man, Energy, Society.* San Francisco: W. H. Freeman.
Derven, Ronald, and Carol Nichols (1976). *How to Cut Your Energy Bills.* Farmington, Mich.: Structures Publishing Co.
Dorf, Richard (1978). *Energy Resources and Policy.* Reading, Mass.: Addison-Wesley.
Duffie, John A., and William Beckman. (1974). *Solar Energy Thermal Processes.* New York: Wiley.
Stoker, H., Stephen, Spencer Seager, and Robert Copener (1975). *Energy: From Source to Use.* Glenview, Ill.: Scott, Foresman.

REVIEW AND DISCUSSION

1. Why is the earth's position relative to the sun considered important?
2. Explain the terms *revolution, rotation, aphelion,* and *perihelion.*
3. Explain the difference between tropical year and sidereal year.
4. Explain Newton's first law of motion: inertia.
5. Why do we have leap years?
6. Has the earth always rotated around its axis at the same speed?
7. What did Foucault prove with his pendulum?
8. Why do earth's weather and communication satellites appear to wander as they orbit the earth?

9. Which direction does the earth rotate: east to west or west to east?
10. What is the plane of the ecliptic?
11. Of what benefit to humanity is the inclination of the earth's axis?
12. What does the tilt of the earth's axis have to do with the changing seasons?
13. Explain the difference between the solstices and the equinoxes.
14. What is the circle of illumination?
15. Why is the sun sometimes likened to a giant furnace?
16. How does our atmosphere protect us from excessive solar radiation?
17. What is corpuscular energy?
18. What are the Van Allen radiation belts?
19. What causes the auroras?
20. Explain the term *insolation*.
21. Discuss the significance of the earth's heat balance.
22. Explain the term *albedo*. Why is it higher at the poles than near the equator?
23. Explain why a large body of water (lake or ocean) can be an important modifier of a region's climate.
24. What does an area's latitude have to do with its heating and cooling?
25. Explain the terms *radiation, reradiation, conduction, convection,* and *compression*.
26. Explain the greenhouse effect.
27. What causes hot drying winds, such as Santa Anas, Afghanets, and chinooks, to occur?
28. Explain the difference between a normal lapse rate and an adiabatic lapse rate.
29. If an air mass were to descend from a mountaintop elevation of 15,000 feet (4500 meters), would its temperature increase or decrease by the time it reached sea level? What would be the amount of change?
30. If an air mass ascended the windward side of a 16,000-foot (4800-meter) mountain, with condensation occurring at the 12,000-foot (3600-meter) level, what would be the temperature increase or decrease at the top?
31. Explain the terms *evaporation, advection, saturation, condensation,* and *dew point*.
32. What causes temperature inversions?
33. Explain the difference between static and dynamic inversions.
34. How do temperature inversions affect life in urban environments?
35. What is smog?
36. Why is solar energy regarded as a nonpollutant?
37. Is solar energy a practical energy resource for the homeowner?
38. Do you believe there is an "energy crisis"?

NOTE

[1] William P. Lowry (August 1967), "The Climate of Cities," *Scientific American* (offprint no. 1215), pp. 15–23.

ADDITIONAL READING

Beiser, Arthur (1971). *The Earth*. New York: Time-Life Books.

Bergamini, David (1969). *The Universe*. New York: Time-Life Books.

Ehrlich, Paul R, Anne H. Ehrlich, and John P. Holdren (1977). *Ecoscience: Population, Resources, Environment*. San Francisco: W. H. Freeman.

Nicks, Oran W., ed. (1970). *This Island Earth*. NASA publication no. SP-2500. Washington, D.C.: Office of Technology Utilization, NASA. A valuable collection of photographs of the earth, taken by astronauts in space and by satellites.

Wallace, Daniel, ed. (1976). *Energy We Can Live With*. Emmaus, Pa.: Rodale Press.

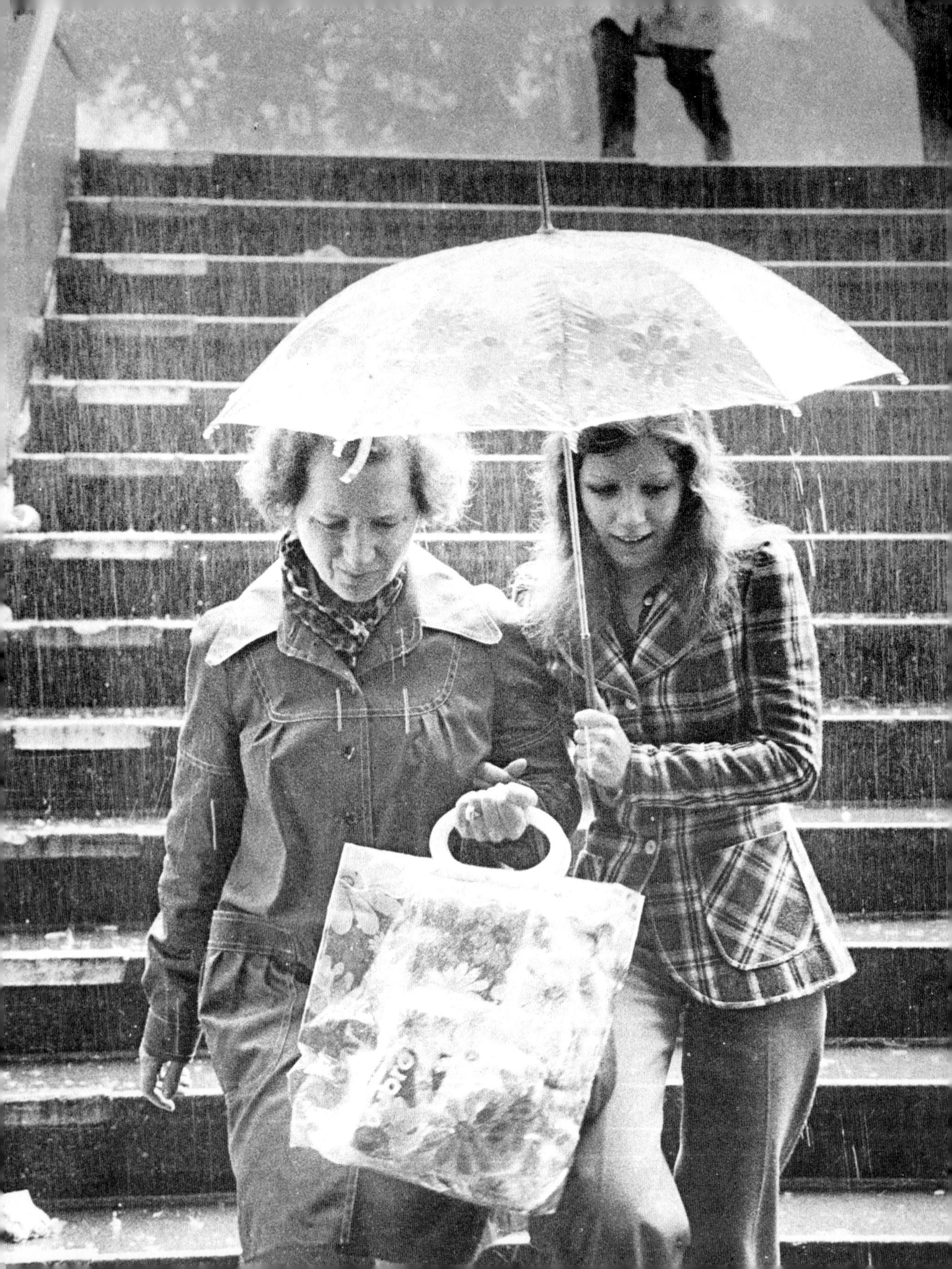

F O U R
EARTH'S DYNAMIC ATMOSPHERE

What is the atmosphere? It is the envelope of air that surrounds our planet. It gives rise to the climates of the world; it gives birth to the weather that confounds us every day. Charles Dudley Warner once said: "Ah! the weather! Everyone complains about it, but no one does anything about it!" True enough for the first half of this century, but true today? No.

Because earth has a dynamic atmosphere, it also has weather and climate. *Weather* and *climate* are two words that seem to mean the same thing and, to some people, are interchangeable. Yet, in a way, they might be likened to the terms *mother* and *daughter*—related, but not the same. In this sense, weather is the mother of climate.

Weather is "what it is like" today or tomorrow. It is the condition of the earth's atmosphere over a specific location for a specified period of time, such as a few hours or days.

Climate, on the other hand, is the weather that is characteristic of a place for a much longer period of time. The climate of a region is based on a generalization of the averages gained from a systematic study of weather data (statistics) gathered over 30 years or more. Numerous elements comprise climate: temperature, moisture, atmospheric pressure, and wind. For example, while the climate records for Fresno, California, show that it averages 11.2 inches (28 centimeters) of rainfall per year, there has been no actual year when that exact amount has fallen.

We would be concerned about an area's weather if we were planning a short vacation there, but we would be more concerned about the climate if we were planning to move there permanently. Weather deals with today or tomorrow—what the temperatures will be, what the chances are for rain, and if there will be wind; climate goes deeper, entailing the evolution of the area's natural vegetation, soils, water supply, and landforms. Climate can tell us why one region on the earth's surface is a vast, empty desert, another is a thick, heavy rain forest, and still another is a wide open grassland.

The study of weather is known as **meteorology** (from the Greek *meteora*, "things out there"); the study of climate is called **climatology** (from Aristotle's *klimata*, "climate regions"). Both the meteorologist and the climatologist investigate and analyze the workings of the earth's atmosphere. They are concerned with the different elements and the way these elements are controlled or modified by differences in latitude; sizes, shapes, and distribution of landmasses and oceans; air temperatures; air pressures and winds; ocean currents; moisture and precipitation; altitude; and the presence or absence of mountain barriers.

We begin our investigation into why earth's atmosphere is so energetic, vital, and dynamic by looking closely at its structure and at the different ways in which meteorologists and climatologists describe its composition.

COMPOSITION OF EARTH'S ATMOSPHERE

A number of ways have been devised to define the composition of our atmosphere. The most popular way was to set up two distinct regions: the "outer" and the "lower" atmospheres (Fig. 4-1). The outer atmosphere was composed of the **exosphere** (from the Greek *exo*, "outer") and the **ionosphere.** The exosphere was supposed to be the outermost shell, or sphere, surrounding the earth, beginning at an altitude of about 500 miles (800 kilometers) and extending outward to about 20,000 miles (32,000 kilometers). It was said to be a distinct part of the atmosphere because it contained charged particles of nitrogen and oxygen. It was difficult, however, to truly fit this region into the atmosphere because no one really knew just where it came to an end. As a matter of fact, the exosphere just fades away into outer space.

The ionosphere is somewhat easier to explain because it is an area of electrical activity, but, again, it becomes confusing when one attempts to set absolute boundaries. Most **aeronomists** have been content to state that the ionosphere varies in depth and in distance

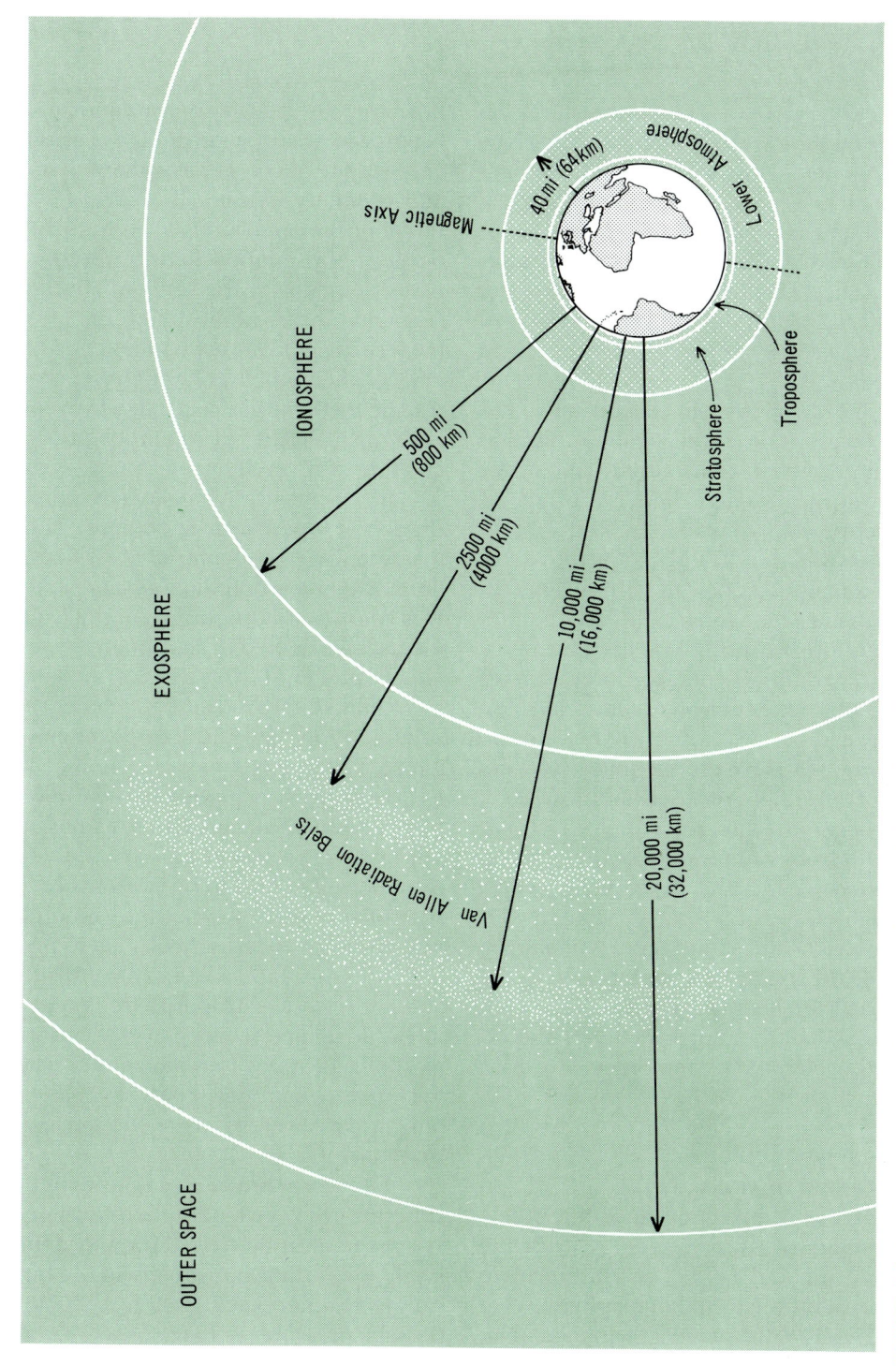

FIGURE 4-1 Earth's outer atmosphere: exosphere and ionosphere (Not drawn to scale.)

91

from the earth. Generally, its lowest level begins about 50 miles (80 kilometers) above the earth and extends outward to about 500 miles (800 kilometers). This region contains a great number of gaseous particles that are under constant bombardment by cosmic rays (solar particles). They become electrically charged, or *ionized,* and form into layers of ionized gases, which serve as a protective shield around the earth.

This shield serves us in several ways. For example, long-distance radio communication would be impossible if transmitted radio waves (which travel in a straight line) were not reflected (bent back) to earth by the ionized layers. Otherwise, they would escape into outer space. We can also be grateful to these ions for aesthetic reasons. As visible light rays bounce from ion to ion during their passage toward the earth, separation occurs, forming what we perceive as blue sky. It always seems bluest directly overhead because when we look up, we only have to "see through" 11 miles (18 kilometers) of atmosphere, whereas when we look at the sky over the horizon, we have to see through hundreds of miles of the densest and most polluted part of the atmosphere, which tends to lighten, or gray out, the blueness of the sky. From a health standpoint we should be thankful for the ionosphere and its electrically charged particles because they also intercept and deflect away from earth many of the incoming ultraviolet and gamma rays.

Structure of the Atmosphere: Gaseous Composition

Another method of atmospheric classification uses the gaseous composition of different levels as the standard. Because the gases of the lower levels are well mixed and uniform in their proportions, the term **homosphere** (from the Greek *homo,* "the same") is applied (Fig. 4-2). Above the 72-mile (115-kilometer) level, however, the gases were found to be in distinct layers, and their proportions were quite different, so the term **heterosphere** (from *hetero,* "different") was used. Both regions are composed of the same kinds of gases (nitrogen and oxygen, plus minor amounts of other types of gases), but not in the same quantities or percentages. In the homosphere the gas particles have sufficient density to form air, but in the heterosphere they are widely separated and, for the tremendous space in which they circulate, insufficient in quantity to be dense enough to form air.

Air is mostly nitrogen (78 percent) and oxygen (21 percent) by volume, with the remaining 1 percent composed primarily of helium, argon, carbon dioxide, and water vapor. It is colorless and odorless, and its very ingredients have helped to sustain life as we know it. Ninety percent of the mass of the atmosphere is contained in the lowest portion of the homosphere, a zone called the **troposphere.** This region might well be called the "home of humanity" because it supplies carbon dioxide for plants, disperses much of the pollutants we send into it, evens out the extremes of hot and cold, permits winds to carry moisture from one part of the world to another, showers us with rain and snow, acts like a blanket to retain heat, and presents fantastic and beautiful sunrises and sunsets. The troposphere is only about 5 miles (8 kilometers) deep over the poles and up to about 12 miles (19 kilometers) in depth at the equator, but, despite its small size, it is the only zone most of us will ever experience firsthand.

The temperature and density of the atmosphere within the troposphere decline with increases in altitude. It is thick and dense at sea level, and it thins out rapidly in the higher elevations to where, at 10,000 feet (3050 meters), the air pressure is halved, and at about 18 miles (28 kilometers) there is virtually no air at

COMPOSITION OF EARTH'S ATMOSPHERE 93

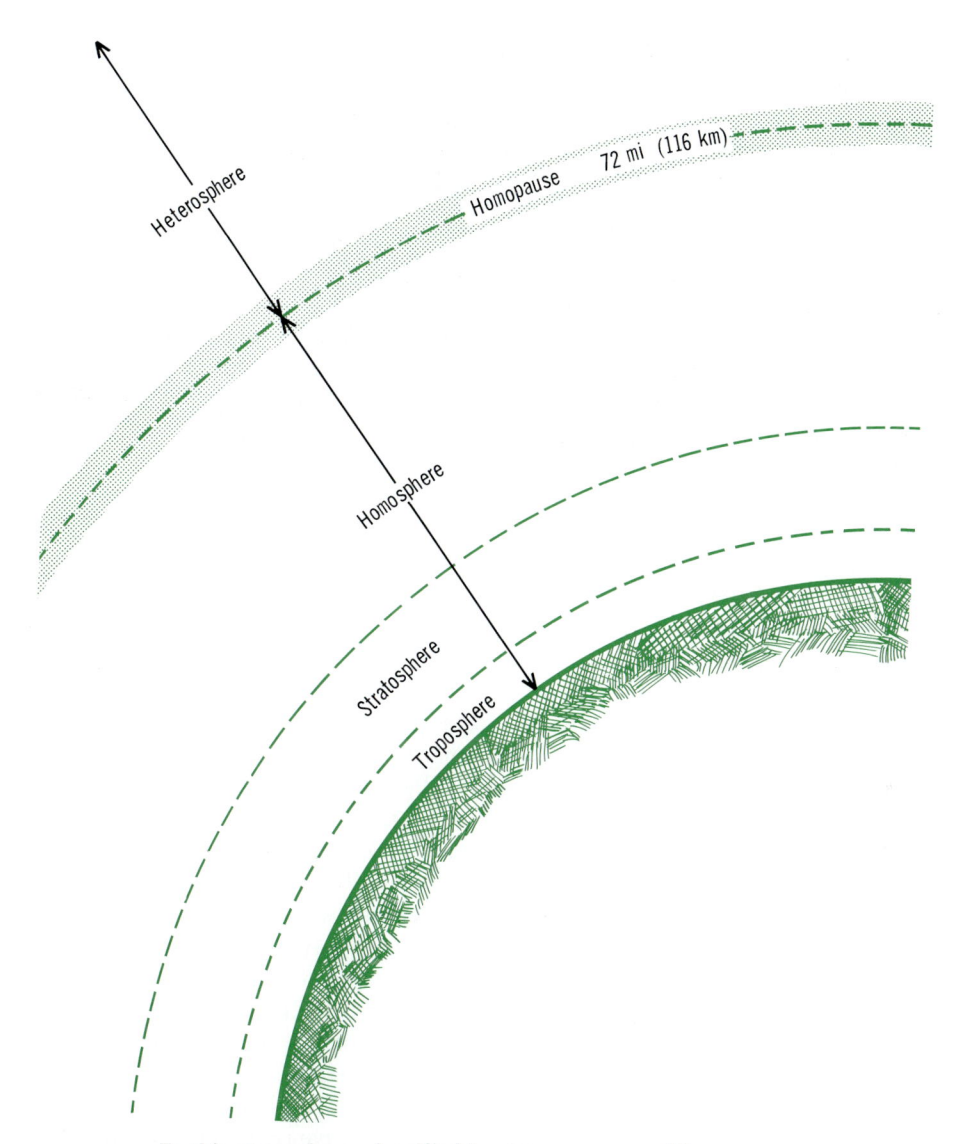

FIGURE 4-2 Earth's atmosphere, classified by gaseous composition

all. You can feel this change every time you ride an elevator in a tall building or drive over a high mountain pass. As the pressure inside your body tries to equalize with the lessening air pressure outside, your ears "pop." Athletes from nations that are at low elevations experienced difficulty adjusting to the atmosphere during the 1968 Olympic Games in Mexico City; they had to compete in a mile-high atmosphere, but their lungs were adapted to the denser air found near sea level. Think of the problem of adaptation faced by the Indians

Modern-day descendants of the Inca, run along the old Imperial Road from Quito to Cuzco.

who migrated to the high Andes of Peru thousands of years ago. Through the process of natural selection (survival of the fittest), they were able to adjust by increasing the size of their lungs and the density of the red blood corpuscles in their systems.

Depending on a region's temperatures, moisture, and the presence or absence of winds, the temperature of the air declines at an average rate of about 3.5°F per 1000 feet (0.6°C per 100 meters), until the upper limit of the troposphere is reached (the **tropopause**). Here the temperature remains fairly constant at about −65°F (−54°C), and there is little or no vertical movement of winds. An interesting feature of the space occupied by the tropopause is the presence of winds that blow in a horizontal direction at tremendous speeds. These high-altitude winds (which we will consider in detail later in this chapter) sometimes reach velocities of 200 miles (320 kilometers) per hour and have a strange, tubular shape. They are called *jet streams*.

Above the tropopause is a region termed the **stratosphere.** This layer within the homosphere extends skyward to an altitude of about 30 miles (50 kilometers) over the equator and slightly lower over the poles. Air still exists here, but it is considerably thinned out. Vertical wind movement is minimal, thus creating little in the way of air turbulence—a fact much appreciated by jet travelers and pilots.

An important feature of the stratosphere is the presence of small amounts of oxygen that are altered by the bombardment of incoming ultraviolet radiation into a heavier oxygen compound called **ozone.** Normally, the oxygen molecule consists of two atoms clinging together like Siamese twins, but the attack of solar energy causes a third atom to be added and to make it *triatomic oxygen*. As the ozone becomes agitated with further ultraviolet radiation, the atoms actually absorb these deadly rays. Without this protective barrier, called the **ozone layer,** much of life on earth would disappear within a matter of a few minutes. The only life forms that would be able to survive would be those that have their skeletons on the outside, such as cockroaches, which have survived similar forms of radiation emitted during nuclear bomb tests. When our astronauts are out "moon walking," they do not have this ozone barrier to protect them, and they would be slowly but seriously burned (like a deadly, delayed-action sunburn) if they did not wear protective suits. It is interesting that some meteorologists consider this ozone layer so important that they would like to term this region the *ozonosphere*.

Structure of the Atmosphere: Temperatures

Another method of classifying the atmosphere is by temperatures. The atmosphere is divided into several layers, or spheres, according to *thermal* (from the Greek *therme*, "heat") characteristics (Fig. 4-3). This does not imply that the various zones or regions have absolute boundaries or completely different characteristics. The earth has only *one* atmosphere, which happens to vary from one area to another. Aeronomists using this newer system have maintained some of the old terms and added new ones. The lowest level is still called the troposphere, and almost all of the earth's weather is contained within its sphere. The stratosphere is next, and its characteristics show how "opposite" it is to the troposphere. The

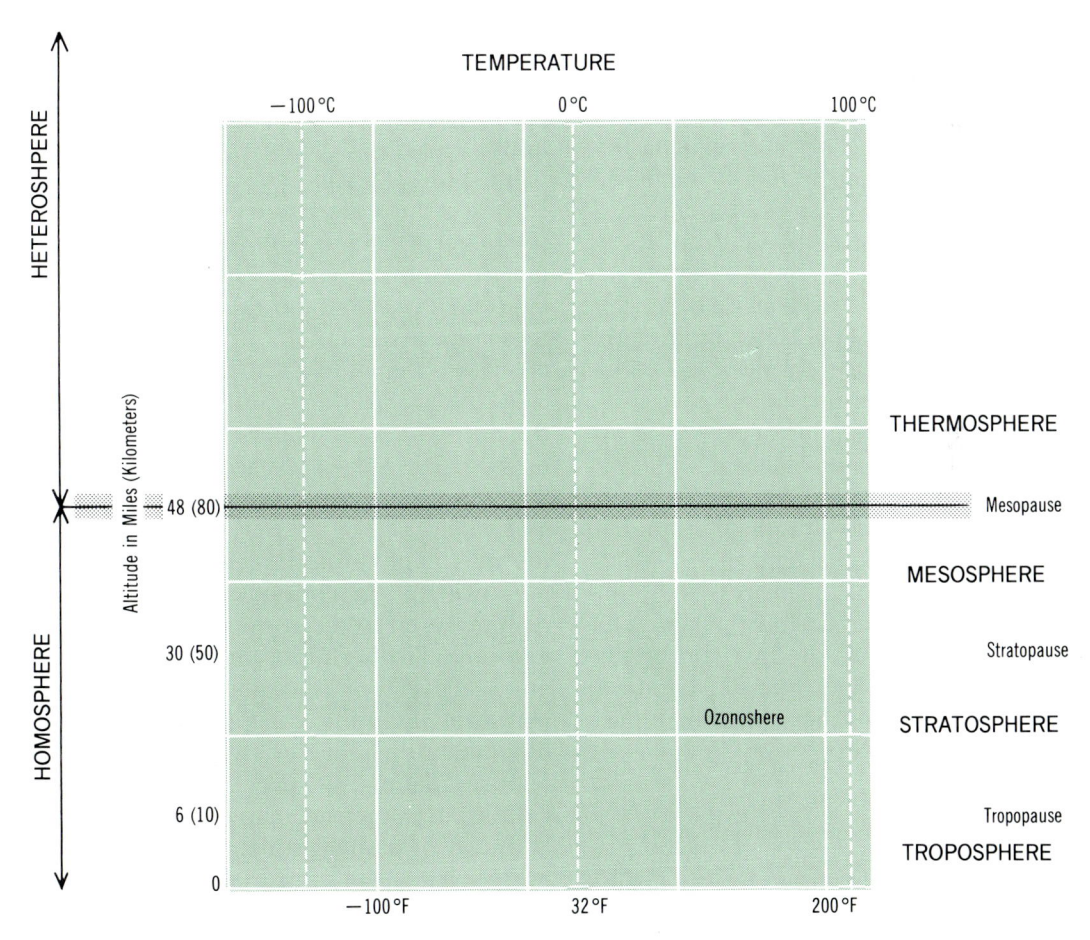

FIGURE 4-3 Earth's atmosphere, classified by air temperature (After A. N. Strahler, *The Earth Sciences*, 2d ed., New York, Harper & Row, 1971.)

temperature of the air declines in the lower region, reaching a low of −65°F (−54°C) at the tropopause, but the temperature through the lower stratosphere remains near that level until an altitude of about 12 miles (20 kilometers) is attained. Then the temperature begins to rise rapidly. By the time a height of 30 miles (47 kilometers) is reached, the temperature has climbed to a level near or not far below freezing.

Beyond the upper limits of the stratosphere (the **stratopause**) another reversal takes effect, and the temperature drops rapidly as the distance from the earth increases. This is the **mesosphere** (from the Greek *mesos*, "middle"), and its characteristics are still being studied. The most striking feature of the middle zone found so far is the sudden and intense decrease in temperature—from near 0°C at the stratopause down to the lowest temperatures ever recorded in our atmosphere, −137°F (−94°C) at an altitude of 50 miles (80 kilometers). The upper limit of the mesosphere (**mesopause**) coincides roughly with the top edge of the homosphere. To many, this is the dividing line between the upper (outer) and lower atmospheres.*

Beyond the mesopause the temperature remains constant until an altitude of around 56 miles (90 kilometers) is reached; then a steady increase in temperature, a process that continues on into space, occurs once again. This region, which includes the previously discussed ionosphere and exosphere, has been given the name **thermosphere** (and might also be called the heterosphere).

---

*To obtain exact figures of the altitudes of the various "spheres" that make up the structure of the earth's atmosphere, consult *U.S. Standard Atmosphere Supplements*, 1966, published by the Environmental Science Services Administration, National Aeronautics and Space Administration, and the U.S. Air Force, U.S. Government Printing Office, Washington, D.C. 20402.

AIR TEMPERATURES

In our discussion of the various methods used to classify the structure of the earth's atmosphere, you undoubtedly have noted that in each method temperature is an important factor. You might also recall from our discussion in Chapter 3 about the movements of the earth that temperatures vary across the face of the earth. This will become more understandable when we have examined the seasonal distribution of air temperatures in the various latitudes.

World Distribution of Air Temperatures

As the earth revolves around the sun in its elliptical orbit, and because of its tilt from the plane of that orbit, not all portions of the earth's surface share equally in the receipt of solar energy. This tells us that the most important controls on our weather and climate are latitudinal position and the angle of the sun's rays because the amount of insolation any area receives is dependent on these factors.

One way to clarify our understanding of the differences in temperature across the face of the earth is to make a map showing these differences. We could call such a map a *world temperature map* or an *isothermal map*. An **isotherm** is a line connecting points of equal temperature and separating regions of higher temperature from regions of lower temperature. Furthermore, an isotherm is based on sea-level (or close to sea-level) temperatures, ignoring abrupt variations caused by mountains, valleys, deserts, lakes, or other features that do not cover a vast expanse. Obviously, an isothermal map of the world, which is drawn to a small scale, could not indicate every change or difference in a region's temperature.

To clarify further, let us consult isothermal maps of the world. Figure 4-4 shows Janu-

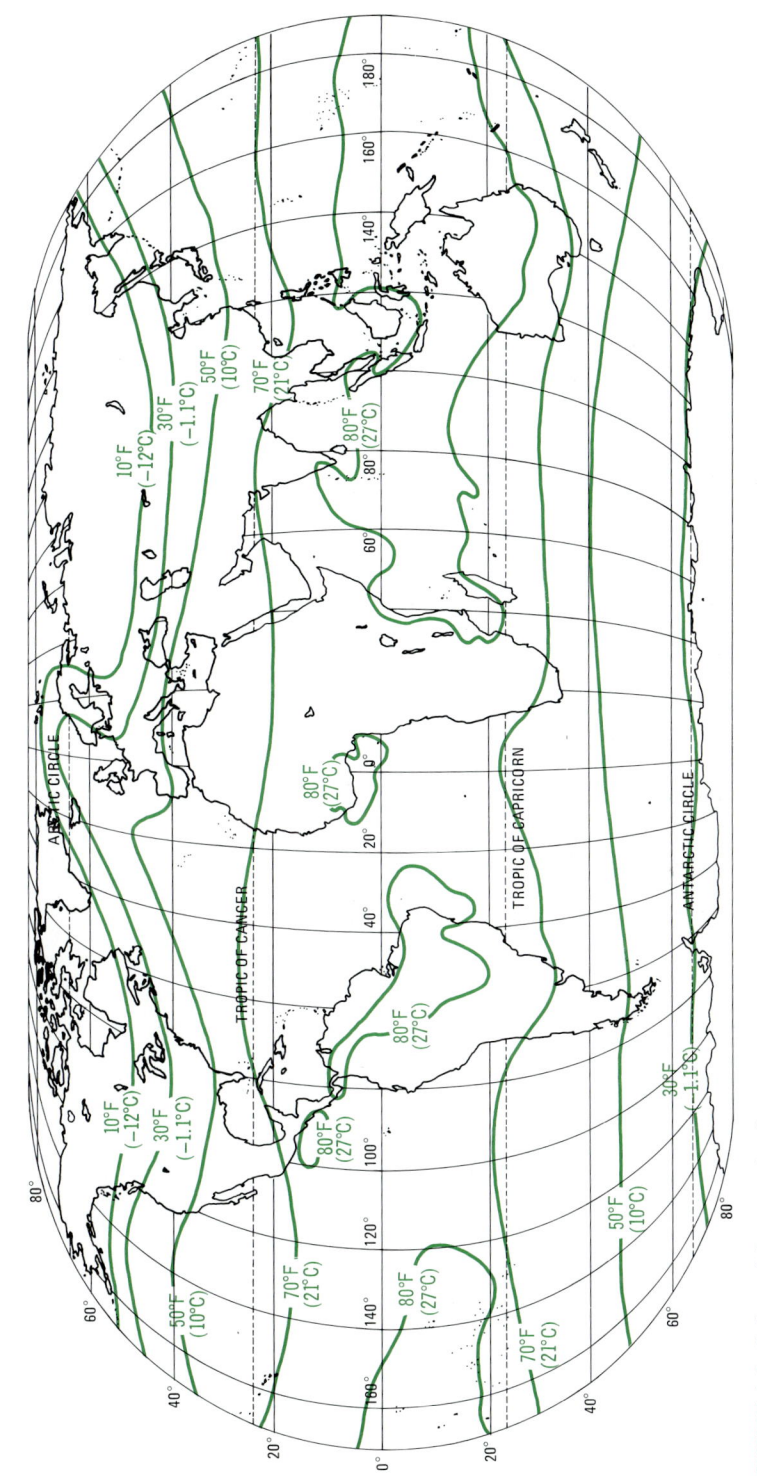

FIGURE 4-4 Selected January isotherms

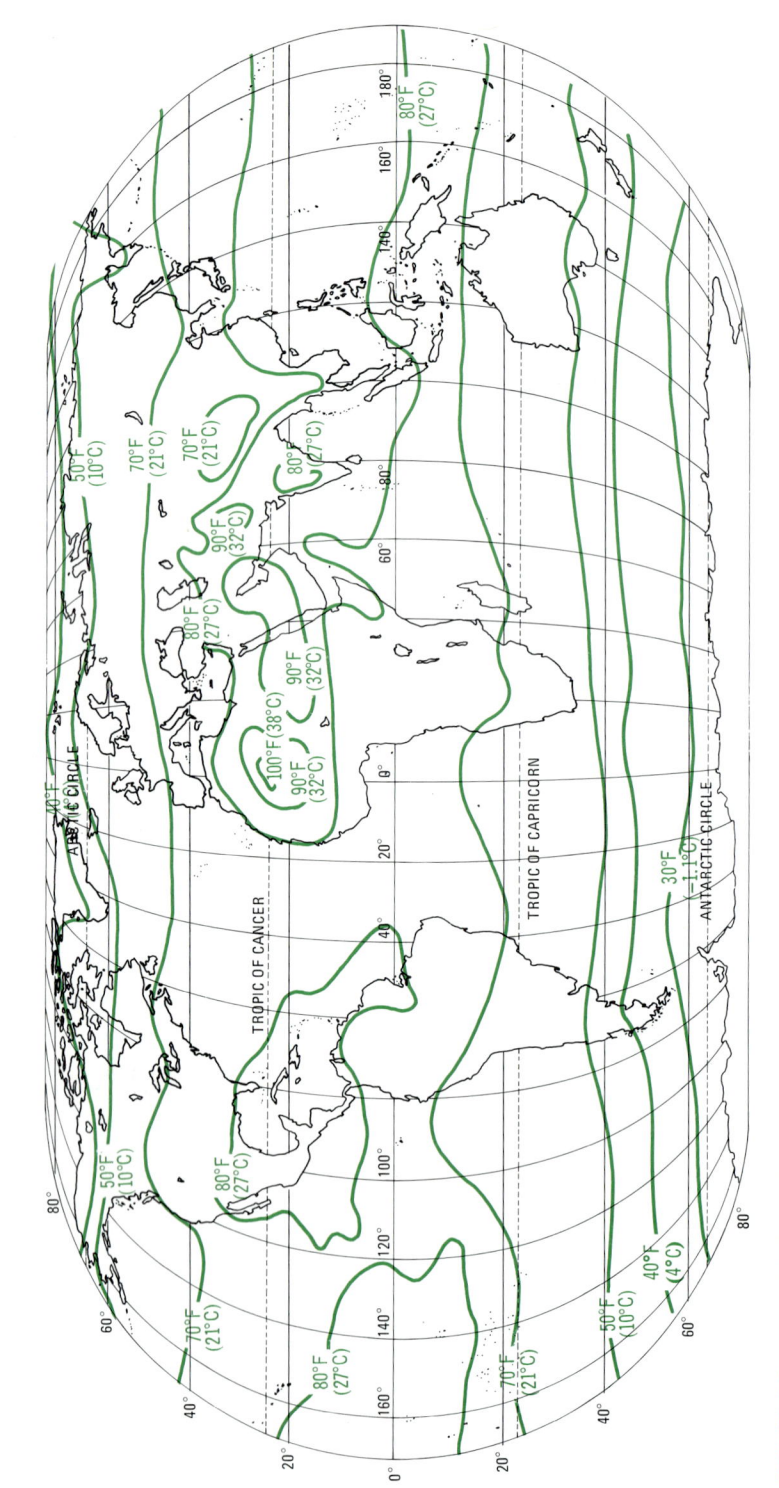

FIGURE 4-5 Selected July isotherms

ary isotherms, and Fig. 4-5 shows July isotherms. Notice how the isotherms are drawn across the face of the earth in a general east-west direction, with the higher-temperature isotherms closer to the equator than the isotherms of lower temperatures. This is proof of the importance of the interrelationship between insolation and a region's latitude. Compare the two maps and you will see that the July isotherms migrate to the north as the Northern Hemisphere tilts toward the sun.

Temperature Gradient Isothermal maps show the **temperature gradient,** which can be likened to the slope of land from high to low elevations, except that it tells us the rate of change (the slope) in air temperatures from one region to another across the earth, whether the change from warm to cold or cold to warm is abrupt (steep slope) or gradual (gentle slope). If two isotherms are close together, it means that the temperature gradient is steep, that temperatures change abruptly from one place, across the isotherms at right angles, to the other place. For example, the January isothermal map (Fig. 4-4) shows isotherms closely packed together over central and southern Asia, yet spread far apart over equatorial Africa. This means there is a rapid change in temperatures (in a north-south direction) over Asia and a gradual change over Africa. If you consult the July isothermal map (Fig. 4-5), you will see how the isotherms are spread far apart over Asia, considerably farther apart than in January, thus denoting a gradual change in temperature from north to south during the summer months.

Land and Water Distribution You may have noticed another peculiarity. The pattern of the isotherms over the continents is quite different from that over the oceans. This indicates to us that the difference in land and water distribution is another important control of the isothermal pattern. For both of the maps, check the isothermal patterns over the landmasses and the oceans. Focus on the isotherms for 30°F (−1.1°C) and 50°F (10°C) in the Southern Hemisphere and for 50°F and 70°F (21°C) in the Northern Hemisphere. The temperature gradient is lower (because the isotherms are more widely spaced) in the Southern Hemisphere than in the north. Furthermore, the isotherms are straighter and lie more in an east-west direction. Why does this happen?

For an answer, compare the extent of the landmasses and the oceans in the two hemispheres. They are not the same. You may remember that in an earlier discussion we said that land and water both assert major controls over temperature (heating and cooling), but that the oceans tended to modify extremes while landmasses tended to exaggerate them. This influence of the presence of large landmasses is termed *continentality*. Again, compare the amount of land south of latitude 20° south with the expanse of land north of latitude 20° north. When you think of that vast stretch of land in the Northern Hemisphere absorbing solar radiation over the summer months, you know that surface temperatures will have to be high. At the same time, the great oceans of the Southern Hemisphere allow only slight changes in temperature from winter to summer.

Because of the intense heating of the continents, the isothermal pattern bends northward over North America and Eurasia during the summer months, then veers southward as the landmasses cool under the grip of winter. Consider the problems such extremes cause the people of Bangladesh. During the summer, interior Asia heats up, creating a large area of ascending hot air. Cooler air over the Bay of Bengal and the Indian Ocean is drawn in to replace the Asian air that has moved up-

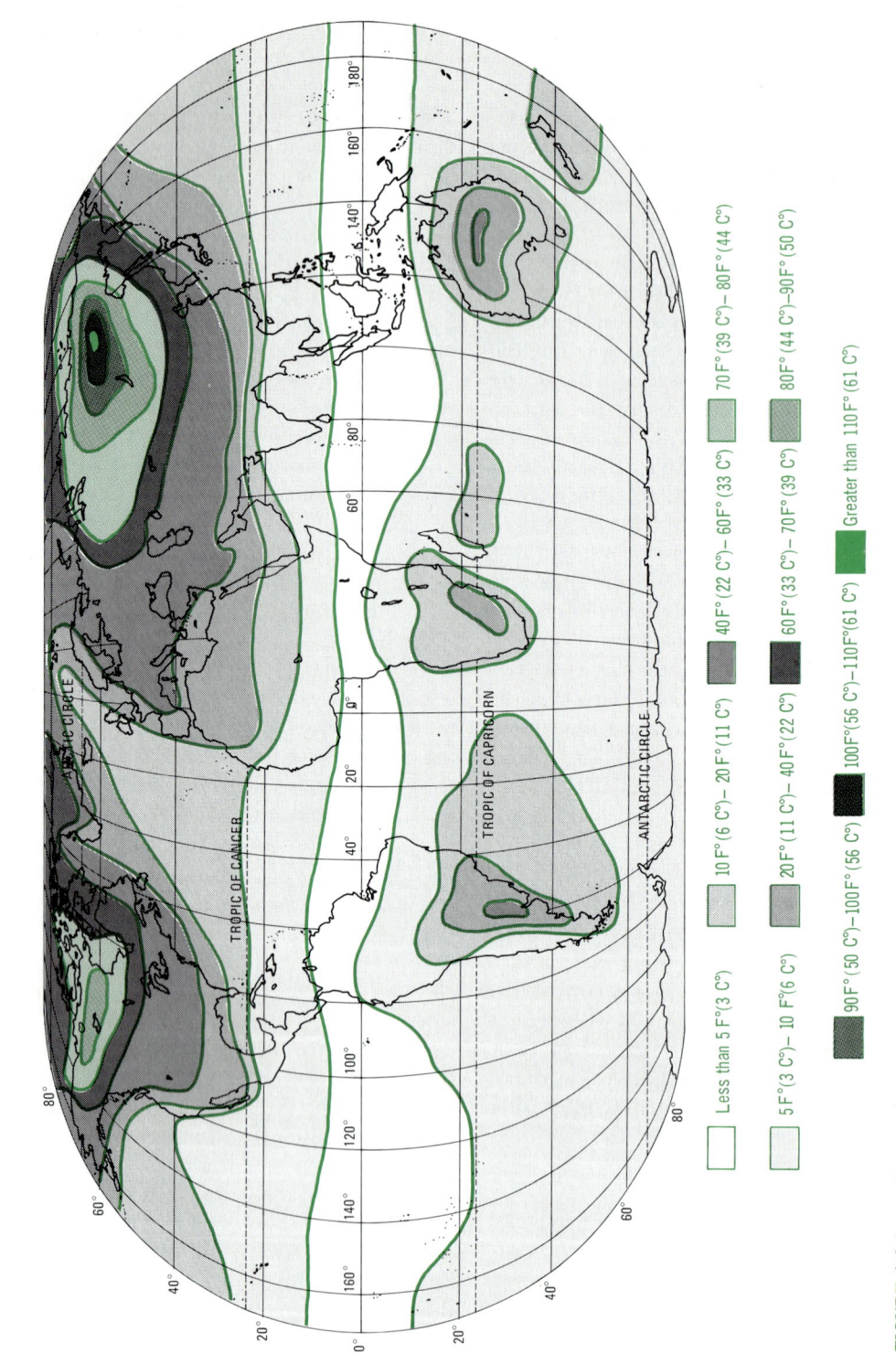

FIGURE 4-6 Annual range in temperature

ward. This cool (not cold) air is moisture laden; as it rises over Bangladesh and strikes the Himalayas, condensation occurs and precipitation follows. These summer monsoons normally bring beneficial rains to the parched farmlands, but sometimes they cause flooding and considerable loss of life. This *monsoon effect* is most noticeable during the summer months; however, as summer fades and winter approaches, the cool air from the continental landmass flows out to replace the rising, warmer air over the oceans, creating a winter monsoon.

Ranges in Temperature Another type of isothermal map that can aid us in understanding the differences in weather and climate across the face of the earth is a map that shows the difference between summer and winter temperatures. Such a map (Fig. 4-6) would indicate which areas of the earth have the greatest ranges. Where would you expect to find the greatest range of variation. Over land? Over the sea? In the Northern Hemisphere or the Southern Hemisphere? After you have studied this map, you will see that nowhere in the Southern Hemisphere does one find as great a temperature variation as in the Northern Hemisphere, and only in the Great Sandy Desert and the *pampa* of Argentina is there a variation as great as 30°F (17°C). It is a different story in the north; in Siberia the range exceeds 110°F (61°C). How would you like living in such a place, where the temperature drops to −40°F (−40°C) in winter, and rises to 70°F (21°C) in midsummer?

Sensible Air Temperature Do you know the difference between the actual air temperature (what the thermometer says) and the way your body *feels* it? This is called the **sensible temperature** and is affected by wind and humidity (moisture in the air) as well as by the temperature. Suppose you and a friend were spending the night in a well-insulated mountain cabin during the winter. You are in the living room sitting close to the fireplace; the other person is outside skiing. You decide to move into the kitchen, leaving the temperature of 70°F (21°C) for a room that is 60°F (15.6°C). Your friend comes into the kitchen from the freezing temperature outside. Will the temperature of the kitchen feel the same to both of you? No; you will feel the coldness of the drop in temperature, but your friend will feel only the tremendous increase. You will shiver; your friend will perspire. The movement of air—wind blowing outside the house or air blown by an oscillating fan—will increase your body's sensitivity. This is dramatically illustrated in Fig. 4-7. Here you can see the effect on a dry-bulb thermometer's reading when the wind begins to blow. You will note that as the wind speed pushes past the 40 mile (64 kilometer) per hour mark, the chilling effect does not increase very rapidly. Obviously, you would appreciate a nice breeze on a hot, humid day, but would you feel the same on a day when the temperature fell to a cool 50°F (10°C)?

AIR PRESSURES AND CIRCULATION

Air is in constant movement throughout the troposphere and all over the earth. It moves because of differences in the weight of air in different parts of the world. It moves horizontally as advective wind, vertically upward as convectional currents, and downward from high altitudes to lower elevations as subsidence. And it moves from high-pressure (heavy air) regions to low-pressure (lighter, less dense air) regions. One might say that the air of our atmosphere moves in circles, and that is why we apply the term *circulation* to the movement of air.

Air pressure means, technically, the weight

MILES PER HOUR	DRY-BULB TEMPERATURE (°F) 35	30	25	20	15	10	5	0	−5	−10	−15	−20	−25	−30	−35	−40	−45
	EQUIVALENT TEMPERATURE OF WINDCHILL INDEX (°F)[a]																
calm	35	30	25	20	15	10	5	0	−5	−10	−15	−20	−25	−30	−35	−40	−45
5	33	27	21	16	12	7	1	−6	−11	−15	−20	−26	−31	−35	−41	−47	−54
10	21	16	9	2	−2	−9	−15	−22	−27	−31	−38	−45	−52	−58	−64	−70	−77
15	16	11	1	−6	−11	−18	−25	−33	−40	−45	−51	−60	−65	−70	−78	−85	−90
20	12	3	−4	−9	−17	−24	−32	−40	−46	−52	−60	−68	−76	−81	−88	−96	−103
25	7	0	−7	−15	−22	−29	−37	−45	−52	−58	−67	−75	−83	−89	−96	−104	−112
30	5	−2	−11	−18	−26	−33	−41	−49	−56	−63	−70	−78	−87	−94	−101	−109	−117
35	3	−4	−13	−20	−27	−35	−43	−52	−60	−67	−72	−83	−90	−98	−105	−113	−123
40	1	−4	−15	−22	−29	−36	−45	−54	−62	−69	−76	−87	−94	−101	−108	−116	−128
45	1	−6	−17	−24	−31	−38	−46	−54	−63	−70	−78	−87	−94	−101	−108	−118	−128
50	0	−7	−17	−24	−31	−38	−47	−56	−63	−70	−79	−88	−96	−103	−110	−120	−128

Very cold

Bitterly cold

Extremely cold

[a]Equivalent in cooling power on exposed flesh under calm condition. Wind speeds over 40 miles (64 Kilometers) per hour have little additional chilling effect.

SOURCE: Mimeographed report, ESSA, Washington, D.C.

FIGURE 4-7 Windchill index

of a column or mass of air. It is measured by instruments called **barometers,** such as the **aneroid barometer** (a metal container that expands or contracts with decreasing or increasing air pressure) and the **mercury barometer** (which is more accurate because it measures the weight of air by the effect of air pressure on the mercury contained in a glass tube). Standard (or normal) atmospheric pressure at sea level is equal to about 29.92 inches (74.8 centimeters) of mercury, which is also expressed as 1013 millibars.* The heavier the air, the higher the column of mercury rises in its tube; if the pressure drops, so does the column of mercury.

**Millibar,* usually abbreviated *mb,* is an international unit for measuring air pressure.

World Distribution of Air Pressure Systems

Like air temperatures, air pressures are not the same everywhere on earth. These variations are caused by differences in latitude, receipt of insolation, seasons, air temperatures of landmasses and water bodies, and by the presence of forests, grasslands, and snowfields. These differences have an important effect on us. Just as we learned more about the effect of temperatures on climate and weather by constructing maps made up of isotherms, we can learn much about the influence of atmospheric pressure by constructing maps composed of **isobars**. An isobar is a boundary line that connects points or places having the same or equal atmospheric (or barometric) pressure and separates areas with differing pressures. Figure 4-8 is a world isobaric chart for the month of January, and Fig. 4-9 is the chart for July. By comparing them, we see that the isobaric patterns are governed by the earth's receipt of solar energy. As we learned from our study of air temperatures, the Northern and Southern hemispheres do not receive the same amount of insolation at the same time. The various pressure systems all migrate north and south about 15° of latitude with the changing seasons.

You will also note that the different pressure regions seem to fit into something similar to zones that stretch across each of the hemispheres in a latitudinal (east-west) direction. A generalized sketch of the world's pressure belts or zones would look like Fig. 4-10 if the isobars were "averaged out," or reduced to what they would be if the localized conditions and differences were omitted from consideration. Note that there is an alternation of high- and low-pressure zones from the North Pole to the South Pole.

The key to the world picture is found in the equatorial regions because there is no letup in insolation there. More solar energy is received on and near the equator than in any other latitude. As the air is heated to an extreme, its molecular activity is increased. The air masses begin to expand and become lighter in weight and density. As they rise, the *equatorial lows* are created. The heated air continues to rise until it reaches the tropopause, where it is forced to split into two masses; one moves north, the other south. This poleward-moving air eventually begins to cool. Some of it descends near the 30° parallels of latitude north and south. As these masses of cooling air subside, high-pressure cells are formed (the *subtropical highs*). On and near this region of high pressure can be found the earth's major desert areas. In the Northern Hemisphere are the Sahara, Arabian, Thar, Sonora (northern Mexico), and Mojave. In the Southern Hemisphere are the Atacama (Chile), the Kalihari and the Namib (southern Africa), and the Great Sandy (Australia).

The remaining masses of upper-level, poleward-moving air begin to subside over the poles. Here, the air is extremely cold because of the scarcity of insolation and a high albedo, and the molecules of the air slow down and begin to compact, forming high-pressure areas. On the earth's surface, the air moves from the poles toward the equator. At the meeting place between the cold, dense air flowing toward the equator and the warmer, poleward-moving air from the subtropical highs the warmer air is forced aloft (since it is lighter and less dense), creating a low-pressure zone (*subpolar lows*). There is a difference in the subpolar lows of the two hemispheres. The Northern Hemisphere, because of its vast land areas, does not allow for one long, continuous belt of low pressure. High pressure builds up during the cold winters because the continents get very cold and the air above compacts into high pressure. Individual low-pressure storms continually are being created in this fluid line of

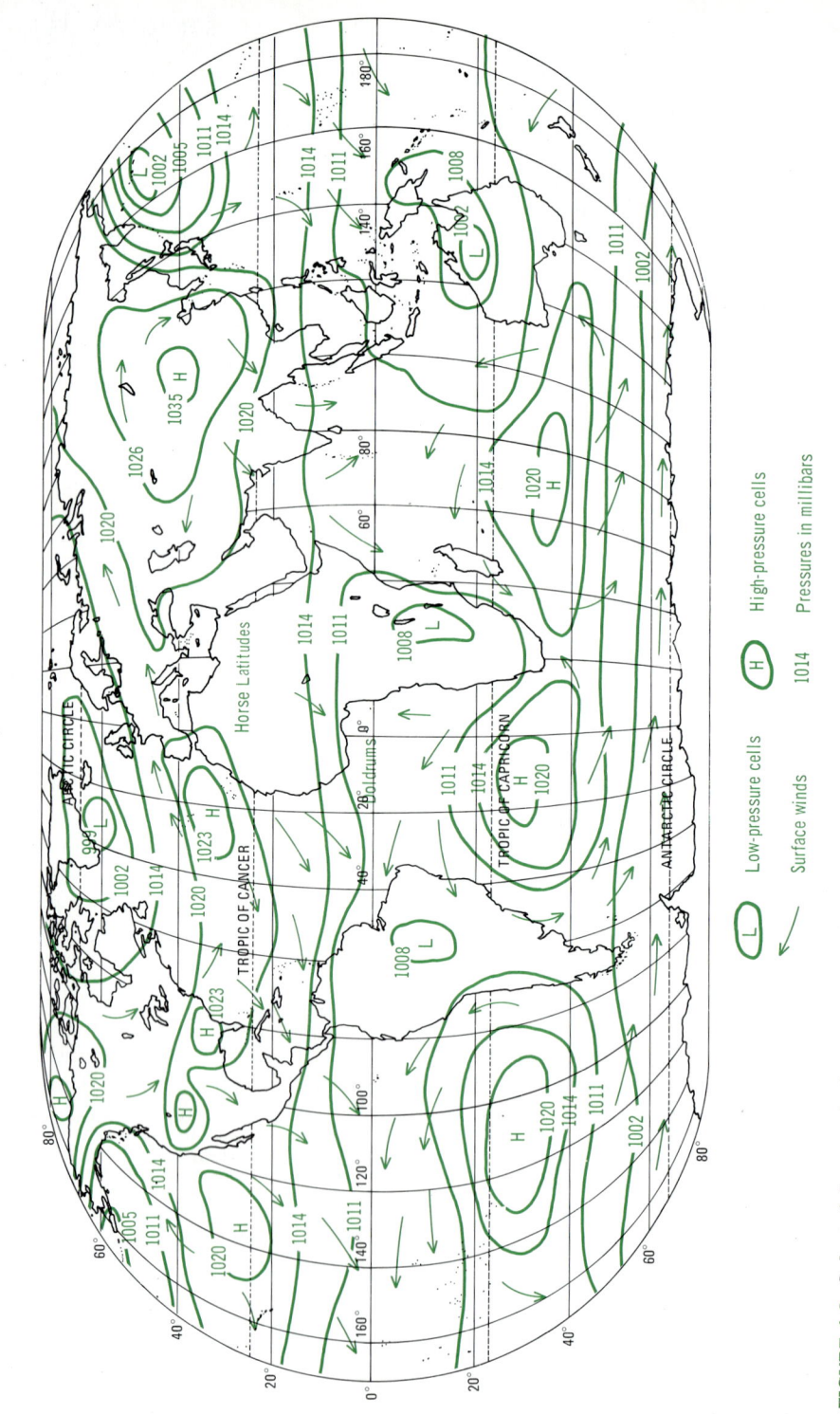

FIGURE 4-8 Selected January isobars

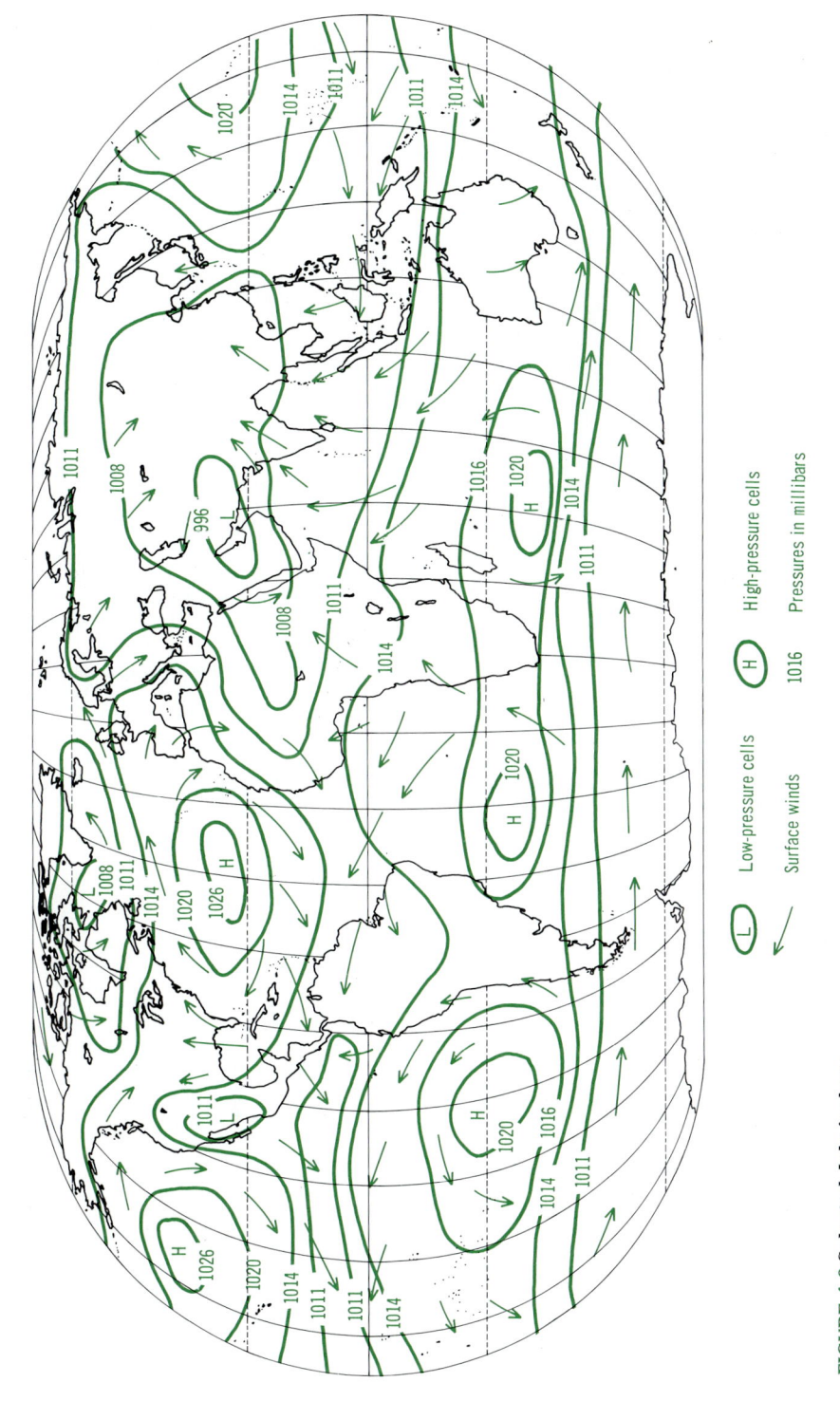

FIGURE 4-9 Selected July isobars

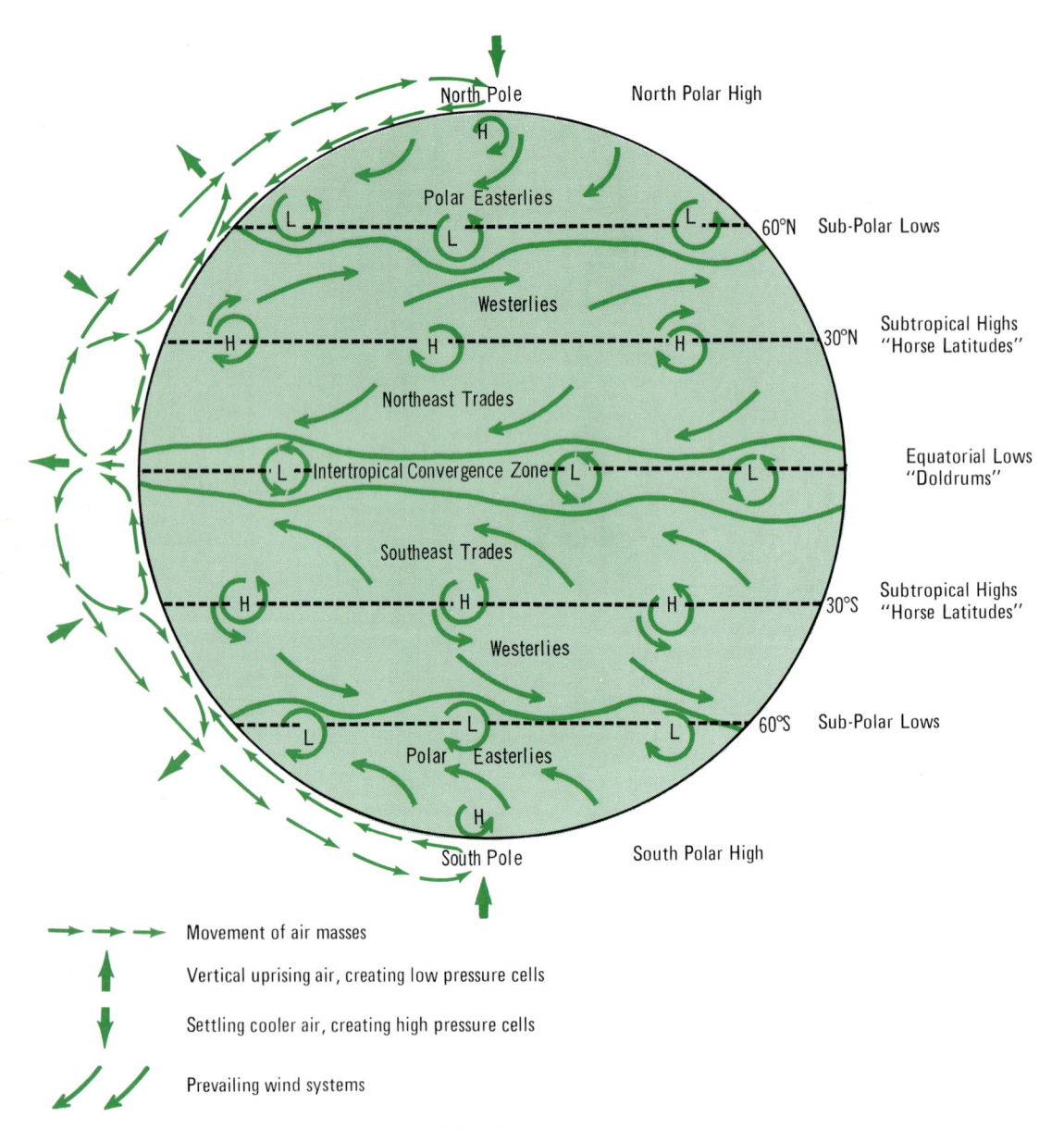

FIGURE 4-10 Generalized diagram of the world's wind and pressure systems

AIR PRESSURES AND CIRCULATION 107

Aneroid barometers The barometer on the right shows the pressure of the moment; the one on the left keeps a permanent record.

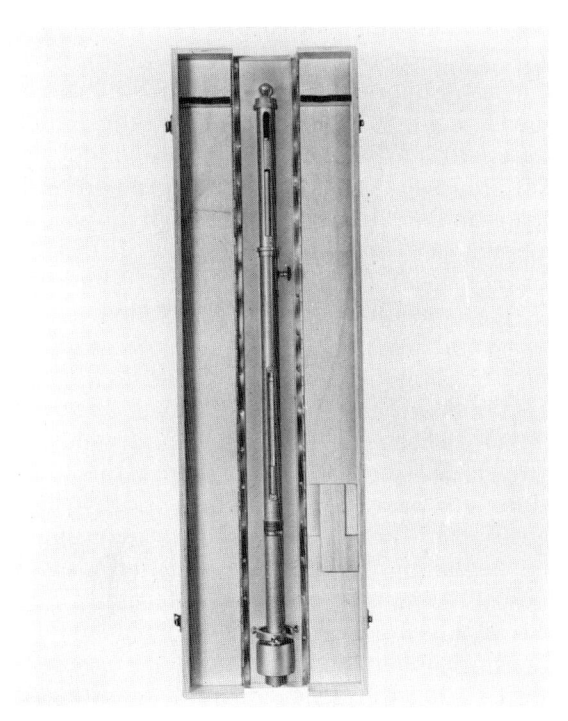

Mercury Barometer

low pressure and, like large waves, they invade the midlatitude regions of North America and Eurasia, creating what has been called the "polar front."*

Over a period of time each of these pressure regions has been given a popular name, or nickname. Each region has its own characteristics and peculiarities that make it stand out as being different.

The equatorial region is a place of high humidity, unstable air masses, permanently fixed low-pressure cells, and the regular occurrence of thunderstorms and waterspouts. The winds that converge within this region are variable, shifting, and generally rising. In bygone days they were not to be trusted by sailors. Ships attempting to cross the Atlantic in the tropics often found themselves wallowing helplessly for days without sufficient wind to move them forward. Climatologists call this re-

*The polar front theory is explained in more detail on page 137.

gion the intertropical convergence zone (ITC) because this is where the trade winds converged, but the sailors used another term, the **doldrums.** Englishmen aboard the becalmed ships became depressed at their enforced idleness and moaned that they were in the "doldrums." The nickname stuck and is now in common usage.

North of the doldrums is another region of variable, unsteady winds that often caused sailing ships much anguish. The subtropical high-pressure zone earned the name **horse latitudes,** and there are a number of opinions as to how this nickname came to be used. The most prevalent one attributes the name to the Spaniards who sailed to the New World with cargoes of horses. If becalmed for a long period of time, unable to take advantage of the winds that did blow, or if the winds blew constantly toward them, making it difficult to proceed, and if the supply of feed and water was running out, the captains would "lighten ship" by throwing some of the horses over the side.

Winds and Wind Systems of the World

During our discussion of the world's air pressure systems we have referred to the winds that move from one pressure region toward another, but we have not explained the reasons behind their direction and speed. Simply, winds are created out of the differences in air pressure, and they move from a high to a low because the air in the high is denser (the molecules are more compacted) than in the low. Winds do not, however, move at a constant speed, nor do they blow in a constant direction, because they are always under the influence of three basic controls: the pressure gradient, friction, and the *Coriolis effect.*

Pressure Gradient
The **pressure gradient** is the determining factor in the initial speed and direction of the wind. It tells us whether the wind will be slight or moderate or a full-blown gale. The pressure gradient indicates the changes from a high-pressure to a low-pressure area and the direction of wind movement. As we have said, pressure builds up to a high-pressure cell (a high) in a region where the air is subsiding and the molecules become packed more tightly together. As this occurs, the air must move out and away, and it does so in a clockwise direction in the Northern Hemisphere. Air rising from a heated surface creates a low-pressure cell (a low), and the air tends to spiral upward and inward in a counterclockwise direction. This inward spiraling is called *cyclonic,* and the outward spiraling is *anticyclonic;* thus lows are also referred to as *cyclones* and highs as *anticyclones.* The direction of spiraling would be just the opposite for lows and highs in the Southern Hemisphere.

The pressure gradient can be found by consulting an isobaric chart such as is shown in Fig. 4-11. On this chart you can see a low-pressure cell bordered by two high-pressure cells. The atmospheric pressure is 1018 millibars in one high and 1015 millibars in the other, while the low has a pressure of only 994 millibars. Note the two arrows; these are pressure gradients and are drawn from the high to the low, crossing each isobar at right angles. If the isobars are close together, the gradient's slope will be steep, the change in pressures obviously is great, and winds blowing along the gradient will be strong (high A to low B). If the isobars are far apart, the slope is gradual since the pressure change is not abrupt, and winds, if any, will be light or moderate (high C to low B). Although pressure differences start winds moving and the pressure gradient determines the beginning speeds and directions, other controls soon take over and affect both, especially if any great distances are involved (either horizontally or vertically).

In order to further our understanding of

AIR PRESSURES AND CIRCULATION 109

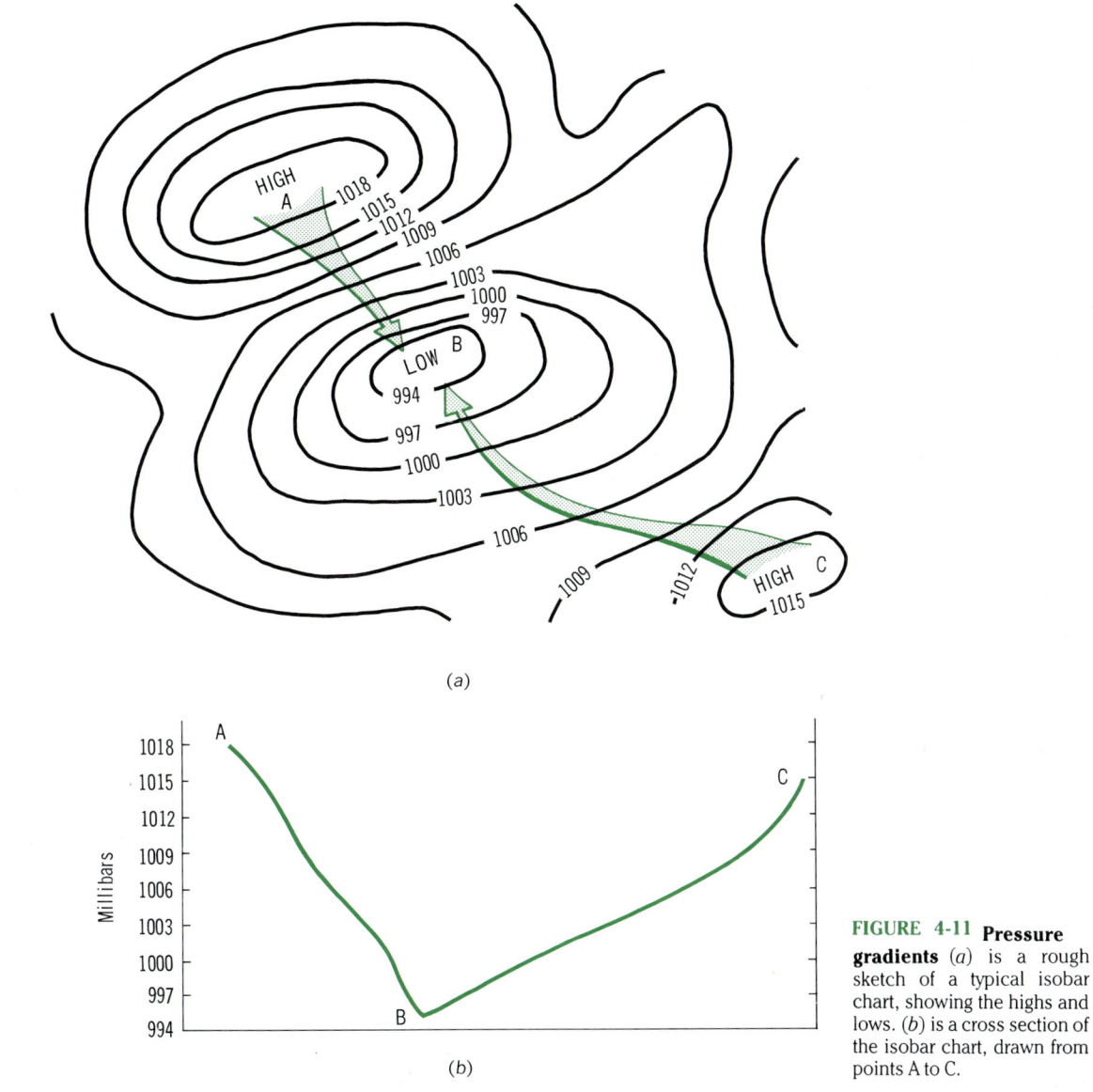

FIGURE 4-11 Pressure gradients (*a*) is a rough sketch of a typical isobar chart, showing the highs and lows. (*b*) is a cross section of the isobar chart, drawn from points A to C.

the pressure gradient force, let us consider the winds called *land and sea breezes*. These are winds that alternate back and forth between bodies of water and the nearby land (Fig. 4-12). People who live on a coast are well aware of these local winds that blow from a lake or ocean to the land during the day and then turn about at night and blow the other way. The reason for this phenomenon is that during the daylight hours, when insolation is at its highest,

110   EARTH'S DYNAMIC ATMOSPHERE

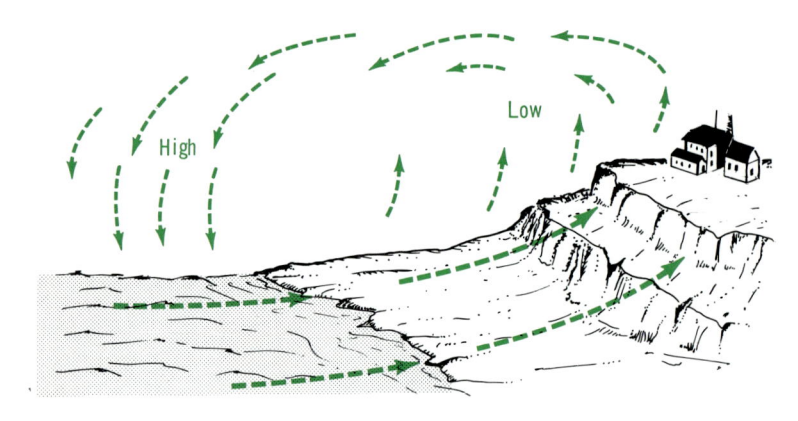

DAY Sea breezes blow onshore.

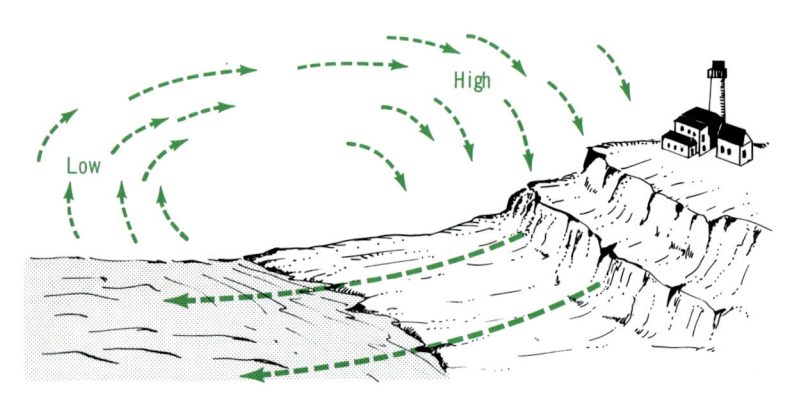

NIGHT Land breezes blow offshore.

FIGURE 4-12 Land and sea breezes

the land warms at a faster rate than the body of water. As a result, the heated air rises over the land, thereby creating a low-pressure area and inviting the cooler air from the lake or ocean to rush onshore. At night the land cools off rapidly, thereby creating a high-pressure area, while the ocean or lake is still warm from its accumulation of heat during the day. Now it is the warm air from the ocean that is doing the lifting and the cooler air from the land that is blowing offshore.

A much larger version of these land and sea breezes is the wind system known as the **monsoon,** a common occurrence in many mid-altitude regions. There are winter monsoons and summer monsoons, and the difference between the two is in the direction in which they move. Winter monsoons may be likened to offshore (land) breezes because the land is much colder than the oceans in winter. Thus a high-pressure cell builds up over the land, and the air moves outward over the ocean, which has retained much of its warmth from the past summer. This is a time of relative stability since the air masses are cold and dry. The summer monsoons, however, are a different matter. In both the southeastern United States and central Asia the land heats up rapidly, thus building a large low-pressure system that pulls in the cooler and moist air from the nearby oceans. Figure 4-13 illustrates the invasion of a summer monsoon into Bangladesh and north-

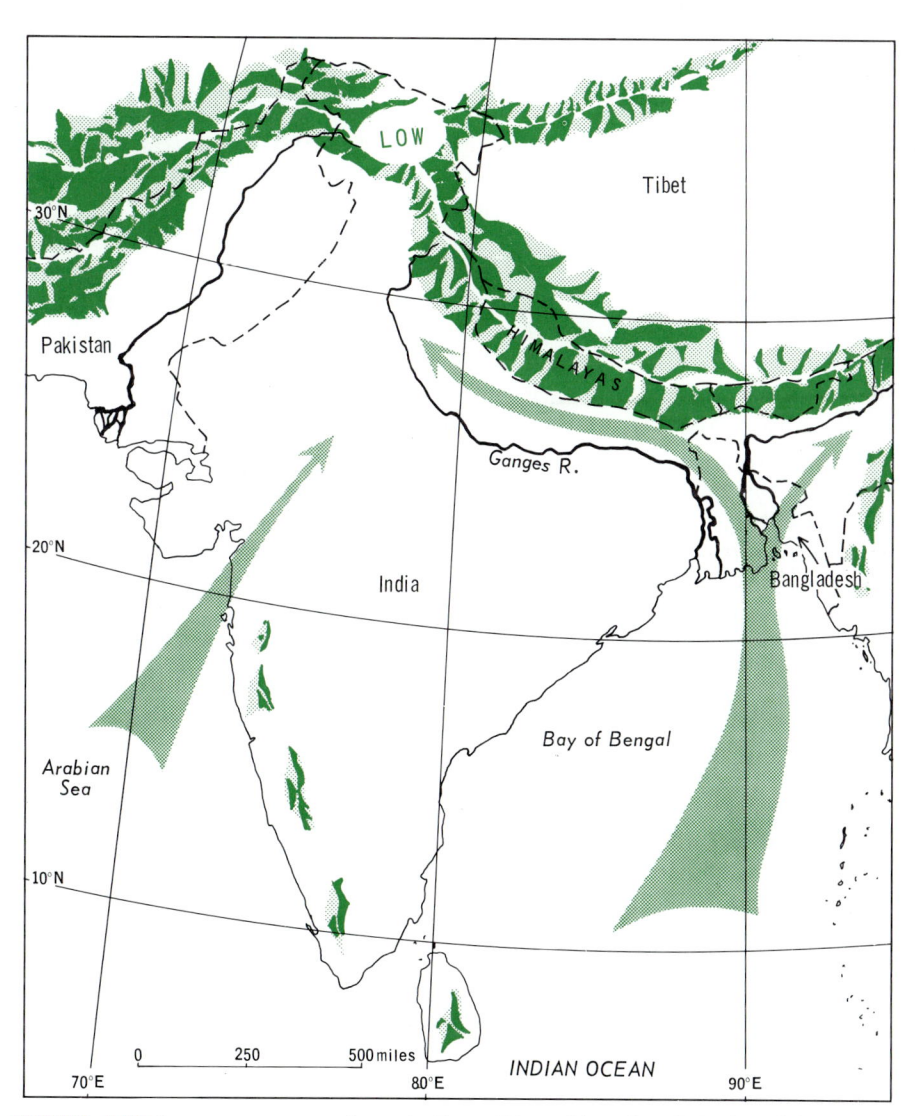

FIGURE 4-13 Summer monsoon from the Bay of Bengal into Bangladesh and India

ern India. As mentioned earlier, the heavy rains resulting from the onshore movement of moisture-laden winds can be both beneficial and disastrous, depending on the severity of the storm and the amount of precipitation that falls.

Friction As winds blow along the earth's surface, a certain amount of resistance is offered by the land or the sea, and the velocity of the wind and the smoothness of its passage will be affected. The amount of *friction,* manifested by the degree of slowing or the creation of turbulence, depends on the unevenness (roughness) of the surface. As might be expected, gently undulating snowfields will not offer as much resistance as high and rugged mountains. A level plain will not impede the wind's flow as will hilly or mountainous terrain. It might be noted that friction does not affect high-level winds; it is normally restricted to altitudes of less than 2000 feet (600 meters).

Coriolis Effect We have shown how winds move from high-pressure to low-pressure areas, governed primarily by the pressure gradient, and the result should be that winds moving from a high-pressure cell move directly toward a low-pressure area. This does not happen, however, where winds travel more than short distances because another control, the **Coriolis effect,** asserts its influence. Coriolis (named after the French scientist who first observed it) has to do with the world's wind systems (Fig. 4-10) will disclose that winds that "should" move directly north or south do not. Instead of blowing from one pressure system to another, they seem to alter their routes, their directions, almost 90° from their intended path: to their *right* in the Northern Hemisphere and to the *left* in the Southern Hemisphere. Why does this happen?

Let us consider winds blowing south from the North Pole. They begin as north winds—and end up coming from the east! They are no longer "north winds"; they are the polar **easterlies.*** Winds moving out of the subtropical highs and heading north seem to change direction and swing to the east. We call these winds the **westerlies,** and they are common to most of the United States. Other winds blowing out of these subtropical highs start off toward the south, but they, too, veer to the right and approach the tropics from the northeast, so we call them the **northeast trade winds** (or northeast trades).*

The same thing is happening in the Southern Hemisphere. The winds alter their directions, but their alteration is to their left. Winds approaching the equator from the southeast are the southeast trades; then, in order, there are the westerlies and the south polar easterlies. This alteration of direction applies not only to winds, but to anything that freely moves over long distances: a migrating flock of birds, a high-flying jet, or ocean currents.

You may now ask this question: Does the wind *really* change its true direction, or does it merely *seem* to do so? For the answer, let us consult Fig. 4-14. This diagram shows the rotating earth and the deflection of its wind systems. The solid arrows indicate the direction the winds would move if the earth did not rotate, and the dotted-line arrows show the path the winds do take in relation to the earth's surface. A wind blowing from the North Pole (point A) does not reach point B, for it appears to be heading directly for point D. Why? Be-

---

*In wind terminology, the direction of a wind is always given in terms of its birthplace, or source. Thus, a west wind is one blowing from the west; an easterly comes from the east; a nor'wester is a wind blowing from the northwest.

*The word *trade* comes to us from the early European traders who used the fairly constant northeast winds to move their sailing ships from Europe to the New World.

AIR PRESSURES AND CIRCULATION **113**

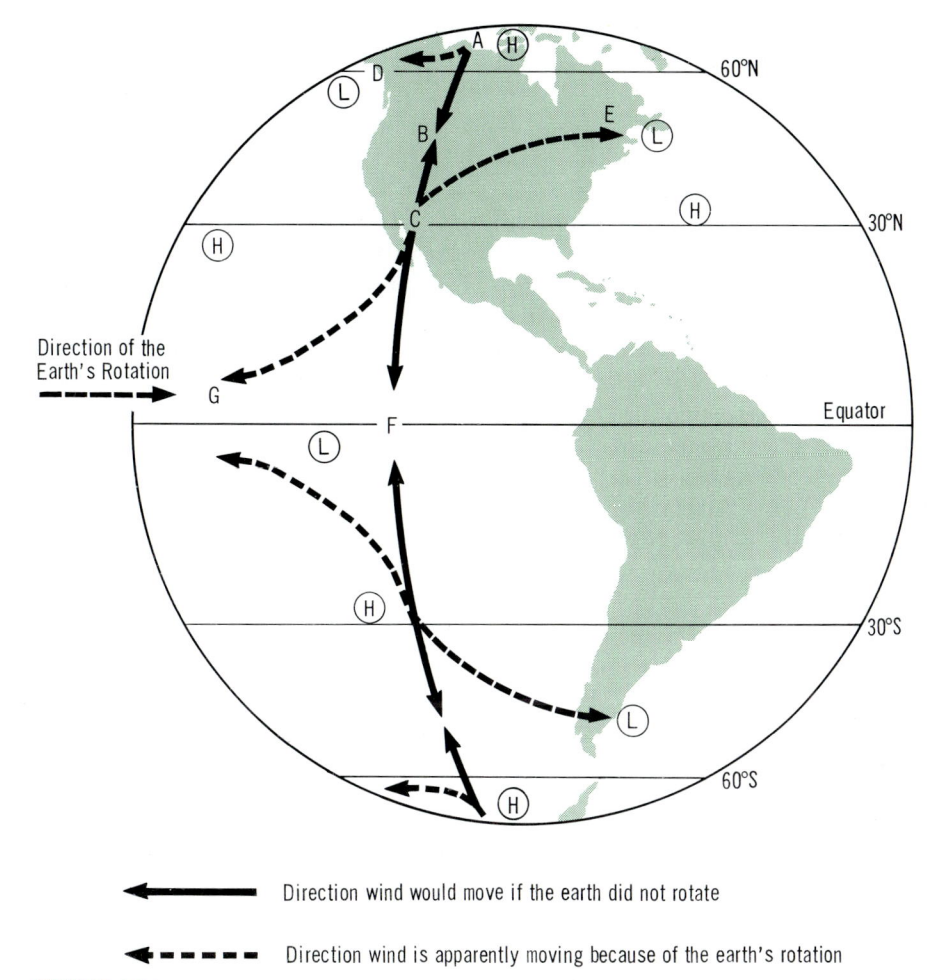

FIGURE 4-14 The coriolis effect

cause the earth rotates from west to east and is moving *under* the flowing wind. Believe it or not, the wind did *not* change its direction. An observer on the ground at point D would take an oath that the wind hitting him in the face was indeed coming from the east, but to a satellite or an observer in the sky (if the wind were colored so that it were visible) the wind would not have changed direction at all. It would have moved from its beginning to the end in a true north-south direction.* The same thing will happen for a wind blowing from point C

*In order to prove this to yourself, take a piece of cardboard and cut it into a circle. Hold a marking pen at the center and have someone rotate the card while you concentrate on drawing a straight line to the outside of the circle. What happens? If the cardboard is rotated in a counterclockwise direction (to simulate the Northern Hemisphere), your line will curve to its right. If the card is rotated clockwise (simulating the Southern Hemisphere), the line will curve to its left.

toward point F: It will end up blowing (apparently) toward point G. What about winds coming north from point C? If the earth is also moving from west to east beneath these winds, why does the wind swing away from B and veer toward E, to become the westerlies? Why does it not bend to the west, as do the polar easterlies and the northeast trades? Well, it does not simply because it cannot. The reason? The earth is rotating faster near the equator, about 1040 miles (1664 kilometers) per hour, than it does at a higher latitude, such as San Francisco (latitude 37°45′ north), where the speed is about 830 miles (1328 kilometers) per hour, or at Anchorage, Alaska (latitude 61°12′ north), where the speed is about 410 miles (658 kilometers) per hour. Now you can understand why there is greater speed at point C than at point A.

Look at it this way. On a calm day the air rotates with the earth as it spins, but we think of the earth and the atmosphere as being still, nonmoving. If the air begins to move over the spinning earth, the situation changes. Because the rotational speed varies with the latitude (greatest at the equator, least at the poles), we can liken this to a freeway in your city. A freeway has several lanes (like parallels of latitude), some for fast-moving and some for slow-moving traffic. When your car (wind) moves from one traffic lane to another and you maintain your original speed, you will be going either faster or slower than the prevailing traffic in the new lane. If you were driving fast and moved into a slower traffic lane, you would immediately flash past the slower-moving cars and, to them, you would be veering to their right. Using the same analogy, if you were going slow and moved into the fast lane, to the speeding cars you would appear to be moving to their left.

It should be noted that the Coriolis "force" is strongest at the poles and decreases in influence at latitudes near the equator. Along the equator the Coriolis effect is at its weakest, and other controls take charge of the speed and direction of winds.

Air Masses

Everything in our atmosphere is tied in with temperature differences and the resulting buildup of air pressure regions across the face of the earth. As we have seen, low-pressure cells form over land or water bodies that have been heated by a heavy receipt of insolation, and high-pressure belts, or cells, form where the air has been cooled and is settling. While the heating and cooling processes continue, the air masses over the regions will assume their special characteristics or properties. For example, consider the northernmost portions of Canada and Siberia, lands that are very cold. Air masses forming over these regions will assume the cold and dry properties of the region, and when they move out, they will move as masses of cold, dry air. An air mass over a tropical desert, such as Africa's Sahara, will be much like the hot, dry land beneath it. Equatorial waters not only heat air above, but also send moisture into the air; the resulting air mass will be warm and moist. So, in order to have a greater understanding of the movement of winds and the varying results of their arrival, we must consider their points of origin, the source regions that gave them birth.

Source Regions Let us first consult Fig. 4-15, which illustrates (in a generalized way) the source regions for the different air masses. Meteorologists have devised a code to simplify matters. The capital letters *A, AA, P, T,* and *E* refer, roughly, to latitudinal position, and the lowercase letters *c* and *m* apply to the actual earth surface, whether land or ocean. The letter *A* stands for *arctic; AA* for *Antarctic; mP, polar maritime,* for source regions such as the waters off the Aleutian Islands or near Green-

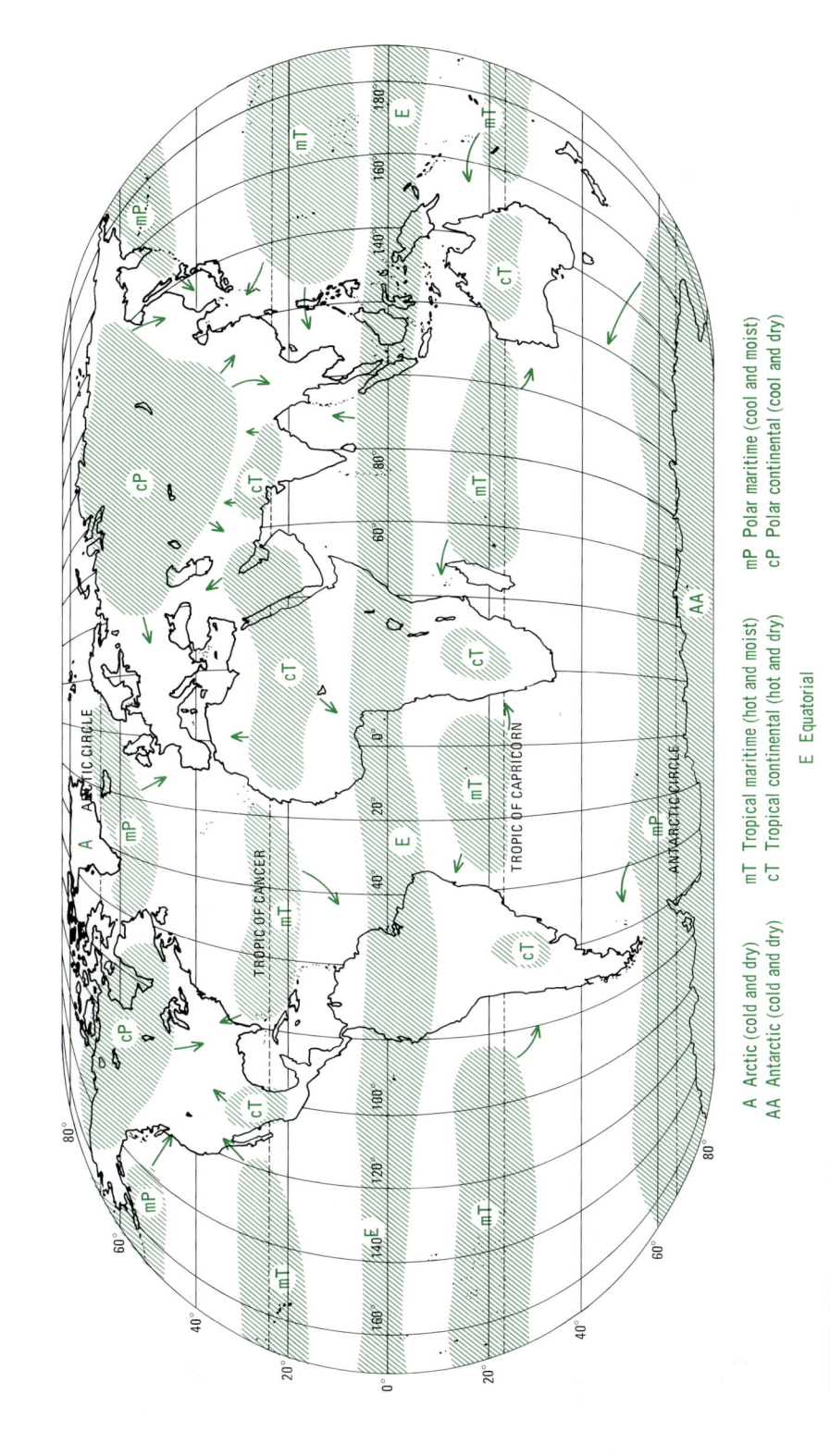

FIGURE 4-15 Source regions of air masses

A Arctic (cold and dry)
AA Antarctic (cold and dry)
mT Tropical maritime (hot and moist)
cT Tropical continental (hot and dry)
E Equatorial
mP Polar maritime (cool and moist)
cP Polar continental (cool and dry)

115

land; *cP, polar continental,* for regions such as the cold interior of Canada or Asia; *mT, tropical maritime,* for tropical waters such as the Caribbean, the Gulf of Mexico, or the mid-Atlantic; and *cT, tropical continental,* for air masses forming over the hot deserts, grasslands, or forests of large land areas within the equatorial region. *E, equatorial,* is a relatively new addition as a source region and is confined to the area between latitudes 10° north and 10° south.

We have previously discussed many of the winds that result when air masses of different properties collide or, at least, when one of them intrudes on another's territory. As an example, the monsoons that periodically sweep over India and Bangladesh are the result of a very hot polar continental air mass building up over interior Asia during the summer, thus inviting the intrusion of slightly cooler tropical maritime air from the Bay of Bengal and the Indian Ocean. The reverse occurs in winter, when the Asian polar continental air mass is cold and dry and sweeps down from Tibet's high plateau and over the Himalayas onto the lands surrounding the Bay of Bengal.

Now let us consider an important and interesting outgrowth of the meeting of differing air masses, as when cool air from maritime or continental polar regions of the high latitudes and warm air marching poleward from the tropical maritime or continental regions converge with air masses that are moving out of the subtropical high-pressure belts. Wind systems are created that are called jet streams.

Jet Streams Many strange phenomena occur in our atmosphere, but one of the least understood is the phenomenon of **jet streams.*** These are high-altitude winds that encircle the earth at elevations of from 4 to 10 miles (6 to 16 kilometers), moving at speeds of up to 300 miles (480 kilometers) per hour. Both the Northern and Southern hemispheres have at least three of these "swiftly flowing rivers of air": near the tropics, near the subtropics, and along the polar fronts. Although these winds have probably existed since the beginnings of the atmosphere, the world did not become aware of them until the 1920s. The first inkling of their presence came when weather balloons were sent aloft through gentle breezes to probe the upper reaches of the troposphere. Imagine the consternation of meteorologists when a balloon traveling at the rate of 15 miles (24 kilometers) per hour and expected to take about two weeks to drift from Des Moines, Iowa, to New York City landed in New York only four days later? The final proof that these racing winds did actually exist came during World War II, when American bombers tried to raid Japan. In order to avoid detection, they climbed into the tropopause and headed for Japan—and discovered they were not traveling at a speed of 450 miles (720 kilometers) per hour, as their speed indicators showed, but only at 250 miles (400 kilometers) per hour. They were heading directly into the subpolar jet stream.

Today, aircraft flying to Europe from North America make use of this jet stream to save fuel and time. If they are heading west, however, it is another story. Planes returning to North America try to avoid the jet streams by flying north or south of the jet stream's errant path.

What are the jet streams? For an answer, let us consider the jet stream that affects North America (Fig. 4-16). It seems to follow the tortuous meanderings of the leading edge of the polar front. The sometimes violent winds are born through the collision of cold polar maritime and continental air pushing down from

---

*For a thorough and interesting explanation of the jet streams, see Elmar R. Rieter (1967), *Jet Streams* (Garden City, N.Y.: Doubleday, Anchor Books).

AIR PRESSURES AND CIRCULATION 117

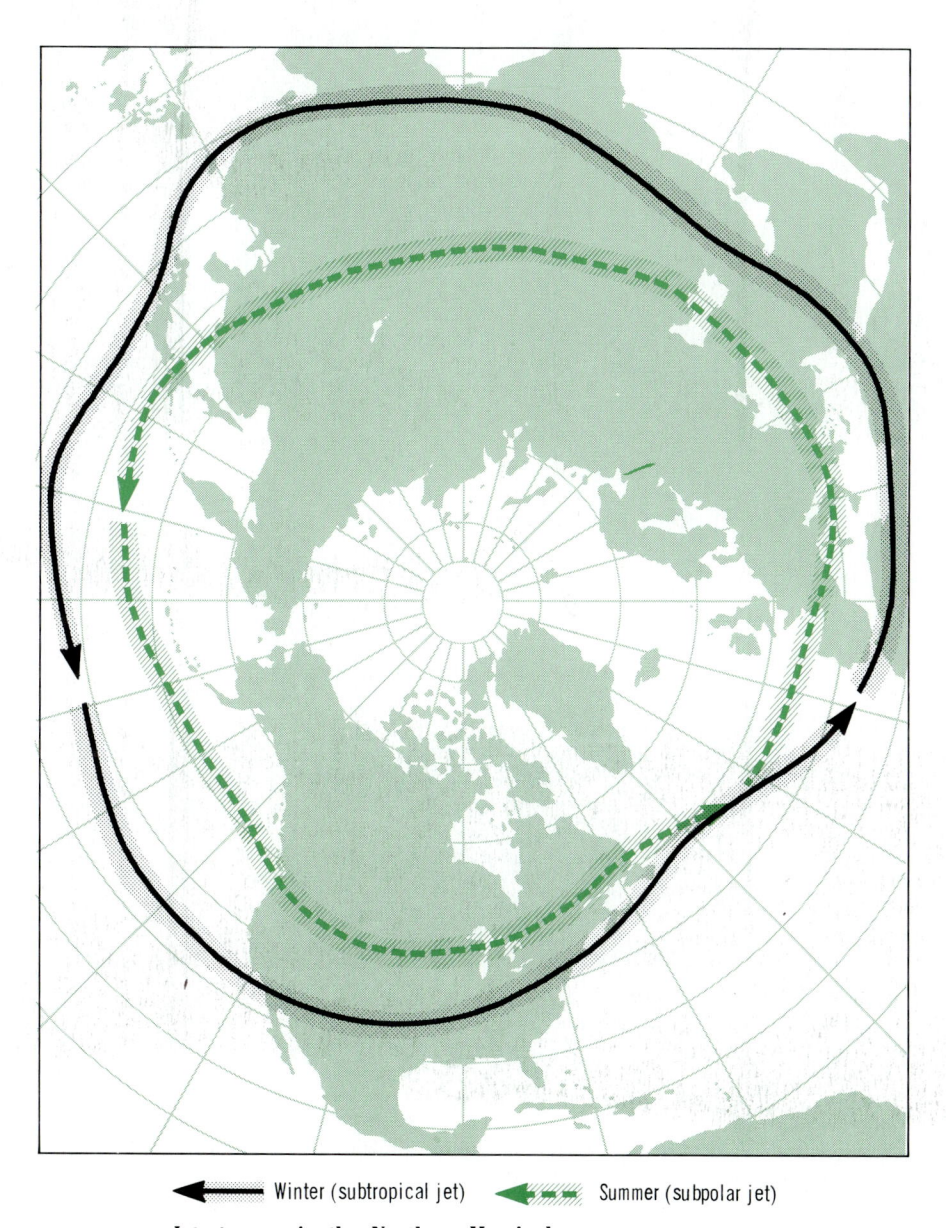

FIGURE 4-16 **Jet streams in the Northern Hemisphere**

Winter (subtropical jet)   Summer (subpolar jet)

Jet stream These long cirrus clouds, occurring about 100 miles (160 kilometers) northwest of Dakar, Senegal, probably marked the location of a subtropical jet stream. They were photographed by a camera aboard unmanned Apollo 6 in early 1971.

the Arctic and Canada against warmer air moving north from the subtropical high-pressure belt (roughly 30° north). Turbulence is created over the upper level of the air masses, and strong winds begin to flow down the pressure gradient. Since the same thing is happening over northern Europe and Asia, the result is a ribbon of rapidly moving air that encircles the globe. It is surrounded on all sides by much calmer air, and some meteorologists liken it to a flattened tube or hose that is about 1 mile (1.6 kilometer) in thickness and up to 100 miles (160 kilometers) in width. Within the "tube" wind speeds vary from 30 to 300 miles (48 to 480 kilometers) per hour, while wind speeds outside may be as low as 10 miles (16 kilometers) per hour, depending on the season and the geographical region. It seems that the greatest velocities occur off the eastern edge of Asia and near Greenland. It should also be noted that the jet stream is not one long, continuous ribbon; it is broken into sections, some short and some stretching out for over 1000 miles.

As for jet-stream terminology, the summer-created jet is usually called the *subpolar jet*, and the one formed during the winter is the *subtropical jet*. Both are primarily westerly winds and exert a termendous influence on our weather. It is believed that these jets propel storms (especially tornadoes) into the mid-latitudes during the winter and early spring. They also influence summertime weather, as when the snakelike wanderings of the jet stream isolate masses of cold or warm air, causing the particular affected region to suffer an unusual cold spell or an unseasonably hot and dry period. This is probably an important spin-off from the phenomena of jet streams: They cause a mixing of cold polar air and warm tropical air.

REVIEW AND DISCUSSION

1. Explain the difference between weather and climate.
2. Define *ionosphere, exosphere,* and *aeronomist.*
3. What makes the sky blue? Why is it bluest overhead?
4. Explain the composition of the atmosphere by its gaseous composition. Explain the terms *homosphere* and *heterosphere.*
5. Explain the difference between the troposphere and the stratosphere.
6. Why do your ears "pop" at high altitudes?
7. What is ozone? What is the importance of the ozone layer?
8. Explain the structure of the atmosphere by temperatures.
9. Explain the terms *mesosphere, mesopause, thermosphere,* and *thermopause.*
10. What is an isotherm? Is there a difference in the patterns of isotherms between January and July?
11. What is temperature gradient?
12. Why is the pattern of isotherms different over landmasses and the oceans?
13. Why do India and Bangladesh experience summer monsoons?
14. What is sensible air temperature?
15. Explain the development of different air pressure regions.
16. What is the difference between an aneroid and a mercury barometer?
17. What is meant by the term *millibar?*
18. What is the effect of the changing seasons on the January and July patterns of isobars?
19. What is the difference between the zones called the doldrums and the horse latitudes?
20. Explain the control of the pressure gradient, friction, and Coriolis effect over the wind systems of the world.
21. What is the ITC?
22. Explain the local wind system known as land and sea breezes.
23. Why do we call a wind blowing from the west a westerly wind and not an easterly?
24. Why is knowledge of air mass source regions important for a meteorologist?
25. Explain the symbols mP, cP, mT, cT, A, AA, and E.
26. Explain the origin, shape, and importance of jet streams.
27. What is the "windchill index?"

ADDITIONAL READING

Blair, Thomas A., and Robert C. Fite (1965). *Weather Elements: Text in Elementary Meteorology.* 5th ed. Englewood Cliffs, N.J.: Prentice-Hall. A text in elementary meteorology.

Dempsey, M. (1965). *The Ocean of Air.* London: Purnell and Sons. Distributed in the United States by Ginn and Company, Boston.

Donn, William L. (1975). *Meteorology*. 4th ed. New York. McGraw-Hill.

Huschke, Ralph E., ed. (1959). *Glossary of Meteorology*. Boston, Mass.: Meteorological Society.

Navarra, John G. (1979). *Atmosphere, Weather and Climate: An Introduction to Meteorology*. Philadelphia: Saunders.

Riehl, Herbert (1978). *Introduction to the Atmosphere*. 3rd ed. New York: McGraw-Hill.

Spiegel, Herbert J., and Arnold Gruber (1983). *From Weather Vanes to Satellites: An Introduction to Meteorology*. New York: John Wiley & Sons.

Starr, V. P. (December 1956). "The General Circulation of the Atmosphere," *Scientific American* (offprint no. 841).

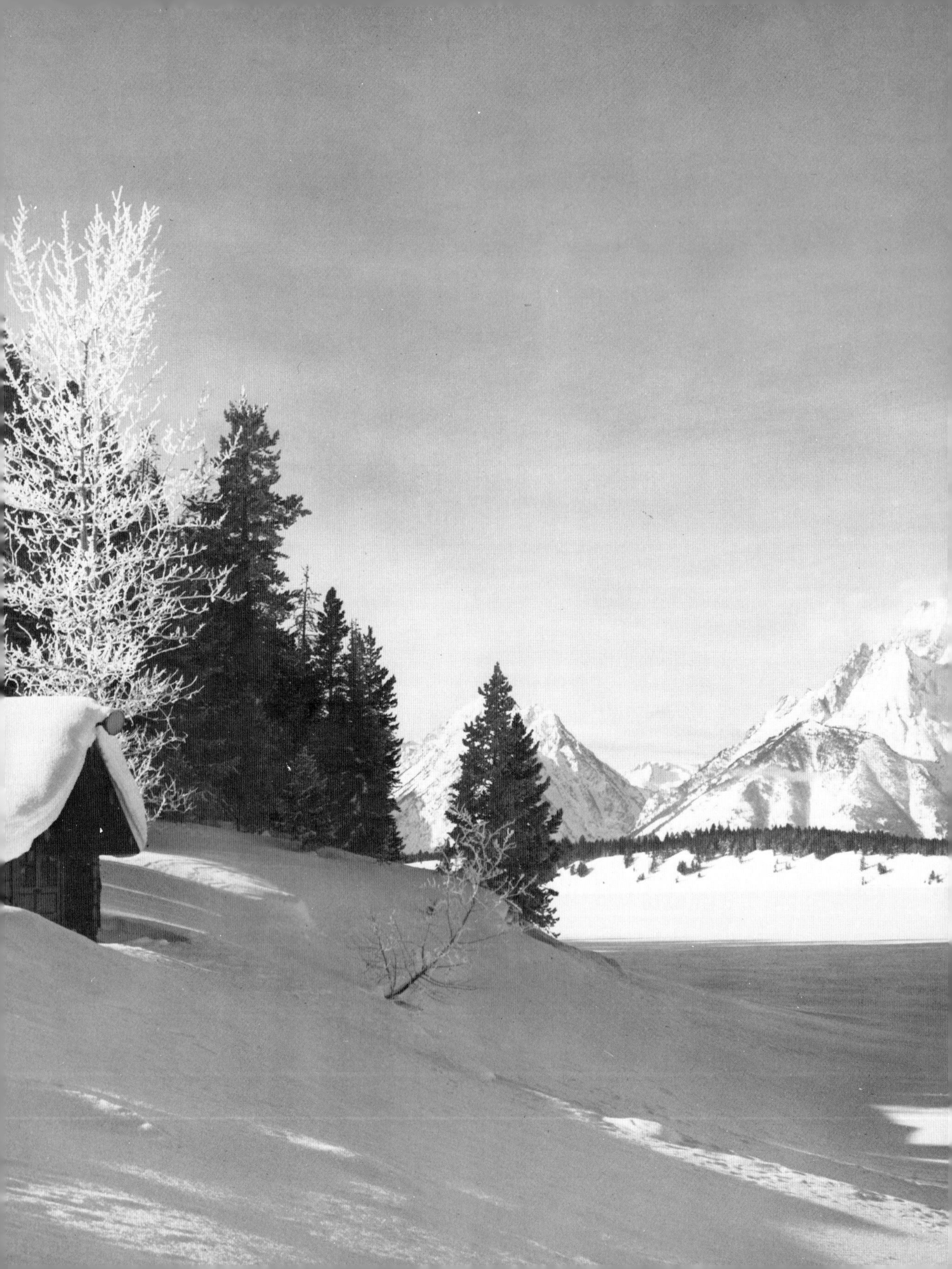

FIVE
MOISTURE: CLOUDS AND STORMS

All the waters run into the sea; yet the sea is not full; unto the place whence the rivers come, thither they return again.
ECCLESIASTES 1:7

Our investigation into the earth's dynamic atmosphere has shown how important water is in contributing to and modifying weather and climate. Without moisture our earth undoubtedly would be uninhabitable, much like the planets Mercury and Mars. Fortunately, however, our earth has sufficient gravitational attraction to hold the atmosphere in place—as it has for more than 2 billion years. Furthermore, the earth is situated at the right distance from the sun to allow moisture to exist in all of its three forms: vapor, liquid, and solid. There are a number of theories as to how the earth received its water supply, but the one most widely accepted suggests that the earth was once a fiery liquid that cooled slowly through countless millions of years. In the beginning the surface was too hot to allow water molecules to remain, and most of them immediately boiled away into the atmosphere. As time passed, the crust cooled sufficiently to allow the falling water to accumulate in great pools.*

At first, there may have been only one ocean that covered the earth's rock mass, but, in time, the earth's crust began to break up and the landmasses rose above the surface of the seas. As the ages passed, solar radiation continued to strike the waters of the oceans, causing some of the moisture to evaporate and ascend to the skies, where it cooled, condensed, and was precipitated. Thus, a process that we might call the "cycle that permits life" was born. Hydrologists call this the *hydrologic cycle*.

THE HYDROLOGIC CYCLE

The word *cycle* can be defined as "a repeating or recurrent series; a succession of events that repeat themselves in an orderly manner over a long period of time." The term *hydrologic* means "pertaining to water." We can, therefore, combine these words and say that the **hydrologic cycle** is the water cycle, which is another way of saying that water is involved in a recurring series of events during which it changes form in an orderly manner.

Evaporation is an important step in the hydrologic cycle (Fig. 5-1); it is the result of the transformation of water from either a liquid (water) or a solid (ice or snow) state to a gaseous (water vapor) state. The water vapor is carried upward by convectional currents and is spread throughout the air masses that lie over the water body. Onshore winds then push the moisture-laden air inland. Condensation occurs through orographic lifting, convection, or frontal uplift in cyclonic storms, and precipitation falls to the surface of the earth. Some moisture falls as snow or hail and is held at the higher elevations (or high latitudes) as a sort of "solid reservoir" until it melts. Thawing and melting add water to the underground supply.

Most precipitation is in the form of rain, and much of that becomes surface runoff, flowing through streams to larger streams, rivers, lakes, and, eventually, oceans. Some water infiltrates downward through porous materials into what is called the zone of saturation, from where it moves by gravity flow to the ocean. Evaporation goes on throughout the entire cycle, and the water, whether in vapor or liquid form, is in constant motion.

EARTH'S WATER BALANCE

How a region fares with regard to its water supply depends on its water balance. Essentially, **water balance** (or water budget) is the ratio between the receipt and the loss of moisture. It takes into consideration not only the additions through the various forms of precipitation, but the amount retained by the

---

*For those who would like to pursue this topic further, see L. B. Leopold and K. S. Davis (1970), "A Sun-Powered Cycle," in Chap. 2, *Water* (New York: Time-Life Books). The authors are hydrologists, and their book is well written, authentic, and interesting.

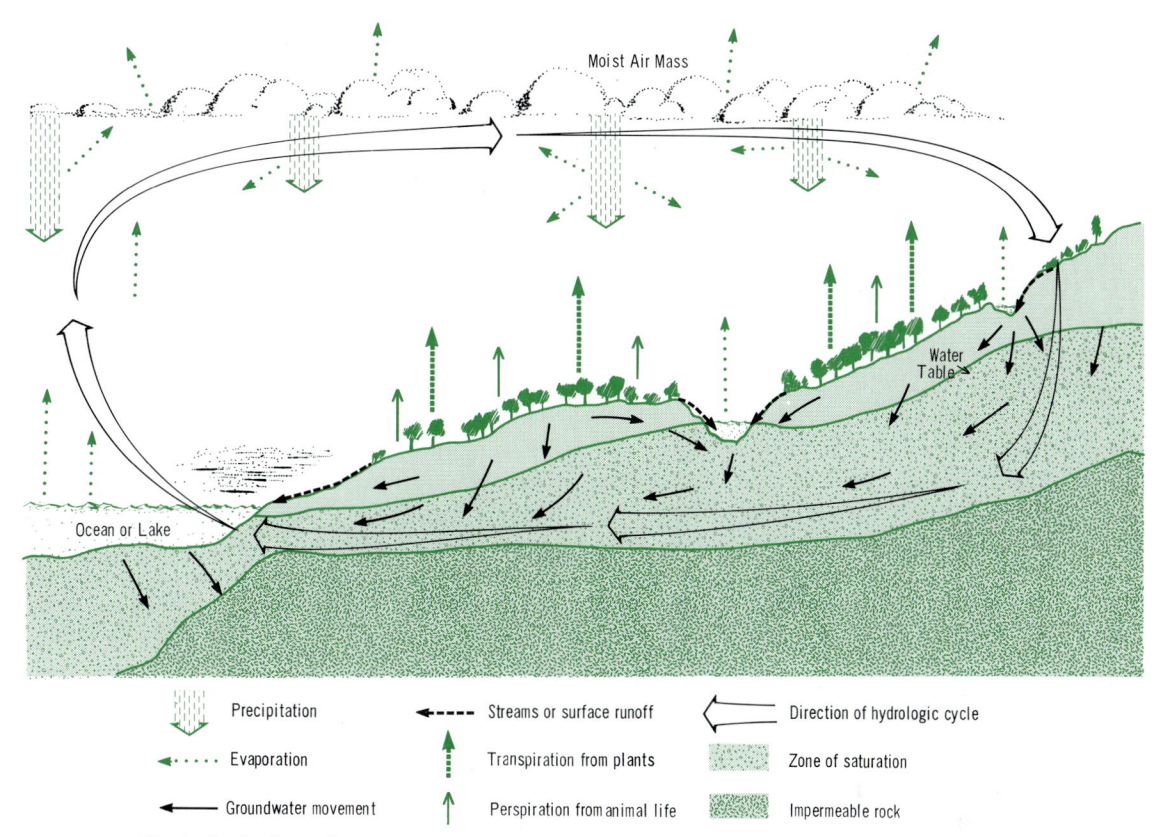

FIGURE 5-1 The hydrologic cycle.

soil, underground reservoirs, vegetation, lakes, streams, and rivers. These totals are then related to the losses by gravity out of the groundwater zone, by evaporation into the air, and by **transpiration** (moisture emitted through leaf tissues) from plants. If the precipitation exceeds the losses, an area is said to have a favorable water balance. If the balance is unfavorable, agriculture may be severely limited, as residents of many areas of North Africa, Australia, central Asia, and the American Southwest can testify. There are a number of contributing factors, such as the region's air temperatures throughout the year, the velocity and direction of winds, the types of vegetative cover and soils, the topography—whether smooth or rugged—population numbers and distribution, the extent of urbanization and, most important, the region's location in relation to the distribution of the earth's waters.

MOISTURE AND PRECIPITATION

Moisture in the form of water vapor is an important part of the troposphere. It is the direct result of the change in form from a liquid (through a process called **evaporation**) or from a solid (through a process called **sublima-**

tion, when heat causes water molecules to separate from their compacted state and rise as vapor without passing through the liquid state). Even in small quantities, water vapor is important. It supplies the earth with moisture in the form of clouds and precipitation, and it absorbs heat that radiates from both the sun and the earth.

The amount of water vapor affects the rate of water loss through evaporation from the earth's water bodies. This rate varies from rapid to slow, depending on the amount of moisture available for evaporation in a given area, the size or extent of the surface area, the salinity (saltiness) of the water, the air pressures, the presence or absence of winds, and the temperature of the air mass. These factors form a kind of "chain," and if one is changed, the entire situation is altered. For example, if there is no wind, evaporation will be less, and water might lie in ponds, lakes, or oceans for a long time, If the air temperature drops too low, the water might freeze, and there would be less evaporation. And, although there is considerable potential for evaporation from any large water body because the supply is there, the rate of evaporation will be greater over tropical waters, because of their consistently higher temperatures, than over polar waters. Ideal conditions for a high rate of evaporation would be a hot, dry, windy day in an area that is plentifully supplied with moisture.

It would be well to note here that the condition of an air mass is the direct result of its *stability*. If an air mass is *stable*, as in the case of a temperature inversion, there will be little vertical air movement. If, however, the cool ground layer is warmed by insolation, it begins to develop ascending air movements. Instability results as the rapidly rising air develops a greater dry adiabatic lapse rate than the air mass into which it is intruding. Storms such as tornadoes or thunderstorms come from an air mass that has become *unstable*.

Humidity

The presence of moisture, dampness, or water vapor in the air around us is termed *humidity*, and there are three ways of expressing this measurement: absolute, specific, and relative humidity. *Absolute humidity* (see Fig. 5-2) measures the actual weight (expressed in grams per cubic foot, or grams per cubic meter) of the water present in a specific volume of air. *Specific humidity*, on the other hand, measures the weight of the water vapor present in a given mass of air relative to the weight of the air; it is usually given as grams of water vapor per kilogram of moisture-laden air. For most of us, however, the term *relative humidity* is used as the normal standard of measurement for the amount of moisture a given mass of air can hold, and it is expressed as a percentage. More technically, we can say that relative humidity is the amount of water vapor in a given mass of air at a given temperature, as compared to the amount of moisture the air mass could hold if it were fully saturated and there were no accompanying change in temperature. If the daily weather report gives the humidity of your locale as 60 percent, it means that the air mass contains only 60 percent of the moisture it could hold at its present temperature. What happens if there is nearby a large body of water that sends moist air into your locale? As the moisture penetrates the air, the humidity rises. If there is no temperature change, the humidity will continue to rise until the air mass is fully saturated, or soaked, with water. If dew point is reached, condensation can occur.

This tells us that a given body of air can hold and retain just so much moisture at a certain temperature, but if the temperature is altered, so is the moisture-holding capacity of the air mass. If the temperature of the air is

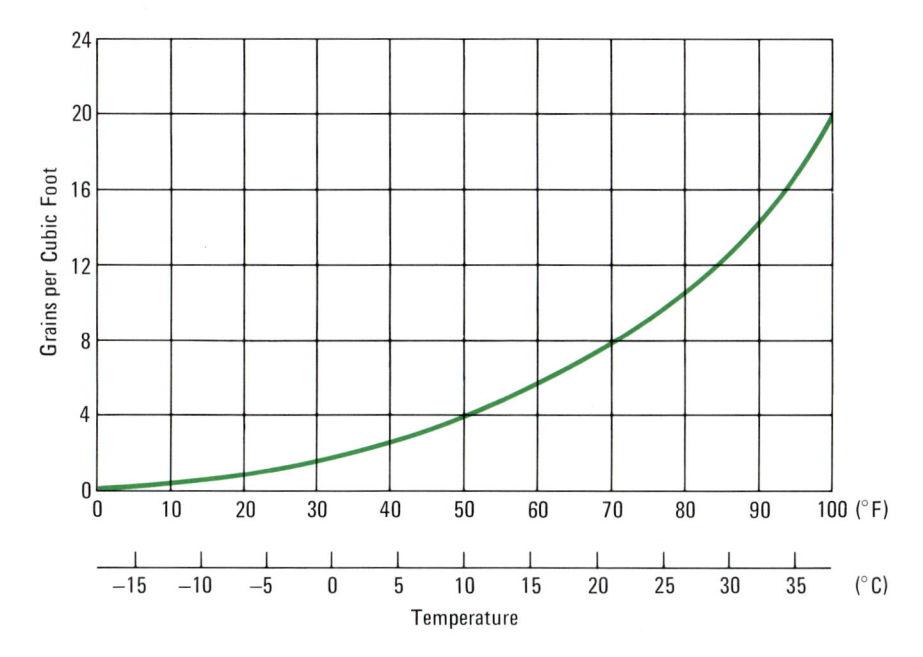

FIGURE 5-2 Absolute humidity The higher the temperature, the more moisture can be held by a given mass (volume) of air. If the temperature of the air mass should decline and fall below its dew point, condensation would occur and would be evident in the form of fog or clouds.

raised through terrestrial radiation, molecular speed within the mass is increased and the air can hold more water vapor. Winds can cause an air mass to move, thus enabling it to expand and increase its moisture-holding capacity. A decrease in temperature on a calm, windless day will lower the saturation point. Dew point can be attained sooner and some form of condensation, such as fog, clouds, or ice crystals, may occur; if the saturation point is exceeded, some form of precipitation, such as hail, snow, or rain, will fall.

It is interesting to note that of all the moisture contained in our atmosphere, more than half lies below the 2-mile (3-kilometer) level.

Forms of Condensation

As we have seen, **condensation** will occur with either one of two factors: (1) if there is an increase in the amount of water vapor in an air mass to its saturation point, or (2) if the air temperature falls to a level where its capacity to hold moisture is reduced. If this **dew point** is reached when the air temperature is at or below freezing, clouds made up of ice crystals will appear. If the dew point is reached at a higher temperature, fog or clouds made up of water droplets will form.

Fog At sea level or in high inland valleys, fog may occur. It appears when a warm, moist air mass comes in contact with a layer of colder air close to the ground. A common fog for inland areas is **radiation fog,** called **ground-inversion fog,** which occurs when the land is chilled during the night through the loss of heat by radiation and the warmer air above has its bottom layers cooled to dew point. Such fogs usually disappear by midday or as soon as the sun's rays can warm the ground and end the inversion.

A fog common to San Francisco is called an **advection (moving) fog.** This type of fog

Forms of condensation: clouds and fog.

Cirrus

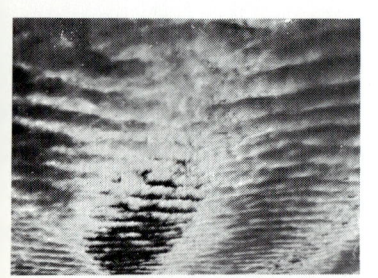

Altocumulus

Nimbus

Cumulus

Stratocumulus

Stratus

Cirrus — 40,000 ft (12,200 m)

Altocumulus

Nimbus

Cumulus — 20,000 ft (6100 m)

Stratocumulus — 5000 ft (1525 m)

Stratus

Fog — 1000 ft (305 m)

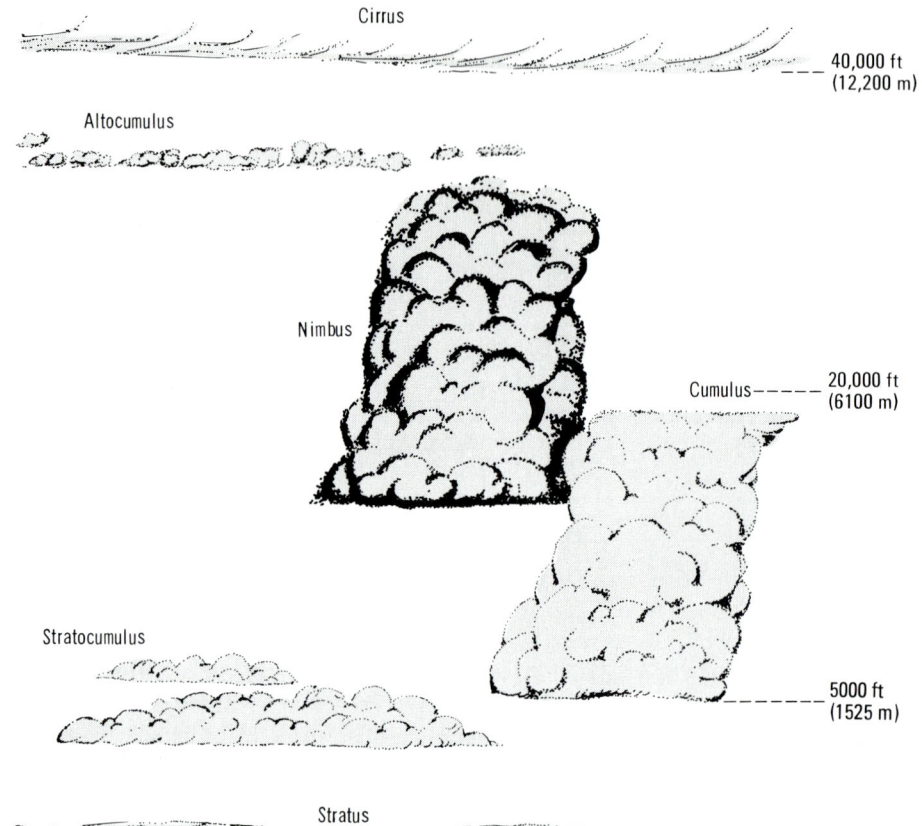

FIGURE 5-3 Forms of condensation: clouds and fog.

Stratocumulus

Stratus

Fog

results when a warm, moist air mass (in this case, pushed by westerly winds from the Pacific Ocean) crosses a cool or cold landmass or a cool ocean current (such as the California current that moves past the Golden Gate), where it is chilled to the point of condensation. Because the warm air mass continues to converge with the cool body of land or water, the thick blanket of fog is constantly being resupplied. In addition, the fog's thickness prevents the sun's rays from warming the land surface. The fog layer, therefore, remains for much longer periods of time than the thinner ground-inversion fog.*

Clouds Probably the most important result of condensation is the phenomenon we call clouds, because clouds can furnish precipitation in the form of snow, sleet, hail, or rain. If there is a rapid upward movement of warmed air, an unstable condition will be created. Clouds will form, and if the air mass is deep enough and contains enough moisture, water droplets will form around minute particles (called condensation nuclei) of dust, salt, or volcanic ash, and larger, heavier droplets will fall as rain or snow.

Clouds may be classified according to their general form and the altitude at which they appear (Fig. 5-3). We use the term **stratus** to mean any kind of cloud formation that spreads out in layers over a wide area, seeming to cover the sky from horizon to horizon. Stratus clouds can be found at different elevations, but they usually form within a few thousand feet of the earth's surface. An "overcast sky" is a reference to decks of stratus clouds that block out the sun. **Cumulus** clouds appear like huge puffs of cotton (or giant gobs of whipped cream). Their flat bottoms seem to rest close to the earth while their heads tower thousands of feet into the sky. Smaller versions of this type of cloud often look like small cotton balls scattered across the sky. **Cirrus** clouds form at high altitudes, stretching out in thin, lacy, or streamerlike wisps near the tropopause.

Both stratus and cirrus clouds result from condensation occurring without any up or down movement; cumulus clouds are the result of convectional uplift. Depending on the altitude of formation and the process, these terms can be combined to explain different or unusual clouds, such as **stratocumulus** (broad layer of thick and puffy cumulus clouds), **altostratus** (a layer of clouds at elevations of thousands of feet), **cirrostratus** (a high but thin layer of stratus clouds), or **cumulonimbus** (from the term *nimbus*, which means "rain bearing"). It is interesting to note that the most common type of cloud the world over is the stratus, especially from about the 30° parallel northward.

Precipitation
Before precipitation can occur, clouds must form. Note, however, that not all clouds yield rain, hail, or snow. Condensation and the resulting formation of clouds is the first step, but before there can be precipitation, the water droplets must combine and become sufficiently heavy in order to fall to the earth.

What forms will precipitation take? If the temperature is below freezing, snow may form as tiny flakes or hail may result from the freezing of coat after coat of moisture around the beginning nucleus of dust particles or icy crystals. If sufficient water droplets coalesce to form a raindrop and the temperature is above freezing, rain will fall.

There are three primary causes of precipitation, and they all have to do with the lift-

---

*San Francisco's famous fogs are usually more intense during the summer months, when warm air from nearby valleys and hillsides encroaches the cold California current, thus creating heavy fogs.

Farm country, Pennsylvania the calm after the storm

ing of warm, moist air masses to a zone of cooler air temperatures, where dew point can be reached and exceeded. The three types of precipitation are orographic, convectional, and frontal (Fig. 5-4).

Orographic precipitation results when a warm, moist air mass moves onshore and is forced to rise, because of an obstruction such as a higher. elevated landmass (mountains, up-sloping plains, or plateaus). This forced change in altitude causes the air mass to lose its heat through the adiabatic cooling process.

Convectional precipitation results when a mass of unstable air rises rapidly from an area that has been heated to extremes, such as a South Pacific atoll, in a tropical or subtropical desert, or around a warm ocean current. Such an updraft is normally quite violent in its movements as the heated air rushes upward and cools at a higher elevation. This cooling creates severe downdrafts, and dew point is quickly attained. Condensation occurs and a cumulus cloud begins to form—a furiously boiling white cloud thrusting up into the sky. Precipitation is short-lived, although sometimes quite heavy, and sometimes thunderstorms are created.

Frontal precipitation (also referred to as cyclonic) is the result of a confrontation between two large air masses, one warm, the

MOISTURE AND PRECIPITATION 131

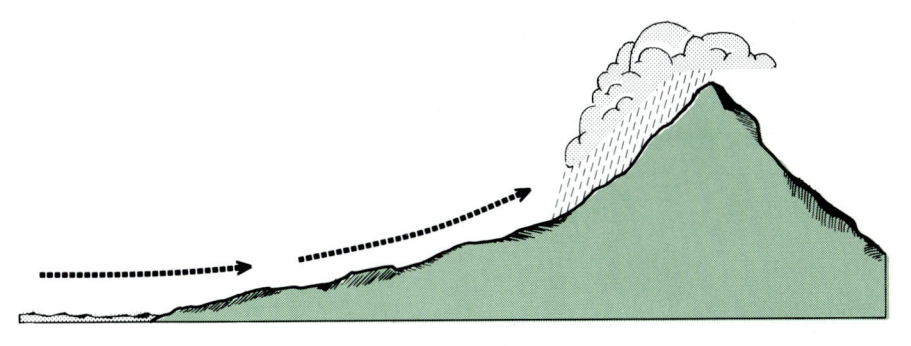

OROGRAPHIC

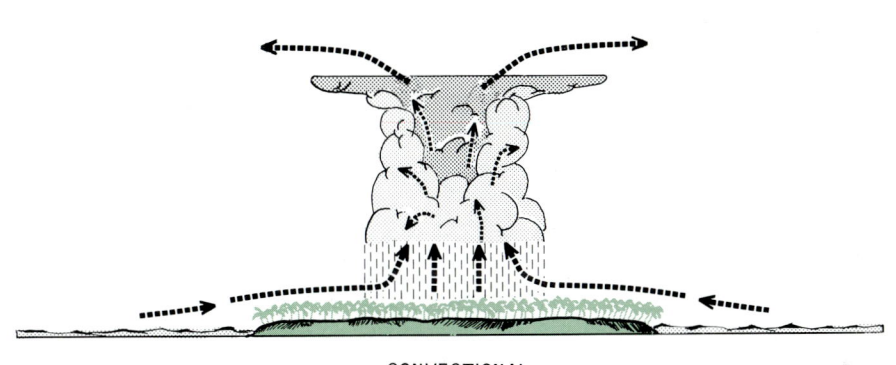

CONVECTIONAL

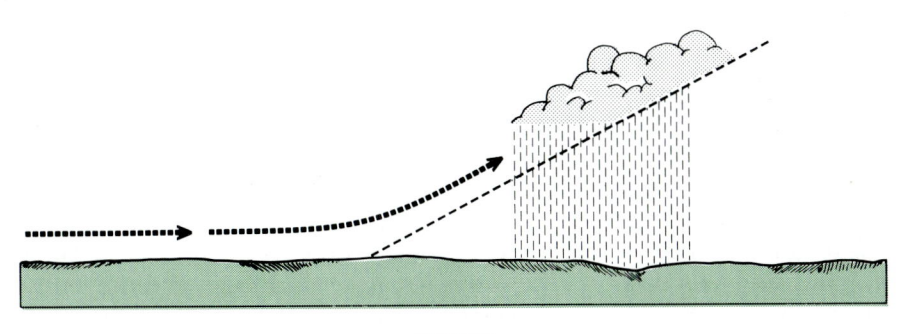

FRONTAL

------▶ Movement of air masses

FIGURE 5-4 Causes of precipitation.

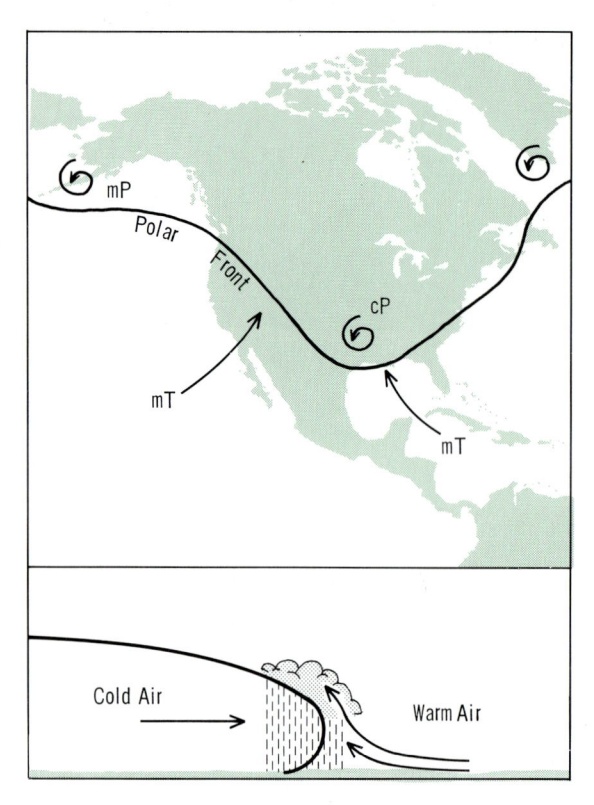

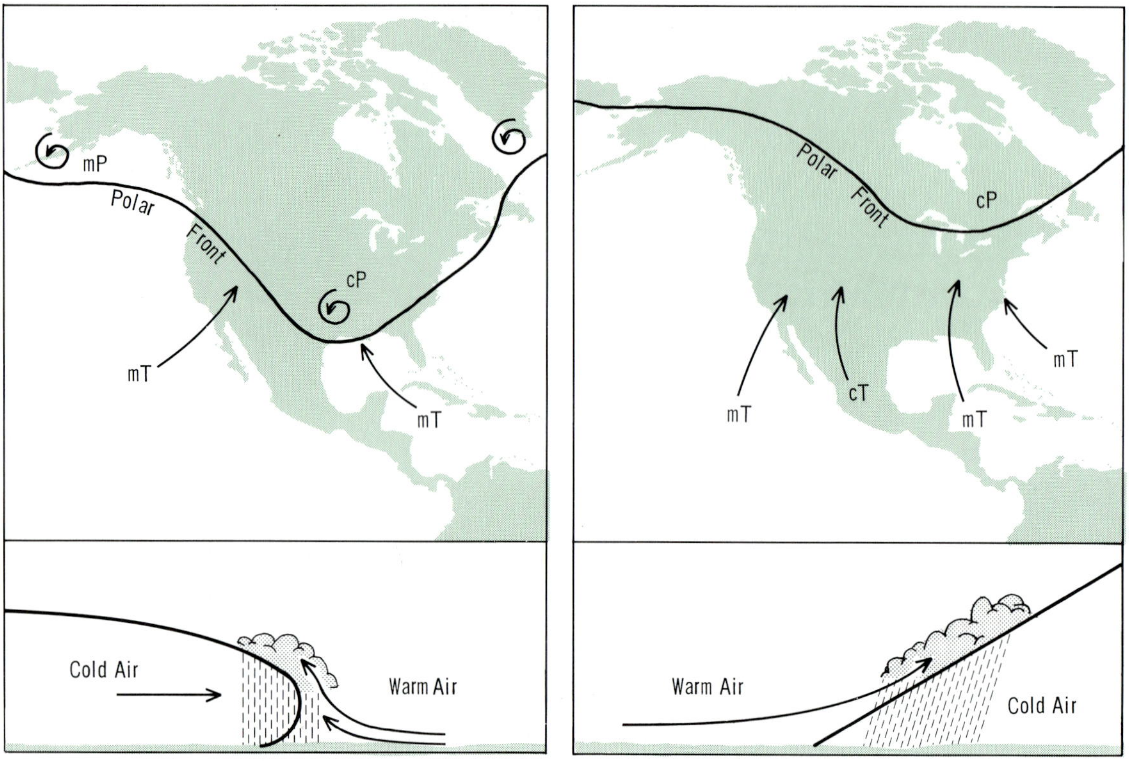

132 MOISTURE: CLOUDS AND STORMS

WINTER Cold fronts usually dominate, and the cold air mass is the aggressor.

SUMMER Warm fronts usually dominate, and the warm air mass is the aggressor.

FIGURE 5-5 Frontal activity in the midlatitudes.

other cool. If the cool air is the aggressor, the result is a **cold front;** if the warm air is doing the moving, a **warm front** develops. Consult Fig. 5-5, which presents two types of fronts that cause precipitation. During the winter the most likely occurrence is for a cool mass of air to move southward out of the polar front. When the cool air collides with the stationary mass of warm air, it wedges under the warm air, much like the sloping blade of a bulldozer. This forces the warm air aloft, followed by the familiar adiabatic cooling process, with condensation and precipitation resulting.

Fronts are common (and quite typical) during the late fall and winters along the Great Plains of the United States. If a warm, tropical air mass is the aggressor and moves into territory already occupied by a stationary and colder air mass, the warm air mass will be forced to override the cold air mass until dew point is reached, condensation occurs, and precipitation results. This is typical of air masses leaving the Caribbean and the Gulf of Mexico to swing over Texas and the southeastern United States, there to collide with cold air masses from the north.

Figure 5-6 explains the development of a cyclonic storm along a weather front. Here, you see the counterclockwise spiraling of the low-pressure cell that converges on a warm air

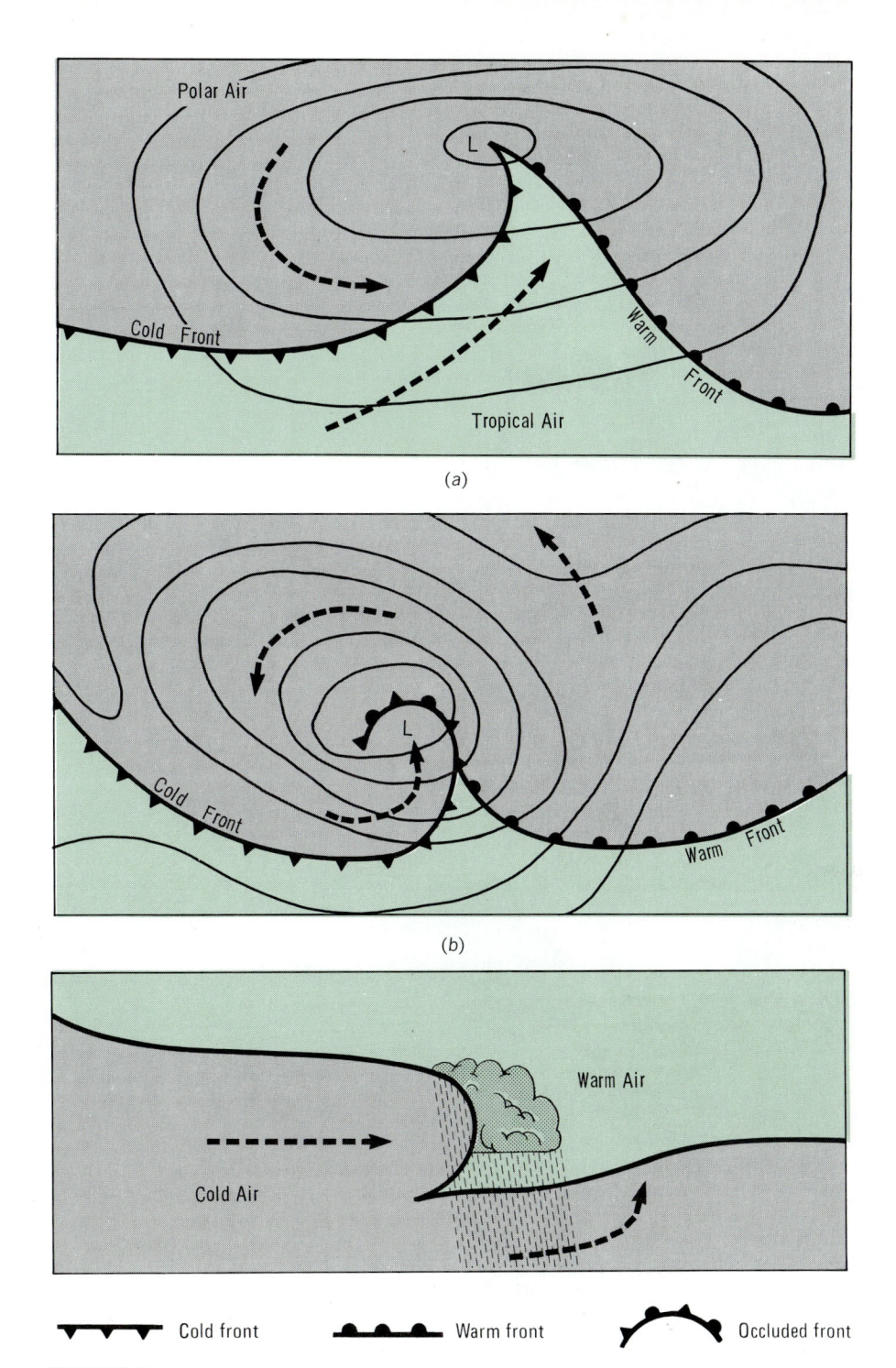

FIGURE 5-6 Development of a cyclonic storm along a weather front *(a)* shows the aggressive advance of a cold air mass on a stationary warm air mass. *(b)* shows how the counterclockwise spiraling cold air has trapped the warm air and forced it aloft, thus creating an occluded front with resulting condensation (clouds) and precipitation.

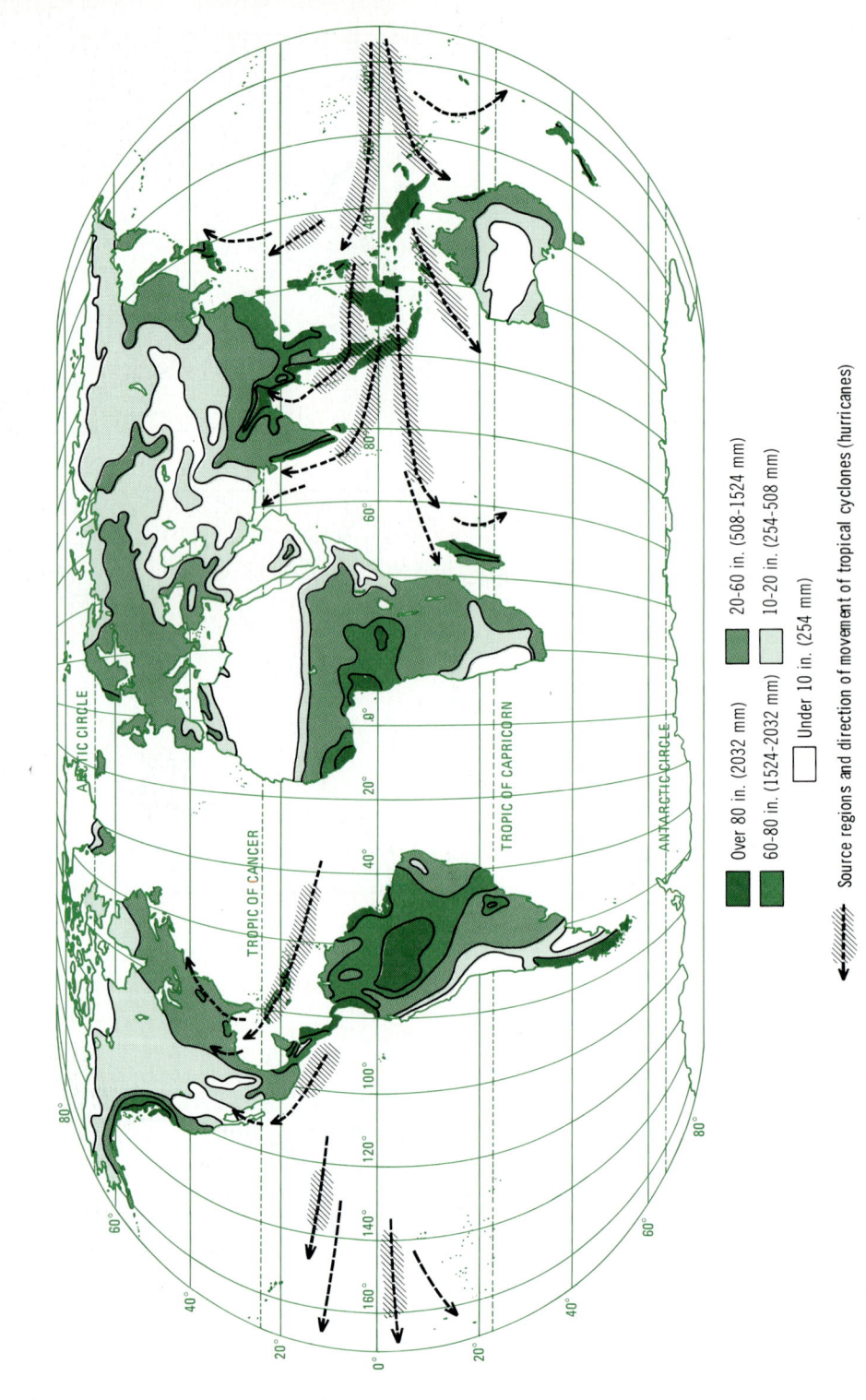

FIGURE 5-7 Annual world rainfall and tropical cyclones of the world.

The windward coast of Oahu

mass. When a warm air mass is forced aloft, completely cut off from its source and the ground, the result is an **occluded front.** You can tell which front is arriving by the temperature change that occurs. If a cold front passes when the air around you is warm, the temperature will drop. The reverse occurs with the passage of a warm front. Precipitation during a warm front will be widespread and heavy, and it may last for hours or, sometimes, for days. Cold-front precipitation, on the other hand, has a shorter life span, although rainfall may be quite intense (thunderstorms).

We might say that orographic precipitation will be most common in regions where barriers (such as mountains) obstruct the free movement of air masses; convectional precipitation will be the normal type for tropical regions because of the excessive insolation received; and frontal activity will be the customary precipitation for the midlatitude regions, which are dominated by polar fronts.

Because rainfall is so important to us, it is necessary for us to know where and how much rain falls to enable us to plan for the future by understanding how to preserve our present water supply. To do this, meteorologists collect precipitation data and then construct maps to show the areas of world rainfall and all other forms of precipitation. By connecting points of equal precipitation with lines (called **isohyets**), we can establish boundaries of the different regions and thus establish a map showing the world's average annual rainfall (Fig. 5-7). On consulting this map, it becomes obvious that the equatorial regions receive the heaviest annual rainfall; many places in the tropics receive more than 80 inches (2000 millimeters) per year. Note also that areas separated from large bodies of water or those on the lee (or rain shadow) side of mountain ranges get little precipitation. Areas that lie in the path of the trade winds, typically the Hawaiian islands, get the heaviest rainfall. On the windward coasts

136 MOISTURE: CLOUDS AND STORMS

of Hawaii, Maui, Oahu, and Kauai, heavy precipitation results in the tropical rain forest types of vegetation; on the downwind, or lee, side the vegetative cover is much like that of Arizona or New Mexico. The same applies to the lee and windward sides of South America's Andes. Regions in the midlatitudes get little of what might be termed consistent rainfall, primarily because of conditions created by the belt of subtropical high pressure, with its descending, compressing air.

Now consider Fig. 5-8, which shows the

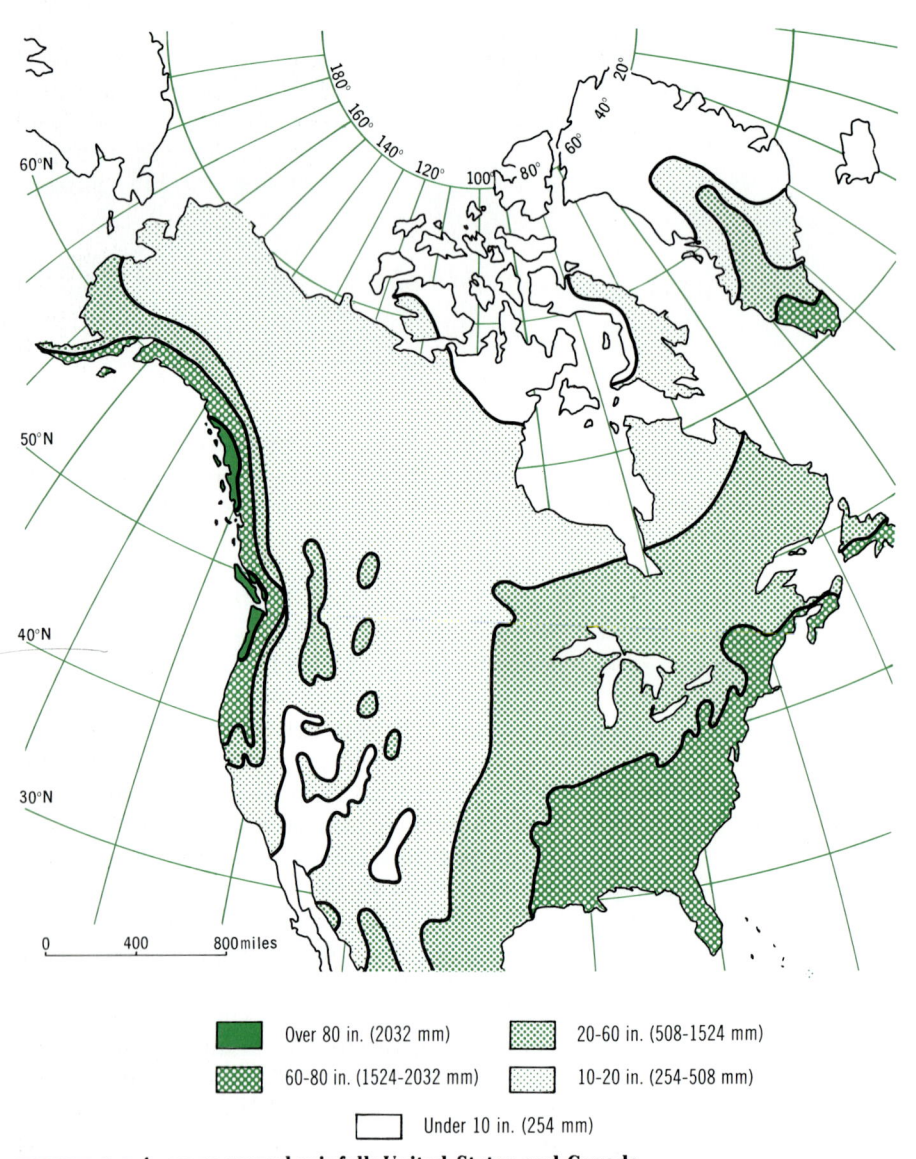

FIGURE 5-8 Average annual rainfall: United States and Canada.

Legend:
- Over 80 in. (2032 mm)
- 60-80 in. (1524-2032 mm)
- 20-60 in. (508-1524 mm)
- 10-20 in. (254-508 mm)
- Under 10 in. (254 mm)

average annual rainfall of Canada and the United States. The southeastern United States receives over 40 inches (1016 millimeters) of rainfall, while most of the remaining states and Canada receive less than 20 inches (508 millimeters). One amazing contradiction may be seen in the coastline from the state of Washington north through British Columbia and Alaska, where 80 inches (2032 millimeters) of rain falls annually. This region receives so much precipitation mainly because of the series of wavelike onslaughts of low-pressure cells coming from the Aleutian islands; moisture-laden winds strike the mountain ranges, and this results in orographic precipitation.

STORMS

Many storms are the result of a collision between two advancing air masses, one cold, the other warm. If you recall our discussion on pressure gradients and on precipitation caused by frontal activity, you probably realize that this type of storm is the result of cyclones and anticyclones coming together. For an example, consult Figs. 5-9, and 5-10, which are weather maps for two days in November 1982. Inhabitants of the American Northwest, the Great Plains, and the Great Lakes, and, it is important to note, the citrus growers of Florida have long been aware of storms brought to them through cyclonic activity, the invasion of polar air masses. During World War II a Norwegian meteorologist, Vilhelm Bjerknes, examined the data showing the continual onslaught of these cool lows. It seemed to him that it was much like a battleground during a war, because frontal storms advanced in a series of waves, much like regiments of soldiers. His "wave theory of cyclones" came to be known as the **polar-front theory** and is still accepted today. Many of the storms that strike the North American continent, such as thunderstorms, tornadoes, and hurricanes, are the result of cyclonic action.

Thunderstorms

Thunderstorms are born out of the upward rush of heated air to high altitudes, which is created out of the differences in air pressure brought on by unequal heating of areas of land and sea. In the American Midwest, a region will be heated until the air begins to rise and outside air moves in; the outside air is then also heated and begins to rise. The low-pressure region sends its air spiraling upward in a continuous cyclonic flow (counterclockwise) to great heights, aided by what has been termed the "chimney effect," where the air at these high altitudes is expanding and moving at faster speeds, thus drawing more air upward from the ground. As condensation occurs, latent heat is released as the towering cumulus cloud is formed, and this increases the violence of the boiling cloud. As the cloud increases in height, sometimes growing as high as 9 miles (15 kilometers), and as temperatures within the cloud decline, rain and snow form and begin to fall, and the top of the cloud spreads out in wispy streamers.

The cloud becomes a thunderstorm when the precipitation creates a downdraft within the cloud as the subsiding air drags air behind it. Winds are violent within the thunderhead, as warm air races past cold air and the electrical field within the cloud becomes highly charged; lightning flashes and the sound of thunder is heard across the land. Cold air pours out of the cloud's base, accompanied by heavy rain and high winds and, sometimes, tornadoes. Eventually, the downdraft destroys the thunderstorm because it has cut off the supply of warm air that was the impetus for the storm's beginning. Without a continuous supply of warm, rising air and the resulting precipitation, the thunderstorm loses its vitality

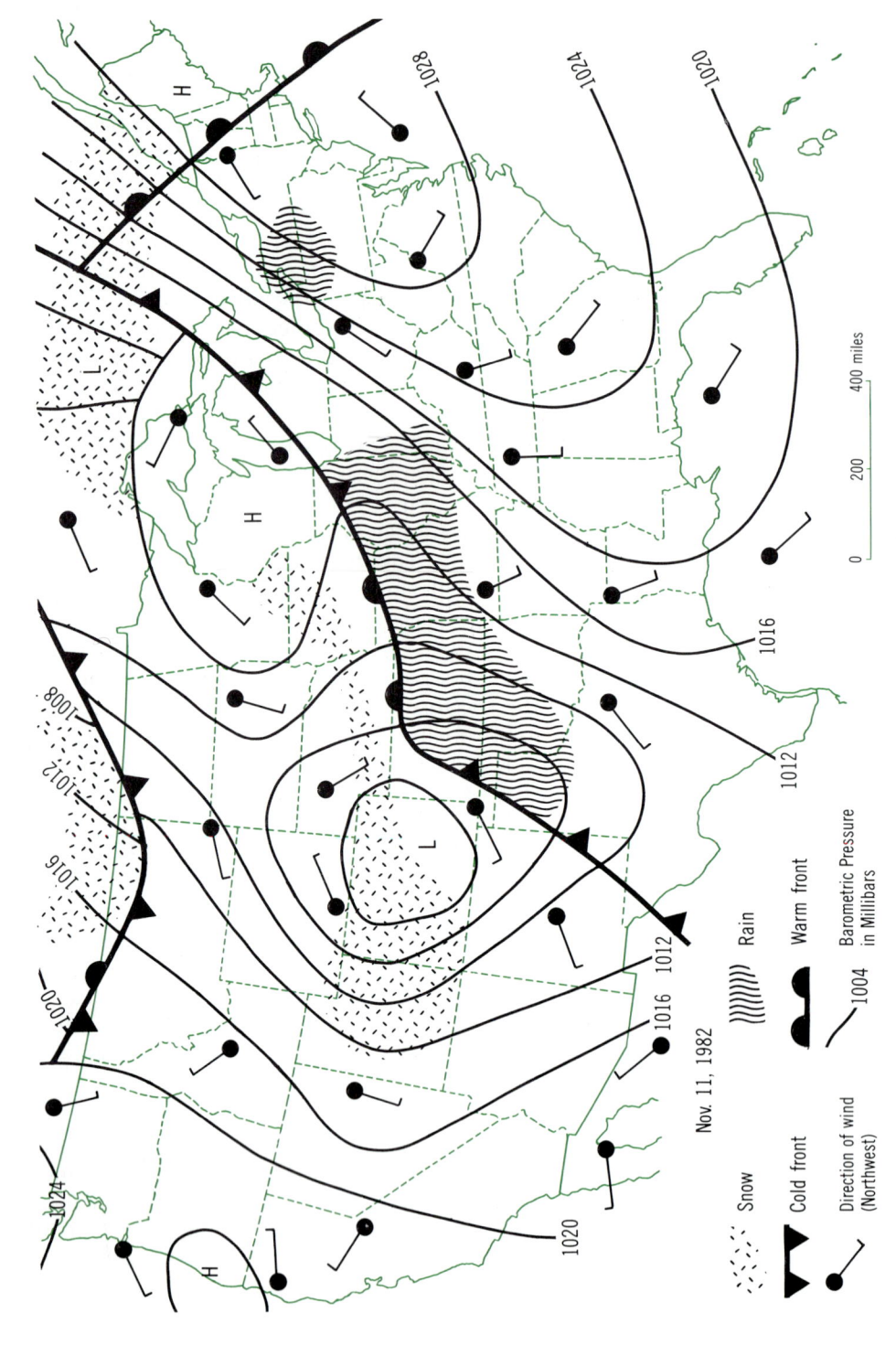

FIGURE 5-9 **National weather forecast, Saturday, November 11, 1982** Prepared in cooperation with the National Weather Service.

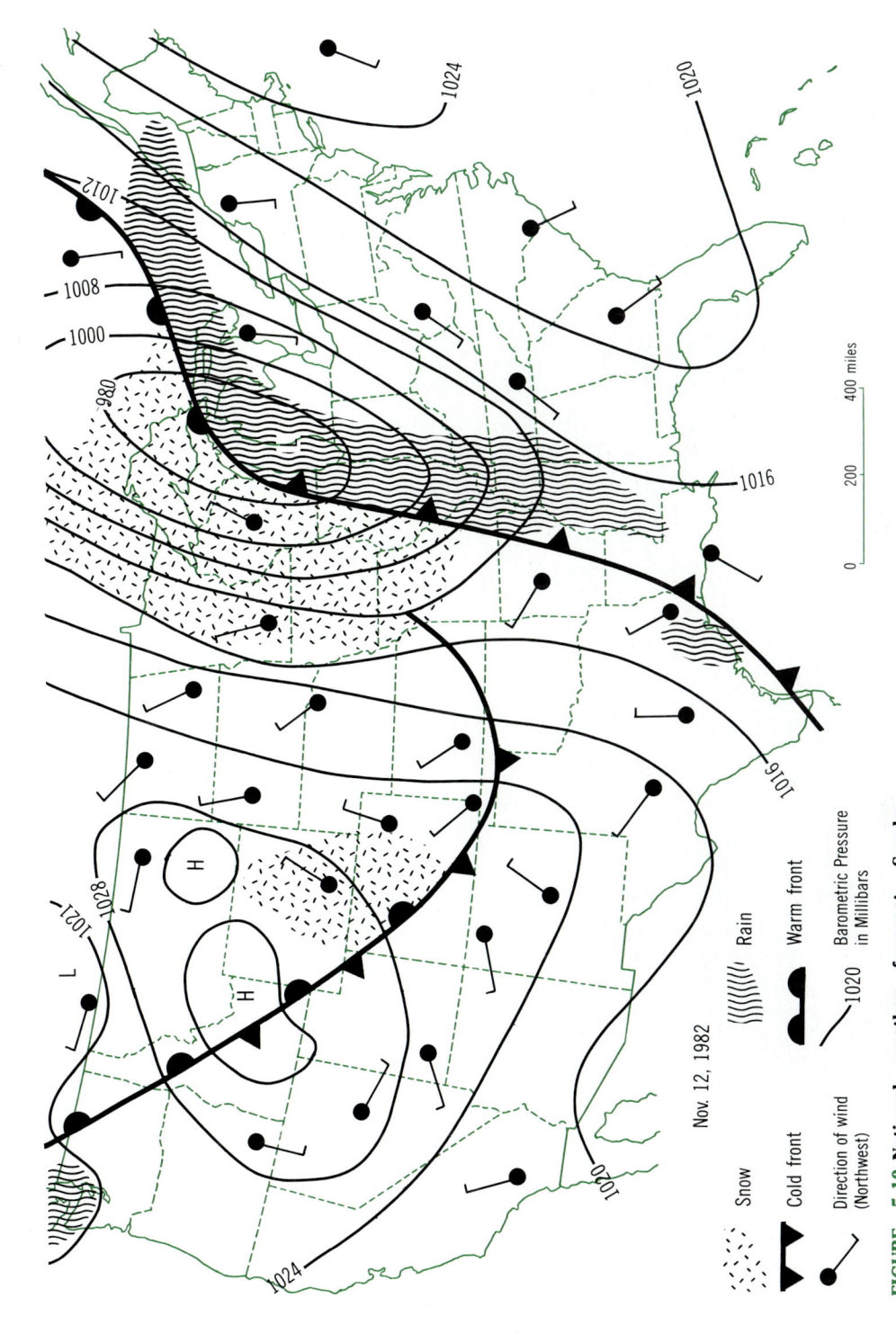

FIGURE 5-10 National weather forecast, Sunday, November 12, 1982 (Prepared in cooperation with the National Weather Service.)

139

and its heat energy and slowly settles into oblivion.

Although thunderstorms normally result from convectional uplift, many evolve out of the interaction between warm and cold fronts, as happens over most of the United States and southern Canada, and from the differences between the surface temperatures of land and water, as is the case in storms that occur throughout the southeastern United States.

Tornadoes

Tornadoes are violent windstorms that "spin off" from thunderstorms; although they are relatively small in size and short-lived, they can be very destructive.* In the beginning the towering thunderhead is dark and foreboding, with a heavy curtain of rain descending to the ground, and it is difficult to distinguish the separation line of cloud and rain. Then, from the bottom of the cloud a funnel appears, circling wildly in a counterclockwise motion, and reaches for the earth. It may appear and disappear once or twice; then, finally, the funnel touches down and the destruction begins.

The storm sometimes moves very slowly, almost seeming motionless, and sometimes at speeds of up to 70 miles (112 kilometers) per hour. One thunderstorm may spawn several tornadoes that form at intervals as it sweeps across the land, and not all of them reach the ground. The path of a tornado is usually quite narrow in width, perhaps 0.25 mile (0.4 kilometer) wide, although some attain lengths of 16 to 300 miles (26 to 480 kilometers).

The phenomena we know as tornadoes are pretty well limited to the United States. Figure 5-11 details the incidence of these storms that have struck the United States for a 27-year period (1953 to 1980). It would seem that if a "tornado alley" does exist, it would include Texas, Oklahoma, Kansas, Nebraska, Iowa, Missouri, Illinois, and Indiana. Florida and the Southeast also have their share of tropical cyclones-turned-tornadoes.

The worst tornado on record "ripped across Missouri, lower Illinois, and Indiana in three hours on March 18, 1925, and killed 689 people."[1] Since "the early 1950's, the tornado death toll has averaged about 120 per year" in the United States.[2] This does not include, of course, the destruction of property that, during this century alone, must amount to billions of dollars.

Hurricanes and Tropical Cyclones

We have concentrated on storms of the mid-latitudes primarily because of our concern with the United States and Canada, which lie in the path of invading polar-front air masses. There is, however, another type of cyclonic storm, the **tropical cyclone,** which originates in the tropics. Depending on which part of the world you are in, the names will vary: *hurricane* (coming to us from the Arawaks of the Caribbean), *typhoon* (from east Asia and the China Sea), *cyclone* (from the Indian Ocean), and *baguios* (from the Philippines). These storms result from the rapid buildup of heated air over the warm equatorial waters and, occasionally, over the broad stretches of land along the equator. As the heated air rises, it develops a counterclockwise spiraling vortex upward, similar to the motion of a thunderstorm or tornado. When the air mass begins to move, speeds of the whirling winds may reach more than 100 miles (160 kilometers) per hour within the storm, and the storm itself may advance across the open sea at a speed of from 20 to 40 miles (32 to 64 kilometers) per hour. Accompanying the violent winds will be thunderstorms and heavy precipitation.

During the summer the southeastern coast of the United States must occasionally face these storms as they sweep up from the tropics and move across the Caribbean and the Gulf Coast. Individual hurricanes have severely punished the land all the way from Texas up to New England, averaging about six per year. Some hurricanes also originate along the western side of Mexico and Central America, but they seldom have the strength to move far enough north to bother California.

An interesting feature of a hurricane is what meteorologists call the "eye of the hur-

---

*For informative and interesting information on both thunderstorms and tornadoes, see the National Oceanic and Atmospheric Administration's publications: *Thunderstorms,* NOAA/PA 70014 (1973), and *Tornado,* NOAA/PI 70007 (1970). They can be obtained from the Superintendent of Documents, U.S. Government Printing Office, Washington, D.C. 20402.

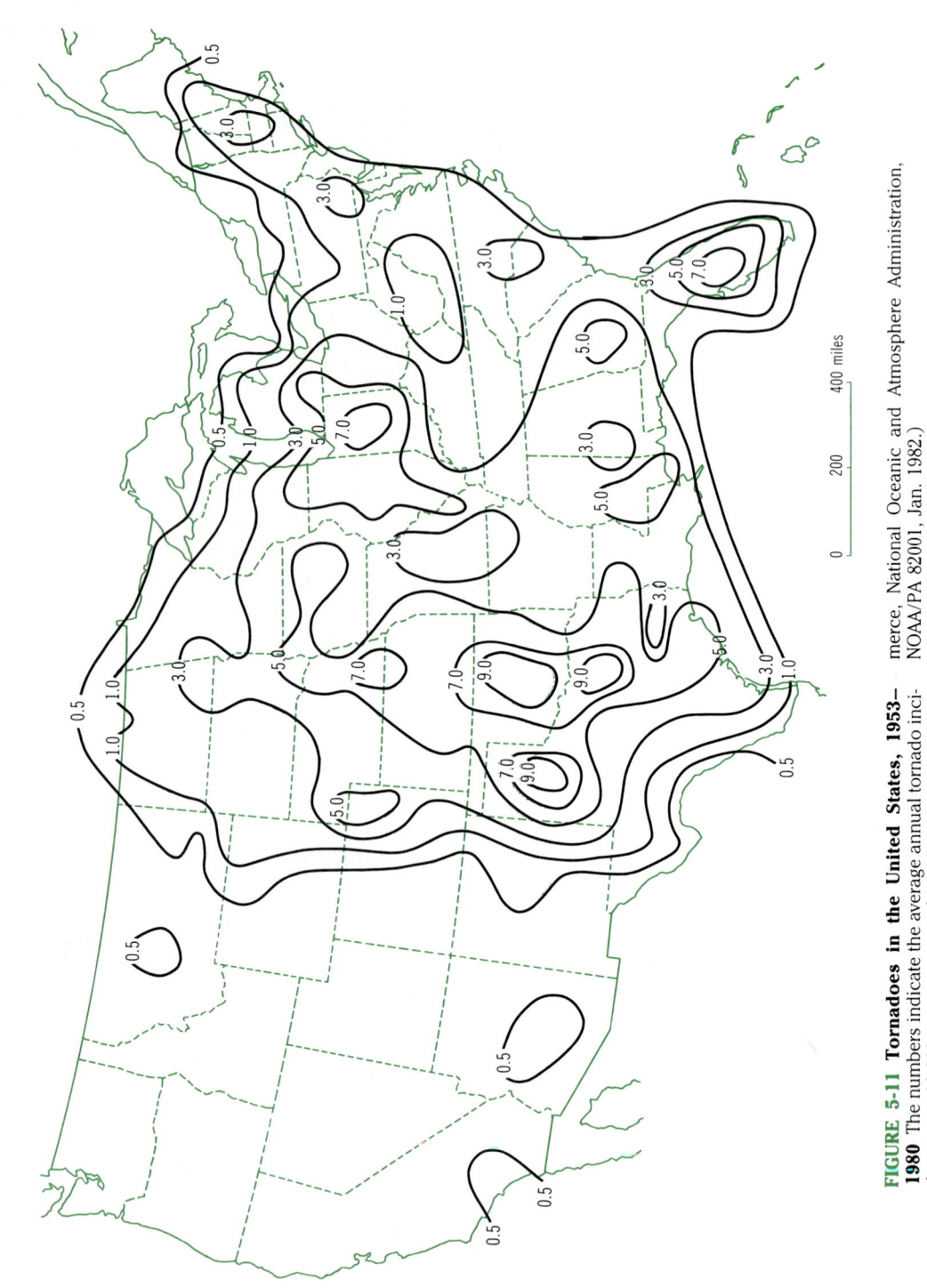

FIGURE 5-11 Tornadoes in the United States, 1953–1980 The numbers indicate the average annual tornado incidence per 10,000 square miles. (U.S. Department of Commerce, National Oceanic and Atmosphere Administration, NOAA/PA 82001, Jan. 1982.)

STORMS 143

Tornado!

The aftermath.

25,000 ft (7625 m)

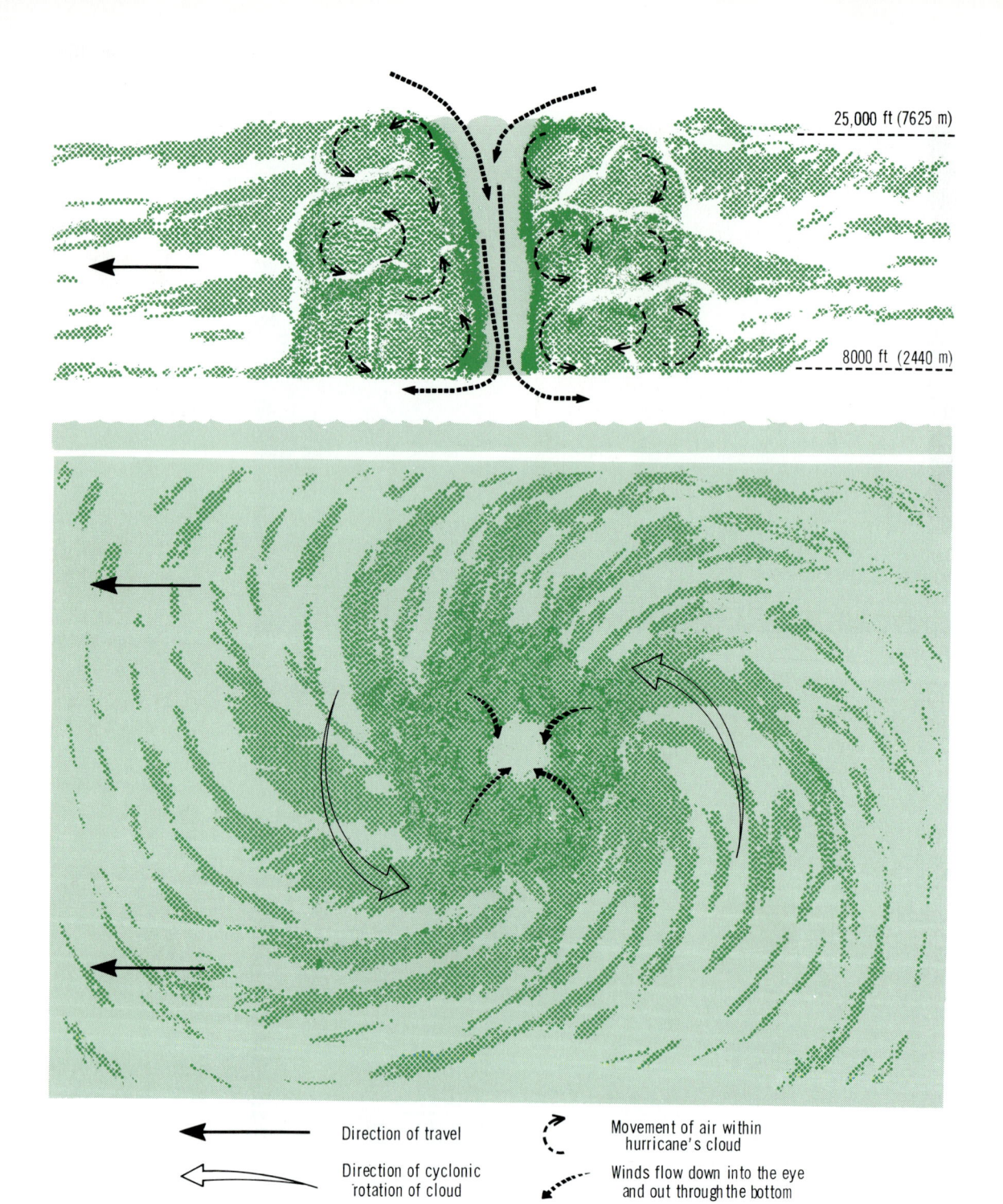

8000 ft (2440 m)

	Direction of travel		Movement of air within hurricane's cloud
	Direction of cyclonic rotation of cloud		Winds flow down into the eye and out through the bottom

FIGURE 5-12 The formation of a hurricane.

Satellite view of two hurricanes. This photo, taken on September 2, 1978, shows hurricane Ella sweeping across the mid-Atlantic Ocean, bringing heavy rain and winds to the east coast, and hurricane Norman swirling off the west coast, sending high humidity to America's southwest.

ricane" (Fig. 5-12). This eye results from the rapid spiraling of the air mass as it travels across the waters; in the eye the air is calm, and flocks of birds may be found flying serenely around in circles, not daring to brave the angry winds all around, waiting patiently for the storm to die. The eye gives a few moments' respite for a storm-tossed ship or a wind-battered city, but then comes the other side of the storm and winds coming from the opposite direction, and the storm-ravaged area goes under siege once more.

Worldwide, tropical cyclonic storms seem to have a regular pattern (see Fig. 5-7). They appear during a particular hemisphere's late summer and fall, and most of them appear on the west sides of the oceans, apparently because these low-pressure cyclones are pushed westward by the trade winds and are affected by the Coriolis effect, and thus swing away from the tropics as they near land. You will note that there are no tropical cyclones in the South Atlantic or along the east coast of South America. This is probably because there is no opportunity for a storm of this nature to build up in the space that is available.

REVIEW AND DISCUSSION

1. Explain the term *hydrologic cycle*.
2. How important is a region's water balance?
3. What is the difference between the processes of evaporation and sublimation?
4. Explain the meaning of *relative humidity*.
5. When can condensation occur?
6. How does fog form? Explain the difference between ground-inversion fog and advection (moving) fog.
7. Explain the difference between condensation and precipitation.
8. How are clouds classified? Explain the differences between cirrus, stratus, and cumulus clouds.
9. What does it mean when the word *alto* is attached to a cloud's classification, as in *altostratus?*
10. What does *nimbus* mean? What does *cumulonimbus* mean?
11. Explain the terms *orographic, convectional,* and *frontal* with reference to precipitation.
12. Explain the difference between cold front, warm front, and occluded front.
13. What does *rain shadow* mean?
14. Explain why the southeastern United States receives more rainfall than the northeastern portion.
15. Explain the polar front theory.
16. Describe the formation of a thunderstorm. How does it differ from a tornado? How does it differ from a hurricane?
17. What parts of the United States are most affected by tornadoes? What parts are most affected by hurricanes?
18. Why is the South Atlantic Ocean spared the onslaughts of hurricanes?

NOTES

[1] Morris Tepper (May 1958), "Tornadoes," *Scientific American* (offprint no. 848), p. 6.

[2] U.S. Department of Commerce (1970), *Tornado*, NOAA Publication NOAA/PI 70007 (Washington, D.C.: U.S. Government Printing Office).

ADDITIONAL READING

Court, Arnold (August 1970). *Tornado Incidence Maps*. U.S. Department of Commerce ESSA Technical Memorandum ERLTM-NSSL 49. Norman, Okla.: National Severe Storms Laboratory.

Court, Arnold, and Richard D. Gerston (1966). "Fog Frequency in the United States," *Geographical Review*, 4, pp. 543–550.

National Oceanic and Atmospheric Administration (July 1972). *The Flood That Strikes in a Flash*. NOAA Publication 511–323/164. Washington, D.C.: U.S. Government Printing Office.

Petterssen, Sverre (1968). *Introduction to Meteorology*. 3rd ed. New York: McGraw-Hill.

U.S. Department of Commerce (1969). *Clouds*. ESSA Brochure 0–355–259. Washington, D.C.: U.S. Government Printing Office.

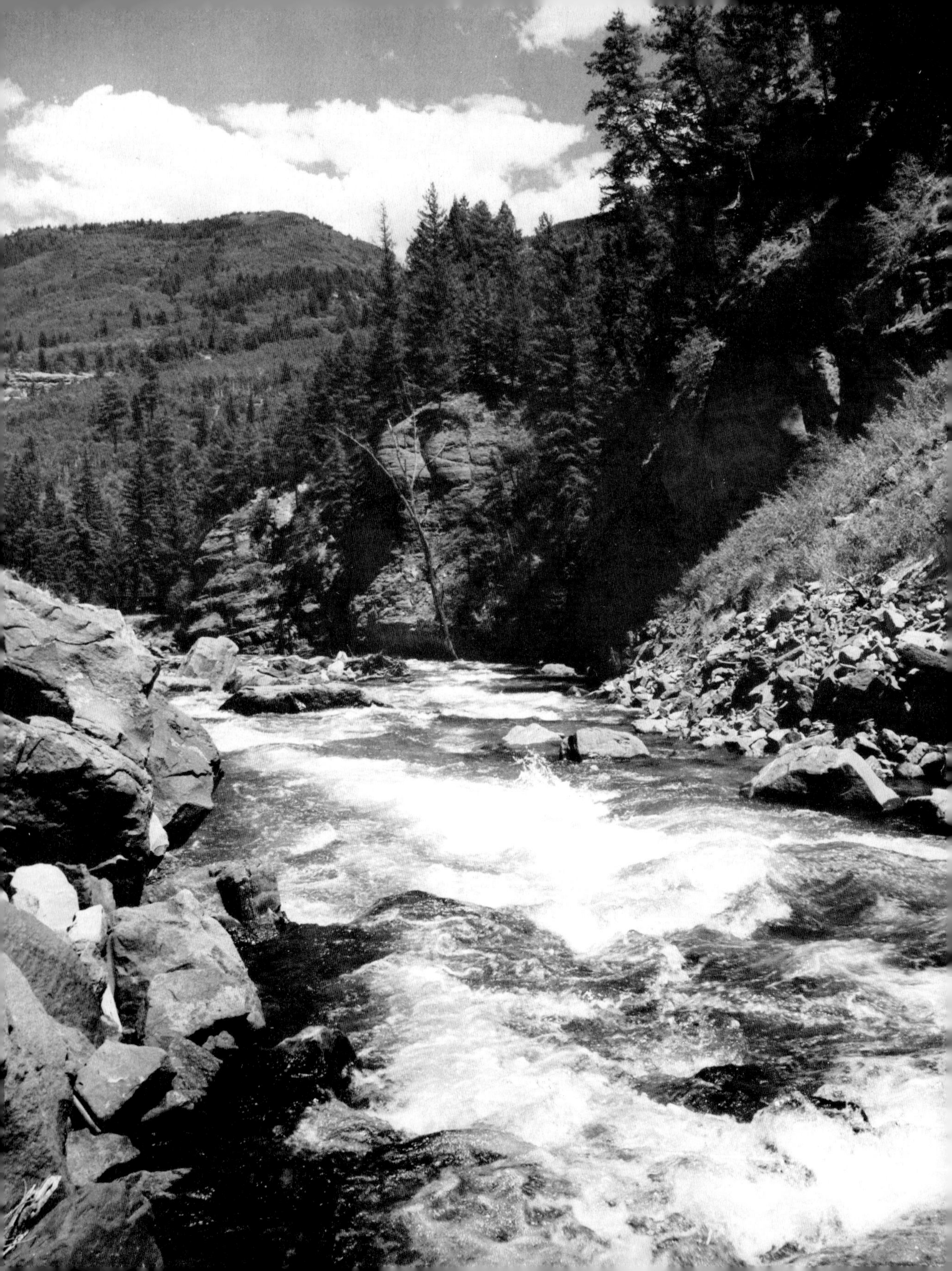

SIX

THE EARTH'S WATERS

Water has three natural forms: liquid, solid, and vapor. With changes in form, it has recycled itself constantly since the creation of the earth.

From the time, hundreds of millions of years ago, when water vapor formed earth's first clouds and the first rain fell, this process of recycling has continued with little change in the quantity of the earth's water.

Every drop of the earth's original water supply is still in use . . . in the earth's atmosphere, on the earth's surface, or beneath it.

RIVER OF LIFE. WATER: THE ENVIRONMENTAL CHALLENGE
U.S. DEPARTMENT OF THE INTERIOR

It is probably obvious by now that all places on the earth are not fortunate enough to have a favorable water balance. The reason for this, as Table 6-1 shows, is the unequal distribution of the waters that make up earth's hydrosphere. Even a brief examination of this table will make it apparent that although we cannot readily drink the salt waters of the oceans, they are the greatest suppliers of the moisture required to feed the hydrologic cycle and thus furnish the earth with the waters it must have in order to survive. Because our future may depend on how we treat them, we cannot ignore the waters of the earth. We must learn more about them so that we can devise methods to prevent abuse and pollution.

THE ATMOSPHERE

At times, when the skies are crowded with storm clouds and rain is falling, it almost seems as if the sky were filled with water. Then the storm passes, and the days that follow are clear and dry. At such times we realize that even though the lower atmosphere does cover a considerable amount of territory—encompassing the earth as it does—it is sadly lacking in moisture. The troposphere contains a small amount of water at any given time, but it does its part in carrying out the hydrologic cycle. One might say that the lower atmosphere serves as a kind of intermediary, because it takes water from the oceans, carries it inland, and then precipitates it.

Moisture in the atmosphere is held in the form of a gaseous vapor that is distributed throughout its tremendous volume. Hydrologists tell us that if all of this vapor were precipitated at one time and fell to the earth as rain, it would form a layer only about 1 inch (25 millimeters) thick. How, then, can Mount Waialeale in Kauai, Hawaii, average about 472 inches (11,800 millimeters) of rainfall per year? The reason is the constant movement of air masses over the island. As one air mass is emptied of its moisture, a new one moves into the area, and the cycle continues. The atmosphere does not have the opportunity to become fully saturated at any one time. It is always giving and receiving. Consider the conterminous United States, which receives an average of 30 inches (762 millimeters) of precipitation every year and loses about 21 inches (530 millimeters) of this moisture by evapotranspiration.* As for the difference of 9 inches (230 millimeters), some of the moisture is obviously absorbed by the air over the land and reprecipitated.

SOIL MOISTURE

If you were to examine soils all over the world, you would find that almost all of them contain some moisture. "Some" moisture, however, is not enough for most forms of vegetation. If an area is deficient in moisture, artificial irrigation methods must be introduced for agriculture to be successful. What happens to the natural vegetation if moisture is lacking? Plants that have adapted to such an arid climate would survive—if the next arrival of moisture from precipitation or through groundwater movement is not too long delayed. Desert plants may flourish in an area that lacks moisture because of their ability to store water in their tissues and because they transpire less than other vegetative forms. Transpiration can seriously affect the availability of soil moisture, and some trees may withdraw and transpire as much as 50 gallons (190 liters) of moisture per day.[1] For natural vegetation, this demands an energetic hydrologic cycle and frequent rains;

Evapotranspiration is the total loss of moisture by evaporation from land and sea and transpiration from vegetation.

TABLE 6-1 World's Estimated Water Supply.

LOCATION	SURFACE AREA Square miles	SURFACE AREA Square kilometers	WATER VOLUME Cubic miles	WATER VOLUME Cubic kilometers	PERCENTAGE OF TOTAL WATER VOLUME
Surface waters					
Freshwater lakes	330,000	858,000	30,000	123,000	0.009
Saltwater lakes and inland seas	270,000	702,000	25,000	102,400	0.008
Average in stream channels	—	—	300	1,229	0.0001
Subsurface waters					
Soil moisture			16,000	66,000	0.005
Groundwater within depth of 0.5 mile (0.9 kilometer)	50,000,000	130,000,000	1,000,000	4,096,000	0.31
Groundwater—deep-lying			1,000,000	4,096,000	0.31
Ice caps and glaciers	6,900,000	17,900,000	7,000,000	28,672,000	2.15
Atmosphere below an elevation of 7 miles (11 kilometers)	197,000,000	512,000,000	3,100	12,700	0.001
World oceans	139,500,000	363,000,000	317,000,000	1,298,432,000	97.2
Totals (rounded)			326,000,000	1,335,600,000	100.0

Source: Adapted from U.S. Geological Survey Publication No. 2401-2189, 1972.

for trees that are a part of our agricultural system, repeated irrigation is required.

Silts and clays and other fine-textured soils with an adequate supply of humus (decaying vegetation) have the greatest potential for retaining moisture. They may take a long time to reach a level near saturation, but they also resist heavy losses. Sands and other coarse soils may reach saturation quickly, but they also lose their moisture rapidly through downward movement by gravity and through evaporation from the surface. There is a great variety of soils, and each responds in a different way to a favorable or unfavorable water balance. In Chapter 9 we will investigate the soils of the earth in depth.

GROUNDWATER

While it is likely that most people have some knowledge of the importance of underground water, it is also probable that they do not understand the true and vital significance of this source of water. Table 6-1 makes it clear that not all of the water we require for drinking and agriculture comes from lakes, rivers, or streams. These waters are important for a number of reasons, but we cannot discount the water supply that is contained in the massive underground reservoirs. Compare the figures given for the water volume. All locations of surface water total a scanty 55,300 cubic miles (227,000 cubic kilometers), while the water volume of the subsurface water to a depth of a mere 0.5 mile (0.8 kilometer) totals 1,016,000 cubic miles (4,162,000 cubic kilometers). Consider also the percentage of surface water in the total water supply of the earth. The oceans, ice caps, and glaciers account for more than 99 percent of the total; of the remaining amount (less than 1 percent), two-thirds is contained beneath the surface.

Groundwater has long been important to us. The ancient Babylonians worshipped a god called Ea, who ruled the domain of "springs that gushed from the earth,"[2] and the Judeo-Christian Bible tells us how Moses "struck the rock with his staff twice, whereupon water in abundance gushed out, and the community and their cattle drank."[3] There is no doubt that fresh water issuing out of the ground was believed to be the gift of God. People residing in arid or semiarid regions could not exist without wells, and history is full of tales of those who tried every means, including magical devices and divining rods, to find water. Only in relatively recent times have hydrologists understood the mysterious movements of groundwater.

The Movement of Groundwater

Moisture penetrates the groundwater area by infiltration from streams, ponds, lakes, melting snow and ice, and stream runoff. It moves by gravity. Water will infiltrate **porous** rocks and move through rocks that are **permeable.*** If a layer of rock has both of these properties and is also underlain by impermeable rocks that prevent the waters from infiltrating farther downward, it is termed an **aquifer.** An aquifer is much like an underground tunnel because the waters may flow freely from high to lower elevations through the many spaces between the rock grains.

When water infiltrates the upper layers of the groundwater zones (Fig. 6-1), it can move in several different directions. Downward flow is caused by a pull of gravity (called **gravitational movement**), and this is a continuing action; as the bottommost layer becomes saturat-

---

*The most porous rock layers are composed of sedimentary rocks, rocks built up from the deposits of sediments from lakes and oceans or from winds and running water. Those composed of fine particles, such as clay or shale, are porous enough to allow water to enter, but the spaces are too small to allow much vertical or lateral movement. Large-grain rock layers, such as sandstone, are both porous and permeable. Volcanic lava, if many cracks and fissures are present, can also be quite porous.

GROUNDWATER 153

Groundwater zones (Not drawn to scale.)

Zone of Soil Water

Zone of Aeration (Vadose Zone)

Capillary Fringe
Water Table (Interface)

Zone of Saturation

Impermeable Rock

↓ Gravitational water

A Capillary (upward-moving) water

↑ Evapotranspiration from plants

⇑ Evaporation from surface

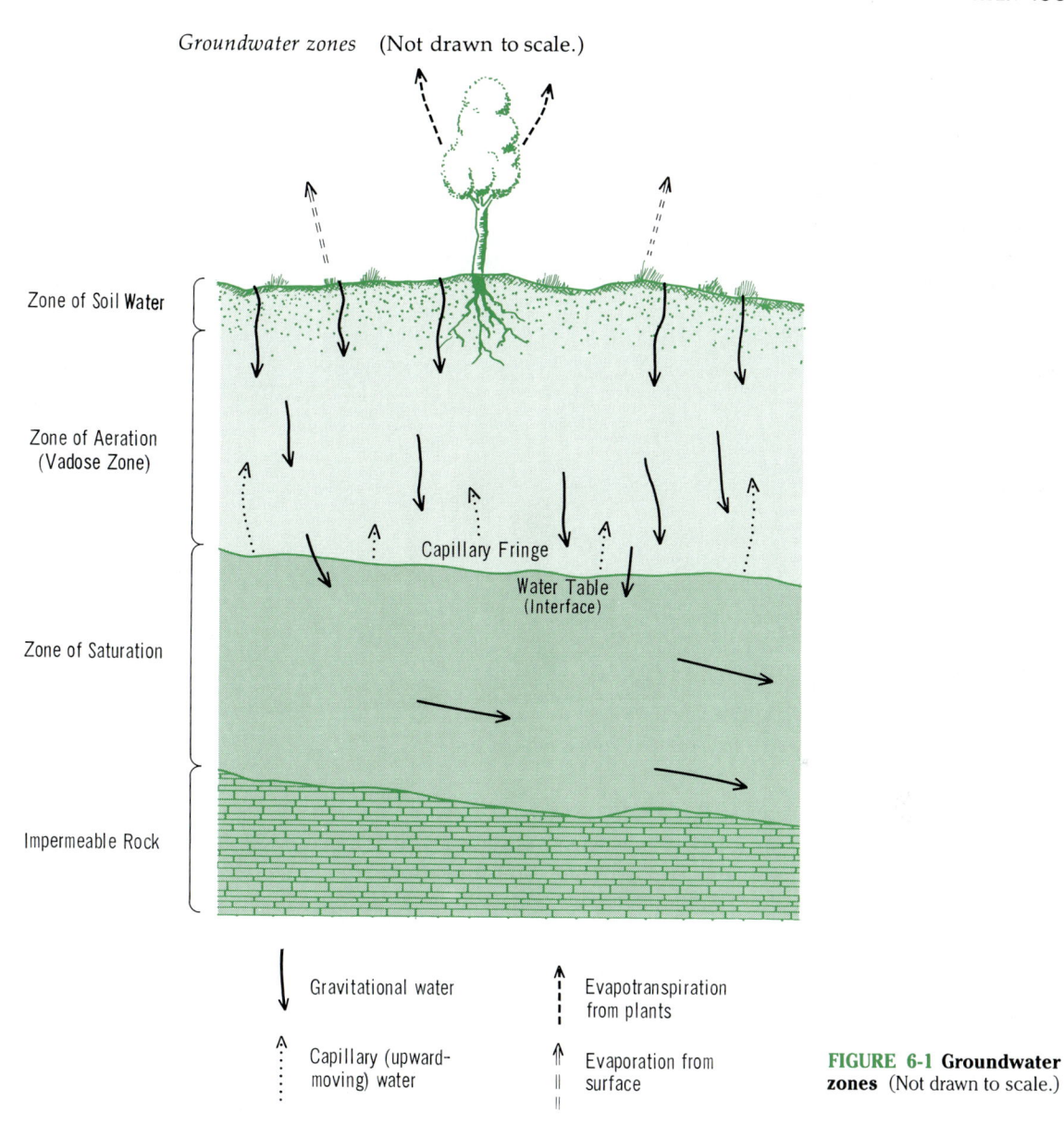

FIGURE 6-1 Groundwater zones (Not drawn to scale.)

ed, the water begins to flow laterally (more or less horizontally) by the force of gravity. Water can also flow vertically upward by **capillary action,** which is caused when surface tension is reduced and the water molecules are drawn up through the tiny spaces within the rock or soil, or when plants reach down with their root systems to tap, or "suck up," the water. Not all of the water molecules will move in either direction; some are held in place by **hygroscopic attraction.** This occurs when the water molecules coat or cover soil particles with a film and therefore cannot move.

The topmost zone, called the **zone of soil**

Mount Waialeale, Kauai, Hawaii: the wettest mountain on earth.

water, supplies most of the short-rooted vegetation with moisture. This is also the area in which most of the plant nutrients can be found. Its moisture content fluctuates constantly, losing moisture through evaporation from the surface and through transpiration from its vegetative cover and gaining moisture during times of precipitation.

As the amount of water increases, the molecules begin to gravitate downward into what might be termed the intermediate zone, but is more properly called the **vadose zone,** or the **zone of aeration.** This zone is aerated; that is, there is room for both water and air in sufficient quantities to support the larger forms of vegetation. Directly beneath this zone is the **water table,** which is the upper boundary of the lowest of the groundwater zones, the **zone of saturation.** The water table can be found close to or near the earth's surface (in humid areas or during seasons of exceptionally high precipitation) or as much as 100 yards (90 meters) deep during dry periods. The zone of saturation is the region where most of the underground water is to be found—about 1 million cubic miles (4,096,000 cubic kilometers) in volume—as Table 6-1 indicates. At the bottom of this zone is, in many places, the earth's bedrock and, in others, dense rocks with only slight permeability. When water penetrates to this depth, it can only fill up the spaces between grains and crystals and eventually move down and away. The water table is sometimes referred to as the **interface,** or **contact surface,** between the region of soil moisture and the saturated layer, and it tends to parallel the ground surface above it.

The slope of the water table from its highest point of elevation, where it receives its initial charge of water, to its lowest point some distance (whether long or short) away, where the water is lost through discharge into a lake, stream, ocean, or plain, is called the **hydraulic gradient** (Fig. 6-2). The amount of water flowing along this gradient is of the greatest importance, but much depends on the height of the source. The higher the elevation, the heavier the flow will be. If the water must travel a long distance, some will be lost en route and the flow will be diminished. The height of the intake point will vary with the seasons, rising during times of maximum precipitation and falling during times of minimum precipitation or drought. The vegetative cover will show telltale symptoms of lushness when moisture is plentiful and drooping and signs of limpness and wilting during the drier periods. As for streams in the area, Fig. 6-2a shows a stream that is supplied primarily by groundwater when the water table is at its highest. Fig. 6-2b shows a streambed that is dry because the water table has dropped too low to replenish it. If there were water running in this stream, it would have to be supplied by surface runoff from some distant place where precipitation still occurs.

Wells

It has been estimated that close to 25 percent of the freshwater supply for our urban areas comes from underground sources. *Wells* are, simply, holes dug in the ground deep enough to go below the water table and into the saturated zone. In dry-land climate regions you may have to dig very deep, and in some humid climate regions you may only have to dig deep enough to form a basin in which the water can accumulate. The water table in China 2600 years ago must have been very deep; Confucius is reported to have said that if you dig a well to a depth of 72 feet (21.6 meters) and stop before reaching the spring, you have, after all, thrown away the well. In the present-day United States the depths of wells are increasing, primarily because of our increasing populations. As more and more water is removed through wells, more wells must be dug, further depleting the water supply. Some places, such as the city of Los Angeles, have had to reach out great distances to find the water needed to replace that taken from the ground. The problem started nearly 70 years ago with the aqueduct that stretched 180 miles (288 kilometers) from Owens Lake, at the foot of the Sierra Nevadas, to Los Angeles. With the increasing demand, water was brought to the Los Angeles area from Hoover Dam by a great canal through the Imperial Valley. Now the Feather River Project has been created to bring in water from the rivers near Sacramento. In addition, Los Angeles hydrologists take advantage of periods of heavy rainfall and divert much of the water from the rivers to catchment basins, where the water is temporarily impounded so that it will infiltrate into the aquifers and replenish, at least for a while, the groundwater reservoirs.

Another kind of a well, the **artesian well** (from the French province of Artois, where the first wells of this kind were dug), does not require pumps to withdraw the water, because the force of gravity and the buildup of hydrostatic pressure behind the confined groundwater causes the water to erupt through the well opening in what sometimes seems to be a spouting fountain. As an example of the importance of artesian wells, consider the vast livestock region of Australia's Great Artesian Basin (Fig. 6-3), which is bordered on the east by the Eastern Highlands and on the west by

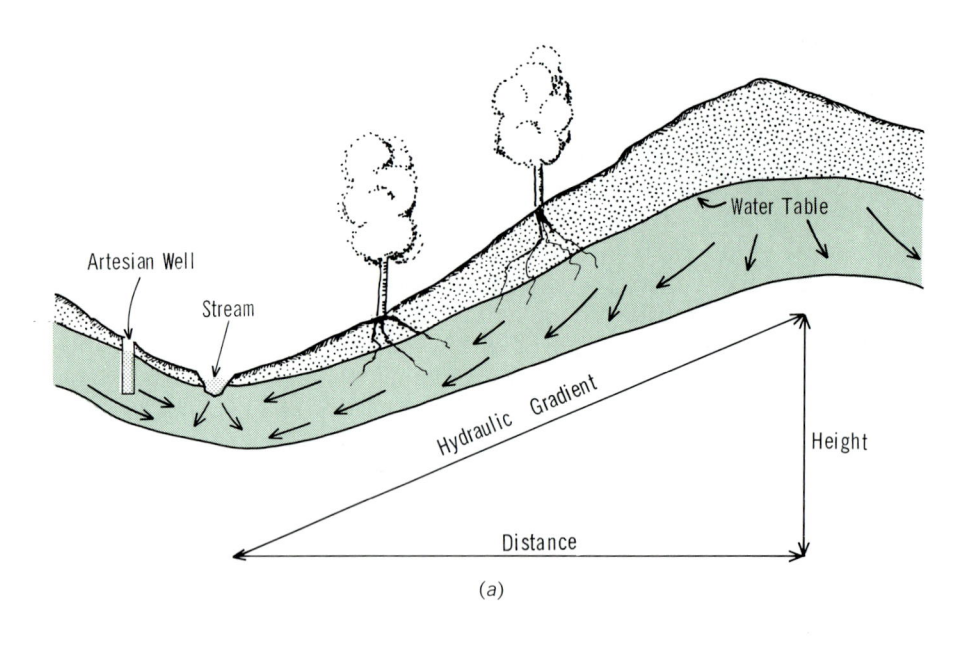

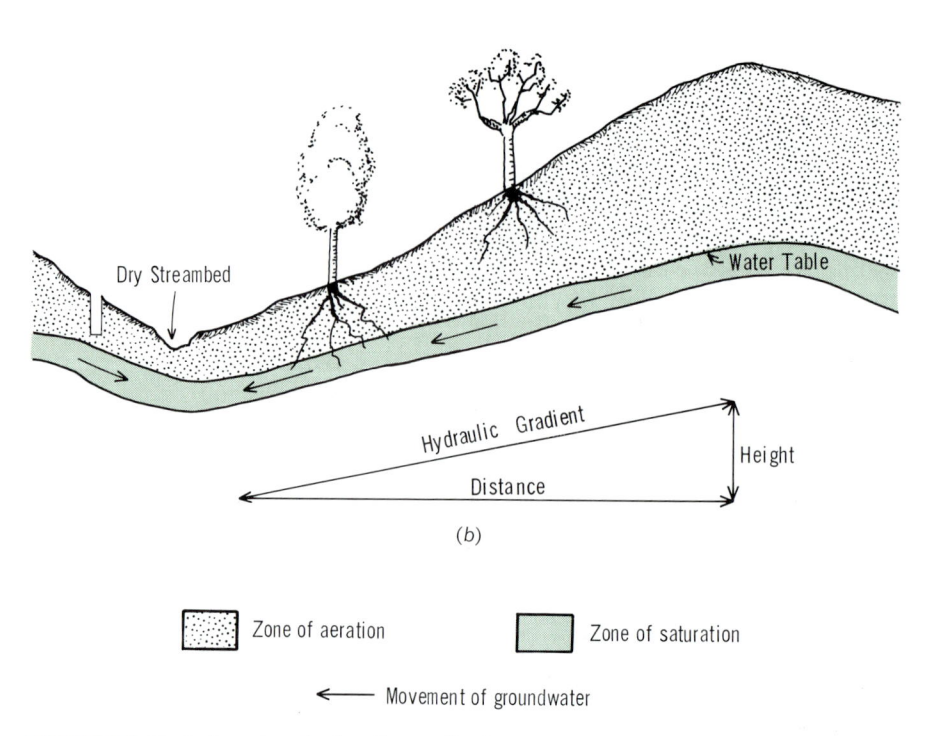

FIGURE 6-2 **Variations in a hydraulic gradient.**

GROUNDWATER 157

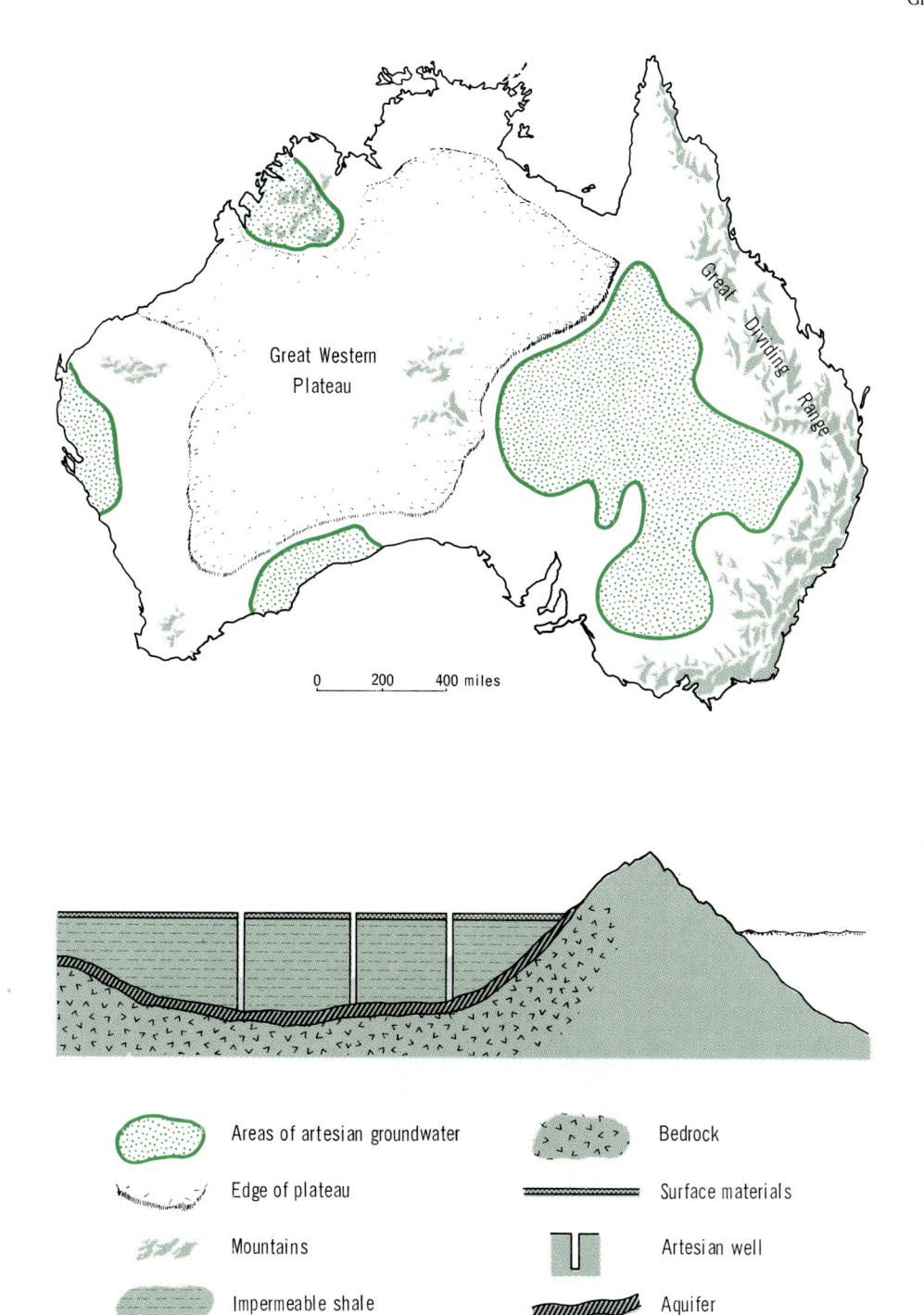

FIGURE 6-3 **Australia's Great Artesian Basin.**

the plateau that contains the Great Sandy Desert. An aquifer composed of sandstone dips downward from the highlands and stretches westward, beneath an overlying cap of impermeable shale, to the borders of the plateau, where it rises. Beneath the aquifer, which is over 1000 miles (1600 kilometers) in width, is impermeable bedrock. Water that falls on both plateau and highlands flows down through the aquifers and collects in the basin. When wells are dug, water escapes upward from its place of confinement.

Springs, Geysers—and Geothermal Energy

When an aquifer bearing flowing groundwater breaches the surface and the water moves down he slope, a spring has come into being. Some of these springs flow continuously, others only intermittently, but they are always a welcome sight to a farmer or rancher or to a backpacker hiking in the wilderness on a hot, dry day.

If, however, the descending groundwater moves through cracks or fissures within the earth's crust and comes into close contact with the superheated rocks at the bottom level of the crust that overlies the extremely hot upper layer of the mantle (see Figs. 6-4 and 10-3), the result is the conversion of relatively cool water to boiling water or steam. As the water is heated, it expands and turns into steam and rises toward the surface. The steam may bubble through standing bodies of water, sending noxious fumes into the air while it stains the surrounding surface rock with a fantasy of

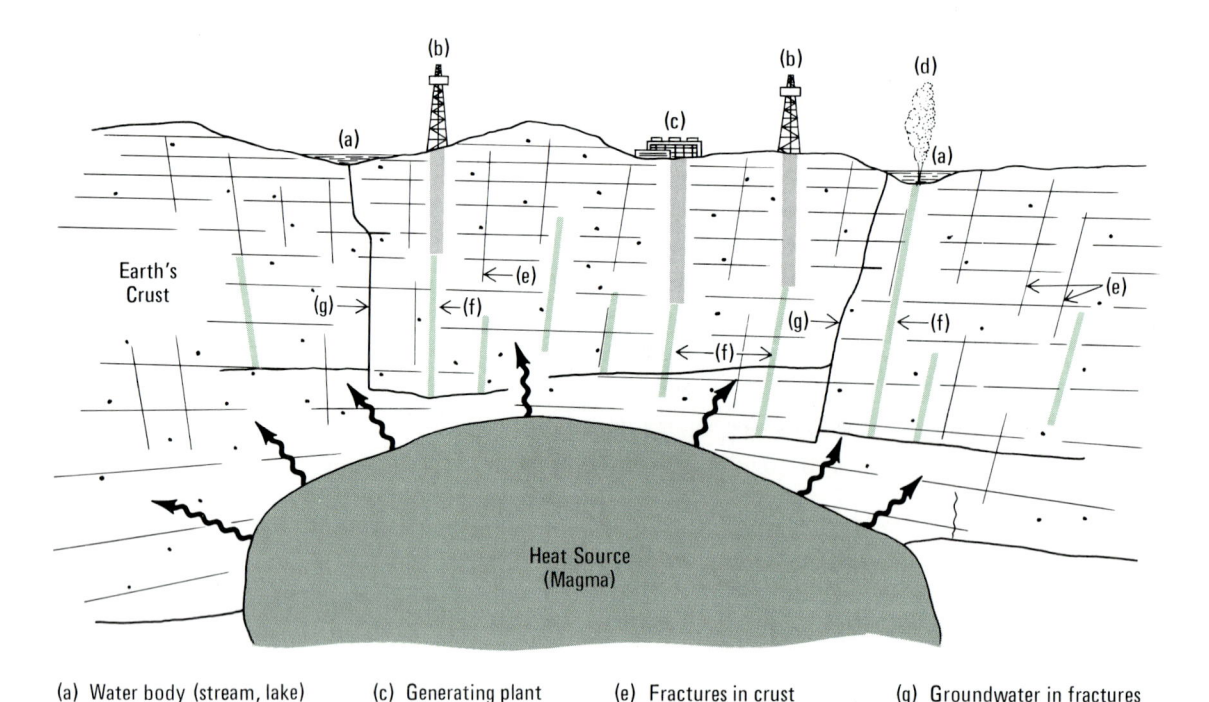

(a) Water body (stream, lake) (c) Generating plant   (e) Fractures in crust   (g) Groundwater in fractures
(b) Drill rig   (d) Geyser   (f) Steam-filled fractures   (fissures)

FIGURE 6-4 Geothermal energy.

GROUNDWATER 159

bright colors. These *fumaroles* can be very interesting, but when the steam erupts violently, sending great spouting fountains of boiling water and steam into the air, the result- known as *geysers*—can be breathtaking. Millions of visitors to Yellowstone National Park can testify to the magnificence of Old Faithful, whose regularity was almost unbelievable.

Hot springs and geysers may appear to have little commercial value, except as health spas or vacation resorts in places such as Hot Springs, Arkansas, but scientists have already found ways to harness the tremendous power that is created as steam expands from the superheated rocks deep within the earth's crust. This new energy resource is called *geothermal energy* (*geo* means "earth," and *thermal* means "heat"), and it has already been harnessed in many countries as an efficient alternative energy source.[4]

Old Faithful in Yellowstone National Park: geyser in action.

CASE STUDY GEOTHERMAL ENERGY

Geothermal energy has long been used by humans who wished to take advantage of the heat that water close to heated rocks produced. Consider the Roman baths and the spas and saunas of Finland, where beds of rocks were heated by fires set underneath them. The steam rising from the heated waters was healthful and relaxing. Since that time, however, technology has greatly advanced; a number of electricity-generating plants have been constructed that do not use up nonrenewable resources (as the use of petroleum, natural gas, or coal does).[5] Although Iceland, New Zealand, and Italy have used geothermal energy efficiently, the world's largest and most productive geothermal site is in northern California, in the area known as The Geysers.

Take a close look at Fig. 6-4, which is a cross section of a portion of the land in The Geysers region. Here you can see a portion of the earth's crust overlying the mantle, with an upward intrusion of extremely hot liquid rock that is called *magma* [1800°F (990°C)]. Throughout this mountainous area many fumaroles exist, but the electricity-generating plants have used drilling rigs to reach deep into the earth in order to capture and use the steam that is expanding as the groundwater encounters the heat that is rising from the magma. As the drill taps the underground reservoir, the pressure is released and the hot water, or steam, rises to the surface—to spin the turbines and produce electricity. It is interesting that the Pacific Gas and Electric Company* is now able to produce enough

---

*Pacific Gas and Electric Company (PG & E) has contracts with firms such as Thermogenics, Inc., Aminoil, USA, and Union Oil Company, which do the actual construction work, to establish addtional generating units.

electricity to satisfy the needs of a city as large as San Francisco. Consider the costs of producing energy by geothermal means as compared to other sources: geothermal—1.7 cents per kilowatt hour; nuclear energy—2.3 cents; coal—2.9 cents; and petroleum—4.5 cents. By this century's end, it is conceivable that other areas of the United States (Fig. 6-5) will be generating geothermal energy.

Finally, geothermal energy produces little in the way of pollution other than "rotten eggs" smell, which is caused by tiny quantities of hydrogen sulfide gas.

Artificial "Aquifers"

For thousands of years, humans have dug wells and constructed lengthy ditches to transport waters from high country to lower-level farmlands, but most people probably think of "aquifers" as the concrete and steel water mains or irrigation canals in use today. Some may recall reading about the aqueducts built by the Romans several thousand years ago, but the most interesting water-transporting conduits are those devised by Persians. These people left the broad and relatively lush steppes of central

Geothermal turbine generators at Pacific Gas and Electric Company's the Geysers Geothermal Power Plant, Sonoma County, California.

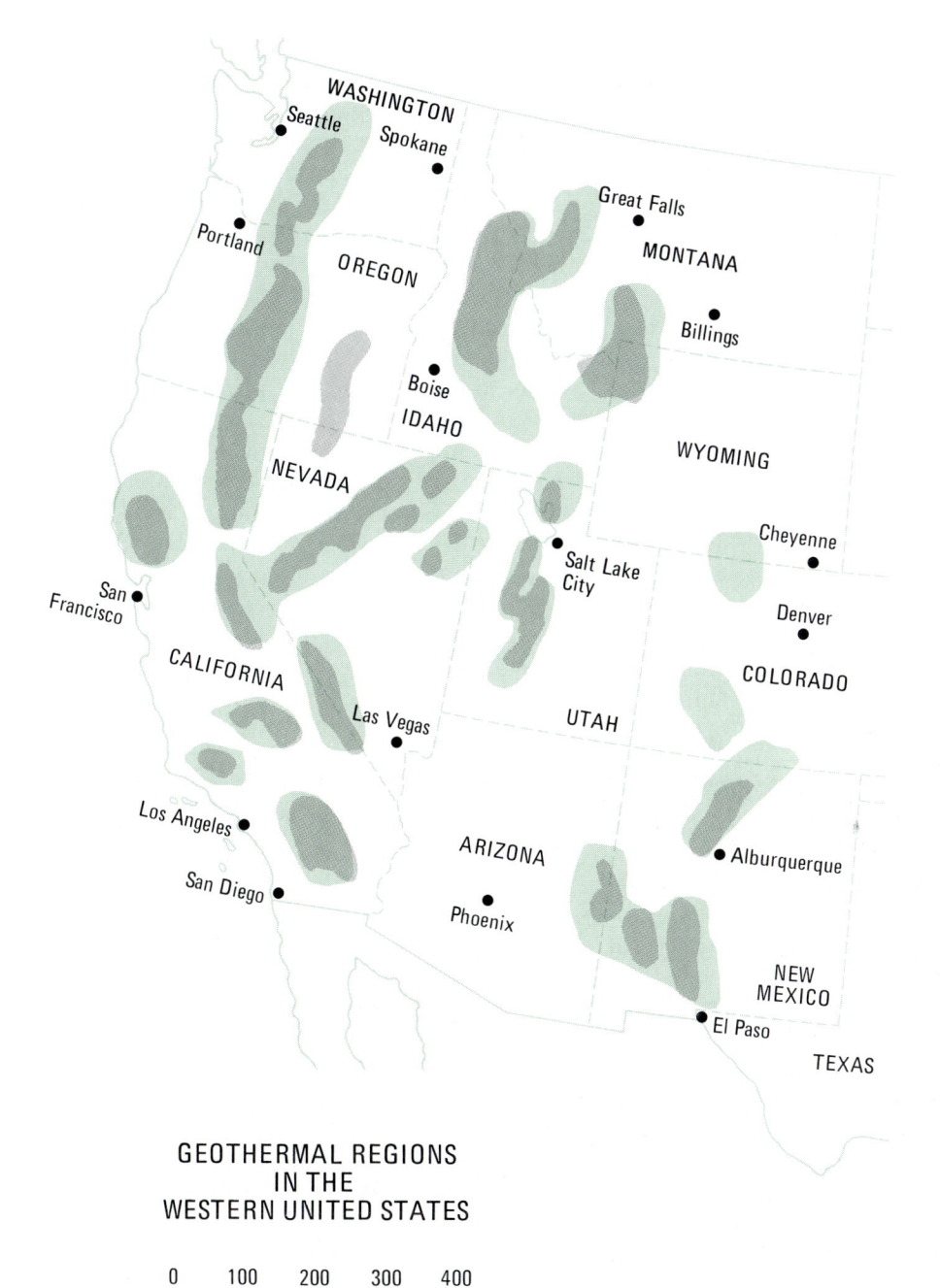

FIGURE 6-5 Geothermal regions in the western United States.

Asia 4500 years ago and migrated to the arid plateau of Persia. They were primarily cattle and horse breeders, but they also practiced primitive agriculture, so they required water for their feed crops and for themselves and their animals. This was a problem in an arid land where evaporation far exceeded precipitation. At first they dug wells, but these too often filled with minerals and salts and were useless after a short time. They tried constructing irrigation ditches from the distant Taurus-Zagros Mountains, but the flowing water either seeped into the parched ground or rapidly evaporated into the dry air. Quite often villages died, and the inhabitants were forced to move on, hoping for new land where water would be plentiful. Then, some of these horsemen of so long ago came up with the answer: the **qanat**.*

Essentially, a qanat (kanat) is a well that is nearly horizontal instead of vertical. In order to get water for a distant village, a team of water-finding experts located a water-bearing aquifer. Then a team of "engineers" and "diggers" dug a well next to the mouth of the aquifer, which was usually located in the upper levels of a broad alluvial fan at the mouth of a valley. This first well was dug to a depth of about 100 feet (30 meters) and was called the mother well. A line was laid out toward the distant village, and the digging began in earnest. Shafts, or wells, were dug at regular intervals, usually about 50 yards (45 meters) apart, to serve as ventilation tubes and to allow the excavated soil and rocks to be removed. At the top of each shaft a mound was formed, giving the land above the qanat a characteristic appearance—like a series of doughnuts laying on the land. Then began the arduous job of connecting the shafts by means of an underground tunnel or conduit (Fig. 6-6). The tunnel, roughly rectangular in shape, averaged 2 to 3 feet (1 meter) in width and up to 6 feet (1.8 meters) in height.

When the qanat was completed, the aquifer was breached, and the water began to flow down through the length of the tunnel to the village. Some of these qanats in use today measure up to 50 miles (80 kilometers) in length and still supply the villages with water. During the rainy season in the western mountains, some qanats may deliver as much as 4200 gallons (16,000 liters) per minute, diminishing to only 400 gallons (1500 liters) per minute during the dry season.[6] One qanat is generally adequate for 200 acres (80 hectares) and is enough to support the average Iranian village.[7] It is estimated that there are between 20,000 and 50,000 qanats still in use today.[8]

STREAMS, RIVERS, AND LAKES

We have seen in Table 6-1 that the earth's groundwater reservoirs contain about 36 times as much water as the surface locations, but those figures do not tell the entire story. It may be obvious that groundwater is vital to us as a supplier of fresh water for plants, animals, and humans, but surface waters are important, too. Consider the relationship of humans to a deep lake that is well stocked with fish and a full-flowing river that provides easy access to the hinterland and to the sea. Consider also the tremendous variety of agricultural, commercial, and industrial activities that depend on surface water for the generation of power. Think of the recreational activities made possible by the construction of giant hydroelectric dams and the resulting artificial lakes that sometimes stretch for miles behind them.

---

*To learn more of this interesting method of water transport, see George B. Cressy (January 1958), "Qanats, Karez, and Foggaras," *Geographical Review*, 48, pp. 27–44; also see Anthony Smith (1953), *Blind White Fish in Persia* (New York: Dutton).

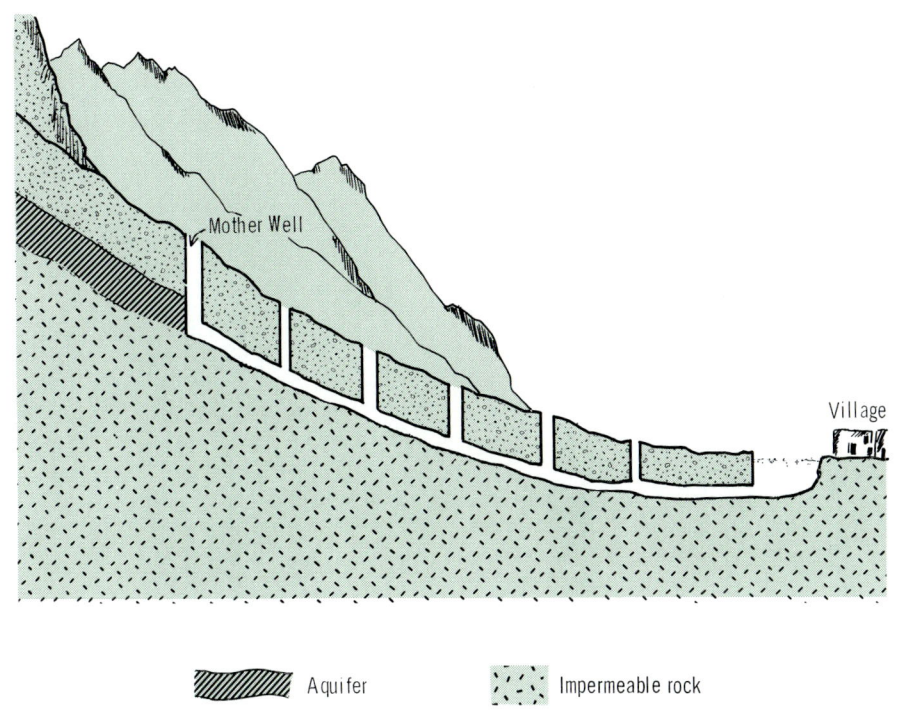

FIGURE 6-6 Cross section of an Iranian qanat.

Another significant consideration is the role played by surface waters in the hydrologic cycle. It has been estimated that the 66 principal rivers of the world discharge about 3720 cubic miles (15,237 cubic kilometers) of water every year. The estimated total from all rivers, large and small, measured and unmeasured, is about 9200 cubic miles (38,000 cubic kilometers) yearly, which is 25 cubic miles (104 cubic kilometers) daily. At any given moment, however, only about 300 cubic miles (1228 cubic kilometers) are physically present in all the world's stream channels, and it seems evident that river channels, on the average, contain only enough water to maintain their flow for about two weeks.[9] It also seems that groundwater comes to the aid of streams, supplying them with water during times of minimum precipitation.

Figure 6-7 shows some of the most important rivers in the world, rivers that made their mark on human history and on the face of the earth. River flow is governed by the amount of precipitation or additions from melting glaciers or snowpacks at the river's source; consequently, a regular flow is never certain, with the possible exception of Egypt's "Holy Mother" Nile. This river, swollen by tropical rains and meltwater from the Ruwenzori Mountains and the Ethiopian highlands, rises by late spring and overflows its banks, flooding both sides for considerable distances. By fall the waters subside, leaving a thick deposit of fertile silt, a "gift of the goddess" to the farmers. So

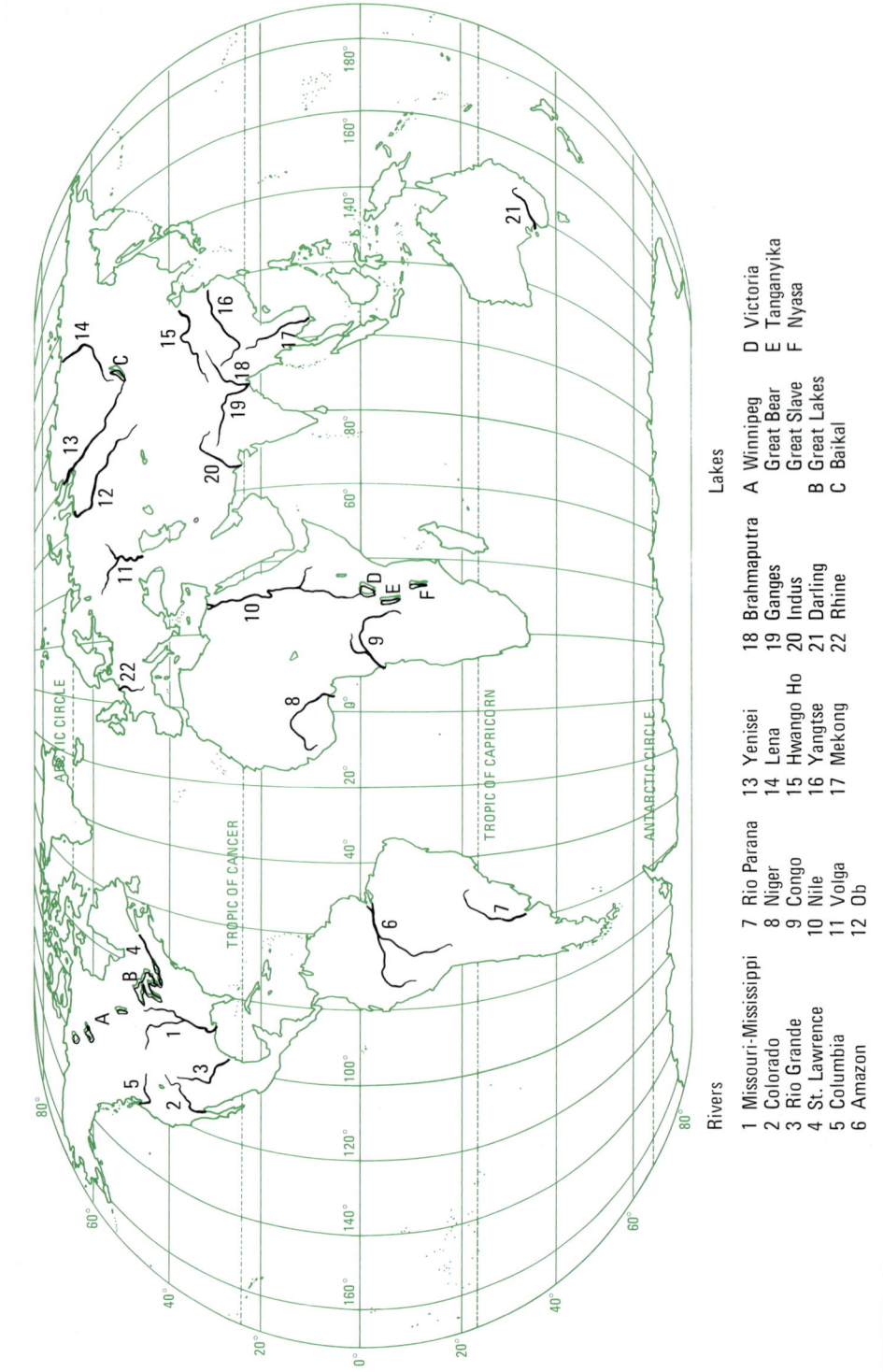

FIGURE 6-7 Major rivers and freshwater lakes.

Rivers

1 Missouri-Mississippi
2 Colorado
3 Rio Grande
4 St. Lawrence
5 Columbia
6 Amazon
7 Rio Parana
8 Niger
9 Congo
10 Nile
11 Volga
12 Ob
13 Yenisei
14 Lena
15 Hwango Ho
16 Yangtse
17 Mekong
18 Brahmaputra
19 Ganges
20 Indus
21 Darling
22 Rhine

Lakes

A Winnipeg
B Great Bear
C Great Slave
D Great Lakes
E Baikal
D Victoria
E Tanganyika
F Nyasa

164

regular was the flooding of the Nile that the ancient Egyptians devised their 365-day year from the Nile's clocklike precision.[10] Since World War II, however, Egypt's expanding population has brought on changes that have turned out to be both beneficial and detrimental. The need to supply the arid southern regions with water for irrigation and to furnish electricity to the spreading settlements caused the construction of several hydroelectric dams (most notably the Aswan High Dam). These dams now force the farmers to import expensive chemical fertilizers to replace the fertile silt once supplied free by the "Holy Nile."

The Nile, however, does not match rivers such as America's Mississippi, which has a drainage area of 1,243,000 square miles (1,988,800 square kilometers) and discharges at an average rate of 620,000 cubic feet (18,600 cubic meters) per second. This amounts to 133 cubic miles (545 cubic kilometers) per year, which is almost 34 percent of the total river discharge from the United States. Compare these figures to those of the mighty Amazon River of Brazil, which discharges almost 1300 cubic miles (5324 cubic kilometers) per year—almost three times that of all the rivers of the United States.[11]

It becomes readily apparent that the tremendous discharge of water in the form of **surface runoff** (movement of water across the land) to the oceans is of considerable importance in maintaining the hydrologic cycle. In addition, the waters from the rivers add to the groundwater reservoir and supply moisture to the atmosphere through evaporation.

Egypt's "Holy Mother" Nile river

Lakes, which number in the hundreds of thousands, are also important in maintaining the hydrologic cycle. Although every lake, large or small, is valuable to someone, the large freshwater lake merits our closest attention. Figure 6-7 brings out an interesting fact. The large freshwater lakes are to be found only in the northern third of North America, in east-central Africa, and in Siberia. This suggests that in the other areas of the world the water is either in rivers or underground. The estimated total volume of all the large freshwater lakes comes to about 30,000 cubic miles (123,000 cubic kilometers), with a combined surface area of 330,000 square miles (858,000 square kilometers). Some of these lakes are impressive in area as well as in volume. Others, such as Siberia's Lake Baikal, may not have the area of the others, but do surpass them in volume (Fig. 6-8). Baikal resulted from a mighty

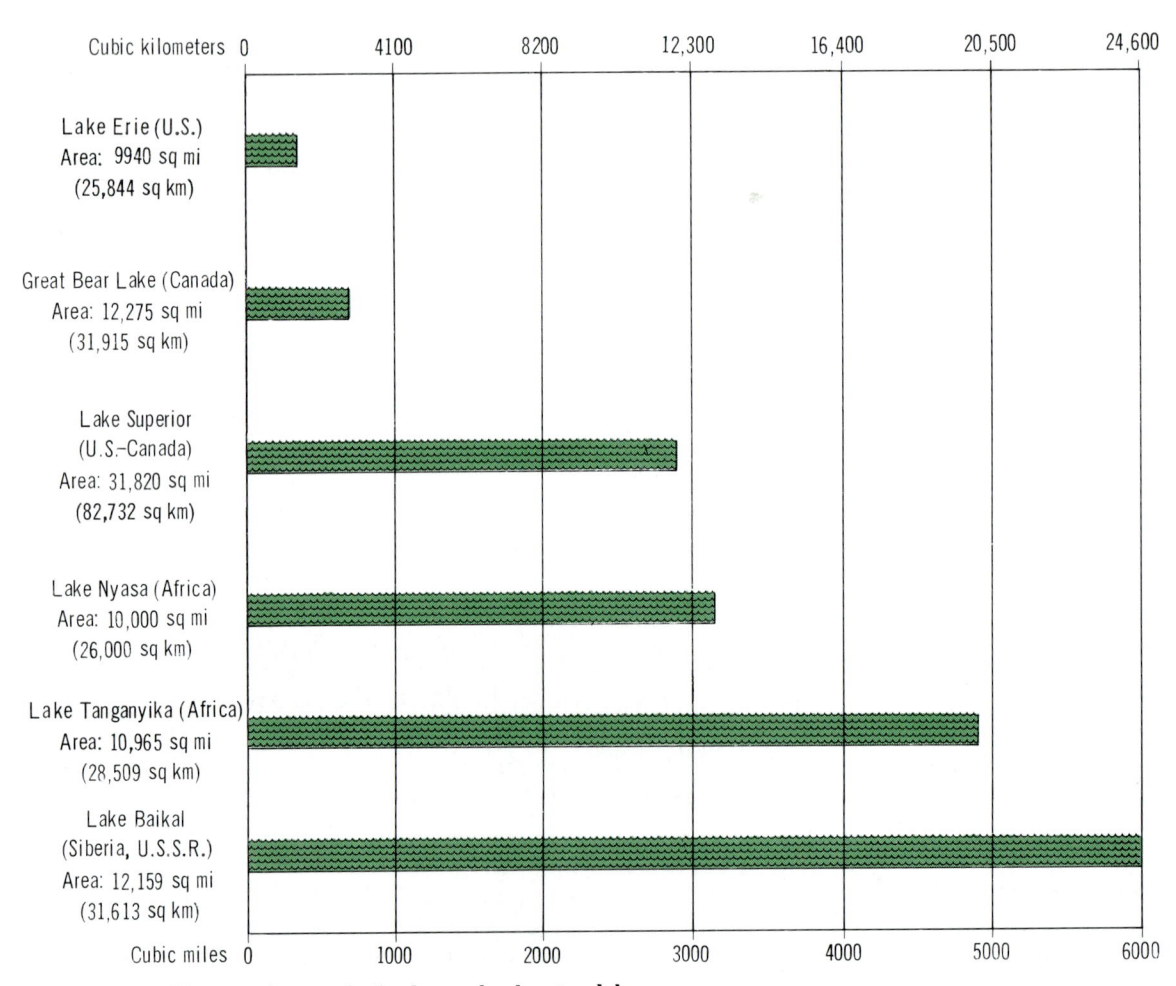

FIGURE 6-8 Water volume of six large freshwater lakes.

fracture of the earth's crust, which opened a deep cleft in the ground (a process we will consider in more detail in Chapter 11). The depth of lake Baikal has been measured to be 4500 feet (1350 meters), making it the deepest lake in the world. However, the five Great Lakes that lie on the border between the United States and Canada, if considered as a body, constitute the largest area of fresh water in the world.

As to the saltwater lakes and their position in the scheme of things, in addition to supplying fish and other products of the sea, their primary service is as waterways. The Caspian Sea is the largest of all, having a surface area of 152,084 square miles (395,418 square kilometers) and a water volume of 19,240 cubic miles (78,800 cubic kilometers); it accounts for over 75 percent of the saltwater lakes. Most of the saltwater (saline) lakes are to be found in Asia. The only one of any significance in North America is Utah's Great Salt Lake, which has shrunk to only 1700 square miles (4420 square kilometers) and has a water volume of 7 cubic miles (29 cubic kilometers).

ICE CAPS AND GLACIERS

Slightly over 2 percent of the earth's moisture is held in glaciers and **ice caps** (Fig. 6-9). This may seem insignificant when contrasted with the 97 percent held by the world's oceans, but when contrasted with the amounts held by streams, lakes, and groundwater (all of which total a mere 0.6 of the world's water supply), it is a significant amount.

Glacial ice comes in many forms, from relatively small glaciers tucked away in remote valleys in high mountains, such as the Alps, the Canadian Rockies, the Andes, the Pamirs, the Tien Shan, and the Himalayas, to huge masses of ice, such as the ice cap that presses down on the island of Greenland, the floating ice sheet of the Arctic, and the vast **continental glacier** that covers Antarctica. The smaller **valley glaciers** (also called **alpine,** or **mountain glaciers**) sculpture the mountainous regions of the world and supply the inhabitants of the farmlands below their terminals with water; they also have a profound influence on local weather conditions.

However, as a major control of weather and climate for immense regions, we must look to the great masses of ice in the polar regions which today have retreated to relatively limited areas. In the past, however, the earth has experienced great ice ages—long periods when ice sheets spread down from the North Pole onto the continents of North America and Eurasia, grinding away at the earth, scouring out huge basins that were later filled with water (giving us the Great Lakes, for example), laying bare the crust's bedrock, sending out powerful rivers that deposited fresh residues of soil and silt over tremendous areas, and lowering the level of the oceans in the process. All of this occurred during times of high temperature and high evaporation that caused the troposphere to become so covered with clouds that incoming solar radiation was turned aside, allowing the temperatures below to drop to very low levels. When precipitation occurred, instead of falling as rain, it fell as snow and ice. The earth has gone through the cycle of ice age and interglacial age again and again; the last of the great ice ages ended 6000 to 10,000 years ago. No one can foretell if and when another will happen.

The world's valley glaciers hold a water volume of about 50,000 cubic miles (205,000 cubic kilometers), which is less than that held by lakes and rivers, far less than that held by the Greenland ice cap—630,000 cubic miles (2,580,000 cubic kilometers)—and a mere fraction of that held by Antarctica—6,320,000 cubic miles (25,800,000 cubic kilometers).

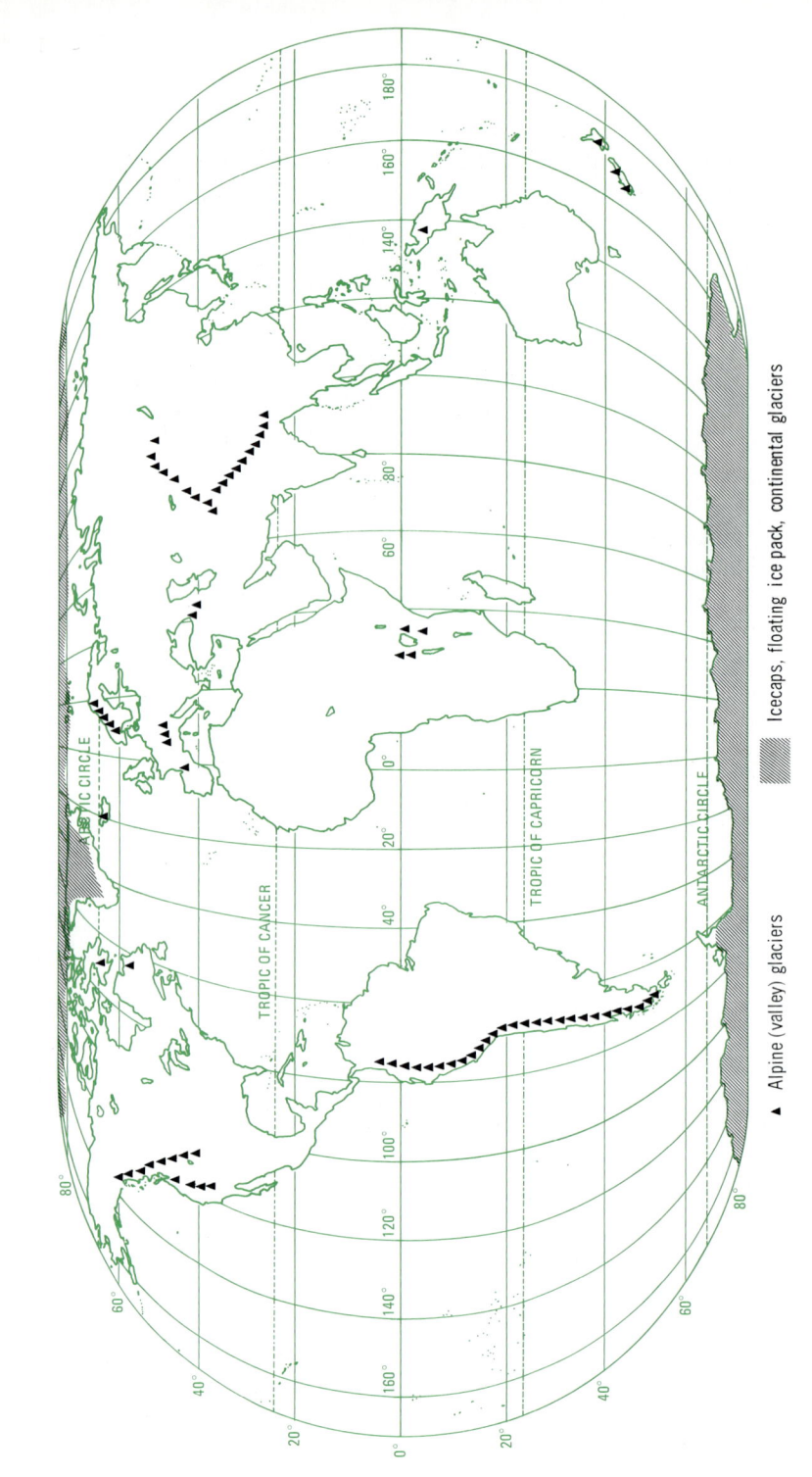

FIGURE 6-9 Earth's ice caps and glaciers.

168

Antarctica: A continental glacier.

What would happen if all the valley glaciers were to melt? If this occurred, the localities below them would experience flooding of their lands and some alterations in their weather and climate. If, however, the Antarctic continental glacier were to melt, the level of the oceans would rise, perhaps as much as 100 feet (30 meters) or more, possibly inundating the lower elevations of cities such as New York, Galveston, London, and Leningrad. Fortunately, such an occurrence is unlikely, at least in our time, but the remote possibility does point out the importance of the immense reservoir of water retained by the world's expanses of ice.

THE OCEANS

One almost has to agree with Leonard Engle, who said that the ancients who gave our planet its name, Earth, were really all wet.[12] Their concern was with the continental landmasses, with little importance placed on the oceans. They thought the earth was all land, with ocean merely meandering about the outer edge. Today, however, we know that the oceans make up almost three-fourths of the earth's surface and account for over 97 percent of the water supply. Obviously, they play a major role in the workings of the hydrologic cycle and exert a tremendous influence on the climates of the world.

Distribution of the Oceans

Examining a globe, you might at first conclude that there is just one big ocean, with a scattering of landmasses here and there. Technically, you would be absolutely correct. There is only one global ocean, but it has been arbitrarily subdivided into four primary oceans. We know

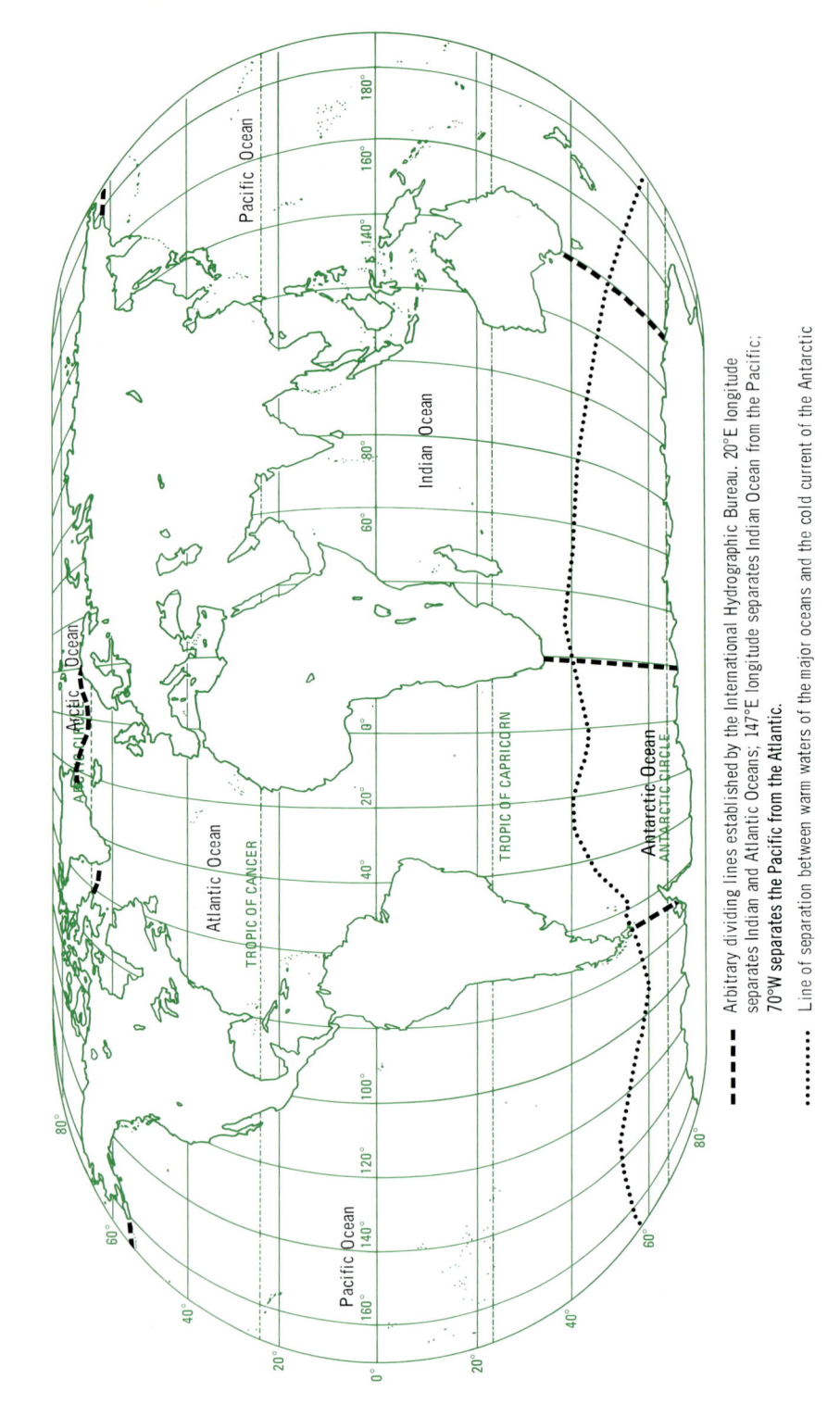

FIGURE 6-10 Oceans of the earth.

these as the Arctic, Pacific, Atlantic, and Indian oceans (Fig. 6-10), but where does each start and stop? The nations of the world, working through the International Hydrographic Bureau (headquarters in Monaco), have devised limits that arbitrarily separate the oceans from each other; these separations, however, are "primarily for the purpose of filing *Notices of Mariners* and have little to do with natural boundaries."[13] Table 6-2 shows the comparative areas covered by the oceans and shows that the Pacific Ocean covers more than twice the area of any other ocean and almost as much as all the other water bodies, including the multitude of seas, bays, and gulfs that surround the continents, combined.

Because each ocean does have its own currents, as well as varying amounts of warm and cold waters, it is possible to go further and subdivide the four oceans into seven bodies of water: North Pacific, South Pacific, North Atlantic, South Atlantic, Indian, Arctic, and Antarctic. As we proceed with our discussion of the oceans, the relevance of such classifications will become apparent—with the possible exception of the Antarctic Ocean.

Most oceanographers and geographers will not accept the body of water surrounding the continent of Antarctica as a separate ocean. Instead, they consider it merely the southern extension of the Pacific, Atlantic, and Indian oceans. However, oceanographic research conducted during the International Geophysical Year (IGY) has suggested that there might be sufficient reasoning for classifying it separately. There is a distinct body of water that is in constant movement about the continent of Antarctica, a powerful cold-water current that circulates from west to east between latitude 40° south and 60° south. This circumpolar current is encouraged (indeed, it is dragged along) by the prevailing westerly winds of the region and by the absence of any obstructing landmasses. When we discuss the circulation of ocean currents, you will see the importance of this whirling mass of cold water and recognize why all the oceans exert such an influence on the climates of the world.* First, however, we must answer the question of *why* the waters of the oceans move from one part to another, much like great rivers. The answer can be

---

*For an interesting report on the oceanographic research conducted in the Antarctic during the IGY (1957–58), see V. G. Kort (September 1962), "The Antarctic Ocean," *Scientific American* (offprint No. 860).

TABLE 6-2 **Comparative Sizes of the World's Oceans.**

OCEAN	AREA IN SQUARE MILES	AREA IN SQUARE KILOMETERS
Pacific	64,000,000	166,000,000
Atlantic	31,700,000	83,000,000
Indian	28,400,000	74,000,000
Arctic	5,400,000	14,000,000
Subtotal (rounded)	129,500,000	337,000,000
Sea, bays, and gulfs not included in above	10,000,000	26,000,000
Totals (rounded)	139,500,000	363,000,000

found by examining the specific causes behind the movement of the waters of the oceans: salinity, temperature, winds, and the Coriolis effect.

The Nature of Ocean Waters

The waters of the world ocean behave in a manner similar to the air masses. They move horizontally and vertically, and they move because of differences in density. **Density** is defined as the weight of a specific volume of water at a specific temperature, expressed in grams per cubic centimeter. For example, water reaches its highest density at a temperature of 39°F (3.9°C). At this point the weight of 1 cubic centimeter of water would equal 1 gram on a balance scale. The density of any ocean water body is determined by its temperature and its salinity.

Salinity In effect, **salinity** means "saltiness." If you have ever gone swimming in an ocean, you know that in comparison to the water of a river or lake, the ocean seemed salty. In actuality, ocean water is almost 96.5 percent "pure." That is, only 3.5 percent of the total is made up of the 57 different kinds of minerals (such as traces of gold, arsenic, nickel, silver) and salts (such as sodium, chlorine, sulfate, magnesium, calcium, and potassium). When sodium and chlorine combine, the result is sodium chloride, more commonly known as table salt. To harvest this important salt, ocean water must be impounded in a collecting basin, and after the water has evaporated, the salt left behind can be gathered and refined. What would happen if all the waters of the oceans were to evaporate? How much salt would we find? The answer shows you just how big the oceans are: The layer of salt would cover the entire earth to a depth of about 160 feet (48 meters)!

Salinity of the oceans varies from place to place, depending on factors such as the amount of precipitation, rates of evaporation, temperatures, and the movements of the waters. Figure 6-11 shows that high salinities can be found in various places along the subtropical high-pressure belt. The Atlantic, especially the region north of the Tropic of Cancer, rates high in salinity, while most of the Pacific, particularly the deeper layers, is low in salts. The lowest salinities are found in the polar regions because of the constant addition of melting snow and ice and a very low evaporation rate. The highest salinities are found in water bodies that essentially are cut off from the general circulation of the oceans, such as the Red Sea, the Persian Gulf, and the Mediterranean. This is because higher salinities will be found wherever the rate of evaporation exceeds precipitation, or wherever fresh or less salty water is prevented from entering the body and thereby diluting it. Generally, in the midlatitude regions lower salinities will be found; although precipitation is down, so is the rate of evaporation. In the tropics, the rate of evaporation is high, yet salinities are low. This is because there is considerable precpitation in these areas. As for the Persian Gulf and the Red Sea, they have high salinities, yet they are situated relatively close to the tropics. Here there is a high evaporation rate, and the narrow inlets to these bodies of water do not allow dilution by waters from the Indian Ocean. Furthermore, those areas are located in the midst of the world's greatest desert region (the Sahara, Arabian, and Iranian deserts), where there is little precipitation and no freshwater rivers.

Contrast that situation with the complex one to be found in the Gulf of Mexico, located at about the same latitude. The Gulf of Mexico averages about 36‰ (3.6‰ salinity) but, near the mouth of the Mississippi River, the

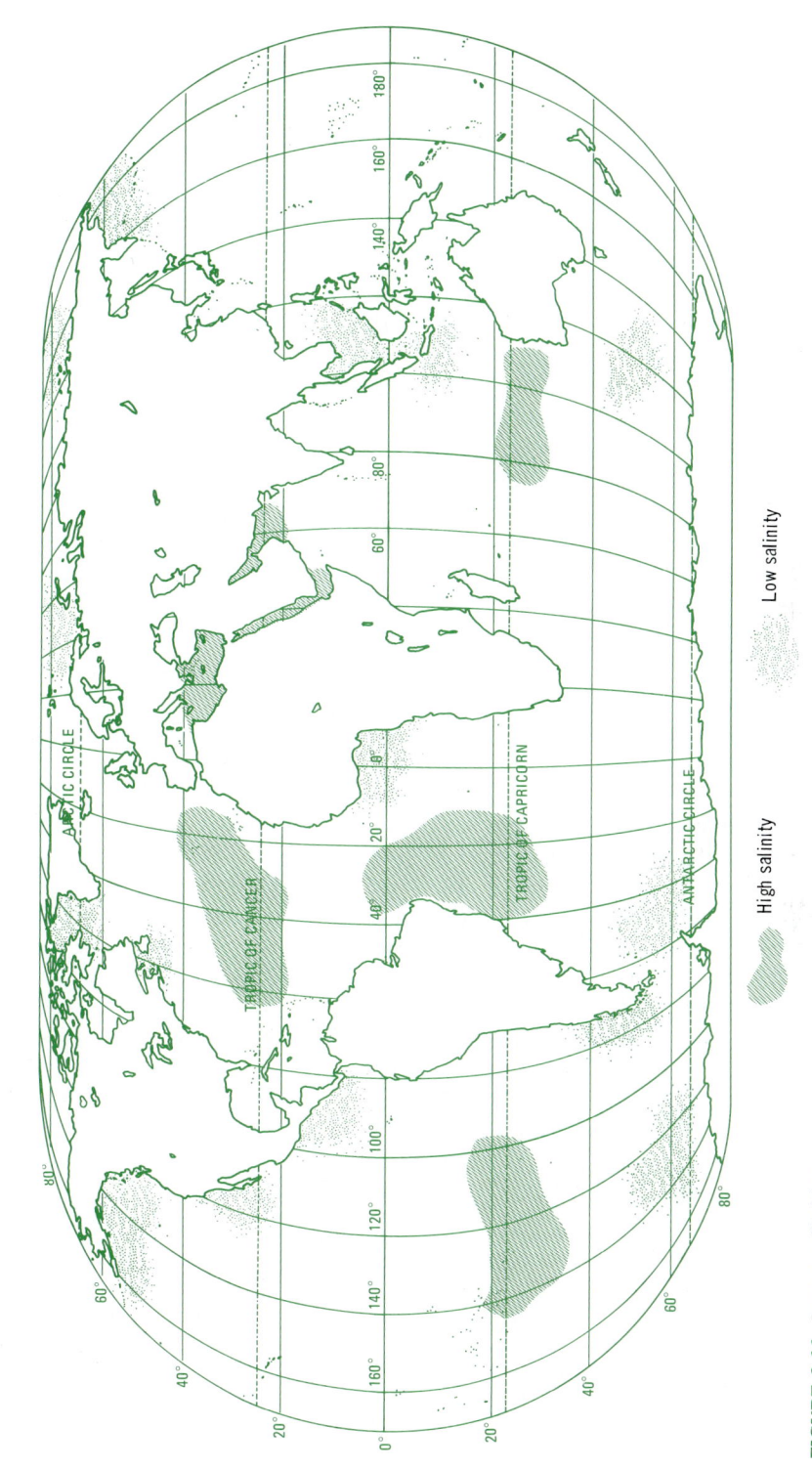

FIGURE 6-11 Ocean salinities.

salinity drops to about 20‰ (2.0‰ salinity).* The reason for this is that the mighty river not only brings salts to the Gulf of Mexico, but also dilutes the saltiness by dispersing the salts throughout the area and adding fresh water to the total supply. Waters that are high in salinity are heavier than less salty waters, and they tend to sink toward the ocean bottom, where they begin to flow out and away from the source. A good example of this can be seen in the interesting, if paradoxical, movements of the waters of the Mediterranean Sea. The waters are high in salinity because of the high rate of evaporation. Thus they tend to sink down to the bottom and flow westward through the only outlet available: the Straits of Gibraltar. At the same time, the less salty, thus less dense, waters of the eastern Atlantic flow eastward into the Mediterranean. The Atlantic waters flow near the surface, passing over the dense waters of the Mediterranean, which flow close to the bottom.

People once believed that the oceans received their salts from the ocean floor; we now know that the salts and minerals come from the land, carried by groundwater and rivers. This may seem like a strange paradox, but the salts arrive within a new supply of relatively fresh water and, although the salts do accumulate through time, they are spread both vertically and horizontally throughout the entire body of water.

Temperature As in the atmosphere, temperature plays an important role in the oceans.

The waters on the surface are heated by insolation and radiation from the atmosphere to a depth of about 6 feet (1.8 meters). Winds and currents cause a mixing of this surface water with waters below, which results in a layer of relatively warm water that measures 600 to 1500 feet (180 to 450 meters) in thickness. This layer is often referred to as the **mixed layer,** a band of constant temperature. Below this mixed layer is a region called the **thermocline,** which means that there is a rapid decrease in temperature, extending downward to a depth of about 3500 to 6000 feet (1050 to 1800 meters). Below the thermocline the water temperature decreases more slowly through the deep waters to the ocean floor.

Where are the warmest waters? Look again at Figs. 4-4 and 4-5, isothermal maps of the world in January and July, and you will see that the receipt of insolation is heaviest on and near the equator and the lower latitudes and lowest in the upper midlatitudes and the polar regions. Compare these maps with Fig. 6-12, a generalization of the world's ocean currents, and you will see that the regions of high insolation send warm water flowing poleward, while cool currents stem from the regions of low insolation and high albedo. Generally, warm currents flow on or near the surface, and cool currents slide along or near the ocean floor. This occurs because the water molecules in the layer of relatively high temperatures separate in their excited movements; thus the water becomes lighter, less dense. In cool or cold waters, the molecules slow down and become tightly compressed, making the waters denser and heavier. Dense water tends to sink and flow outward from the source, and warm water tends to rise in a convective movement similar to that of heated air. Then the winds, aided by the Coriolis effect, take over and steer the currents in the patterns you saw in Fig. 6-12.

---

*Salinity is expressed as a ratio, in terms of the "parts per thousand" of the dissolved salts as related to a specific volume of water. For example, the average salinity of sea water is 35 parts per thousand (35‰), which means that there are 35 grams of dissolved salts for every 1000 grams of water. A high salinity would be 40‰; a low salinity would be 33‰.

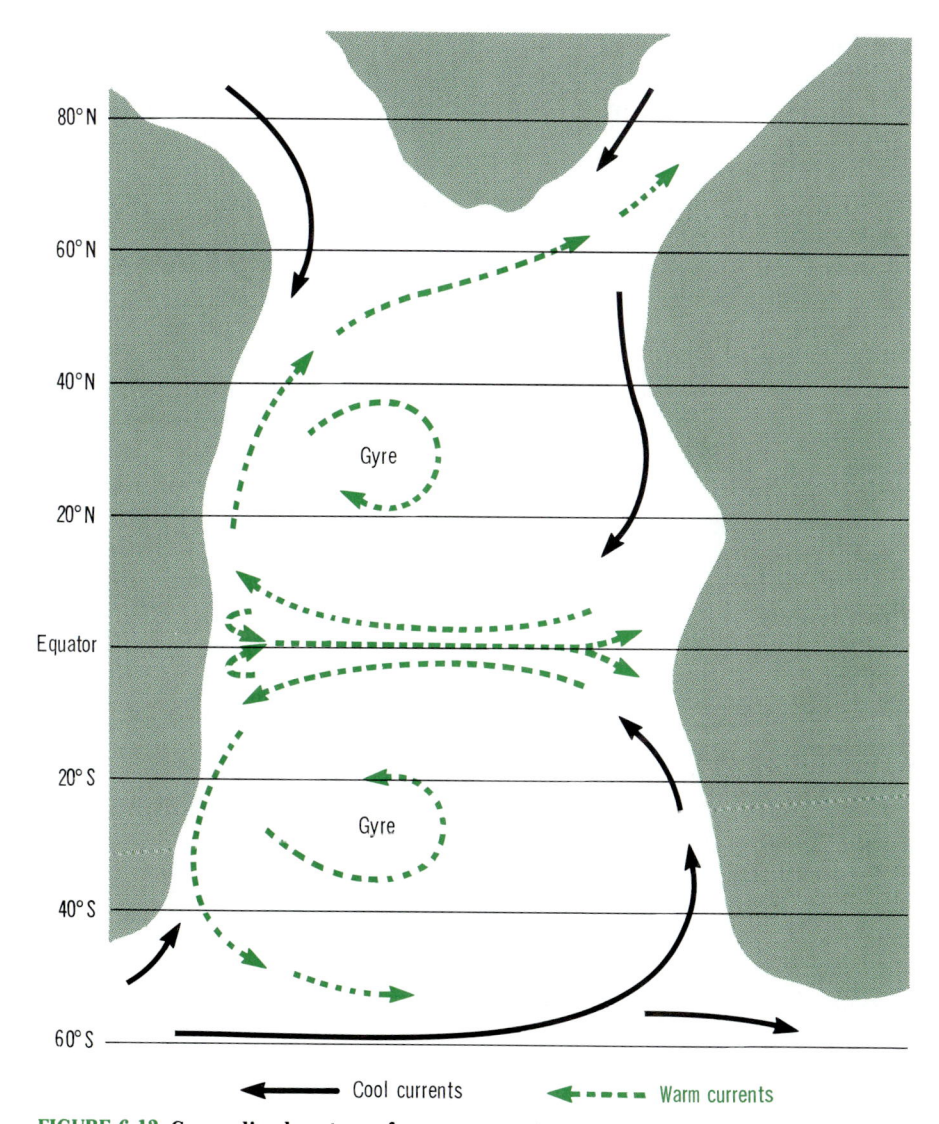

FIGURE 6-12 Generalized system of ocean currents.

Movements of Ocean Waters

Although differences in temperature and salinity have an important influence on the movement of ocean waters, the wind is most responsible for the major ocean currents. We are not referring to local winds, but to the major and prevailing wind systems. The relatively steady blowing of the trade winds, both north and south of the equator, starts the waters moving from east to west, but, like the winds themselves, the ocean currents are deflected by the Coriolis effect, and currents in the Northern Hemisphere tend to veer (curve) to their right (Fig. 6-12), while those in the Southern Hemisphere veer to their left. A circling pattern is set up—aided by other factors, such as a continent that gets in the way—that forces the waters to pile up and to seek avenues of escape. Opposing currents, those that run in the opposite direction of the prevailing system, may be observed flowing primarily in the equatorial regions.

The idea that such equatorial *countercurrents* exist has long been recognized, but only in recent times has oceanographic research provided sufficiently accurate details. For example, consider the Pacific's equatorial countercurrent, called the Cromwell current. This warm current, as a direct response to the high level of water in the region around New Guinea and the Philippines and the unusual weakening of the normally powerful trade winds, flows eastward at speeds up to 5 knots, about 125 miles (200 kilometers) per day. When these extremely warm waters approach South America, near Peru, Ecuador, and Colombia, they curve to the right, creating the phenomenon called *El Niño*,* which overrides the upwelling colder waters of the northerly moving Humboldt current. The result is a dramatic change in the region's climate, with heavy rains devastating the coastline while the increase of the water's temperature over 7°F (4°C) upsets the ecosystem and destroys the abundant sealife. In addition, El Niño has periodically created changes in the climate of the southwestern United States by raising the amount of precipitation to about five times normal.[14]

El Niño means "The Child," in reference to the nativity and the fact that this alteration of the climate normally begins about Christmas and endures until midspring.

The Atlantic also has a countercurrent that flows back out of the Caribbean and along the northeastern coast of South America toward Africa. It is believed that there also is one in the Indian Ocean, but this has not yet been proved.

Currents Oceanographers often compare the circulation of the oceans to that of the earth's wind systems; just as ascending or descending air masses cause lateral movement of the air masses they displace, so do moving bodies, or **currents,** of water. As an example, note that warm currents sweep westward under the equatorial belt of low pressure, urged along by the trade winds, then part, sending one warm current northward and one southward, toward the poles. At the same time, cool waters from the polar regions flow toward the east and then swing toward the equator. Warm waters move along the western sides of the oceans, cool waters along the eastern sides. As an example, consider the circumpolar Antarctic current. This cold-water current has an average surface temperature of around 28°F (-2.2°C) because insolation is low and the albedo is high in the south polar region. The cold waters sweep around the continent, and "rivers" branch off the main stream and slide northward beneath the warmer waters farther north. The cold waters eventually rise, or upwell, near latitude 30° south, along the coasts of three continents. This creates deserts such as the Atacama of Chile and Peru, and Namib of South-West Africa, and the Australian deserts. Cool currents originating in the Arctic also have a hand in the formation of deserts in the Northern Hemisphere.

The waters travel great distances in their circumnavigation of an ocean, and a state of balance is reached, with cold water moving out of the polar region and warm water entering somewhere farther along to replace it. There is constant interaction of warm and cold waters, and these moving masses of water act much like moving weather fronts. Just as low-pressure cells race out of the polar fronts to bring chilly water and storms to the midlatitudes, so the cool ocean currents influence the weather and climate of places like Greenland, Iceland, Alaska, Siberia and, somewhat surprisingly, the Sahara Desert of North Africa. As tropical cyclones, developed from the warm, moist air masses of the equatorial regions, drive north to enter the Caribbean and the southeastern United States, so warm ocean currents flow along the eastern sides of the continents, bringing warmth and moisture to what otherwise might be parched, cold lands.

An interesting sidelight to the pattern of current flow in the Atlantic Ocean is the existence of the **gyres** (sometimes called **gyrals**) that are situated under subtropical high-pressure cells. These gyres result from the inward spiraling of currents that spin off warm, poleward-moving currents (such as the Gulf Stream) and circle around, dragged by the winds of the subtropical high. Not all the proof is in, but the waters of these gyres are believed to "bulge" up to a level 4 feet (1.2 meters) higher than the waters surrounding them.

The North Atlantic gyre is famous in myth and legend as the "graveyard of lost ships" and is a swirling region of thick masses of seaweed. Portuguese sailors named it the Sargasso Sea. An intriguing and mysterious portion of this gyre is a somewhat triangular section of the West Atlantic, which stretches west from Bermuda to the East Coast of the United States, then south to the Virgin Islands. Since 1866, 12 ships and 60 aircraft have disappeared, leaving no traces—no debris, no oil slicks, no survivors, and no explanations.[15] Some have theorized that they may have encountered sudden storms at sea, powerful turbulence in the sky, magnetic interference, attacks by German U-boats (during wartime), or "unidentified flying objects." John Godwin, in *This Baffling World*, reports that there was never any radio communication that implied the presence of danger, and the trouble always "seemed to stem *from direction and location,*"[16] in this area called the Bermuda Triangle.

Major Ocean Currents The major ocean currents are shown in Fig. 6-13, and the pattern we saw earlier in Fig. 6-12 can be seen repeated again and again, from ocean to ocean: a clockwise spiraling in the water bodies of the Northern Hemisphere and a counterclockwise spiraling in the Southern Hemisphere. When we discuss the earth's vegetation, you will see more clearly the tremendous impact of the warm and cool ocean currents. As an example, the eastern portions of Asia and North America are under constant onslaught by low-pressure cells, emerging from the polar front, that could leave the land cold and dry. Fortunately, however, Asia has the Japan current (also known by the Japanese name *kuroshio,* "black current"), which develops in the north equatorial current of the Pacific. This warm current sweeps along the Asian coastline, modifying the climate as it goes, supplying plentiful moisture and warm temperatures to eastern China, Korea, and the islands of Japan. A similar situation exists in the North Atlantic, with the north equatorial current giving birth to the Gulf Stream, which warms the eastern United States and which, as the North Atlantic drift, brings a pleasant climate to the British Isles and Scandinavia.

The portion of the Indian Ocean south of the equator matches the patterns of the other oceans, but the region north is noted for its complexities. The currents in the Arabian Sea and the Bay of Bengal circle in a clockwise motion during the summer, but they reverse their directions in the winter. This rhythm generally

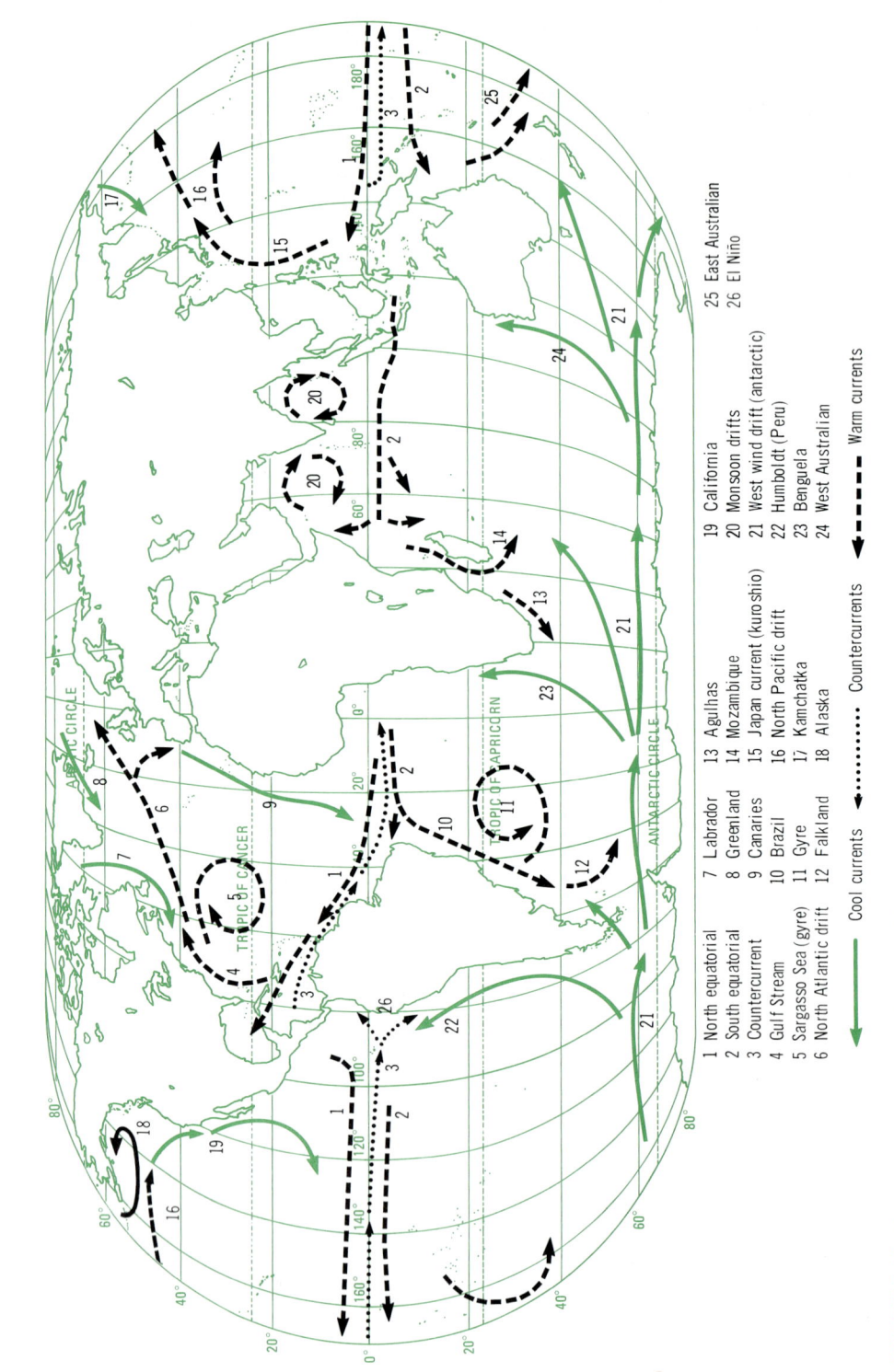

FIGURE 6-13 Major ocean currents of the world.

Wind waves and whitecaps.

follows the prevailing wind system of the region; thus, the currents are called monsoon currents.

Wind Waves The most dramatic and exciting of the ocean movements are the **wind waves** (with the possible exception of the seismic sea waves resulting from earthquakes occurring in the ocean floor, which we will discuss in Chapter 14). Anyone who has been to the seashore knows that the waters of the ocean are in constant, restless motion. The most obvious movements are the waves that come rolling in toward the beach to end up as foaming breakers. These waves are rightly called wind waves because they are caused by the passage of wind over the water's surface. The resulting transmission of wind energy to the water disturbs the once-placid waters, stirs them into dynamic action, and starts them rolling.

There are different kinds of wind waves, depending on a number of factors: depth of the water body, closeness to a shoreline, and wind velocity and duration. If the starting point, or source region, is the open sea, where the waters are very deep, wave movement follows a definite and rather regular pattern (Fig. 6-14). If you were on a ship in the open sea, the waves would appear to be "marching" in a particular direction, wave after wave. In actuality, however, the water is moving up and down in a circular movement. What you see as a moving wave is not the water that is moving, but the wave form and energy. To prove this point, take a 20-foot (6.2 meter) length of rope and tie one end to a post. Now back off until the

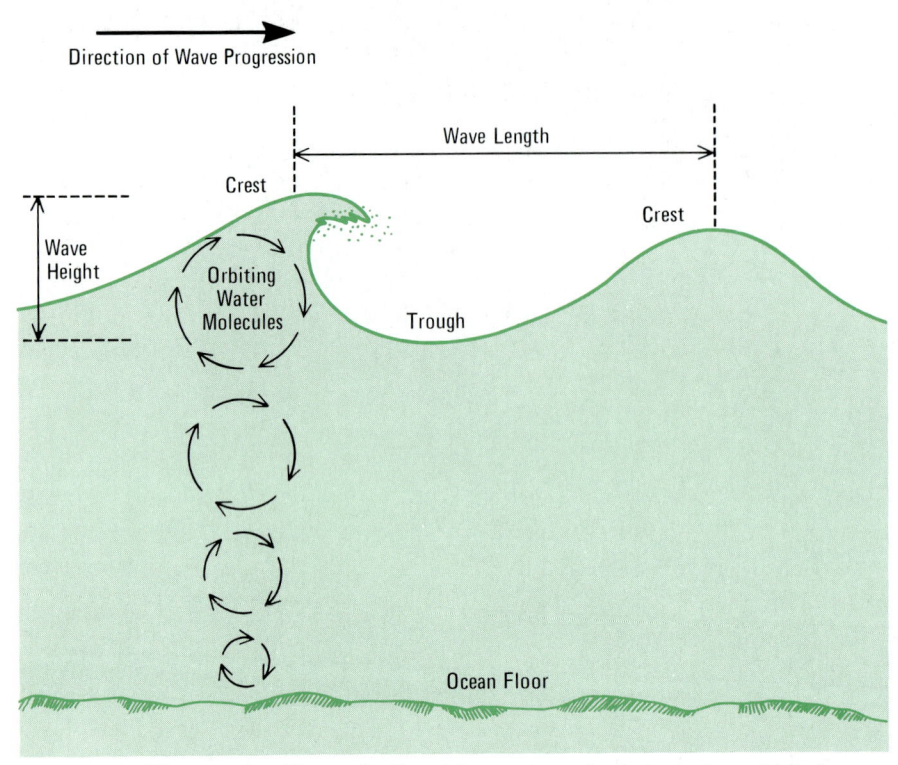

FIGURE 6-14 Wave motion The curving line of the sea shows the rhythmic rise and fall of a wave as it is driven across the ocean by the blowing wind. Water molecules move in a circular path as they orbit about their focal point: up, forward to the crest, down, and back to their starting point. The water mass itself does not move any appreciable distance forward; only the wave energy and form move. An object on the surface (a leaf, a branch, or a patch of oil) is physically transported ahead, pushed by the wind, but the water molecules remain behind, circling about their focal points. The diameter of the orbital path decreases at the ocean floor. Actual forward movement occurs only in shallow waters, when the wave pattern is disrupted and the wave crest surges forward, carrying the mass of water molecules onto the beach.

rope is fairly straight but not too tight. Snap the rope as if you were cracking a whip, and a "wave" will travel all the way through the rope to the post. What you saw was the wave of energy you transmitted to the rope.

Wind waves begin because of a disturbance (storm) at sea. Once the wave process is set in motion, either by the actual pushing by the wind or by friction (sometimes referred to as skin drag), the waves will proceed as long as either of the two forces continues to supply energy to the movement. The orbiting movement of the water particles gives added impetus to the speed of the wave because, as the circling energy rises to the crest and momentarily surges forward before it makes its descent, the wind's push and pull is increased. As a result, you may sometimes see waves that seem to be moving much faster than the wind. *Whitecaps*, tiny breakers at the top of the waves, appear if the wind blows at a faster rate than a moving wave. When the wind waves have moved a great distance from their place of origin and from the winds that gave them birth, *swells,* which may move in a direction different from that of the local winds, result. In calm regions the marching waves lose speed and height and begin to "die." No longer will you see sharply pitching waves studded with whitecaps; instead, you see gently rolling waves that are rounded and smooth.

The height and length of a wave depend primarily on the initial speed and continuing velocity of the wind, how long it lasts, and the amount of ocean expanse to be covered (this distance is called the **fetch**). If the wind is strong and steady and the fetch is long, the waves may build to great size. There are verified reports of waves attaining heights of 90 feet (27 meters) and, on rare occasions, heights of 112 feet (34 meters).[17] During World War II two destroyers were overturned and sunk by high waves in a storm east of the Philippines. Mariners have a rule of thumb that says the wave height will not go above double the wind velocity. If this is true, the winds would have to reach a velocity of at least 50 knots to create waves 100 feet (30 meters) high. Wind waves in the tropics are relatively low and short in length, except during storms that produce winds of gale force.

One of the most interesting of all the earth's regions, with reference to wind waves, is the area between 40° and 50° south latitude, an area aptly termed the "roaring forties." With no landmasses to interfere, the westerly winds attain speeds up to 80 knots! Imagine the sizes of the waves produced, and think of what Magellan, Sir Francis Drake, and Captain Cook, and all the other intrepid mariners of the days of sailing ships experienced when they were forced to "round the Horn" from the Atlantic to the Pacific.

Tides Gigantic waves that overturn ships and drown coastal cities are not the result of tides. Instead, **tides** are the rise and fall of the sea level, as influenced by the moon's gravitational pull against the waters of the earth. The moon may be 240,000 miles (384,000 kilometers) distant, but it does have a vital influence on conditions on earth. The moon orbits around the earth (Fig. 6-15) and exerts a pull that is difficult to perceive in reference to the landmasses, but its presence is felt as far as the oceans are concerned. If the moon did not revolve around the earth, or if the earth did not rotate, we would have no tides at all. Fortunately, both of these do happen, and we have four tidal changes ever 24-hour period: two high tides, two low tides, alternating every 6 hours (approximately) from high to low to high to low. The highest **tidal ranges** (the difference in height between high tide and low tide) occur when the earth, moon, and sun are in line with each other (called *syzygy*). These are called the **spring tides.** The lowest tidal ranges occur when the moon is "out of line" (at right angles to the line between the sun and the earth); thus the pull of the sun tends to cancel out the pull of the moon. These are called **neap tides.**

Solar eclipse—when the earth, moon and sun are in conjunction, (syzygy) and the moon blocks our view of the sun.

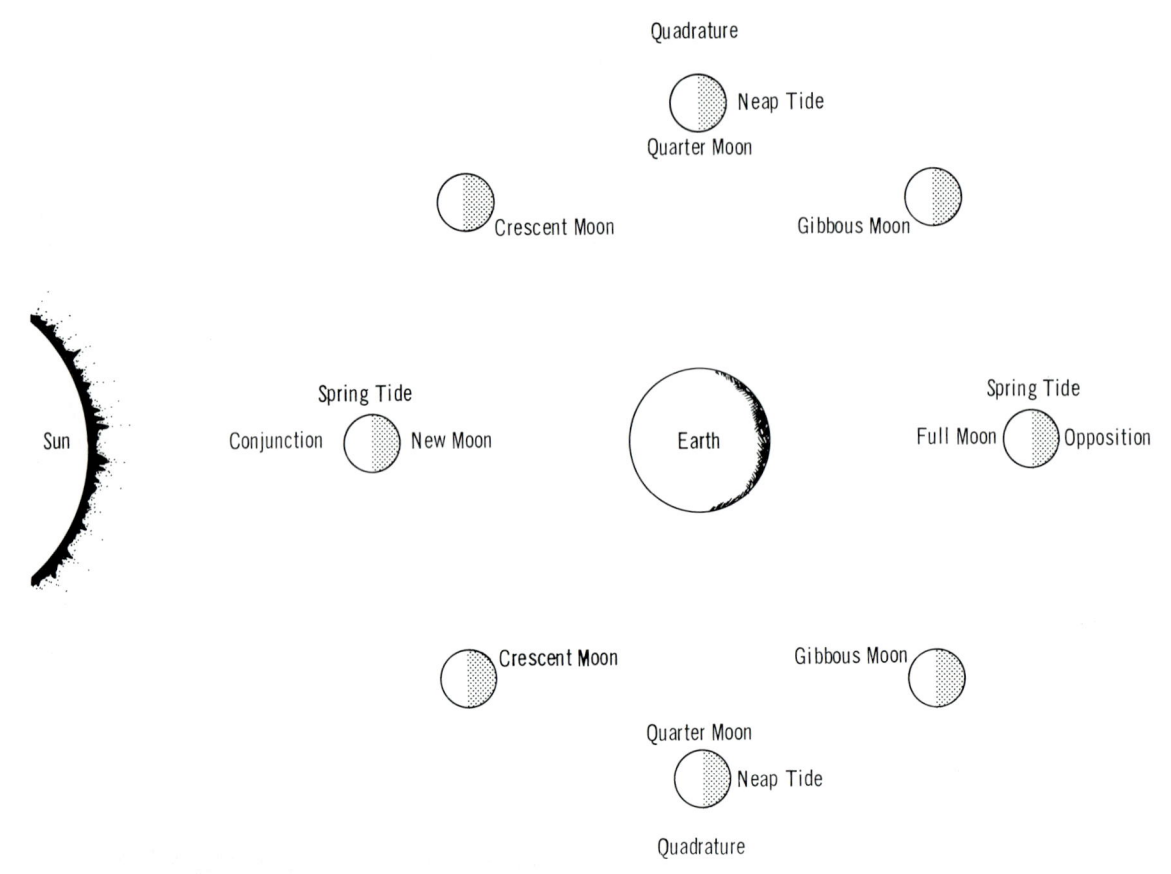

FIGURE 6-15 Positions of the moon and their effect on the tides.

The changing of the tides may not seem to affect our weather or climate, but they do play an important role in many people's lives. When the earth is at perihelion (its closest approach to the sun), the pull of the sun on the waters of the earth is intensified. If this should occur during a spring tide, the resulting tidal range would be dramatically increased. If, also, a storm occurred at sea, sending powerful wind waves in to shore, the resulting surge would be catastrophic for people living along the shoreline. This did happen in January 1983 along California's lower coastline when the extra-high tides combined with storm-tossed waves to cause considerable damage to the houses along the shore. On the beneficial side is the prospect of constructing hydroelectric power plants in narrow river mouths to produce energy by harnessing the tidal flows.

There is much more to be said about the influence of tides, and we will pursue this further in Chapter 14, when we study the role of the oceans in sculpturing earth's landmasses.

EARTH'S WATERS: A SECOND LOOK

The waters of the earth are a global concern, or they should be, because the hydrologic cycle is not confined to any one area. It is a worldwide phenomenon that ties all lands and all peoples together. Consider the long journey of a drop of water that rises from the Pacific Ocean as vapor, precipitates over the Rocky Mountains, flows into the Gulf of Mexico, and is transported by the Gulf Stream to Europe, where it continues the cycle: evaporating and precipitating, carried by groundwater movement, running streams, and blowing winds until, at last, it rejoins the Pacific and the cycle begins anew.

What one nation does to the waters, through carelessness or callousness, affects all of us. The human population has increased to such an extent that the hydrologic cycle has been seriously affected. Think of the growing number of large cities, huge urban areas covered with houses, buildings, streets, and freeways, and consider what this has done to the once-massive groundwater reservoirs, the atmosphere above the cities, and the streams, lakes, and seas nearby.

We have considered the beneficial yet detrimental effect of the Egyptian construction of the Aswan High Dam, but must we merely shrug and sigh at the harmful effects of that dam? Or might we see hope in the discovery (September 9, 1977) of a groundwater reservoir located in the Qatar district, west of the Nile, that is estimated to contain almost 900 million cubic yards (700 million cubic meters) of fresh water? Think of the vast amount of land that could be put into useful cultivation—and the possibility that such a resource might lessen tensions in the Middle East.

Then there is the Soviet Union's scheme to divert the northward flow of western Siberia's great rivers to the south. This plan may surely help the arid regions of Soviet Central Asia, but what will it do to the world's climates? Perhaps it will encourage the refilling of the shrinking Caspian Sea, but what effect will such a proposal have on the Siberian hydrologic cycle?[18]

Consider also the devastating pollution of the once-bountiful Baltic Sea, brought on by pollution poured into that beautiful body of water by the increasing population and industrialization of Scandinavia. If an answer can be found for this problem, then perhaps the Finns and the Swedes will not have to sail the great distances to the North Atlantic to compete with other nations for the declining sea life.

Think also of the drought-stricken region of western Africa, which encompasses the nations of Mauritania, Upper Volta, Mali, Niger, and Chad. If the advanced, industrialized nations of the world could construct desalinization plants along the African coast, would it not free them of the need to supply those nations with millions of tons of food annually, and would it not allow those people the freedom to stand on their own?

It is, perhaps, time for us to act, not just react, and work a little harder to save—and use more wisely—one of our most precious resources, water. Thus far in our discussion we have considered how rivers have been dammed to create hydroelectric power, and we have seen how groundwater has been harnessed to provide us with geothermal energy, but let us now consider the possibility of obtaining useful energy from the waters of the oceans.

CASE STUDY   ENERGY FROM THE OCEANS

For several thousand years, humans have used the flowing waters of rivers to produce power (hydropower). Huge waterwheels sited along fast-moving streams or rivers enabled these early agriculturalists to spin grinding wheels; flour was produced from the ground grain in large quantities and with a minimum of human labor. As time passed, these waterwheels were put into use to power saws, crush ores, and pump out flooded mines in England.[19] As the years passed, waterwheels gave way to giant hydroelectric dams that, if the hydropower potential were fully utilized across the face of the earth, could produce as much energy as the world now consumes by the burning of fossil fuels.

The idea of the waterwheel did not die easily, however; scientists located giant waterwheels in estuaries, where rivers met the oceans. More than eight centuries ago, mills used the flow of the incoming and outgoing tides to spin the paddlewheels, thus creating a very useful mechanical power source. Tidal power was a very important energy resource, but with the advent of hydroelectric power, petroleum, and natural gas, the cost of utilizing tidal movement was too high. Today there are only a few tidal power installations in operation, and the only truly productive tidal power plant that profitably converts tidal movement into electrical energy is located in St. Malo, France, on the northern coast of Britanny. La Rance's water turbines are built into a huge seawall, which gives it the appearance of a giant hydroelectric dam. Plans for the future use of tidal power include Nova Scotia's Bay of Fundy and the Cook Inlet on the coast of southern Alaska. One disadvantage of tidal power is that the plants must be located along the coastline, and the energy produced is usable only for cities located close to the estuary.

Perhaps you have stood on a beach and watched the ocean waves pound inward through the surf zone or stood on the deck of a ship that was fighting its way against a powerful ocean current, such as the Gulf Stream. You may have wondered whether the energy within these forces could be profitably harnessed. If so, you are not alone. During the past century, several scientists have tried to find ways to utilize these energy resources gainfully. Still, the cost of producing electricity by developing installations that could withstand the corrosion of the saltwater is too high for the amont and availability of the energy produced.

The oceans offer us another potential energy source: the ocean temperature gradient (OTG), which produces sea thermal energy (STE). Simply, solar energy heats surface waters, mostly near the tropics, and causes the warmed waters to move toward the poles. Also, the cold waters near the poles move toward the equator. This movement could be harnessed, but, unfortunately, the nations that need this form of energy cannot afford the high costs of putting it into action.[20]

As we consider the alternative energy resources offered to us by the oceans, we have no option but to consider the facts. The power is there for us to use, but we can use it only when we can afford to do so.

Le Rance Tidal Power Station. The tide that rushes up the estuary of the river Rance is one of the highest in the world, 45 feet (13.5 meters).

REVIEW AND DISCUSSION

1. How does the earth's atmosphere compare to surface and underground water as a supplier of fresh water?
2. How important is soil moisture to humanity?
3. What is the difference between porosity and permeability?
4. What is an aquifer?
5. Explain the difference between gravitational water and capillary water.
6. Make a sketch of the groundwater zones and explain the difference between the zone of soil water and the zone of aeration.
7. Explain the term *hydraulic gradient*.
8. What is an artesian well?
9. Explain the use of the Persian qanat.
10. What is meant by the term *surface runoff*?
11. Which is the deepest lake in the world? Which is the largest?
12. Explain the difference between the terms *valley glacier*, *ice cap*, and *continental glacier*.
13. Which of the oceans is the largest? The saltiest? The coldest?
14. What does salinity have to do with the density of ocean water?
15. Explain the term *thermocline*.
16. Explain the term *gyre*.
17. What does *upwelling* mean?
18. What is meant by the term *countercurrent*?
19. How are ocean currents influenced by the Coriolis effect?
20. What makes the *kuroshio* (current) important to the Japanese?
21. What causes ocean waves?
22. Explain the terms *fetch*, *whitecap*, and *swells*.
23. What causes tides? What is meant by *tidal range*? Explain the difference between spring tides and neap tides.
24. Compare the energy potential we might expect from harnessing the power of incoming ocean waves with the changing of the tides to the energy we might obtain from differing ocean temperatures.
25. What is meant by *geothermal energy*?

NOTES

[1] U.S. Geological Survey (1972), *Waters of the World* (leaflet no. 2401–2189) (Washington, D.C.: U.S. Government Printing Office), p. 14.
[2] Richard Aldington and Delano Ames, trans. (1959), *Larousse Encyclopedia of Mythology* (London: Paul Hamlyn), p. 53.
[3] Numbers 20:11.
[4] Garry D. McKenzie and Russell O. Utgard, eds. (1972), U.S. Geological Survey "Geothermal Energy," *Man and His Physical Environment: Readings in Environmental Geology* (Minneapolis, Minn.: Burgess), pp. 232–234.
[5] Donald E. White, "Resources of Geothermal Energy and Their Utilization," in McKenzie and Utgard, pp. 235–239.
[6] George B. Cressy (January 1958), "Qanats, Karez, and Foggaras," *Geographical Review*, 48, p. 39.
[7] Cressy, p. 37.
[8] Cressy, p. 39.
[9] U.S. Geological Survey, p. 8.
[10] Luna B. Leopold and Kenneth S. Davis (1970), *Water* (New York: Time-Life Books), p. 121.
[11] U.S. Geological Survey, p. 6.
[12] Leonard Engel (1962), *The Sea* (New York: Time-Life Books), p. 9.
[13] Harold W. Dubach and R. W. Taber (1968), *Questions About the Oceans*, National Oceanographic Data Center publication G-13 (Washington, D.C.: U.S. Naval Oceanographic Office), p. 10.
[14] Harry Siegal (August 17, 1983), "El Nino: The World Turns Topsy-Turvy," *Los Angeles Times*.
[15] John Godwin (1968), *This Baffling World* (New York: Hart), pp. 235–254.
[16] Godwin.

[17] Willard Bascom (August 1969), "Ocean Waves," *Scientific American* (offprint no. 828), p. 5.
[18] Robert C. Toth (April 20, 1977), "Soviet Goal: Reversing Its River Flows," *Los Angeles Times.*
[19] D. S. Halacy, Jr. (1977), *Earth, Water, Wind, and Sun. Our Energy Alternatives* (New York: Harper & Row), p. 40ff.
[20] William D. Metz (June 22, 1973), "Ocean Temperature Gradients: Solar Power from the Sea," *Science,* 180, pp. 1266–1267.

ADDITIONAL READING

Busalacchi, Antonio, J., Kensuke Takeuchi, and James J. O'Brien, "Interannual Variability of the Equatorial Pacific—Revisited," *Journal of Geophysical Research,* Vol. 88, No. C12, September 20, 1983, pp. 7551–7562.

Chorley, Richard J., ed. (1971). *Introduction to Physical Hydrology.* New York: Methuen, Inc.

Chorley, Richard J., ed. (1971). *Introduction to Fluvial Processes.* New York: Methuen, Inc.

Chorley, Richard J., ed. (1971). *Introduction to Geographical Hydrology.* New York: Methuen, Inc.

Cowan, Robert C. (1971). *Frontiers of the Sea: The Story of Oceanographic Exploration.* Garden City, N.Y.: Doubleday.

Cromie, W. J. (1962). *Exploring the Secrets of the Sea.* Englewood Cliffs, N.J.: Prentice-Hall.

Darwin, Charles (1959). *The Voyage of the Beagle.* Abridged and edited by Millicent E. Selsam. New York: Harper & Brothers.

Erickson, David, and G. Wollin (1967). *Ever Changing Sea.* New York: Alfred Knopf.

Highsmith, R. M. (July 1965). "Irrigated Lands of the World," *Geographical Review,* 55, 382–389.

Horsfield, Brenda, and Peter B. Stone (1972). *The Great Ocean Business.* New York: Coward-McCann & Geoghegan.

Ley, Willy (1962). *The Poles.* New York: Time-Life Books.

Olson, Ralph E. (1970). *A Geography of Water.* Dubuque, Iowa: Wm. C. Brown.

Sayre, A. N. (November 1950). "Groundwater," *Scientific American* (offprint no. 3818).

Smith, Anthony (1953). *Blind White Fish in Persia.* New York: Dutton.

Thomas, H. E. (1951). *The Conservation of Ground Water.* New York: McGraw-Hill.

Todd, D. K. (1980). *Ground Water Hydrology,* 2nd ed. New York: Wiley.

Tolman, C. (1951). *Ground Water.* New York: McGraw-Hill.

Weyl, Peter K. (1970). *Oceanography: An Introduction to the Marine Environment.* New York: Wiley.

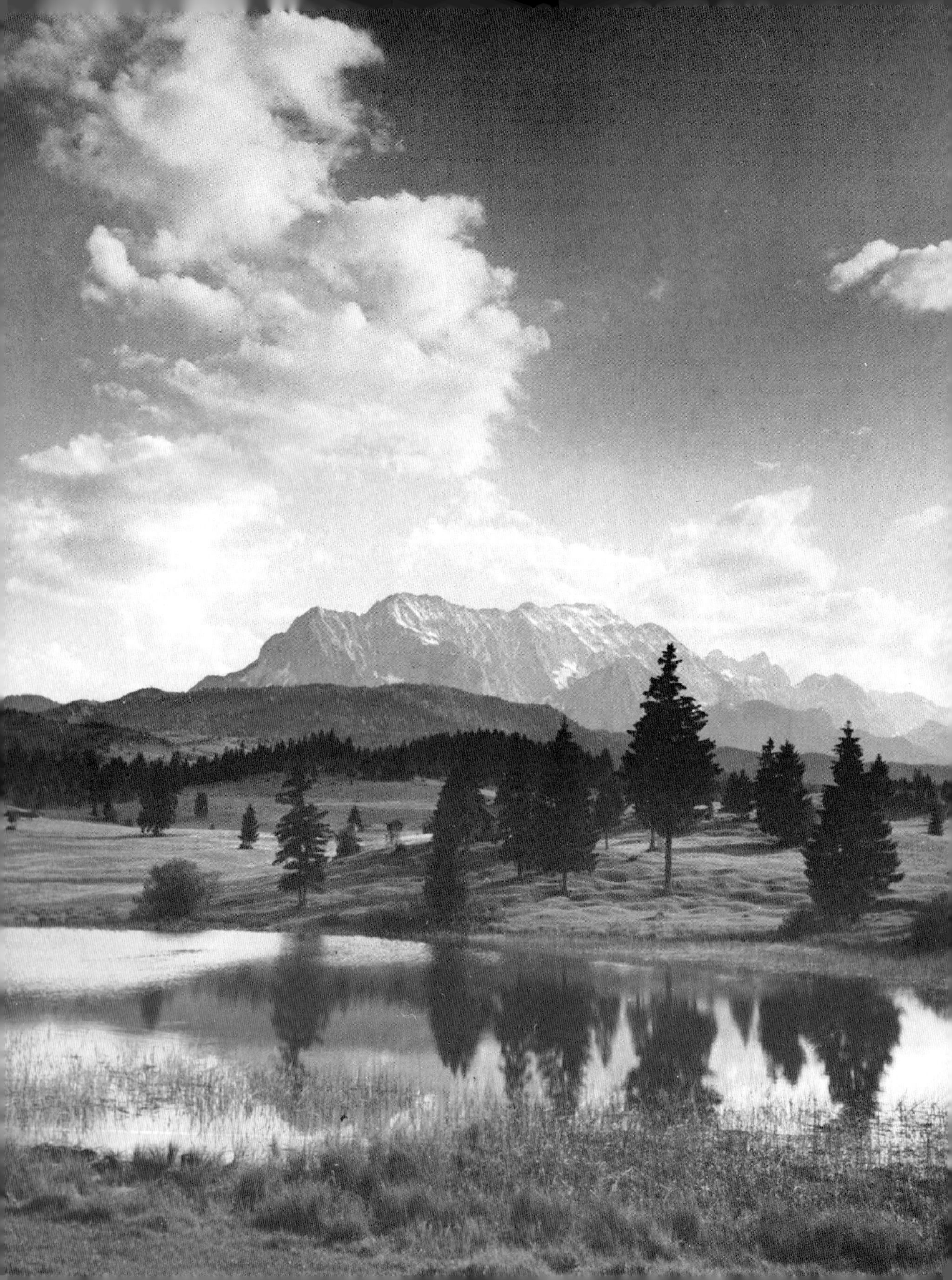

SEVEN
EARTH'S CLIMATES

Aristotle, the noted Greek philosopher of ancient times, composed a simple classification system for the earth's climates. He came up with three zones: torrid (hot), temperate (mild), and frigid (very cold). Unfortunately, he ignored some important governing influences, such as a place's latitude and altitude, its location relative to oceans or other bodies of water, the dominant vegetation of the region, its position on a landmass, wind systems, and the presence or absence of storms. The student of climate comes soon to the realization that climatic regions are like people: No two can ever be truly identical.

For more than 2000 years since Aristotle's attempt to develop a workable classification of climates, scientists of many disciplines have tried to go beyond the attempt to understand why human cultures vary so greatly across the face of the earth. As knowledge of the earth increased, so did agricultural production and the human population. A British economist, Thomas Robert Malthus, foresaw disaster in that the population was increasing faster than the rate of agricultural production. His dire predictions, first published in 1798, fortunately were not fulfilled, although anyone who has studied the increasing population problems of our time realizes there is merit in his theories. When one considers the spread of people over the earth and the desecration of the soils, vegetation, waters, and air, it becomes apparent that we must do more than merely complain. We must learn more about the intimate relationship between humanity, the climates, and vegetation and then actively attempt to find solutions to the problems. As we now begin our investigation into the earth's tremendous variety of climates, we will follow in the footsteps of Wladimir Köppen, the botanist who first endeavored to explain the variations of the earth's climates by studying the connection between vegetation and climate. Even though there have been numerous attempts to upgrade his system, it still stands as a masterpiece. You will find it interesting to continue your research into the newer classification systems and compare Köppen's system to them.

KÖPPEN'S CLASSIFICATION OF CLIMATES

As an experienced botanist of the nineteenth century, Köppen was alert to the gradual changes in the vegetative cover from one place to another. As a young man he studied vegetation maps devised by other botanists and climatologists who had scientifically examined the changes in vegetation at differing elevations, latitudinal positions, and locations on the European landmass. Although he realized the close relationship of a region's vegetation to its climate, he found himself faced with the same problem every mapmaker faces: How does one pinpoint the exact boundary line between regions of different values? He finally concluded that the line marking the dividing place between two climatic regions was, in reality, a zone of flexibility. He believed that this zone was determined by factors such as prevailing seasonal air temperatures, average nighttime and daytime temperatures, amount and variability of precipitation, rates of evaporation, and the area's location in relation to large water bodies, winds, ocean currents (cool or warm), and mountainous regions.*

Although Köppen's classification, published in 1918, was greeted with some skepticism in North America, it soon became the standard system. His climatic map (Fig. 7-1) has been modified several times since then by R. Geiger (A German climatologist) and Glenn T. Trewartha (a geographer-climatologist). By comparing this climatic map with Fig. 7-2 the relationship of Köppen's system to the earth's vegetation becomes apparent.

Köppen's classification is based on an alphabetical code (Fig. 7-3). There are five major climate groups.

A   Moist and hot
B   Dry
C   Moist and warm
D   Snowy and cold
E   Polar; icy

---

*To gain a greater understanding of the problems that confronted Köppen, see C. W. Thornthwaite (April 1943), "Problems in the Classification of Climates," *The Geographical Review*, 33 (2), pp. 233–255. Also available as Bobbs-Merrill reprint no. G-227.

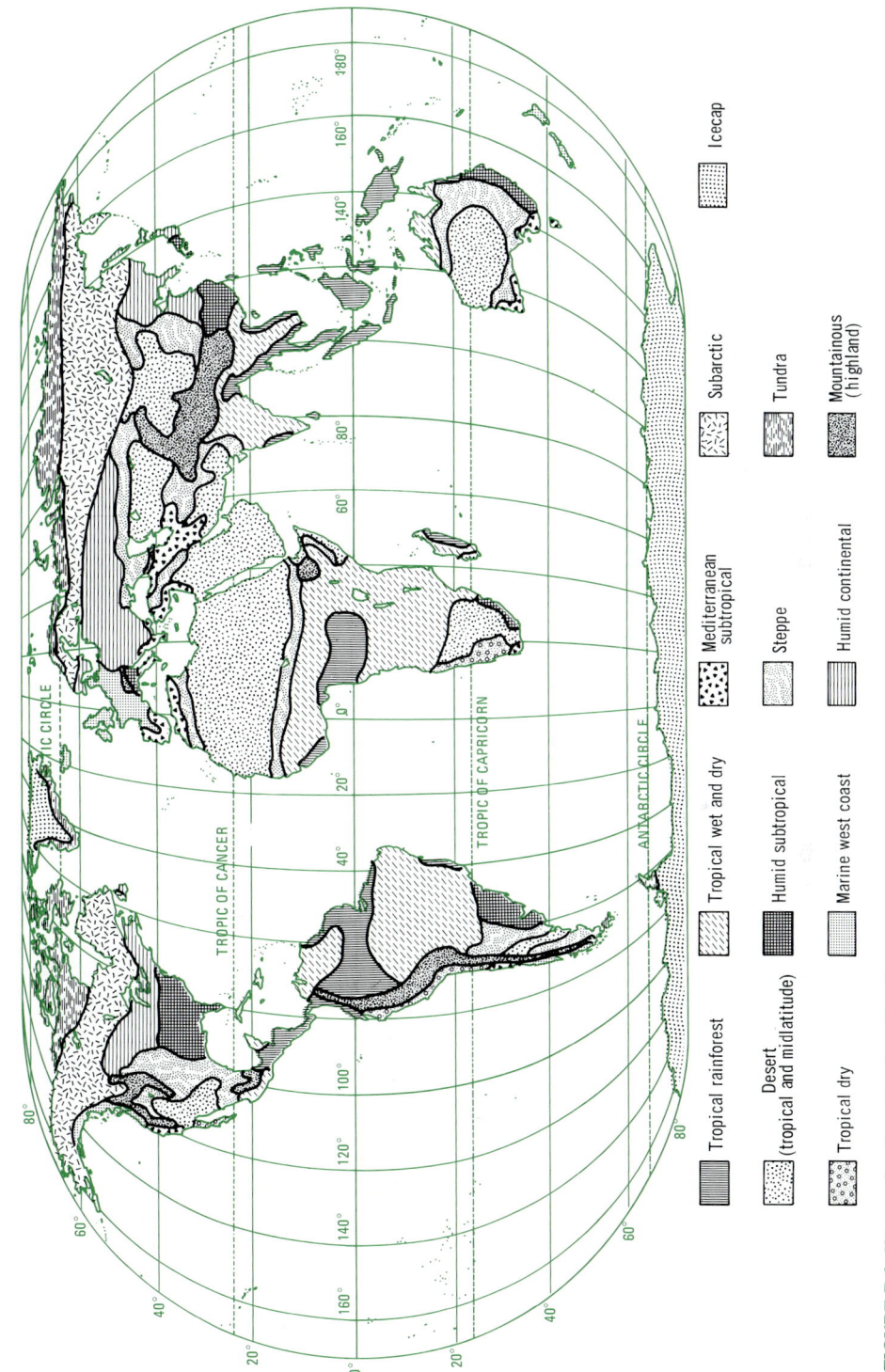

FIGURE 7-1 Köppen's climatic classification.

191

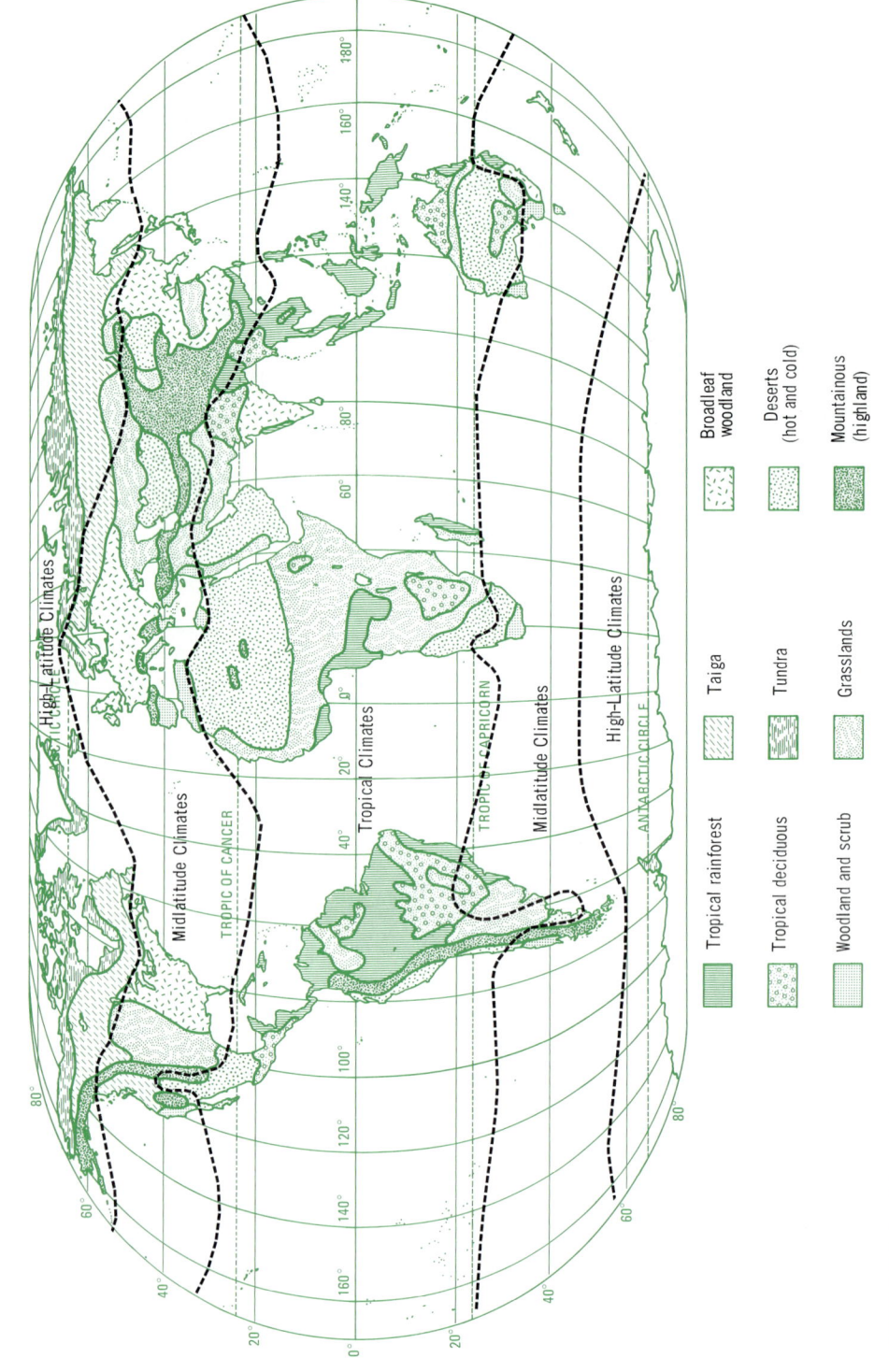

FIGURE 7-2 Earth's vegetation and climate regions.

192

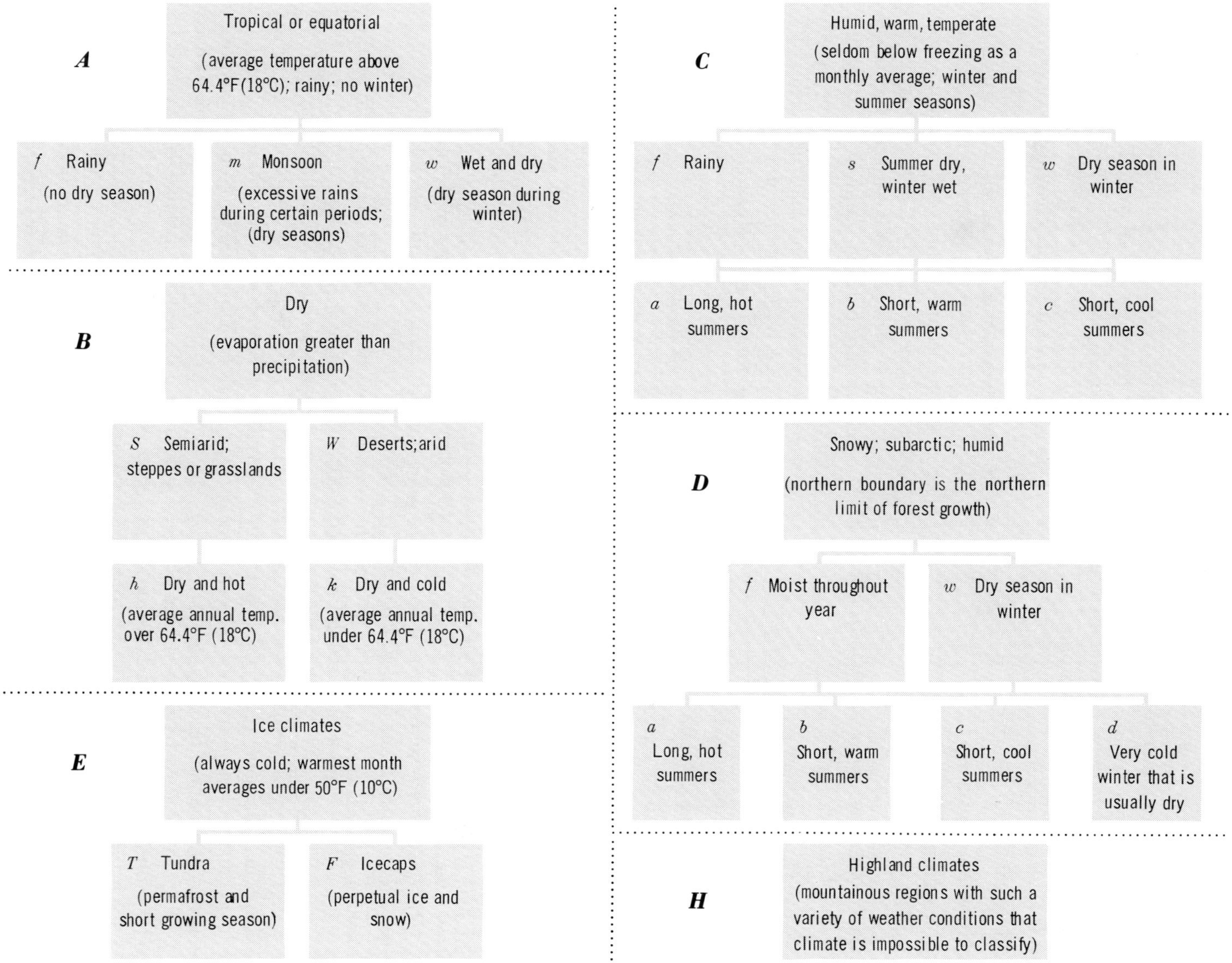

A Tropical or equatorial (average temperature above 64.4°F (18°C); rainy; no winter)
- *f* Rainy (no dry season)
- *m* Monsoon (excessive rains during certain periods; (dry seasons)
- *w* Wet and dry season (dry season during winter)

B Dry (evaporation greater than precipitation)
- *S* Semiarid; steppes or grasslands
- *W* Deserts; arid
  - *h* Dry and hot (average annual temp. over 64.4°F (18°C))
  - *k* Dry and cold (average annual temp. under 64.4°F (18°C))

C Humid, warm, temperate (seldom below freezing as a monthly average; winter and summer seasons)
- *f* Rainy
- *s* Summer dry, winter wet
- *w* Dry season in winter
- *a* Long, hot summers
- *b* Short, warm summers
- *c* Short, cool summers

D Snowy; subarctic; humid (northern boundary is the northern limit of forest growth)
- *f* Moist throughout year
- *w* Dry season in winter
- *a* Long, hot summers
- *b* Short, warm summers
- *c* Short, cool summers
- *d* Very cold winter that is usually dry

E Ice climates (always cold; warmest month averages under 50°F (10°C))
- *T* Tundra (permafrost and short growing season)
- *F* Icecaps (perpetual ice and snow)

H Highland climates (mountainous regions with such a variety of weather conditions that climate is impossible to classify)

FIGURE 7-3 Köppen's climate-classification system.

In order to differentiate between subdivisions within these major groups, secondary letters were added, which are to be combined to furnish a more complete explanation of the *A*, *C*, and *D* climate regions.

f Moist all year
m Monsoon conditions
s Dry summer, wet winter
w Dry winter, wet summer

A third letter was added to include information about temperatures.

a Long, hot summer
b Short, warm summer
c Short and moderate or cool summer
d Very cold and dry winter
h Hot and dry most of the year
k Cold and dry most of the year

The dry climates were assigned the letters *W* and *S* to differentiate between desert (*BW*) and steppe (*BS*) climate and vegetation conditions. For further explanation, the letters *k* (cold) and *h* (hot) are combined with the *BW* or *BS* climates.

The polar and subpolar (subarctic) regions begin with the letter *E*, to which are added either *T* (for tundra vegetation and conditions) or *F* (for regions perpetually covered with ice and snow).

Geiger added the letter *H* to account for highland (mountainous) regions that were difficult to include in Köppen's system, and Trewartha added a tertiary code, *n*, for climatic regions plagued by fog. At first glance, this system may seem confusing, but in actual practice it works out well. Table 7-1 provides further clarification. In this table the climate regions are listed along with the proper alphabetical codes, the typical latitudinal location, dominant air masses, temperatures, annual precipitation, vegetation, soils, and geographic regions representative of the particular climate. To help you further to understand the earth's climates, consult Fig. 7-4, which is a generalized picture of the climatic regions on a hypothetical landmass that extends into both the Northern and Southern hemispheres. Here you can see the importance of continental position, distance from cool or warm water bodies and ocean currents, and latitude.

Because our concern is with the future of the relationship between humans and the environment, we now examine some of the major climate regions to learn what the earth's climates have offered people and how people have responded to this offering.

TROPICAL CLIMATES

Anthropologists tell us that humans first appeared somewhere within the tropical climate region 2 million years ago or more, so perhaps this is a good place for us to begin our investigation into the influence of climate on humans. Figure 7-5 shows the three major tropical climates, and it is apparent why they are called tropical or equatorial climates: They envelope the equator and extend north and south about 30° to 35°. Because of their latitudinal position and their relationship to the heavy receipt of solar energy throughout the year, temperatures are consistently high.

We do not find any one of the three tropical climates dominating over these regions because of the differences in continental position, the presence of cold or warm ocean currents, and the wind systems (compare with Figs. 4-8 and 4-9). Humans may have originated somewhere within this equatorial region, but they apparently had little desire to remain there. If you consult the map of world population distribution (Fig. 1-1), you will note that while there are some tropical regions with rela-

TABLE 7-1 Climatic Regions of the World.

CLIMATE REGION	CLIMATIC SYMBOL	LATITUDINAL POSITION	AIR MASS	TEMPERATURE	PRECIPITATION ANNUALLY	VEGETATION	SOILS	TYPICAL REGIONS
Tropical rain forest	Af Am	10°N–10°S to 20°N	mT mT	Day: 90–95°F (32–35°C) Night: 70°F (21°C)	65–100 in. (1651–2540 mm)	Selva: broad-leaf, evergreen trees	Latosols	Amazon, Meso-America, Guinea coast, Congo, Ceylon, South-East Asia
Tropical dry	BWh	15°–35°N,S	cT	Above 32°F (0°C)	Less than 10 in. (254 mm)	Xerophytic shrubs	Sierozems	Sahara, Arabian, Thar, Australian, Sonoran deserts; Kalihari, Atacama
Tropical wet and dry	Aw	5°–25°N,S	mT	Winter: 70°F(21°C) Summer: 80–90°F (27–32°C)	Summer rainy	Tropical savanna	Latosols	Sudan, veldt, llanos
Mediterranean subtropical	Csa	30°–45°N,S West coasts	mP (summer) mT (winter)	Summers: 70°F (21°C) Winters: 50°F (10°C)	15–20 in. (381–508 mm)	Chaparral, maquis: sclerophyll	Reddish chestnut; reddish brown	Mediterranean region; southern California; central Chile; tip of South Africa
Humid subtropical	Cfa	20°–35°N,S East coasts	mT	Warm months average above 71.6°F (22°C)	40 in. (1016 mm)	Broad leaf, mixed broad leaf	Podzols	Southeast United States; Southeast China; southern Japan; eastern Australia
Marine west coast	Cfb	40°–60°N,S West coasts	mP	Warmest months average below 71.6°F (22°C)	23 in. (584 mm) low 200 in. (5080 mm) maximum	Coniferous forest, broad leaf	Podzols	Northwest United States; Europe; southern Chile

(continued)

195

TABLE 7-1 (continued)

CLIMATE REGION	CLIMATIC SYMBOL	LATITUDINAL POSITION	AIR MASS	TEMPERATURE	PRECIPITATION ANNUALLY	VEGETATION	SOILS	TYPICAL REGIONS
Midlatitude dry (steppe, desert)	BSk, BWk	35–50°N,S	cT (summer) cP (winter)	Less than 64.4°F (18°C)	10–20 in. (254–508 mm)	Short-grass prairie	Brown; reddish	Patagonia, Gobi, Takla, Makan, western Great Plains
Humid continental	Dfa	35–60° north only	cP (winter) mT (summer)	32°F (winter) (0°C) 72°F (summer) (22°C)	30–50 in. (762–1270 mm)	Mixed hardwoods	Gray-brown podzols	Northeast United States and southern Canada, upper Midwest, Soviet Union, eastern China
Continental subarctic (boreal)	Dfc	50–70°N	mP (summer) cP (winter)	Less than 4 months with average above 50°F (10°C)	10–25 in. (254–635 mm)	Taiga: needle-leaf evergreen	Permafrost	Alaska, northern Canada, northern USSR and Siberia
Tundra	E, ET	60–75°N	mP	32–50°F (0–10°C)	Less than 10 in. (254 mm)	Sedges, lichens, shallowroot grasses	Lithosols	Northern portion North America, Asia
Polar ice cap	EF	Ice caps at both poles	Arctic; Antarctic	No above-freezing month average	Less than 10 in. (254 mm)	None	None	Greenland, northernmost tip of Antarctica
Highland	H	At all latitudes	Too variable to classify	Large diurnal range	Orographic	Too variable to classify	Lithosols on ridges and high valleys; alluvium in low valleys	Mountainous areas of world

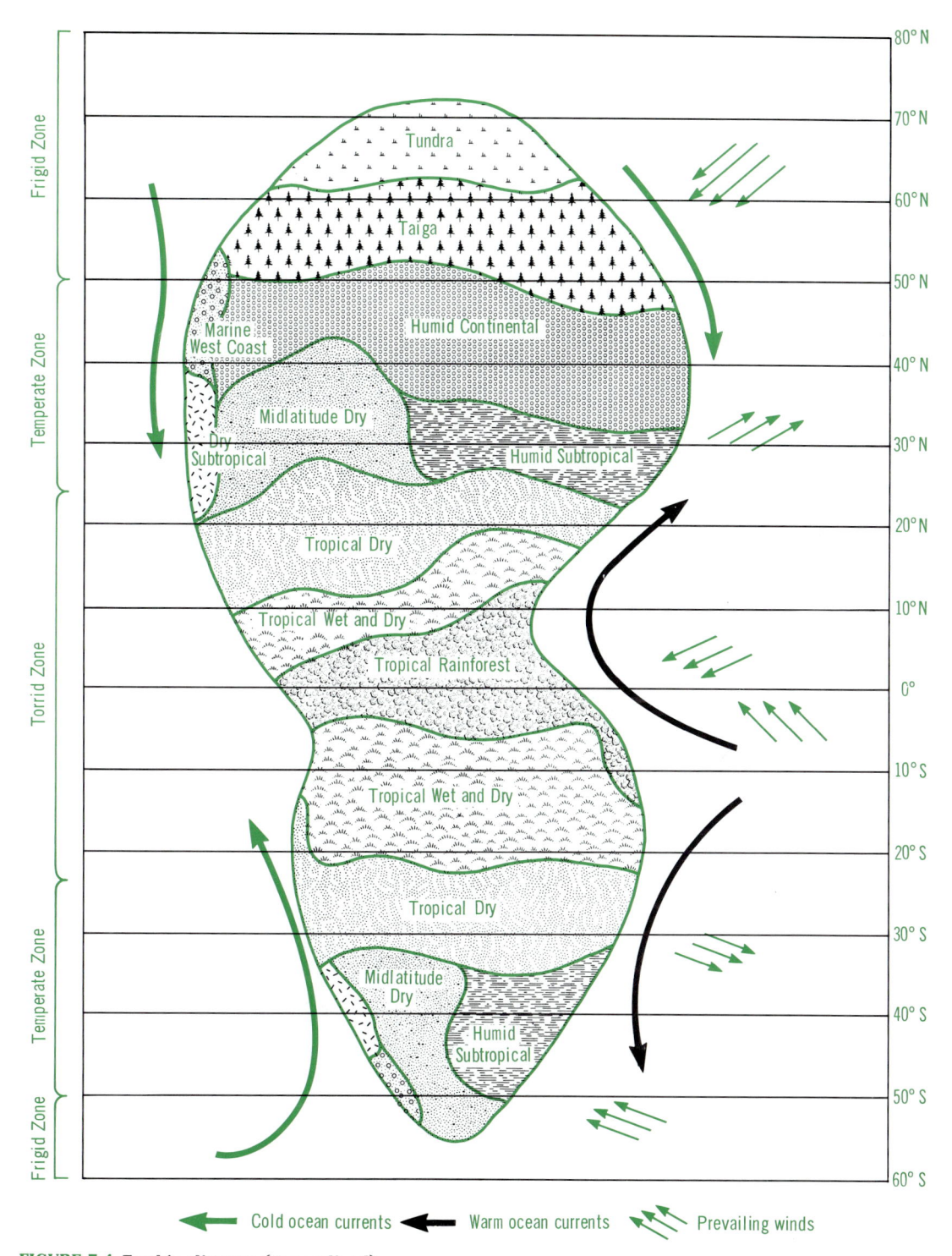

FIGURE 7-4 Earth's climates (generalized).

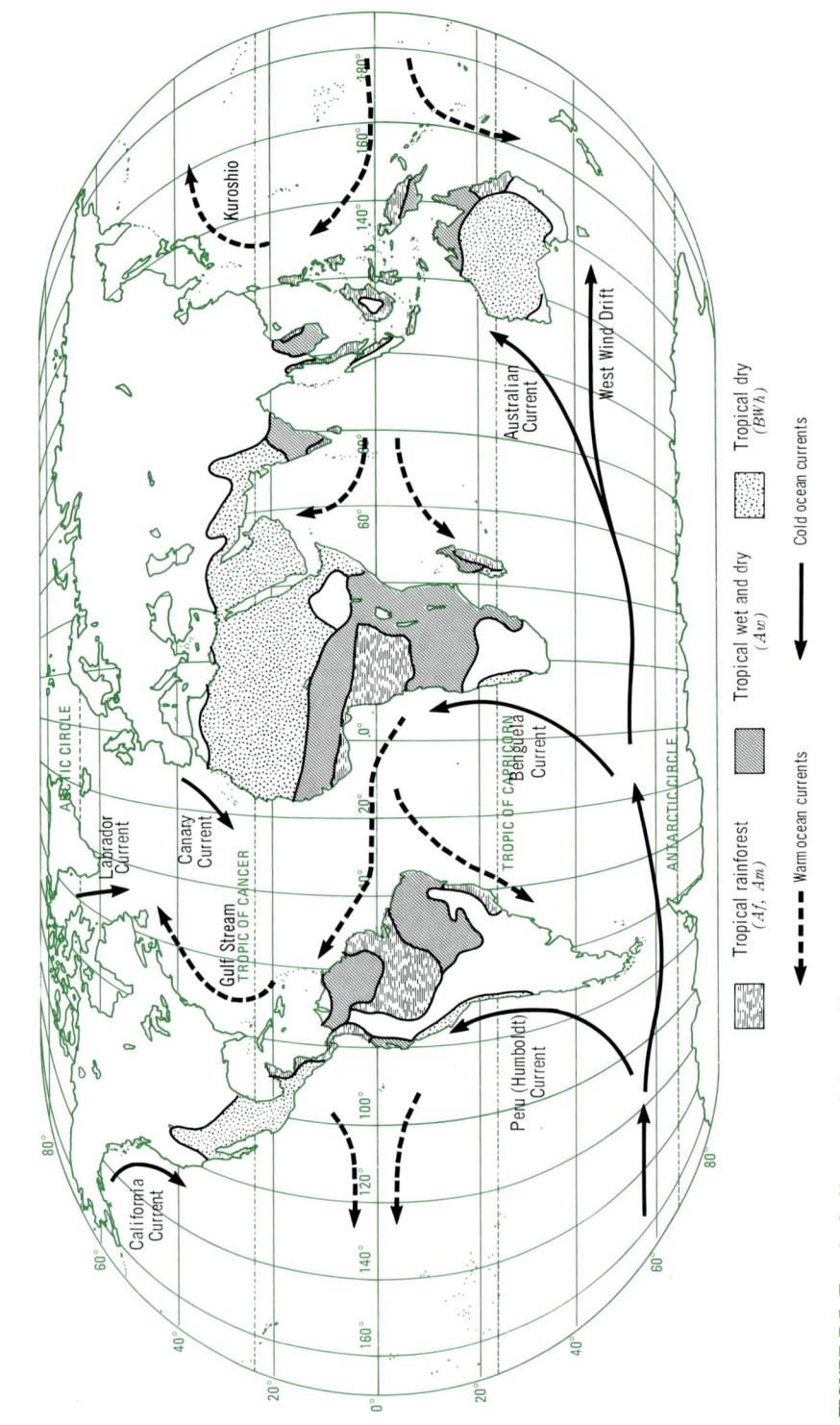

FIGURE 7-5 Tropical climates of the world.

tively high population density, most are sparsely populated.

Tropical Rain Forest

The tropical rain forest climate region is often referred to as equatorial rain forest or tropical rainy, and the letters *Af* are used for its classification code. The largest expanses of this climate will be found within 10° north and south of the equator. The average daily temperatures range between 80° and 90°F (27° and 32°C) (Fig. 7-6) and seldom fall below 64.4°F (18°C) at night. Because this region is a part of the equatorial low-pressure depression, the convectional uplift of the moist, heated air creates daily thunderstorms. The average annual precipitation is 80 to 100 inches (2032 to 2540 millimeters), and there is no dry season. Someone once said that in these tropical rain forest areas there are only wet and less wet seasons.

The average daily pattern is for cumulus clouds gradually to form as the land heats up. By early afternoon the clouds become cumulonimbus (rain bearing) and begin precipitating as they move slowly westward, pushed by the trade winds. If a mountain stands in the way, the precipitation will be heavier, as we saw in Chapter 6 when we considered the rainfall for Mount Waialeale in Hawaii, which average 471.66 inches (11,980 millimeters) of rainfall annually. Such areas of extremely heavy precipitation occur along windward coasts of landmasses, where warm ocean currents make available a plentiful supply of warm and moist air masses.

Even though the tropical rain forest regions are not attractive to people for permanent residence, they have supplied us with valuable products such as rubber, cocoa, medicines, and timber (mahogany, teak, ebony). The constant rainfall also exerts a tremendous influence on the soils, washing away the dissolvable minerals and the little humus that does manage to escape the increased activity of bacteria. The resulting soils, called latosols (from the Latin *later*, "brick"), are able to support the luxuriant stands of trees, but agriculturists (both primitive and advanced) have found that after the forest is cleared, the soils lose their fertility within a year or two and fertilizers become a necessity.

Tropical Monsoon

The tropical monsoon climate region (*Am*) seems to be concentrated in South-East Asia (Bangladesh, Burma, and Malaysia), the East Indies (Sumatra, Java, New Guinea, and Celebes), the Philippines, and the areas near the mouth of the Amazon River of Brazil. Monsoon climate regions have heavy rainfall, caused by monsoon conditions (see Fig. 4-13) during the summer periods. Intense heating of the interior of the landmasses draws in moist air from the nearby oceans, and the resulting rains often cause serious damage. Cherapunji (in the Indian state of Assam) once recorded 1041.87 inches (24,464 millimeters) of rain from August 1860 to July 1861. So great is this problem to the people of Bangladesh that in the area most often hit by monsoon rains, the Bengalis have begun building mounds to serve as a place of refuge when the waters rise too high. Not all monsoons are killers, however, because they do bring water to a normally parched land every summer. Without this rain, the millions of people who must exist on extremely low-nourishment diets could not survive.

Tropical Wet and Dry

Surrounding the rain forest climates are the great expanses of savanna grasslands, a climate region known as tropical wet and dry (*Aw*). Here are the tall-grass areas, such as the Sudan (south of the Sahara Desert), the veldt (East and South Africa), the campos (Brazil), the

FIGURE 7-6 Tropical rain forest (AF): Changuinola, Republic of Panama The dense foliage and giant ferns that line this river in the interior are the result of high monthly temperature averages and heavy precipitation. Climatic data can be presented in any one of three different ways: *(a)* photograph, *(b)* climographic visual display of the data included in the climate chart *(c)*. (Data taken from U.S. Department of Commerce. *World Weather Records, 1951–60*, 1965.)

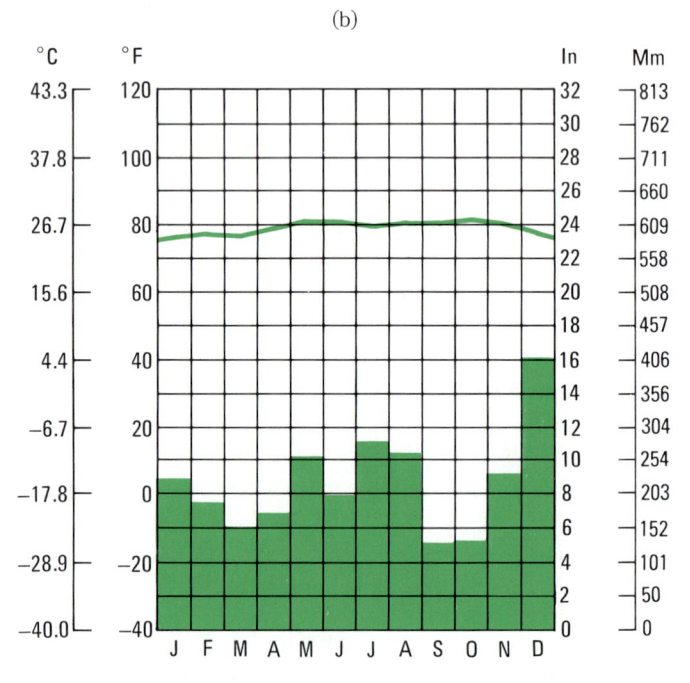

CLIMATE: Af STATION: CHANGUINOLA, PANAMA
ELEVATION: 53 FEET (16 METERS) LAT.: 9°29′N LONG.: 82°29′W

	Jan.	Feb.	Mar.	Apr.	May	June	July	Aug.	Sept.	Oct.	Nov.	Dec.	Year
Temperature													
°F	79.3	78.6	79.5	80.4	81.3	81.7	80.4	81.0	81.1	81.3	80.1	79.0	80.2
°C	26.3	25.9	26.4	26.9	27.4	27.6	26.9	27.0	27.3	27.4	26.7	26.1	26.8
Precipitation													
Inches	8.9	7.5	6.1	6.9	10.2	7.9	11.1	10.4	5.2	5.3	9.2	16.1	104.7
Millimeters	226	190	155	175	259	200	282	264	132	135	234	409	2659

(c)

llanos (Venezuela), and the savannas of India, South-East Asia, and northern Australia. Savannas receive their rain during the summer as the nearby oceans are heated and tropical maritime air masses are dominant. The winters are dry, with average precipitation of less than 2 inches (50 millimeters). The climograph for Fig. 7-7 shows that Darwin, Australia, averages about 81.5°F (27.5°C) for the year, with the yearly rainfall amounting to about 65 inches (1651 millimeters). But see how this precipitation is spread out through the year. The winter months (May to September) average only 0.28 inches (7 millimeters), and the summer months (October to April) average almost 8.5 inches (216 millimeters). Soils in tropical wet and dry regions are not considered to be very fertile, and the grasses seem to be too tough and tall for successful programs of animal husbandry (sheep and cattle raising), so people are discouraged from migrating to these areas. Note, however, that African wildlife seems to flourish in these regions.

Tropical Dry

The tropical dry climate regions are called **desert** (*BW*) and extend 15° to 35° north and south of the equator. They surround the tropical wet and dry regions. These deserts are under the dominating influence of the subtropical high-pressure areas of descending air masses, which bring little precipitation. In regions such as North Africa's Sahara, Australia's Great Sandy, India's Thar, and the Arabian Desert, evaporation usually exceeds precipitation, and the average annual precipitation is less than 10 inches (254 millimeters). The tropical dry regions always have a west coast frontage, extending eastward in some cases for great distances. Some deserts, such as the Atacama of Peru and Chile, are narrow strips bounded on the east by high mountain ranges. The characteristics of the tropical dry regions are also affected by the nearby presence of cool ocean currents, which are subject to less evaporation. Westerly winds blowing across these cool waters are not able to receive or retain much moisture, much less transport it over the parched lands beneath them.

Figure 7-8 tells you that desert lands suffer considerably from the combination of warm temperatures and low precipitation. Beni Abbes, Algeria, has an average yearly temperature of about 73.2°F (22.9°C) and a yearly total of only 2.47 inches (62.7 millimeters) of rain. Summer months show a zero amount, and the total for the "rainy" winter comes to only 1.3 inches (33 millimeters). Desert regions have become noted for setting records for extremes in the temperature ranges. The official world's record of 136.4°F (58°C) as the highest temperature ever recorded was set at El Aziza, Libya, a city located about 20 miles (32 kilometers) inland from the Mediterranean Sea. The record for the greatest diurnal (daily) range was set at In Salah, midway between the Atlas and Ahagger mountains of the Sahara, when during one 24-hour period the temperature ranged from 26°F (−3°C) at night to 126°F (52°C) the next day, a range of 100°F (57°C).

SUBTROPICAL CLIMATES

The earth's subtropical climates are situated on the poleward sides of the tropical climates and in these climate regions some of the world's heaviest concentrations of population are to be found (Fig. 7-9). There are two subtropical climates, the west coastal dry subtropical, also known as Mediterranean subtropical, and the east coastal humid subtropical, also called humid subtropical.

CLIMATE: Aw	STATION: PORT DARWIN, AUSTRALIA ELEVATION: 89 FEET (27 METERS)						LAT.: 12°26'S		LONG.: 130°52'E				
	Jan.	Feb.	Mar.	Apr.	May	June	July	Aug.	Sept.	Oct.	Nov.	Dec.	Year
Temperature													
°F	82.2	81.7	82.9	82.8	81.0	77.7	76.8	78.4	81.9	84.4	84.6	83.8	81.5
°C	27.9	27.6	28.3	28.2	27.0	25.4	24.9	25.8	27.7	29.1	29.2	28.8	27.5
Precipitation													
Inches	15.0	13.7	9.8	6.3	0.68	0.04	0.11	0.02	0.41	3.09	6.02	9.5	64.6
Millimeters	381	348	249	160	17	1.0	2.8	0.5	10.4	78.5	153	241	1641

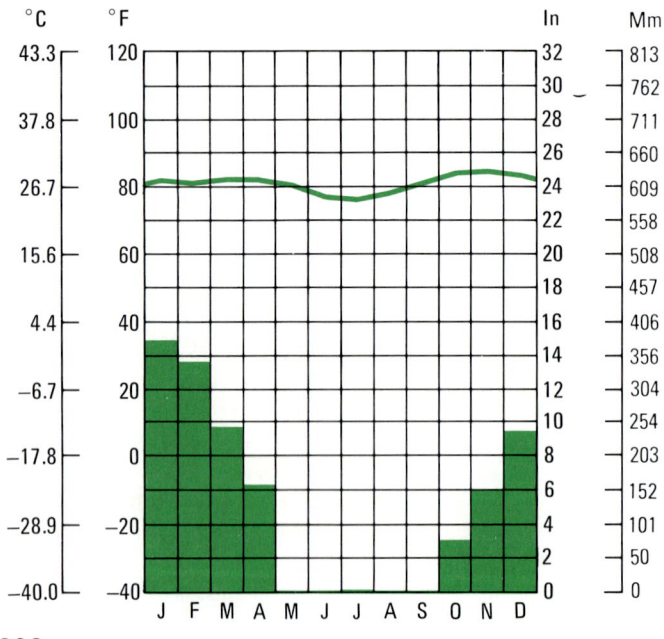

FIGURE 7-7 Tropical wet and dry (AW): Port Darwin, Australia (Data taken from U.S. Department of Commerce, *World Weather Records, 1951–60*, 1965.)

CLIMATE: BWh	STATION: BENI ABBES, ALGERIA ELEVATION: 1643 FEET (501 METERS)					LAT.: 30°08'N		LONG.: 02°10'W					
	Jan.	Feb.	Mar.	Apr.	May	June	July	Aug.	Sept.	Oct.	Nov.	Dec.	Year
Temperature													
°F	51.4	56.5	64.4	72.1	80.1	89.96	96.4	94.1	86.2	73.4	61.9	52.7	73.2
°C	10.8	13.6	18.0	22.3	26.7	32.2	35.8	34.5	30.1	23.0	16.6	11.5	22.9
Precipitation													
Inches	3.73	0.09	0.27	0.32	0.04	0.001	0.02	0.08	0.12	0.41	0.26	0.49	5.831
Millimeters	94.7	2.29	6.86	8.13	1.02	0.03	0.51	2.03	3.05	10.4	6.6	12.4	148.02

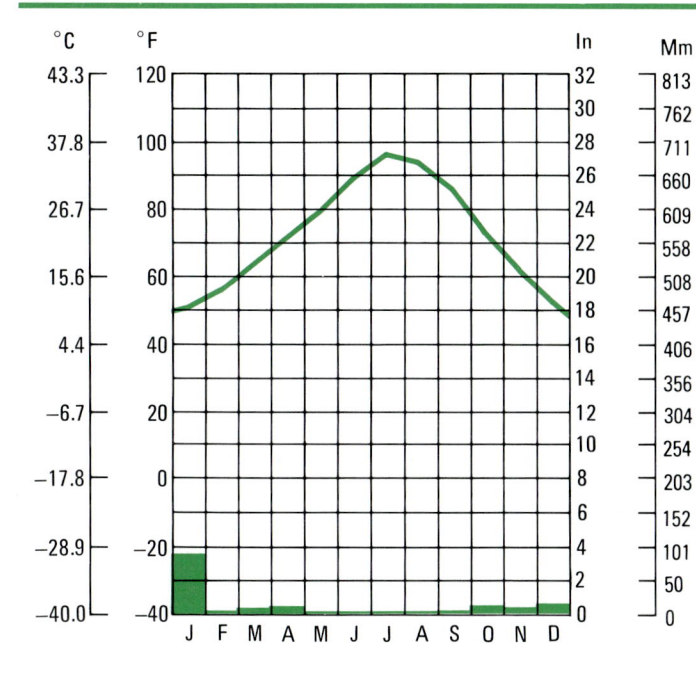

FIGURE 7-8 Tropical dry (BWh): Oasis of Timimoun near Beni Abbes, Algeria The oasis is located in the Grand Erg Occidental (Great Western Sandy Desert). (Data taken from U.S. Department of Commerce, *World Weather Records, 1951–60*, 1965.)

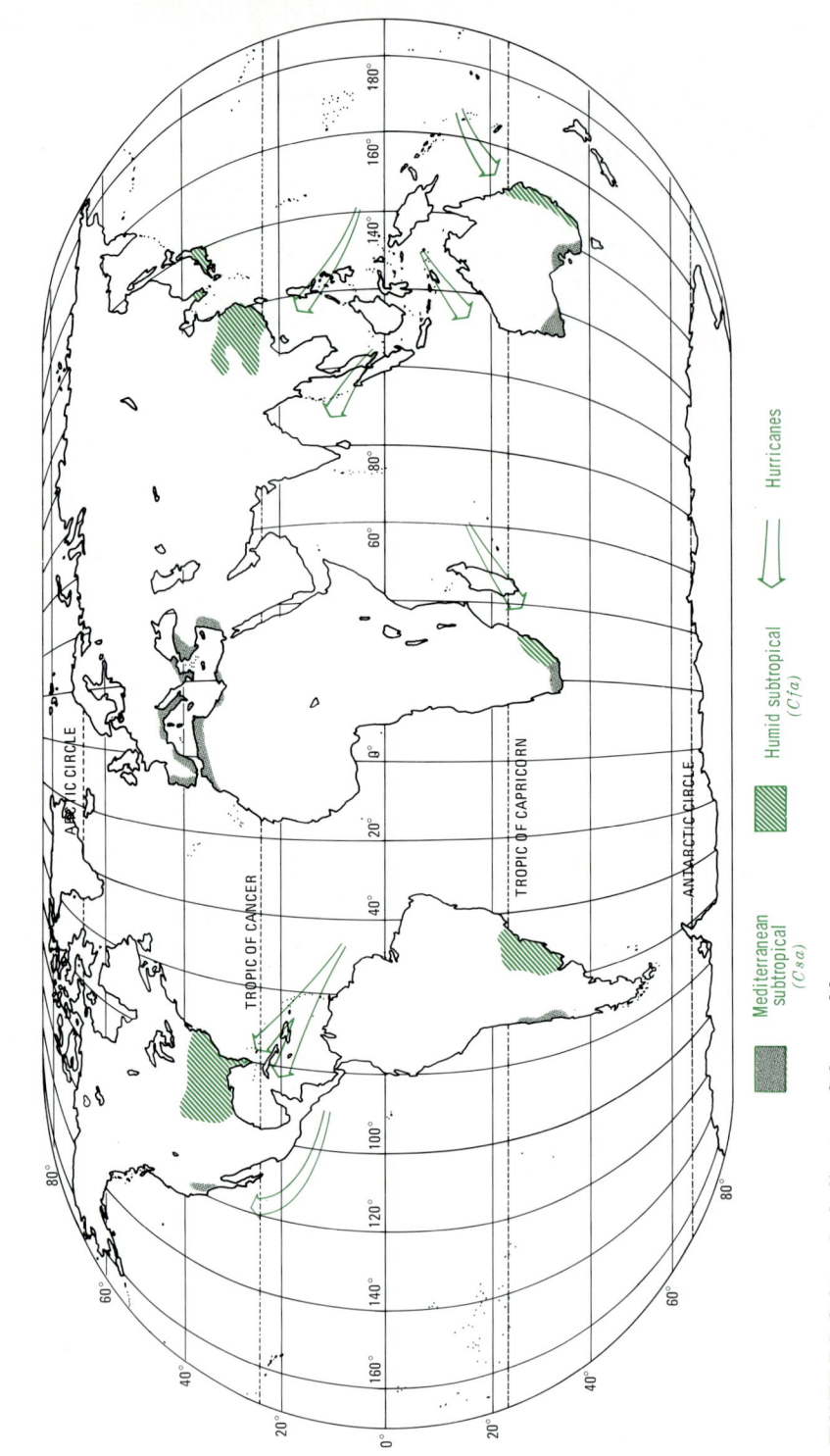

FIGURE 7-9 Subtropical climates of the world.

West Coastal Dry Subtropical

The **Mediterranean dry subtropical** climate (Cs) regions are always found on the west coasts (or extremities of a landmass that face the west, such as the two southernmost portions of Australia), between latitudes 30° and 45° north and south of the equator. They reflect the close proximity of deserts, cool ocean currents, and low-pressure systems from the poles. Added to these factors is the presence of the subtropical high-pressure belts in both hemispheres from which warm and drying air descends (Fig. 7-10).

CLIMATE: Csa	STATION: ROME, ITALY ELEVATION: 432 FEET (131 METERS)						LAT.: 41°48′N		LONG.: 12°36′E			
	Jan.	Feb.	Mar.	Apr.	May	June	July	Aug.	Sept.	Oct.	Nov.	Dec.	Year
Temperature													
°F	45.5	47.1	50.7	55.9	63.3	70.9	75.4	75.4	69.9	61.2	53.4	48.6	59.7
°C	7.5	8.4	10.4	13.3	17.4	21.6	24.1	24.1	21.1	16.2	11.9	9.2	15.4
Precipitation													
Inches	2.56	2.5	2.2	2.56	2.03	1.16	0.98	1.07	2.61	3.07	3.79	3.8	28.4
Millimeters	65.0	63.5	55.8	65.0	51.5	29.5	24.9	27.2	66.3	77.9	96.3	96.5	721

FIGURE 7-10 West coastal dry subtropical (Csa): Wine country north of Rome, Italy (Data taken from U.S. Department of Commerce, *World Weather Records, 1951–60*, 1965.)

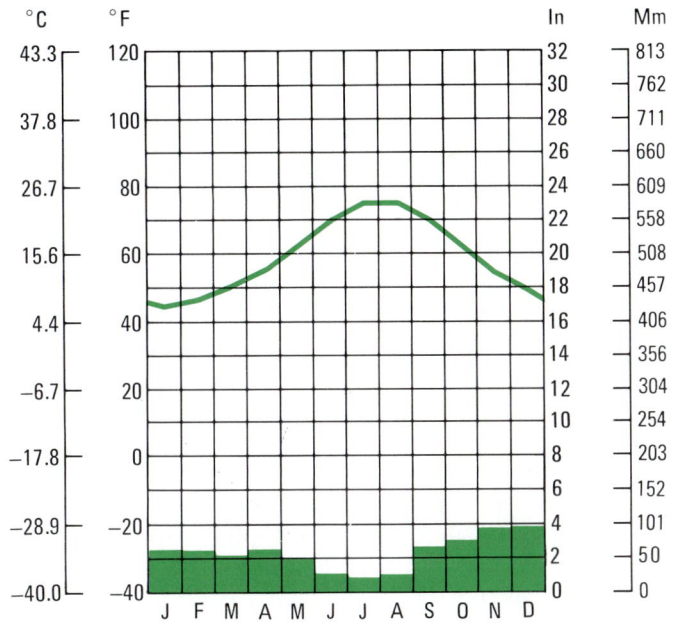

Southern California's climate is greatly influenced by the high-pressure cell far out in the Pacific (called the Pacific or Hawaiian high), and the Mediterranean is affected by the Azores high, located out in the Atlantic off the coast of Portugal. The Hawaiian high tends to "protect" the coast of California from the invasions of cyclonic storms that often lash the coastline farther north during the summer months. Of the 15 inches (381 millimeters) of rainfall received during a year, less than 1 inch (25.4 millimeters) falls during the summer. The average yearly temperature does not seem very warm, since it is only 62.4°F (16.9°C), but the temperature average for any month does not fall below 54.6°F (12.6°C). It is evident that frosts are not common in the Los Angeles lowland, which is one of the reasons why citrus fruits do so well in that region. The long period of low precipitation does favor drought conditions, and this is evidenced by the chaparral (woodland scrub and grasses) that comprises the natural vegetation of the area.

East Coastal Humid Subtropical

On the opposite side of the continents, at the same latitudinal position as the dry subtropical climate, there are regions that are moist and humid because of the warm seas and currents nearby (such as the Caribbean Sea and the Gulf Stream adjacent to the southeastern United States). Figure 7-11 presents climatic data for Nashville, Tennessee, a city within this **humid subtropical** (*Cf*) climate region. You can see that although the annual temperature average is only 60.1°F (15.6°C), the summers are considerably hotter than the winters, staying close to 80°F (27°C). Because of the constancy of the rainfall, heavy irrigation is seldom required. The monthly averages for precipitation amount to about 3.8 inches (96.5 millimeters). Because the region is temperate and humid, broadleaf forests are common.

One might say that the prime difference between these two subtropical climates is that the humid subtropical has sufficient moisture all year while the dry subtropical does not. Another and perhaps more important difference is the prevalence of hurricanes in the humid subtropics. Moist maritime tropical air masses are driven into the region, all too often accompanied by the tropical cyclones. Then, too, during the winter months there are frequent collisions between cold fronts coming down from central Canada and warm fronts moving up from the Gulf of Mexico or the Caribbean. It is interesting to note that the southeastern coast of South America around Buenos Aires is seldom struck by hurricanes. This is undoubtedly because there is insufficient space for a tropical cyclone to build up to devastating fury, and also because the variable winds of the intertropical convergence zone seldom reach this far south.

MIDLATITUDE CLIMATES

The climate regions situated in the middle latitudes, between 30° and 60° north latitude and 30° and 50° south latitude, are generally temperate climates with high precipitation, except for the interior regions, which are quite dry. Figure 7-12 maps the locations of three midlatitude climate regions: marine west coast, humid continental, and midlatitude dry.

Marine West Coast

The term *west coast* does not apply to **marine west coast** climate (*Cfb*) regions in South Africa or Australia, but only to southern Chile in South America, a narrow strip along the western North American coast from the state of Washington on through British Columbia to the southern tip of Alaska, and most of northwestern Europe. This climate region normally receives considerable precipitation, as Fig. 7-13

CLIMATE: Cfa	STATION: NASHVILLE, TENNESSEE ELEVATION: 584 FEET (178 METERS)						LAT.:36°07'N		LONG.:86°41'W				
	Jan.	Feb.	Mar.	Apr.	May	June	July	Aug.	Sept.	Oct.	Nov.	Dec.	Year
Temperature													
°F	39.9	42.1	49.1	59.5	68.5	77.4	80.2	79.2	72.9	61.5	48.6	41.4	60.1
°C	4.4	5.6	9.5	15.3	20.3	25.2	26.8	26.2	22.7	16.4	9.2	5.2	15.6
Precipitation													
Inches	5.5	4.6	5.2	3.74	3.7	3.27	3.7	2.87	2.87	2.32	3.27	4.17	45.9
Millimeters	139	117	132	94.9	93.9	83.1	93.9	72.9	72.9	58.9	83.0	106	1166

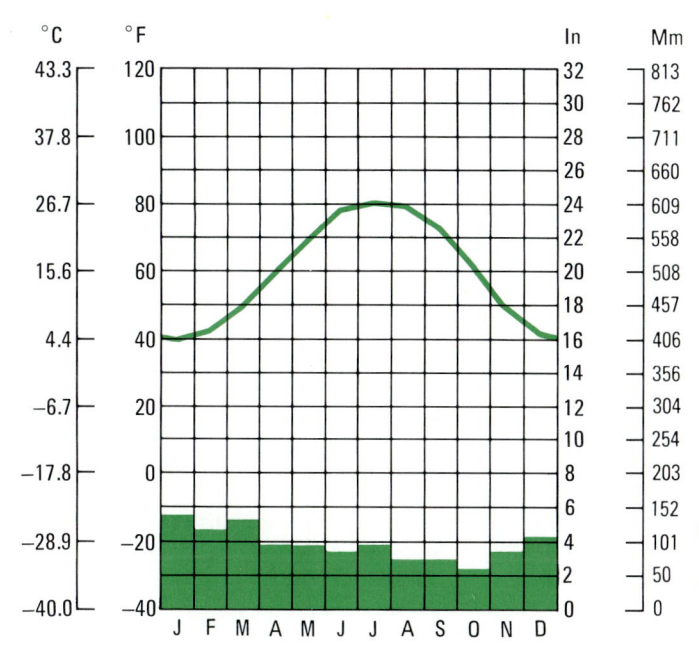

FIGURE 7-11 Humid subtropical (Cfa): Hurricane Mills, southwest of Nashville, Tennessee (Data taken from U.S. Department of Commerce, *World Weather Records, 1951–60,* 1965.)

207

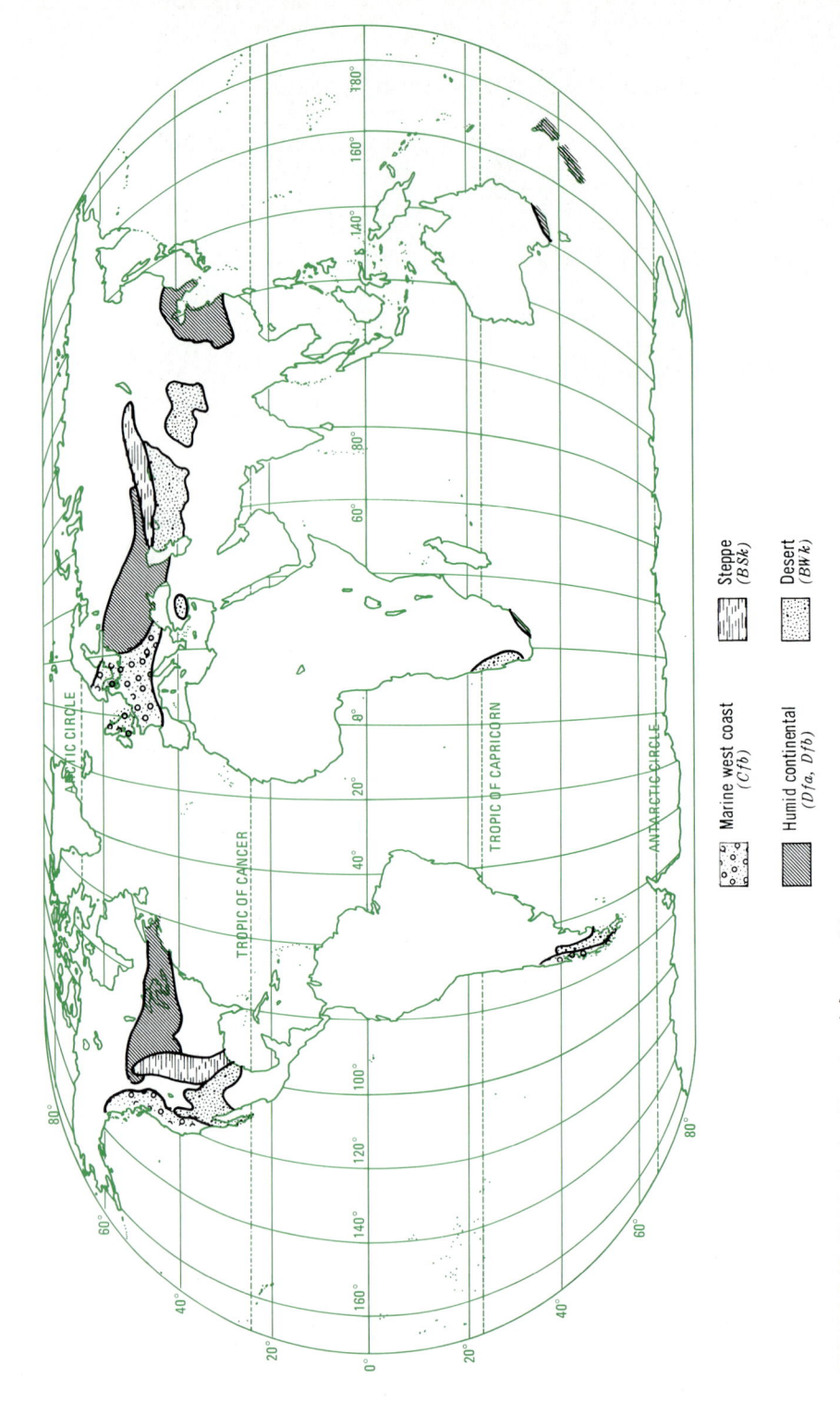

FIGURE 7-12 Mid-latitude climates of the world.

CLIMATE: Cfb	STATION: VANCOUVER, CANADA ELEVATION: 17 FEET (5 METERS)						LAT.: 49°11′N			LONG.: 123°10′W			
	Jan.	Feb.	Mar.	Apr.	May	June	July	Aug.	Sept.	Oct.	Nov.	Dec.	Year
Temperature													
°F	36.1	39.6	42.4	48.4	54.7	59.4	63.7	62.6	57.7	50.2	42.8	39.0	49.7
°C	2.3	4.2	5.8	9.1	12.6	15.2	17.6	17.0	14.3	10.1	6.0	3.9	9.8
Precipitation													
Inches	5.56	4.84	3.78	2.36	1.89	2.01	1.02	1.42	2.2	4.68	5.68	6.24	41.9
Millimeters	141	123	96	60	48	51	25.9	36	56	119	144	158	1064

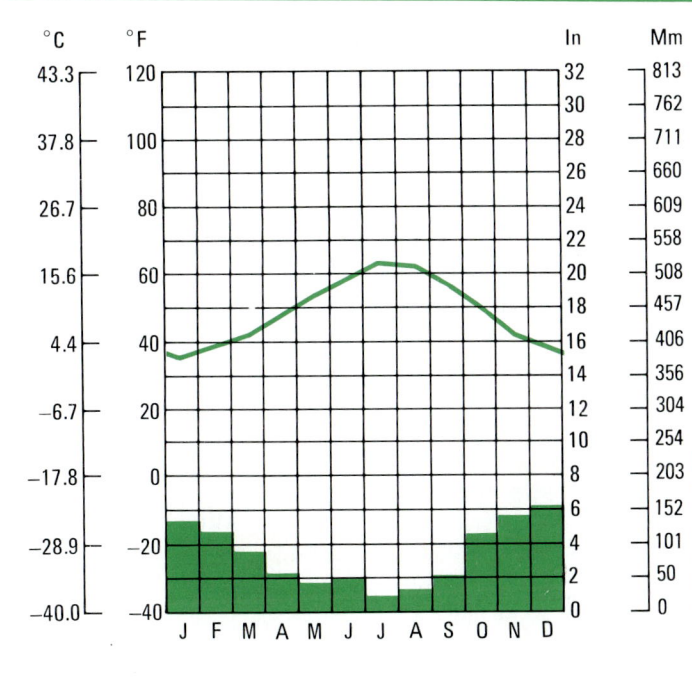

FIGURE 7-13 Marine west coast (Cfb): Kitsilano Park, Vancouver, British Columbia, Canada During the summer sunbathers enjoy the cooling waters off Vancouver, the skyline of the city, and the mountains of the Coast Range in the background. (Data taken from U.S. Department of Commerce, *World Weather Records, 1951–60*, 1965.)

shows. Annual precipitation in the Vancouver area ranges from a low of 23 inches (584 millimeters) to an occasional high of around 80 inches (2032 millimeters). Yearly temperature averages are about 49.7°F (9.8°C), and the amount of precipitation is heaviest during the winter months, totaling 41.9 inches (1064 millimeters) from September through May, and averaging 4 inches (101.6 millimeters) per month. This results from the generation of low-pressure systems that move out of the polar front of Alaska. Windward slopes of the coastal mountain ranges receive considerably more annual precipitation than low-lying areas such as Vancouver, as is evidenced by the record set at the Olympic Mountains of Washington—an average annual precipitation of 150.73 inches (3828.5 millimeters).

One might think that the general climate of this region would be rather inhospitable, but this is not the case because of the proximity of the ocean. The oceans exert a valuable moderating influence on the land, keeping down any extremes in temperature so that the range between the warmest and coldest months averages about 27°F (15°C). This may be a region of hunters and fishers, but farmers can have a profitable agricultural business raising fruits and vegetables during the relatively long growing season (approaching 200 days). In addition, the great expanse of coniferous and broadleaf forests supplies timber.

Humid Continental

The cold and snowy **humid continental** climates (*Dfa, Dfb*) are to be found only in the Northern Hemisphere, extending from 35° to 60° north latitude. In North America the western boundary of this climate region is located in the center of the continent, and from there it stretches to the Atlantic coast, taking in the southern one-third of Canada and the upper Midwest of the United States. As for Europe and Asia, the humid continental climate is found in western Siberia, much of Russia and the Ukraine, and the southern portions of Finland and Sweden. Like many other climate regions, this climate witnesses the battles issuing from the polar front during the winter and from the movement northward of tropical maritime air masses during the summer months.

Winters can be very cold, as shown in Fig. 7-14. The average annual temperature is a low 47°F (8.3°C), ranging from a January low of 22°F (−5.5°C) to a July high of 71.2°F (21.8°C). Snowy winters are common and often encompass 9 or 10 months of a year, especially in the northern sections, which have cooler summer (*Dfb*). Incidentally, the northern border of the *Dfb* climate is the northernmost limit for conventional forms of agriculture. Precipitation totals about 31 inches (787 millimeters) for the year, with a monthly average of 2.7 inches (68.6 millimeters). These conditions create problems for farmers, but they are advantageous for people in the timber business because the humid continental regions are noted for the mixed forests of hardwoods and softwoods, broadleaf deciduous and needle-leafed evergreens.

The southern portion of the humid continental climate region (*Dfa*) receives heavier amounts of precipitation, with monthly averages advancing to the 3- and 4-inch (76.2 and 102-millimeter) levels. Summer rainfall accompanying tropical cyclonic storms can be quite severe, as is evidenced by the 31-inch (787-millimeter) downpour experienced at Smethport, Pennsylvania, on July 18, 1942, during a 4½-hour period.

Midlatitude Dry

The drier regions of the middle latitudes (*BSk, BWk*), called **midlatitude dry** climates, are similar in many respects to those of the lower lati-

CLIMATE: Dfb	STATION: TORONTO, CANADA ELEVATION: 383 FEET (117 METERS)							LAT.: 43°40′N		LONG.: 79°24′W			
	Jan.	Feb.	Mar.	Apr.	May	June	July	Aug.	Sept.	Oct.	Nov.	Dec.	Year
Temperature													
°F	22.0	21.1	32.3	44.4	55.8	66.2	71.2	69.6	61.5	51.1	39.7	28.9	47.0
°C	−5.5	−6.0	0.2	6.9	13.2	19.0	21.8	20.9	16.4	10.6	4.3	−1.7	8.3
Precipitation													
Inches	2.64	1.93	2.64	2.6	2.76	2.48	2.91	2.4	2.56	2.36	2.48	2.4	31.0
Millimeters	67	49	67	66	70	63	74	61	65	60	63	61	787

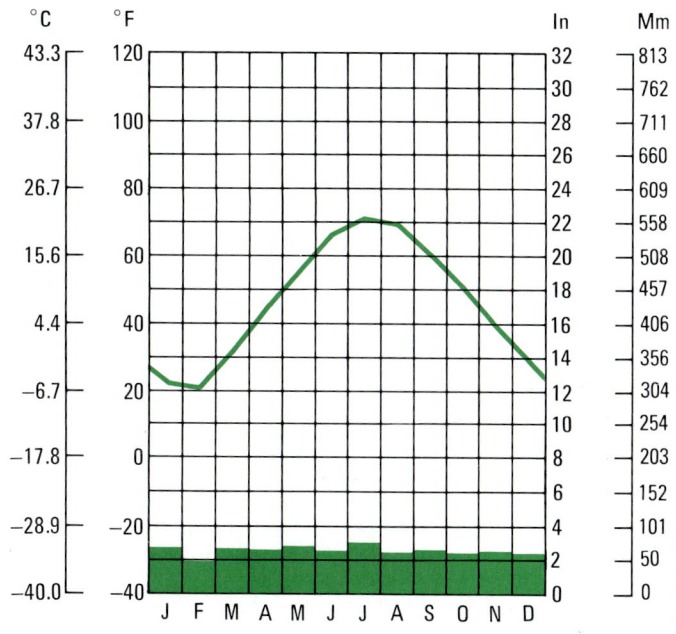

FIGURE 7-14 Humid continental with cool summer (Dfb): Toronto, Ontario, Canada (Data taken from U.S. Department of Commerce, *World Weather Records, 1951–60*, 1965.)

211

CLIMATE: BWk	STATION: ANHSI, CHINA ELEVATION: 3900 FEET (1189 METERS)						LAT.: 40°43′N			LONG.: 95°57′E			
	Jan.	Feb.	Mar.	Apr.	May	June	July	Aug.	Sept.	Oct.	Nov.	Dec.	Year
Temperature													
°F	17.6	26.0	40.6	55.6	66.4	75.2	78.8	76.5	65.1	50.5	32.0	16.9	50.2
°C	−8.0	−3.3	4.8	13.1	19.1	24.0	26.0	24.7	18.4	10.3	0.0	−8.4	10.1
Precipitation													
Inches	0.08	0.04	0.16	0.22	0.12	0.21	0.63	0.44	0.1	0.04	0.06	0.19	2.28
Millimeters	2.03	1.02	4.06	5.59	3.05	5.33	16	11.2	2.54	1.02	1.5	4.8	57.9

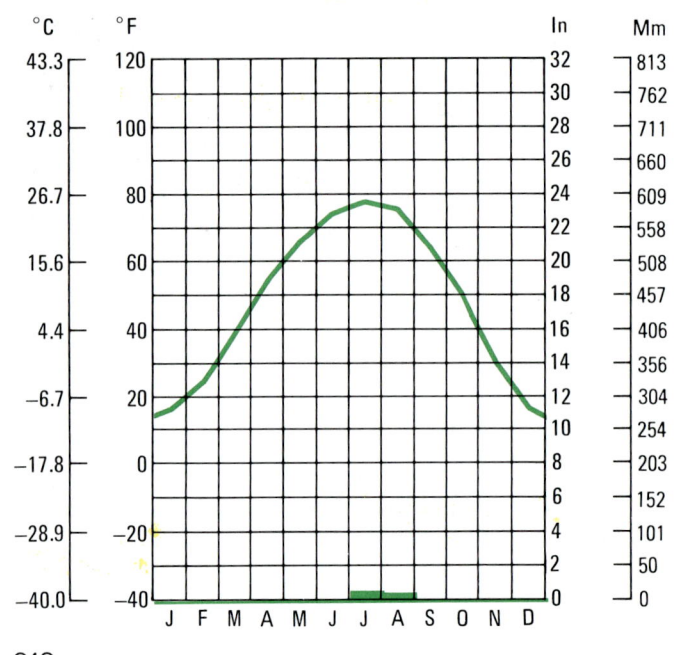

FIGURE 7-15 Midlatitude desert with cool summer (BWk): Anhsi, China This station is situated on the eastern margin of the Tarim Basin, close to the Gobi Desert. (Data taken from U.S. Department of Commerce, *World Weather Records, 1951–60*, 1965.)

tudes. Grasses and desert shrub are the dominant vegetation types, along with scattered clumps of broadleaf deciduous trees. The prime difference is the temperature. Tropical deserts, for example, have monthly temperatures that remain above the point of freezing, while midlatitude deserts often have one or two months when the temperature falls below 32°F (0°C). The reason is the latitudinal position (relative to the angle of insolation) and the presence of high-pressure systems over Canada and Siberia. Air drawn from the colder northern regions is forced down into the deserts during the winters, causing a steep decline in temperature (Fig. 7-15). The yearly temperature average for Tashkent, center of the cotton-growing region of Soviet Central Asia, is a low 56°F (13.3°C), while the summer months see a temperature average of around 80°F (27°C).

The midlatitude steppes and prairies (*BSk*) sometimes seem barren, yet these short-grass regions are vitally important. If you will look at Fig. 8-3, you will note the areas made famous as grazing lands for beef cattle: the Great Plains of the United States, the pampa of Argentina, and the Kirghiz steppes of central Asia and the Ukraine (Fig. 7-16). The veldt of South Africa, notably the steppes of the Kalihari, is now becoming more extensively used for pasturage. The Australians are justifiably proud of the grassy "outback" region that lies west of the Eastern Highlands.

These interior regions are cut off from moisture-laden warm air masses; thus they remain cool throughout the year. Consider also the situation for Cheyenne, Wyoming, where the yearly temperature average reaches 44.5°F (6.9°C), and at least three months of the year have average temperatures below freezing. As for precipitation, the annual total of less than 15 inches (381 millimeters) is an average of slightly more than 1 inch (25.4 millimeters) per month. The chinook winds that sweep down the rain-shadow side of the Rockies during the winter months are drying winds, devoid of moisture. Maritime tropical air masses sweep northward from the Gulf of Mexico and the Caribbean during the summer months, and as they move over the heated land, convectional precipitation results. The four summer months account for 7.7 inches (196 millimeters) of the yearly total, and the remaining eight months account for only 6.7 inches (170 millimeters).

HIGH-LATITUDE CLIMATES

The high-latitude climatic regions (Fig. 7-17) are to be found poleward of the 50° parallel and include the continental subarctic (*Dfc*, *Dwd*, and *Dwc*), tundra, (*ET*), and the ice caps (*EF*). You may recall from our earlier discussion (p. 99) that the presence of huge landmasses interferes with the moderating influence of the oceans. This effect, termed *continentality*, is the prime reason why these climate regions are so severe in their extremes.

Continental Subarctic

The **continental subarctic,** or **boreal,** climate (*Dfc, Dwc, Dwd*) region is the most important commercially of the three high-latitude climates, primarily for the vast expanse of timber available from the taiga, a mixed forest of hardwoods. This very cold region is dominated by polar air masses throughout the year and is constantly being swept by cyclonic storms. Traditional forms of agriculture are not practiced here. Figure 7-18 details conditions in Irkutsk, Siberia. This city is situated about 36 miles (40 kilometers) from Lake Baikal and is a center for the timber industry of the region. The average temperature for the year approaches 31°F (−0.8°C), but the short summer sometimes attains temperatures in the 50°–60°F

CLIMATE: BSk	STATION: ROSTOV-ON-DON, UKRAINE S.S.R. ELEVATION: 254 FEET (77 METERS)						LAT.: 47°15′N		LONG.: 39°40′E				
	Jan.	Feb.	Mar.	Apr.	May	June	July	Aug.	Sept.	Oct.	Nov.	Dec.	Year
Temperature													
°F	22.5	23.2	31.8	48.9	62.2	69.6	74.3	72.1	61.5	48.2	36.3	27.1	48.2
°C	−5.3	−4.9	−0.1	9.4	16.8	20.9	23.5	22.3	16.4	9.0	2.4	−2.7	9.0
Precipitation													
Inches	1.52	1.64	1.28	1.56	1.44	2.32	1.96	1.48	1.28	1.76	1.6	1.48	19.3
Millimeters	38.6	41.6	32.5	39.6	36.6	58.9	49.8	37.6	32.5	44.7	40.6	37.6	490

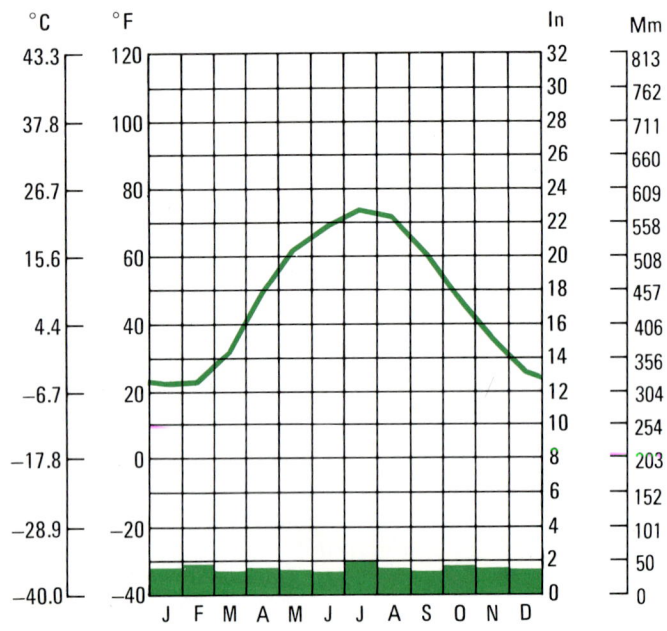

FIGURE 7-16 **Midlatitude steppe (BSk): Rostovon-Don, Ukraine S.S.R.** (Data taken from U.S. Department of Commerce, *World Weather Records, 1951–60*, 1965.)

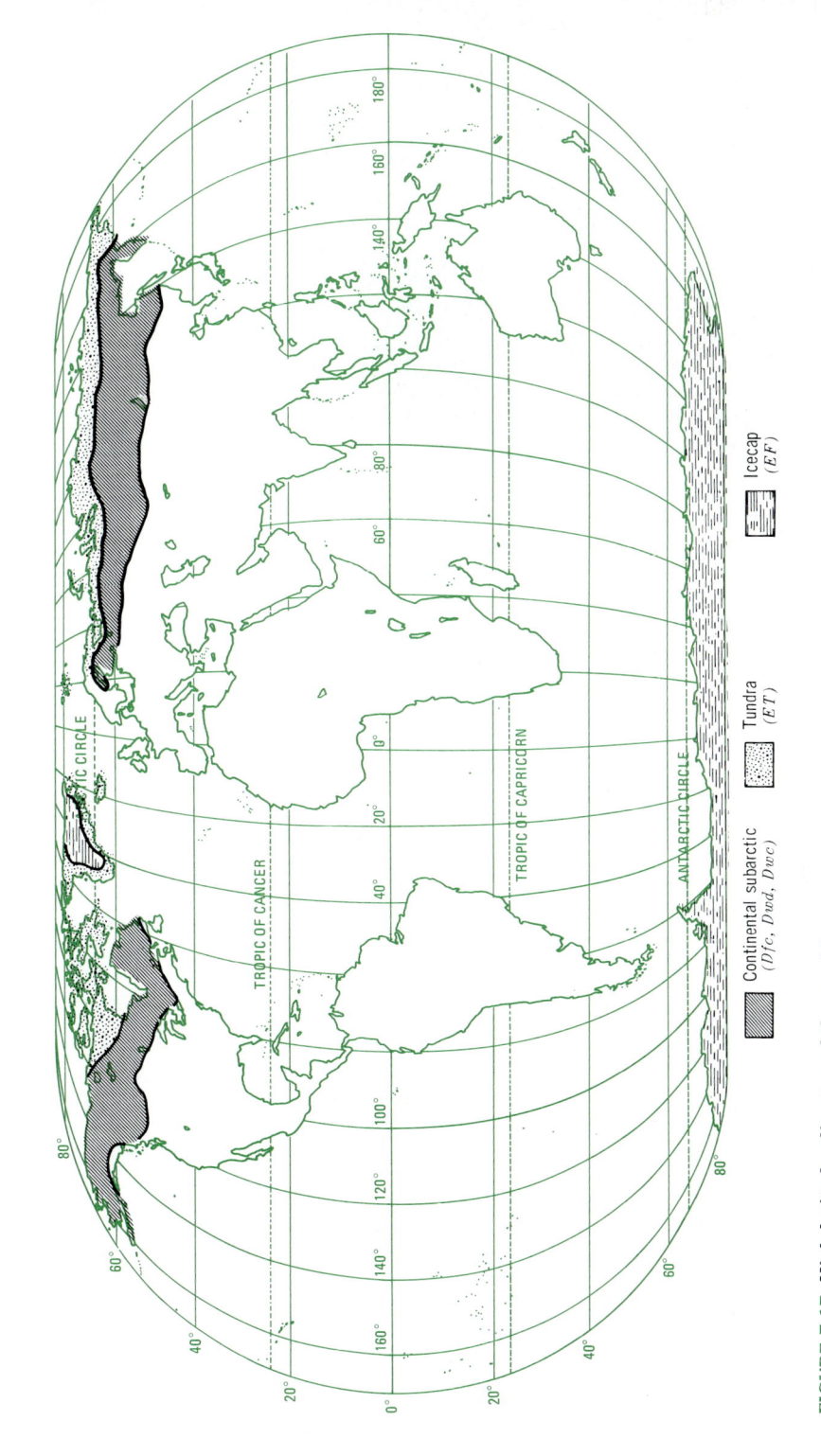

FIGURE 7-17 High-latitude climates of the world.

215

CLIMATE: Dwc	STATION: IRKUTSK, U.S.S.R. ELEVATION: 1930 FEET (588 METERS)								LAT.: 52°16′N		LONG.: 104°21′E		
	Jan.	Feb.	Mar.	Apr.	May	June	July	Aug.	Sept.	Oct.	Nov.	Dec.	Year
Temperature													
°F	−5.44	−0.04	15.26	34.9	47.8	59.7	64.2	59.2	46.8	33.9	12.56	−1.3	30.6
°C	−20.8	−17.8	−9.3	1.6	8.8	15.4	17.9	15.1	8.2	1.1	−10.8	−18.5	−0.8
Precipitation													
Inches	0.48	0.32	0.36	0.6	1.16	3.32	4.08	3.96	1.96	0.8	0.68	0.6	18.3
Millimeters	12	8	9	15	29.5	84	104	101	49.8	20	17	15	465

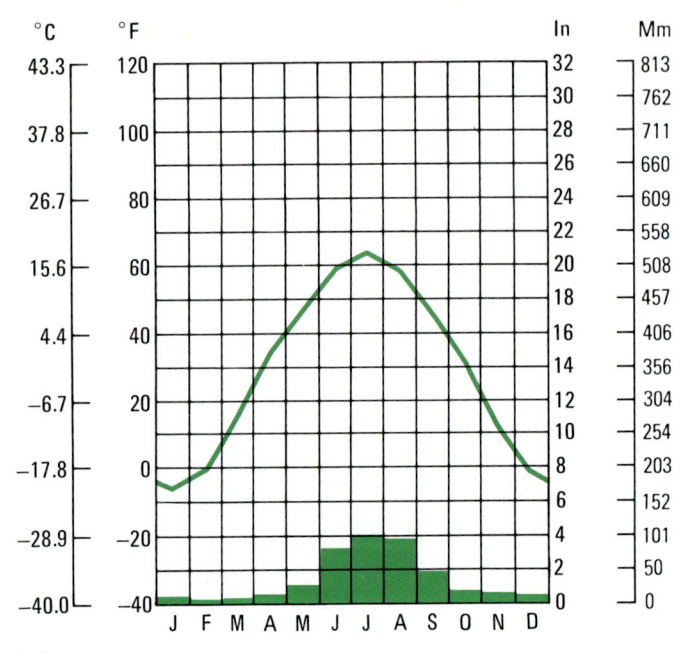

FIGURE 7-18 Continental subarctic (Dwc): Irkutsk, Siberia This station is the center for the Russian timber industry. (Data taken from U.S. Department of Commerce, *World Weather Records, 1951–60*, 1965.)

216

(9°–18°C) range. Precipitation is concentrated in the summer months (almost two-thirds of the total).

Tundra

North of the boreal forest, or taiga, is the usually drab region called the **tundra** (*ET*). Here, maritime polar air masses are dominant throughout the year. Precipitation (see Fig. 7-19) in Point Barrow, Alaska, amounts to about 5 inches (127 millimeters) per year and is concentrated in the period of late fall. Temperatures are below freezing for most of the year; only in July and August do temperatures creep over that point. Although the ground is permanently frozen to depths of hundreds of feet in most places, there is vegetation to be seen: sedges, mosses, lichens, and tiny dwarf willow trees. The vegetation springs to life during the short summer and presents a dazzling display of colorful flowers.

The permafrost presents problems to the tundra's human inhabitants, especially if they construct houses or buildings that are large and without adequate insulation in or under the floors. A visitor may be surprised to see a house tilted from the horizontal, one side having sunk down into the upper 10 inches (254 millimeters) of thawed soil. If a building were constructed on a steep slope in this region, it would have to be firmly anchored deep in the permafrost or it would slide down the hill when the summer thaw arrived.

Ice Cap

The **ice cap** (*climate* (*EF*) region embraces most of Greenland, the floating ice pack of the Arctic, and the great continental glacier that covers the continent of Antarctica. There is no true soil in this region, because the temperatures are too cold, and there is little bacterial activity.

The tremendous amount of water encased in these ice regions has an important bearing on our supply of water and on the temperature of air masses that move out to other regions of the earth. Antarctica contains about eight times as much ice as all of the Arctic Ocean and Greenland combined, and the ice is 1 mile (1.6 kilometers) thick in many places. Consulting Fig. 7-20, you may wonder how the ice caps can build to such depths when there is so little precipitation—less than 5 inches (125 millimeters) per year. The answer is found in the high albedo (reflectiveness) of the ice fields and in the low evaporation. What precipitation an ice cap receives, it retains (except when the outer sections break off and float away on the river or ocean current).

HIGHLAND CLIMATES

For highland climate regions (*H*), we must make the same statements we made when discussing the precipitation of mountainous areas. There are too many variables to permit generalizations. The climate is influenced by determinants such as whether there are high or small mountains, broad ranges or narrow ones, deep and wide valleys or shallow and narrow ones, and whether or not there are interior plateaus. In addition, of course, an important variable is a region's latitude.

CLIMATE AND HUMANITY

So far in our discussions we have attempted to understand the intimate relationship between the earth's vegetation and its climates, but we have not delved very deeply into how this relationship affects humans. Perhaps if we were to make a close study of the map detailing the world distribution of human population (Fig. 1-1) and then scrutinize the maps that portray the various climate regions, we might be able to gain a clearer understanding of why human migration patterns have generally trended toward the areas of "best" climates, the ones that

CLIMATE: ET	STATION: PT. BARROW, ALASKA ELEVATION: 30 FEET (9 METERS)						LAT.: 71°18′N		LONG.: 156°47′W				
	Jan.	Feb.	Mar.	Apr.	May	June	July	Aug.	Sept.	Oct.	Nov.	Dec.	Year
Temperature													
°F	−16.2	−18.2	−14.6	0.14	18.3	33.1	39.0	37.9	30.6	16.5	−0.76	−11.2	9.68
°C	−26.8	−27.9	−25.9	−17.7	−7.6	−0.6	3.9	3.3	−0.8	−8.6	−18.2	−24.0	−12.4
Precipitation													
Inches	0.2	0.16	0.12	0.12	0.12	0.36	0.8	0.92	0.64	0.52	0.24	0.16	4.4
Millimeters	5	4.06	3	3	3	9	20	23.4	16	13	6.1	4.06	112

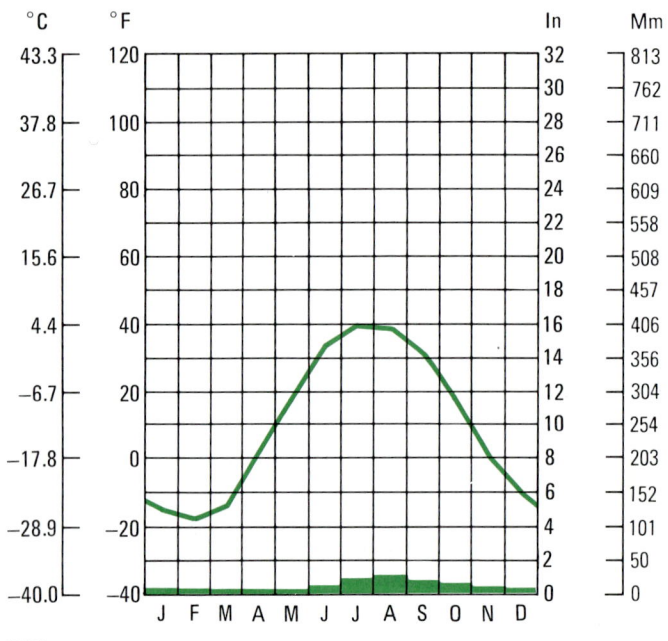

FIGURE 7-19 Tundra (ET): Pt. Barrow, Alaska This area is noted for its permafrost, which thaws at its surface each spring, creating numerous rivers and lakes that are often filled with floating ice. (Data taken from U.S. Department of Commerce, *World Weather Records, 1951–60*, 1965.)

CLIMATE: EF	STATION: McMURDO SOUND, ANTARCTICA ELEVATION: 150 FEET (46 METERS)							LAT.: 77°50′S		LONG.: 166°30′E			
	Jan.	Feb.	Mar.	Apr.	May	June	July	Aug.	Sept.	Oct.	Nov.	Dec.	Year
Temperature													
°F	25.5	16.3	−1.12	−8.5	−9.58	−12.3	−16.9	−20.2	−10.1	−4.72	14.7	25.3	−0.22
°C	−3.6	−8.7	−18.4	−22.5	−23.1	−24.6	−27.2	−29.0	−23.4	−20.4	−9.6	−3.7	−17.9
Precipitation													
Inches	0.44	0.14	0.24	0.25	0.5	0.18	0.18	0.43	0.5	0.3	0.25	0.26	3.67
Millimeters	11	3.6	6.1	6.4	12.7	4.6	4.6	10.9	12.7	7.6	6.4	6.6	93.2

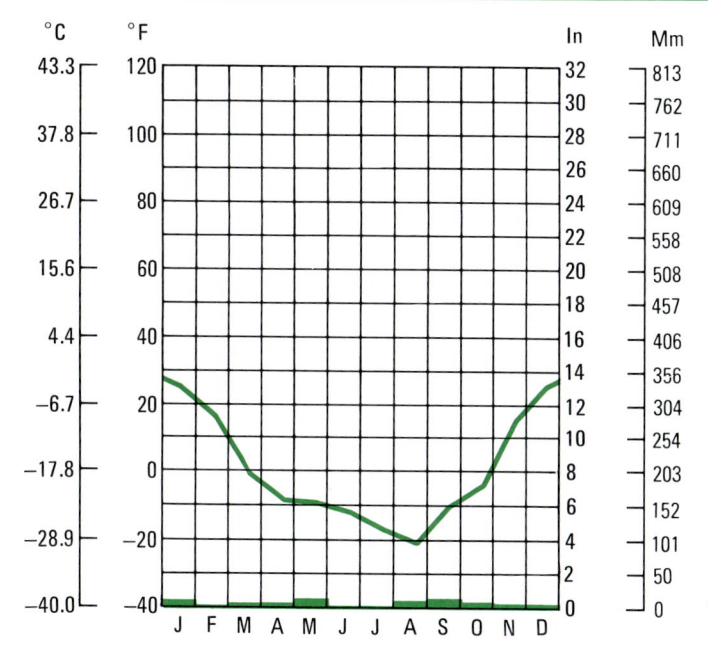

FIGURE 7-20 Ice cap (EF): McMurdo Sound, Antarctica This station is in a land of perpetual ice and snow. (Data taken from U.S. Department of Commerce, *World Weather Records, 1951–60,* 1965.)

are the most "comfortable" and offer better opportunities for industry and agriculture. When we consider that the earth's expanding human population is facing insufficient means of subsistence, we have to concede that although the physical environment does not *determine* the destiny of human beings, it does exert considerable influence on their ability to cope with the problems. Even though many people insist that human ingenuity will discover solutions to these problems of energy shortages, pollution of soils, air, and water, and reduction of the world's forests and animal and sea life, there is also the danger that climatic changes might thwart our advancing technology. This could leave us helpless to do more than struggle weakly to keep from falling behind in the battle for survival.

Emmanuel LeRoy Ladurie, a Frenchman and historian of agricultural problems, has pointed out the many changes in world climate and weather that have significantly affected human existence during the past 1000 years.* There have been a number of times when entire populations have been forced to leave formerly hospitable areas because of localized alterations in the weather. He cites, for example, the case of Europeans who were forced to move from Scandinavia because the weather during the eleventh century turned too cold. Another example occurred near the end of the thirteenth century when Indians of what is now Arizona and New Mexico were decimated by a drought that lasted for 60 years. Fluctuations in climate have been an integral part of the earth's solar budget about which, until possibly the present time, humans could do little. Ladurie makes a strong point in favor of a relatively new approach to the study of agricultural problems, a scientific discipline called **agricultural meteorology.**

An agricultural meteorologist (or agricultural climatologist) studies the important food crops to discover which climate regions are most suitable for them. If cereal grains are under investigation, the scientist will determine what temperatures (day, night, and seasonal) are best for growing bountiful crops of wheat, corn, rye, or barley. Then he will search the world over for the places that fit the requirements. When scientists from all countries of the world cooperate in this endeavor, it may be possible to provide food at more than mere subsistence levels for all people.

*Emmanuel LeRoy Ladurie (1971), *Times of Feast, Times of Famine: A History of Climate Since the Year 1000*, Barbara Bray, translator (Garden City, N.Y.: Doubleday).

REVIEW AND DISCUSSION

1. Explain Wladimir Köppen's classification system and how he arrived at this method.
2. Explain the basic differences between Köppen's five major categories: A, B, C, D, E.
3. Using Table 7-1, explain the following climatic alphabetical codes: Aw, Csa, ET, BSk, Dfc, Cwa, BWh.
4. In which climate region is the northern limit of forest growth?
5. Why are highland climates difficult to classify or examine in detail?
6. Which climate regions are the most preferred for human settlement? Why?
7. What air masses dominate (a) tropical savannas, (b) continental subarctic, and (c) midlatitude steppe and prairie?
8. Explain the differences between the climates of Dallas, Texas, Los Angeles, California, and Montgomery, Alabama.
9. How do humid subtropical regions differ from humid continental regions?

10. Explain why the geographical positions of the Mediterranean dry subtropical and the humid subtropical climates cause them to be different.
11. What is the main difference between tropical dry and midlatitude dry climates?
12. Explain why cold or cool ocean currents have such an effect on climates.
13. Locate on a map the most famous savanna regions of the earth. Do the same for steppe and prairie regions.
14. What are the reasons for the different vegetation forms of the tundra and the taiga?
15. The Great Plains of North America lie in what climate region?region?

NOTE

The extremes in temperatures in various climate regions cited in this chapter were taken from the following interesting and informative article: Glenn Cunningham and James Vernon (December 1968), "Some Extremes of Weather and Climate," *The Journal of Geography*, 67 (9), pp. 530–535.

ADDITIONAL READING

In addition to the citations for Chapters 4, 5, 6, and 8, consult the following:

Appleman, Philip (1965). *The Silent Explosion.* Boston, Mass.: Beacon.

Broek, Jan O. M., and John W. Webb (1973). "Problems of Population Growth," *A Geography of Mankind*, Chap. 20. New York: McGraw-Hill.

Burton, Ian, and Robert W. Kates (January 1964). "Slaying the Malthusian Dragon," *Economic Geography*, 40 (1).

Court, Arnold (1974). "The Climate of the Conterminous United States," in R. A. Bryson and F. K. Hare, *Climates of North America*, Chap. 3. Vol. 11 of *World Survey of Climatology*. Amsterdam: Elsevier Scientific Publishing.

De Blij, Harm J. (1974). "Infinite Humanity: The End of Evolution?" *Man Shapes the Earth*, Chap. 3. Santa Barbara, Calif.: Hamilton.

Higbee, Edward (1960). *The Squeeze: Cities Without Space.* New York: William Morrow.

Johnson, Cecil, ed. (1970). *Eco-Crisis.* New York: Wiley.

Johnson, Warren A., and John Hardesty (1971). *Economic Growth vs. the Environment.* Belmont, Calif.: Wadsworth.

Manners, Ian R., and Marvin Mikesell, eds. (1974). *Perspective on Environment.* Publication No. 13. Washington, D.C.: Association of American Geographers. See especially Werner H. Terjung, "Climatic Modification," Chap. 5.

Thornthwaite, C. W. (1943). "Problems in the Classification of Climates," *The Geographical Review*, 33 (2), 233–255.

Trewartha, Glenn T. (1981). *The Earth's Problem Climates.* 2nd ed. Madison: University of Wisconsin Press.

Went, Frits W. (June 1957). "Climate and Agriculture," *Scientific American* (offprint no. 852).

Went, Frits W. (1963). *The Plants.* New York: Time-Life Books. See especially "The Moulding Forces of Climate," Chap. 6.

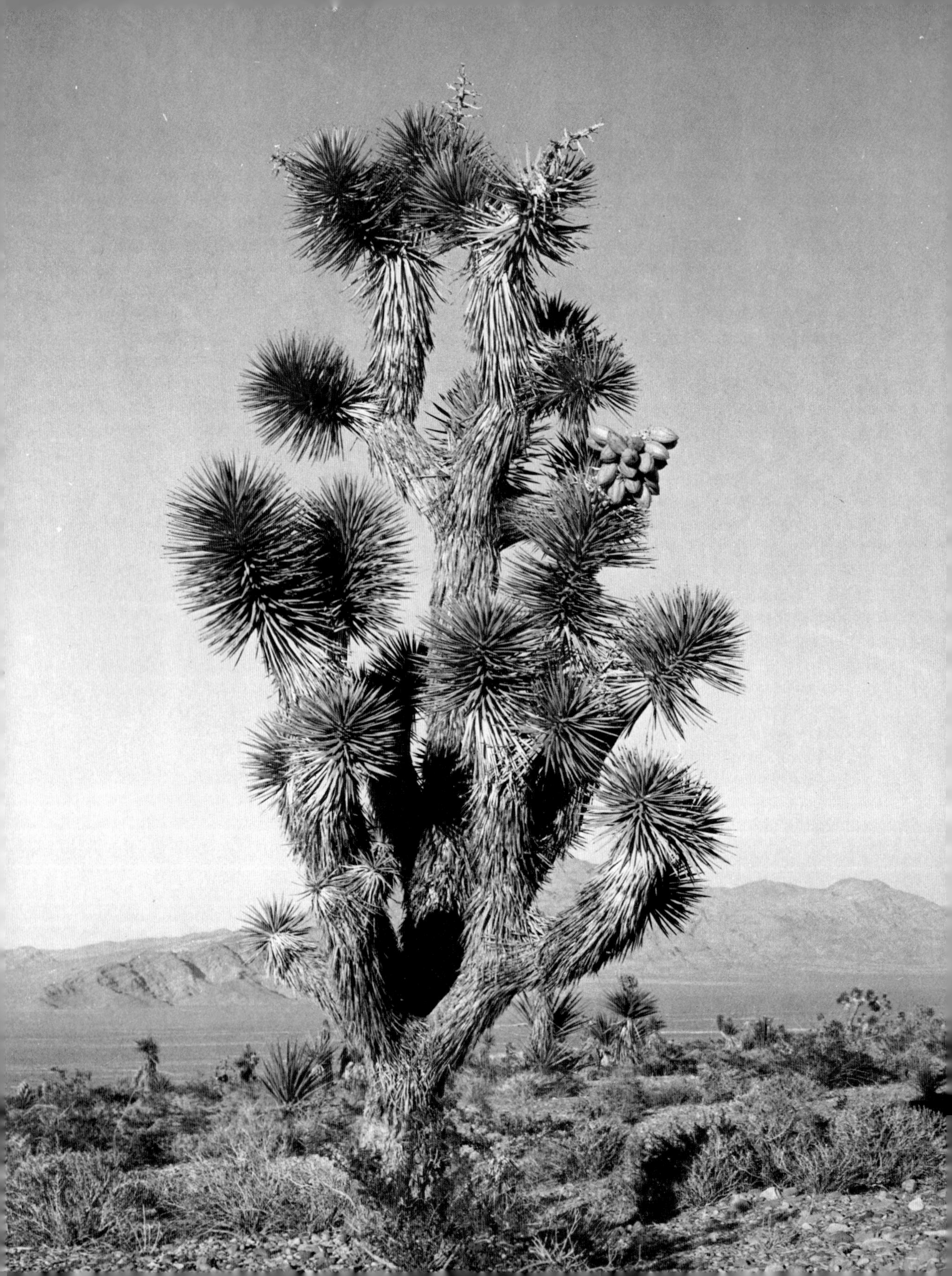

EIGHT
EARTH'S VEGETATION

Climate exercises a very strict and often direct control on the development and distribution of plants, animals and soils. . . . On the other hand, climate itself is profoundly affected by the nature of the vegetation cover; wind speed, temperature regime and atmospheric humidity just above the surface of the ground beneath the shade of dense forest are remarkably different from those at just the same height above soil level in nearby grassland.

VEGETATION AND SOILS: A WORLD PICTURE,
S.R. EYRE

So far, we have considered a number of important cycles: the cycle of the changing seasons as the earth revolves in its orbit about the sun; the cyclic heat balance reached between the sun and the earth's atmosphere; and the hydrologic cycle. These cycles are important to the continued operation of the systems that affect our environment: the atmosphere (air) and the hydrosphere (water). We will now examine the third great system, the biosphere.

The **biosphere** (from the Greek *bio,* meaning "life") is sometimes referred to as the life zone; it is here that earth's life forms are able to exist. Its base begins somewhere within the upper reaches of the earth's crust, and it extends an unknown distance out through the lower atmosphere. There is a close relationship within this zone among all the elements—soils, vegetation, moisture, air temperature, and the varied life forms—and earth scientists refer to the biosphere as a **biotic complex,** or **ecosystem.**[1] This means that a change affecting one element of a system can affect all the others. Our immediate concern is with one element of the ecosystem: vegetation and its cycle of life.

THE VEGETATIVE CYCLE

The **cycle of vegetation** is, to many people, a wondrous phenomenon. In early spring you may have paused to wonder at the new buds bursting forth from the once-naked limbs of a tree in your yard. When this happened, you may have discovered within you a growing feeling of awe, a feeling of reverence—a feeling that you were witnessing the creation of life. Then, after a long, hot summer, comes the fall season. Commercial crops are harvested, and the natural vegetation around you begins to show signs of weariness. Grasses and shrubs wilt and leaves turn color and fall from the trees. You may then have experienced a feeling of sadness, as though you were watching death reach its hand over the landscape. Did you then realize that you had observed the beginning and the end of a life cycle? Well, if you experienced these feelings, you were not alone, because such sentiments have been an integral part of the life experiences of many individuals throughout history.

Religion and Vegetative Cycle

The religions evolved by the peoples of ancient times speak plainly of their reverential attitude toward the grasses, shrubs, and trees they knew and of their belief that each person was a part of nature, not above it. As an example, think of the cattlemen and horsemen of the steppes of central Asia who roamed those wild pastures some 4000 or 5000 years ago. They called this land "the sea of grass," and it meant life to them and their herds. If the sky-father did not send sufficient rain, or if the sun-god did not provide enough days of warmth, the grasses died and, then, so did the cattle and horses. Many times, when conditions of drought lasted for long periods, the tribes were forced into long marches, searching for greener pastures. They soon came to associate the bountiful produce issuing from the earth as "gifts" from the earth-mother. Perhaps they did not truly understand the relationship between sky and earth or the reasons behind the changing of the seasons and the death and rebirth of the sun, or the difference between day and night. To them it was all a mystery related to the capriciousness of gods and goddesses, a capriciousness that must, somehow, be placated. The Slavic ancestors of the Russians believed that a diminutive god named Leshy ruled the forests. The Babylonians worshipped a sun-god called Shamash and hoped that he would take care of them and the world about them. The peoples of the hot and humid rain forest regions of Africa, South America, and

THE VEGETATIVE CYCLE 225

Religion and the vegetative cycle. (a) A rendering of Leshy, forest divinity of the ancient Slavs of Russia. (b) Shamash, Babylonian judge of the heavens and the earth, god of the sun, worshipping the sacred tree of life. (c) Balder, revered by Scandinavians and Germans as the messenger from God, lies dying, struck down by an arrow made from mistletoe. Among those gathered around Balder are warriors of Valhalla, the great king Odin, and evil Loki, the Prince of Darkness.

the East Indies were more concerned with the spirits of the rivers and the trees than with a supernatural figure who ruled the skies. Peoples of middle and high latitudes were forced to endure harsh winters, and it seemed to them that nearly everything died during that period when the icy north winds screamed down on their tents and villages. There was, however, one exception. The sturdy evergreen trees withstood everything the storms threw at them and so came to be a symbol of eternal life.

An ancient and beautiful legend tells the story of Balder, son of the great god Odin. Balder the Beautiful, the best beloved of the Germans and Scandinavians, was the intermediary between the sky-gods and humanity, but he was slain by the evil Loki, Prince of Darkness, who had discovered that the only living thing that had not pledged fealty to Balder was the humble parasite mistletoe. A fire-hardened dart of this normally limp and fragile plant killed Balder, sending him into the underworld. Odin's plea to the god of Hel that humanity needed Balder was only partially granted: Balder could return to the earth's surface at the end of winter, but he must return to Hel each fall.*

Similar stories throughout all the religions point to the idea that the rebirth of life each spring is associated with the rebirth of a god or, at least, the result of the will of a god. Primitive agriculturists also recognized the changing of the seasons; life was always more comfortable and easier for them during the time of spring, summer, and early fall. Some plants seemed to have more persistence than others, and these came to have religious significance, such as the aforementioned evergreen of the Germans, and lotus of the Indian Buddhists, and the maize that was reverenced by the Aztecs and Maya of Meso-America.

The earth's natural vegetation has changed considerably since those early times, changes made by human hands. Hopefully we can regain some of the old concern in the years to come.

Succession and Climax Vegetation

As we learned in Chapter 7, the type of vegetation that abounds in any given area is determined by the physical environment, which, in turn, is the result of a number of factors: the parent material from which the soil is derived, the topography of the land, the latitude, and the climate (amount of precipitation, rate of evaporation, insolation, and air temperatures). If no drastic changes have been wrought by people, the hooves and teeth of animals, or vagaries in the climate, a particular plant community will evolve. This vegetation will be "just right" for the region. It will be the **climax,** the culmination of a series of plant communities that strove to adapt to the environment. The climax vegetation will have established a balance, or state of equilibrium, with the environment.

The term **succession** refers to the changes in the plant communities, from the first types that appeared in the region to the climax. As an example, let us consider one of the Hawaiian islands, the so-called "big island" of Hawaii. The island is volcanic in origin, so all of the soils have the same beginning—but not the same ending. The northeastern slopes are on the windward side and are exposed to heavy precipitation; the southwestern slopes are deficient in precipitation but receive higher amounts of insolation. Suppose seeds, an equal amount on each side, are dropped by high-flying birds. Will the two plant communities evolve in the same way? Not likely; if the seeds

---

*One of the most interesting discourses on the close interrelationship between humanity and vegetation is Sir James George Frazer's (1960), *The Golden Bough* (New York: Macmillan).

are from plants that require considerable moisture but do not need great amounts of sunlight, those dropped on the windward slopes will thrive, and those dropped on the leeward slopes will die out. As time passes, more seeds are either washed ashore or deposited by birds. Perhaps this time the seeds are from plants that need high temperatures and can exist on little moisture. The seeds deposited on the southwestern slopes will take root; the other seeds will not be able to adjust and will disappear.

If, however, we consider an area that is bare of all other forms of vegetation, new types will try to adapt. They will take root, struggle, and die out. Then, in succession, other types will try to become established. Some will succeed and some will not. Eventually, only those that can survive in the climate will remain. Thus, the climax is reached.* The competition among plants for living space is severe, and some plants that might have survived are crowded out or "strangled" by invading plants that are better able to make a rapid adaptation to the climate and the soils. A good example of this can be seen in the Tuolumne Meadows area of Yosemite National Park. Forest fires and the intrusion of civilization reduced certain sections of coniferous forests (pines and spruces) to charred rubble. Grasses immediately appeared and filled the vacant areas. In time, however, the conifers rebounded and reasserted their "right" to the meadows. Now the meadows are receding and the forest is advancing on all sides.

What happens when humans enter the picture? When the first Viking set foot on the soil of New England and gazed about him at the thick forests, was he seeing the original climax vegetation of that region? What of the vegetation that confronted the ancestors of the Australian aborigines when they first arrived in the land "down under"? Archaeologists tell us that the Sahara of northern Africa may seem barren today, but it was not always this way. Until 1500 B.C. it was covered with great expanses of grasses and forested areas. Perhaps in this instance humans should not be held responsible; other factors may have been more instrumental in changing the vegetation that existed in those regions.

Environmental Controls

The main controlling factors of environment are: moisture, air temperature, topography, soils, and biological activity—including that of people. All are integral parts of the biosphere, and it is difficult to classify them in order of importance.

Moisture A potent factor in the development of plants is moisture. Plants must adapt over time to varying conditions, but the presence or absence of water is a strong determinant. Some plants, such as seaweed, water lilies, and sedges (called **hydrophytes,** from the Greek *hydro,* "water," and *phyton,* "plant"), can live under water and can survive at depths of about 15 feet (4 to 5 meters) because of their ability to utilize sunlight at those depths and thus allow the process called **photosynthesis** to proceed.* Other plants can live in seemingly impossible regions, such as deserts, where the precipitation is almost zero. If moisture-loving plants are transplanted to those regions, they will

---

*Some plant geographers use the term *climatic climax vegetation,* thus emphasizing the influence of climate in the establishment of the true natural vegetation of any given area.

**Photosynthesis* is the process used by plants to manufacture the food they need. They take carbon dioxide out of the air—or the seawater—and, with the aid of sunlight, use the chlorophyll within their tissues to transform the carbon dioxide into carbohydrates (sugar and starch) they require.

(a)

(b)

(c)

(d)

The effect of water on the evolution of plant communities. (a) Hot, wet and dry grasslands of the savanna type: average annual rainfall over 10 inches (250 millimeters). Cover: grasses, occasional shrubs and trees. (b) Hot, dry desert: average annual rainfall less than 5 inches (125 millimeters). Cover: occasional plants after rainshowers. (c) Hot, semiarid desert: average annual rainfall from 5 to 10 inches (125 to 250 millimeters). Cover: succulents such as cactus, Joshua trees, small and scattered shrubs. (d) Hot and moist forests, average annual rainfall over 45 inches (1150 millimeters). Cover: permanent stands of trees with little undergrowth.

quickly die, even though the deserts do have occasional rainstorms in which rain falls in a hurried downpour and quickly soaks into the ground or rushes off through *wadis* (normally dry streambeds similar to gullies). Sometimes there are seeds just waiting for this opportunity. They have been lying underground, dormant and apparently lifeless. With the addition of this new moisture, they quickly germinate, and the desert blooms for a short period. Seeds are produced that fall into the soil and wait their turn for next year's brief showers.

Deserts lying within reach of warm, moist maritime air masses gain a little more precipitation, and this allows a considerably different type of vegetation to evolve. The Sonoran Desert of northern Mexico and the American Southwest may seem rather barren at first glance, but a closer look discloses an exciting variety of plants, the giant saguaro cactus, the Joshua tree, and many other species of cacti, that can thrive in a hot desert that has only 5 to 10 inches (120 to 254 millimeters) of precipitation.

Around the edges of the dry deserts are grasslands called prairies and savannas. Although the air temperatures in these regions are generally high, precipitation of over 10 inches (254 millimeters) is able to support vast expanses of varying types of grasses and, primarily in the savanna regions, occasional clumps of trees or shrubs.

Forested regions require plentiful precipitation, and some of them receive over 45 inches (1120 millimeters) of rainfall annually. Where the forests are composed of towering trees with heavy foliage, sunlight will not be able to reach ground level, and undergrowth will be scanty or nonexistent.

Air Temperature In conjunction with the availability of moisture, air temperature operates to determine the size of plants, their leaf structure, their root systems, and whether they can grow in clumps, in large stands, or singly. Temperature is important in controlling the growth of all plants because it affects the chemical activity (movement of nutrient-bearing sap and the process of photosynthesis), periods of dormancy, production of seeds, and amount of time needed for an adequate growing season. If temperatures are high, evaporation from the soils will be high, as will transpiration from the plant itself. If moisture is lacking, plants in this type of area will be forced to adapt by developing leaves that are thick and spongy in order to retain the moisture. If both temperature and precipitation are high, plants will cover the earth surface, and the leaves may be thin, perhaps filmy, and transpiration rates will be high.

Temperature can also limit the spread of certain types of plants. The July 50°F (10°C) isotherm (see Fig. 4-5), for example, is the limit of tree growth in the Northern Hemisphere. Beyond this line is the tundra region, a land of lichens and sedges that can survive in soil that is permanently frozen (**permafrost**) beneath the surface of the ground. The same control is asserted in mountainous regions where the **timberline** reflects the temperature boundary above which trees cannot grow.

Topography The shape of the land, its **topography,** is also important as a control in the evolution of plant communities, especially with regard to the **slope** and the **aspect** of the area. Many areas of the earth are fairly flat and level, but most have some slope. If the terrain is steep and rugged, soils have a difficult time evolving because of their inability to retain water, which tends to run off swiftly. If slopes are exposed to constant winds, there will be an increase in evaporation and thus less moisture retention. Flat bottomlands may be unable to rid themselves of excess moisture, and swamps or bogs may result.

Aspect refers to the direction a slope is facing in relation to the sun. Slopes facing the sun's rays will receive more insolation than slopes facing in the opposite direction. As an example, consider the accompanying photograph, which shows a section of mountainous terrain in the Northern Hemisphere. The area facing the south has a light cover of grasses and clumps of shrubs; the area facing north is covered with heavy vegetation. If both areas receive about the same amount of precipitation, why does one slope have heavier ground cover than the other? For one thing, moisture is more easily retained on the north-facing slope because there is less insolation and thus a lower rate of evaporation.

Near the equator or in the lower latitudes, slope is more significant than aspect because all slopes will receive about the same insolation. If precipitation is heavy, soils may be washed down from higher elevations, thus limiting the growth of plants along ridges and creating steeper slopes. In tropical areas with high temperatures as well as high precipitation, the cover of vegetation may be rather dense on all mountain slopes. If a timberline exists, it will be at a much higher elevation than on mountains located in the mid- or high-latitude areas. Low-lying areas in hot, moist climatic regions will be swampy or marshy if there is not enough slope to carry away the excess moisture. In such an area the water table is usually high and close to the surface, so the soils tend to be waterlogged.

Topography as a factor in the development of plant communities. Notice the difference in aspect: The north-facing slope has heavy vegetation; the south-facing slope has very light vegetation.

Soils Another effective determinant in the development of plant communities is the soil (a subject that will be covered in detail in Chapter 9). If a soil has had time to evolve from its beginnings as a part of the earth's bedrock in a humid region, and if it contains sufficient amounts of **humus** (decaying organic material) or nutritious chemicals, it will be able to retain more moisture and, thus, support a luxuriant vegetative cover. On the other hand, soils evolving in arid or semiarid areas (where evaporation exceeds precipitation) may have a considerable amount of humus, but they are loose-grained and therefore more porous and allow moisture to drain away rapidly. Many plants in dry regions are drought-resistant and are referred to as **xerophytic** (from the Greek *xeros*, "dry").

Biological Activity The term **biological activity** refers to the action of living organisms, such as bacteria, worms, rodents (such as gophers and moles) and, quite definitely, humans. Humus, mentioned earlier as necessary for a good vegetative cover, results from the activity of bacteria, which cause decomposition of leaves, blades of grass, limbs (and even the entire trunks) of trees, and bodies of dead animals. When precipitation occurs, the water soaking into the ground carries the humus particles with it, thus enriching the soil. Worms and small burrowing animals move about through the uppermost layers of the soil, manufacturing air spaces and making channels for water to infiltrate by gravity to deep-lying roots of shrubs and trees. All of these factors contribute to the evolution of soils, but the same cannot always be said for human activities. With the ever-increasing population and the subsequent spread of urbanization, streets and highways are paved, houses and buildings are erected, the native water supply is depleted, and the soils are sometimes excessively robbed of their nutrients. As an example, cattle ranchers moved into the Great Plains of the United States, stripped the land of its climax vegetation, and broke up the *sod cover* (the upper soil layer that is tightly knit because of the extensive root systems of the native grasses). These "sodbusters" of the period from 1870 to 1930 practiced serious overgrazing of the land, which resulted in the "dust bowl" of Kansas and Oklahoma.[2]

Another problem facing humanity today is the increasing amount of air pollution created by the extensive burning of fossil fuels. Sulfur and nitrogen oxides are emitted into the atmosphere, where they change into strong sulfuric and nitric acids, which result in *acid rain*. Recent scientific studies have proven that there has been serious damage to the soils and vegetation of Europe, Canada, northeastern United States, and regions surrounding the giant cities of the world. No one has a definitive answer to the problem at this time, but one possible solution would be a change to alternative forms of energy, thus lessening the potential for disaster.*

THE STRUCTURE OF PLANTS

Natural vegetation is "wild" vegetation, not yet tamed or altered by human intervention. It is to be differentiated from domesticated varieties (which we might term cultural vegetation), and it varies from region to region all around the world. We have discussed the controlling factors of the environment, and it would seem that climate is the most important of all. The climate of one place is not neces-

---

*For an interesting and in-depth look at the problems created by our energy crisis, read Ehrlich, Paul R., Ann H. Ehrlich, and John P. Holdren (1977), *Ecoscience: Population, Resources, Environment* (San Francisco, W. H. Freeman and Company).

(a)

(b)

(c)

(d)

(e)

(f)

Vegetation life forms. (a) Algae formed on a body of still water. (b) Giant sequoias (redwoods). (c) Grasslands in Denmark. (d) Ferns and lianas in Corkscrew Swamp, Florida Everglades. (e) Bird's-nest fungus in the Hoh rain forest. (f) Moss-covered maple on the Hoh River Trail (Olympic rain forest, Washington).

sarily the climate of another; other factors must be considered, such as the position of an area relative to mountains, deserts, plains, rivers, lakes or oceans, and winds. All of these controlling factors combine to create plant forms that differ from each other in the structure, shape, and texture of the leaves and their coverage, size, and periodicity.†

Life Forms of Plants

There is great variation in the life forms of plants, from trees such as the towering *Sequoia gigantea* (more commonly known as the giant redwood) down to the minute, simple-celled systems called algae and fungi. Trees are the largest of all plant systems and are not difficult to distinguish from other plants since they have a single trunk from which many branches are usually put forth. Shrubs are tough, woody plants that are hardy, low-growing systems. They are not limited to one trunk but have several stems sprouting from the nodule and root base. Another type of woody plant is the liana, a plant system that has many climbing vines. Included in this group are wisteria, cucumbers, peas, poison ivy, and beans. Some of these climbers pull themselves up with stems that operate like tiny fingers, others use thorns as grappling hooks, and others wind their tendrils around the plant they are ascending and so move higher and higher.

The next group of plants, grasses, are uncomplicated systems with simple leaf structures and tubular stems. Some grasses, including corn, wheat, rye, oats, and barley, which might be called food grasses for humans, and some, such as fescue and Kentucky bluegrass, are used for animal food and lawn covers. Whether annual or perennial, grasses become dormant after flowering and producing seeds. Related to the grasses are herbs, plants that also lack woody stems and may be either annual or perennial in their life cycles.

Plants that grow and thrive on top of other plant forms are called **epiphytes** (from the Greek *epi,* "upon"). Wind-carried seeds become lodged on branches or in the bark of trees and, in time, the seeds germinate. Epiphytes move through their cycle completely out of touch with the ground, yet they are not parasites; that is, they do not take nourishment from their host tree. The most common epiphytes are orchids, ferns, mosses, liverworts, and lichens, and they are usually found in the humid tropics.

At the very bottom of the plant scale are the fungi and algae, members of the simplest of all plant forms and systems, the **thallophytes** (from the Greek *thallos,* "young shoot or frond"). Many of the thousands of life forms within this system are single-celled and, thus, have no need for a vascular network (such as tiny blood vessels) to carry nutrients. Because algae live in water, they are able to absorb the required moisture directly through the walls of the cell. The most common forms of algae are diatoms and seaweed. Fungi are represented by, for example, bacteria, yeasts, mildew, and mushrooms and are considered parasitic; that is, they live on and draw nourishment from dead or living plants and animals. Neither algae nor fungi have root systems or stems. When algae and fungi live together, almost as one plant, the result is called a lichen.

Plant size is another way to identify different vegetation life forms; plants range in size from tiny, almost invisible bacteria to gigantic trees. For example, grasses vary greatly in size, from the kinds we use as lawn covers, which

---

†For a detailed description of plant classification by structure, see Pierre Dansereau (1957), *Biogeography: An Ecological Perspective* (New York: Ronald Press).

can grow to heights of 2 or 3 feet (610 to 914 millimeters) if left unattended, to the savanna grasses of the African veldt, which can grow tall enough to hide the movement of a herd of elephants. Consider also the grass called bamboo, which can attain heights of 100 feet (30 meters).

Coverage and Periodicity

Coverage can also be used to classify plants. Some trees grow singly, such as East Africa's acacia, while other grow in clumps surrounded by grasses. Still others flourish in great stands as, for example, the teak forests of Burma, where the trees grow very close together, crowding out other plant forms. Climate sets the boundaries for the different systems, but competition for sunlight and space is also an effective determinant.

The term *periodicity* is used to express a plant's response to the cycle of the seasons. For example, some lawn grasses never seem to die but continue growing (and needing to be mowed every week) throughout the year, while others grow rapidly through spring and summer, than die back when fall arrives. This also applies to shrubs, herbs, and trees. Some go dormant and lose their leaves; others seem to retain their leaves from season to season. The period of dormancy can be likened to a period of sleep, when the life processes slow down because of lack of sunlight or precipitation, or when the temperature drops too low. Trees that drop their leaves during this time are called **deciduous.** Trees that seem to keep their leaves are called **evergreen.** We say that they only *seem* to retain their leaves because while the tree always has leaves, there is a continual shedding of old leaves to make way for the new. Some forests may appear to be evergreen but actually have trees that take turns in dropping leaves throughout the year. Such forests are classified as *semideciduous.*

Leaf Types

Leaves of different plants vary considerably, providing another way to classify plant systems. There are two basic leaf types: **broadleaf** and **needle leaf** (Fig. 8-1). Needle leaves are found on conifers (cone-bearing trees) and certain species of cactus. In the case of the conifer, the needle is so constructed that a single vein runs through its center. This vein collects moisture and carbon dioxide from the air and also transports plant sugars to other sections. Needle leaves also resist excessive losses of moisture during periods of high temperatures and evaporation and are not subject to freezing. Broad leaves are found on flowering plants and come in many different sizes, shapes, and textures. The most common type of broad leaf is termed *membranous* because it is a thin, flat layer of tissue of average thickness and is soft in texture. In contrast to the needle leaf, the broad leaf has many veins—main, secondary, and small netted veins—and all perform the necessary task of manufacturing food for the plant. Oxygen and moisture (by transpiration) are emitted by the leaf, and it has been estimated that 1 acre (0.4 hectare) of corn will transpire more than 300,000 gallons (1.1 million liters) of water during a single season.[3]

Other types of broadleaf plants, such as ferns, have thin, almost transparent leaves, which are called *filmy*. Plants such as holly and mesquite exist in a dry habitat, and the lack of plentiful moisture causes their leaves to be tough and thick. These are termed **sclerophyllus** (from the Greek *scleros,* "hard"). The thick, spongy leaves of a cactus are called **succulents** (from the Latin *succulentus,* "full of juice") because they hold a considerable amount of water. These types evolve as a response to an area of light-to-scanty rainfall and a high rate of evaporation. To protect themselves from moisture loss, the plants evolve tough, reflective skins and spongy interiors.

Pine   Joshua   Hemlock

NEEDLE LEAF

Sugar Maple   Magnolia   Cinquefoil

BROAD LEAF

Beavertail Cactus   Prickly Pear Cactus

SUCCULENTS

FIGURE 8-1 Leaf shapes and textures.

DISTRIBUTION OF NATURAL VEGETATION

As you now face a consideration of the tremendous variety of plant life spread across the earth, you may be wondering how this natural vegetation can be classified. You may be mentally comparing and contrasting the vegetation of your own neighborhood with other plants that somehow exist on hot, burning deserts, the frozen permafrost lands of the Arctic, or in the steaming jungles of the tropics. If you were a plant geographer intent on serious research into the biosphere, you would adhere to a very precise vegetation classification system that might well be closely related to Köppen's climate classification. However, in order to obtain a simplified, general overview of our earth's natural vegetation we will create only three major divisions—forests, grasslands, and deserts—and then formulate further subdivisions as needed.*

---

*As we proceed in this discussion of earth's vegetation, it would be well to check back from time to time to the climate maps of Figs. 7-1, 7-2, 7-5, 7-9, 7-12, and 7-17.

Forests

Figure 8-2 is a map of the earth's major forest regions: tropical rain forests, tropical deciduous, Mediterranean woodland and scrub, broadleaf summer deciduous, midlatitude coniferous, and the taiga.

Tropical rain forests, sometimes referred to as equatorial rain forests, are the most prolific, luxuriant, and extensive of all the forests on and near the equator. These forests are commonly referred to as *selva* (the Brazilian term for this type of woodland) and are typical through much of middle America (from Yucatán southward); in Venezuela, Colombia, Ecuador, and Brazil of South America; in portions of West Africa and Zaire; on the east coast of Malagasy; on the west coast of southern India; in Bangladesh and South-East Asia; in the East Indies to New Guinea; and on the east coast of Australia.

Most of the trees are of the broadleaf variety. They appear to have evolved a workable system of stratification; that is, they "layer" themselves, and each level is allowed so much room and just so much sunlight. Therefore,

Tropical rain forest in Puerto Rico. This area receives 150 inches (3750 millimeters) of rainfall annually.

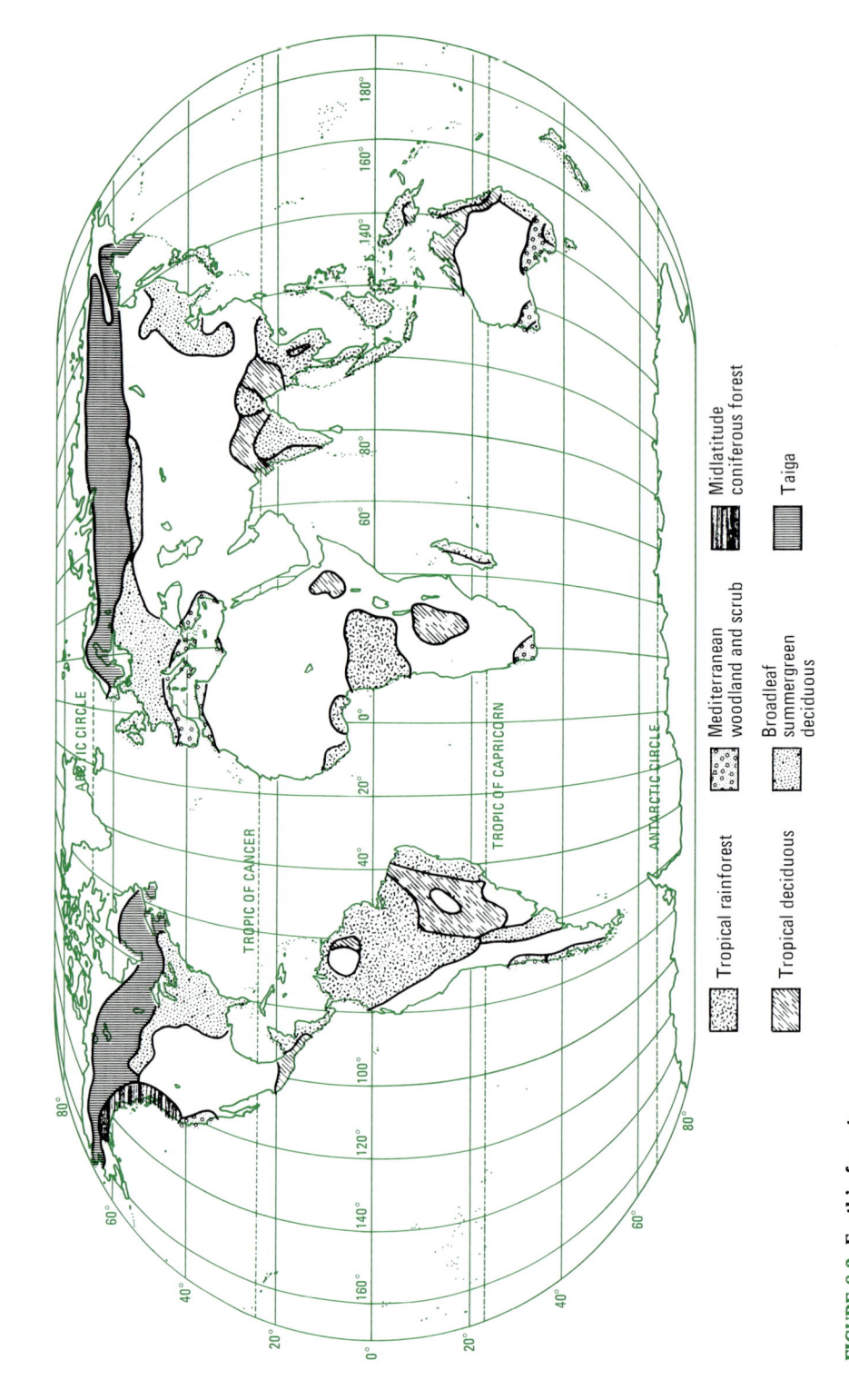

FIGURE 8-2 Earth's forests.

adapting to its situation, a tree will grow to a particular height and no more. While there are a number of vines, lianas, and creepers at the higher levels, there is little in the way of undergrowth at ground level because of the intense overcrowding and the leaf cover, which prevents sufficient amount of sunlight to reach incipient shrubs. You may question that statement if you recall "jungle" movies that you have seen, but the film showing an explorer hacking his way through tangled undergrowth is not, in reality, portraying the true selva of the tropical rain forest. What you have seen is the shrubbery that thrives among the trees alongside the bank of a river or a highway in the midst of a rain forest. In this situation, small plants are given life because they are exposed to the warming rays of the sun. This heavy, nearly impenetrable "jungle" is more properly termed a *galeria* forest.

The tropical rain forest endures because of heavy precipitation (the convective type), which ranges from 80 to 200 inches (2032 to 5080 millimeters) annually. This creates problems because the soils are always wet and thoroughly leached (drained) of nutrients. Thus, human inhabitants of these regions who clear the land for agriculture must contend with infertile soils.

Tropical deciduous forests are found where precipitation and evaporation engage in seasonal combat; the regions are sometimes wet and sometimes dry. There is too much rainfall for a broad grassland, yet not enough for a true rain forest. There are some grassy areas, but trees dominate. Because of the periodic dry seasons, the trees experience dormant periods and so are not evergreen. One of the most commercially usable forests of this type is to be found in Burma, where large stands of teak and mahogany and other important timber products cover much of the land.

Mediterranean woodland and scrub is the typical woodland to be found around the Mediterranean Sea, central Chile, the southern tip of Africa, the southern extremities of Australia, and the coastal area of southern California. Plants in these regions include olive, fig, chestnut, and live oak trees, as well as low-growing woody bushes and shrubs, called *maquis* by the peoples of the Mediterranean countries and *chaparral* by the Californians. Since these are areas of semiaridity because of their location on the western sides of the continents,

Mediterranean woodland and scrub, called chaparral in California.

the presence of cool ocean currents that pass offshore, and the dominating influence of the subtropical high-pressure belts, the trees and other plants have had to adapt to relatively high temperatures and low precipitation.

Midlatitude coniferous forests contain cone-bearing trees typical of many areas, but nowhere are they so dominant as in the region stretching from the northwestern United States through coastal Canada into Alaska. Here, there are commercially usable stands of redwood, Douglas fir, cedar, and spruce.

Broadleaf summergreen deciduous forests are located in southern Canada, Europe, China and Japan, southeastern Australia, Argentina, and much of the southern, midwestern, and northeastern sections of the United States. The climate of these areas, which creates long, hot summers, brings forth a wide variety of deciduous trees; hickory, birch, maple, and walnut are the most typical.

Taiga forests (sometimes referred to as **boreal**) are found in the northern two-thirds of Canada in North America and from Sweden in Europe across Russia and Siberia to the Pacific coast. This vast expanse of conifers is the earth's largest forest. In the northern reaches it is a mixed forest comprised mostly of pine, spruce, and fir, and also including birch and maple. The climate is not conducive to human occupancy, which you will see if you contrast the taiga regions with the distribution of population shown in Fig. 1-1. The northern boundary of this region, where the timber resources are virtually untapped, roughly follows the 50°F (10°C) isotherm and blends into the tundra grasslands bordering the Arctic Ocean. There are no trees beyond this line; the winters are too long, the summers too short, and there is insufficient precipitation.

Grasslands

The earth has many areas where grasses, the savannas, prairies, steppes, and tundra (Fig. 8-3), are the predominant form of plant life.

Savanna grasslands might more properly

Taiga of central Alaska.

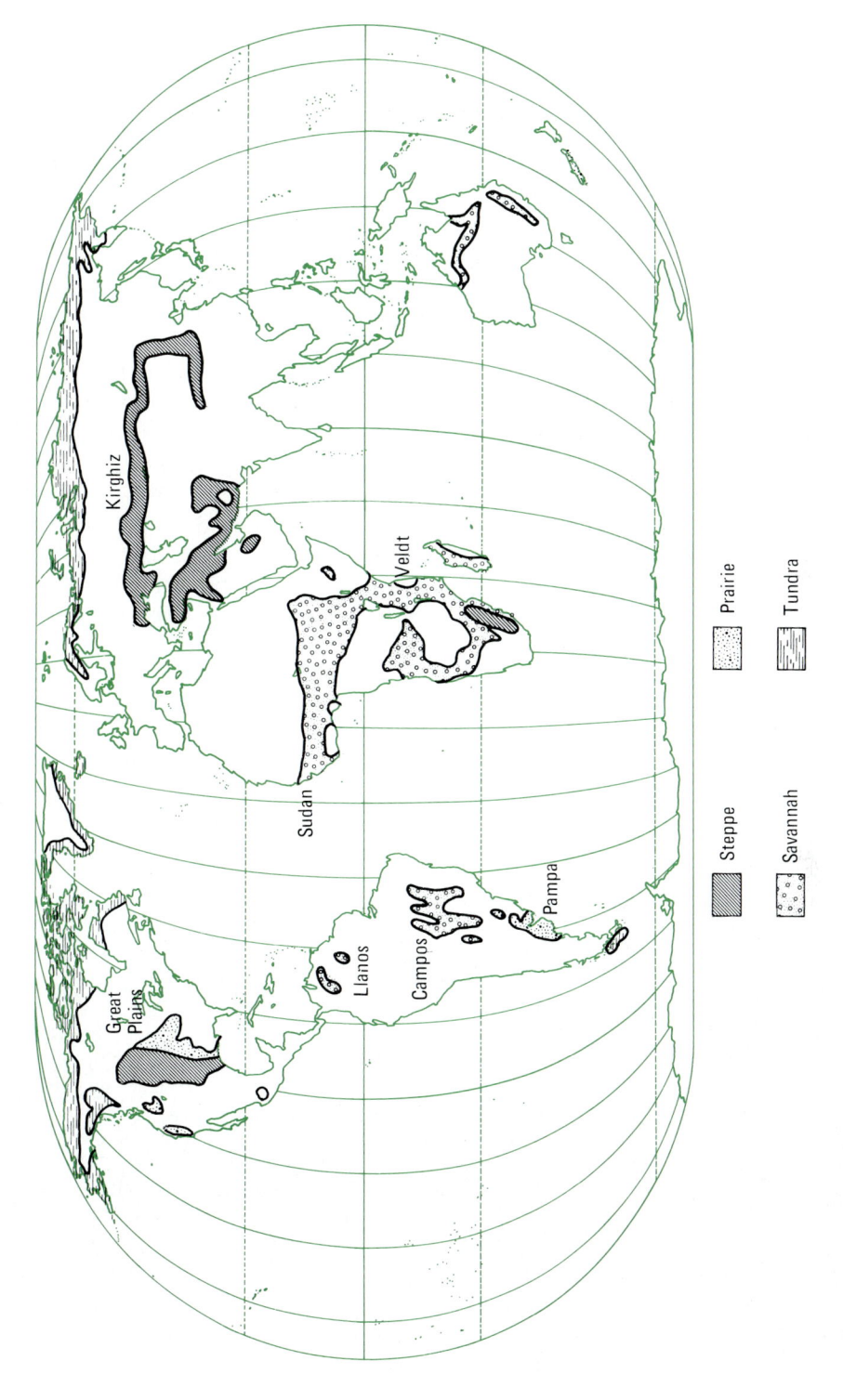

FIGURE 8-3 Earth's grasslands.

241

be called tropical grasslands because they exist roughly from 20° north to 20° south latitudes, and they are included within climatic regions that alternate between wet and dry periods. Trees are present, but usually in small clumps or singly. One might find the umbrella-shaped acacia, palm, and banana trees, as well as shrubs and bushes, especially along streams or around water holes. Savanna grasses are tall and are normally very tough and not well liked by domesticated grazing animals; thus they have not yet been commercially used to any great extent—a fortunate circumstance for the wild herbivores of Africa. The most famous of the earth's savannas are the Sudan, which stretches across Africa south of the Sahara, the llanos of southern Venezuela and eastern Colombia, the campos of south-central Brazil, and the veldt of East and South Africa.

Prairies and **steppes** are regions of the midlatitudes where short grasses are dominant and where trees and large shrubs are found only along streams. Grasses are shorter in steppe regions than in the prairies, but both are quite suitable for grazing animals (if the grazing is not overdone). It is interesting to note that the steppe regions of northern Turkestan and the Ukraine of the Soviet Union have been used for pasturage for 6000 years, and the very rich, black soils have not suffered excessively. Wheat growing was begun centuries ago in the Ukraine with great success, but the conversion of the grazing lands in the eastern steppes to grain lands in 1954 has not fared as well. This is because this region suffers from an unreliable rainfall, periodically creating poor harvests.

Tundra (Russian for "frozen land") is the name applied to lands bordering the Arctic Ocean, a vast region located north of the sixtieth parallel that is permanently frozen beneath the surface, in some places down to depths of 700 feet (210 meters). This land of permafrost cannot support large forms of tree life, but a number of plant forms manage to

The steppes of the Ukraine: harvestime.

Typical tundra landscape during the short summer growing season at Cape Dorset, Canada's Northwest Territories.

exist in spite of (or because of) the scanty precipitation and the great extremes in temperatures. There are short grasses, sedges (grasslike plants that have stems without joints), mosses, lichens, and tiny 6-inch (152-millimeter) willows, which remain so small because of the very short (two months) growing season each year.

Deserts

The earth's **deserts** (Fig. 8-4) do not support large populations (compare with Fig. 1-1) simply because most people prefer cooler and moister climates. In ancient times, the few nomads who managed to exist in these inhospitable regions tended to favor the rain-god over the sun-god—for good reason. More recently, the discovery of oil reserves in some of the most isolated and desolate spots of the Sahara has brought people into the drilling stations and, with the aid of air conditioners, they are able to survive with some comfort.

The prevailing vegetation forms are called desert shrubs; these are deciduous plants that are generally leafless until rain showers force them into immediate flowering, when leaves burst forth and seeds are expelled onto the dry soil. The few evergreen plants have shiny, leathery, and very small leaves and extensive root systems; they include cactus, Joshua trees (given that name by viewers who were reminded of the Biblical prophet who always spoke with outstretched arms), mesquite (able to survive drifting sand dunes because of its far-reaching branches that manage to stay above the dune surface), sagebrush, and the creosote bush (most common desert shrub of the Sonoran Desert of the United States). Spring in the desert can be a time of beauty, when a great variety of wildflowers such as daisies, dandelions, and sand verbena dots the landscape.

Highlands

The vegetation to be found in the higher elevations of the earth, such as the Rocky Mountains

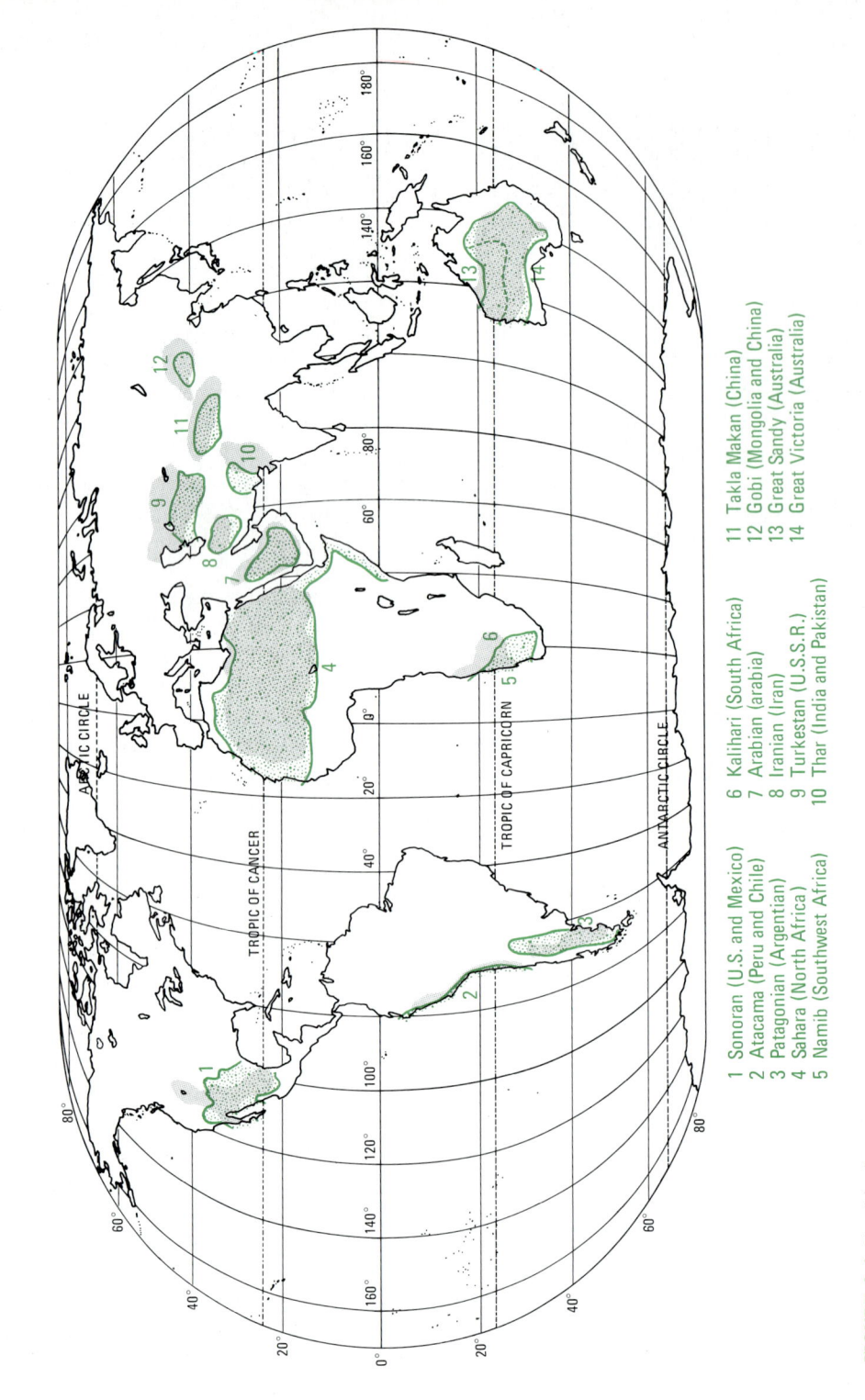

FIGURE 8-4 Earth's deserts.

1 Sonoran (U.S. and Mexico)
2 Atacama (Peru and Chile)
3 Patagonian (Argentian)
4 Sahara (North Africa)
5 Namib (Southwest Africa)
6 Kalihari (South Africa)
7 Arabian (arabia)
8 Iranian (Iran)
9 Turkestan (U.S.S.R.)
10 Thar (India and Pakistan)
11 Takla Makan (China)
12 Gobi (Mongolia and China)
13 Great Sandy (Australia)
14 Great Victoria (Australia)

of the United States and Canada, the Andes of South America, the Alps of Switzerland, or the Tien Shan of central Asia, varies considerably with elevation above sea level and latitude. The timberline, for example, is over 10,000 feet (3000 meters) on or near the equator, but it is down to less than 6000 feet (1800 meters) in the Canadian Rockies. Above this boundary one finds vegetation so similar to that of the tundra regions that it has been termed alpine tundra vegetation.

As you move from the foot of the mountains upward, you move from plant community to different plant community through what plant geographers term "vertical zones of vegetation," almost as if you were moving from the subtropics northward toward the Arctic. For example, let us consider the massive Ruwenzori Mountains of central Africa. On the west side of these mountains tropical rain forests approach the foothills, but on the eastern side the savanna grasses encroach, along with scattered clumps of deciduous trees. If you continue up this eastern slope, you leave the scrub woodland, chop your way through bamboo thickets, hike through pine forests, and eventually emerge into the tundralike grasses to stare up at the snow-capped peak, which is devoid of any vegetation.

If you were to pass a deep valley on the way, you would notice that the ecosystems change with decreases in elevation down the sides of the valley. These abrupt, lateral changes that occur within short distances make it almost impossible to include the variations in the highland vegetation on a map of the earth's vegetation. The vegetation of vast areas can be averaged out and generalized for most atlases, but the highland regions will be either ignored or "blacked out."

VEGETATION AND CLIMATE

In our discussion of the earth's weather patterns and climates we emphasized that weather is a short-term phenomenon and climate is the weather of a region over a long period of time. If you were to travel to a region far from your home, you would know immediately what the weather in the new place was like, but you could not know what climate prevailed unless you had studied the climatic data of the region prior to your arrival or you had knowledge of the vegetation typical of the climate.

Climate exerts a rigorous control over the evolution of any biotic ecosystem, and by carefully examining an area's vegetation, you can learn about its climate. Vegetation will tell you about the ratio of precipitation to evaporation, air temperatures, and moisture retentiveness. With this information you might be able to determine climate with a reasonable degree of accuracy. Remember how Köppen used this procedure in 1918 to devise the first acceptable map of the earth's climates?

We might now pose the question, Does vegetation exert any appreciable influence on any area's climate? We know that plants participate in maintaining the hydrologic cycle and the water balance by absorbing moisture from the soil and air and returning it to the air through the process of transpiration. We also know that plants remove carbon dioxide from the air and replace it with oxygen. But the question remains: Does vegetation affect climate, and if so, how much is its influence felt?

We do not yet have the answer, although humanity has interfered with the earth's vegetation in many areas, eliminating some climatic climax communities and replacing them with domesticated varieties. We may obtain some of the answers when we have studied the results of wide-ranging plans, such as the one recently introduced by the Russians into western Siberia.

This plan is to open the vast taiga land that lies between the Ural Mountains and the Ob River to agriculture and livestock production.[4] As one considers the effect of the taiga on the region's climate, a number of questions arise. Will the loss of 25 percent of the taiga affect the climate? Will constant irrigation affect the permafrost condition of the land? Will the temperature extremes of Siberia be modified? Will the summertime monsoon-producing central Asian low-pressure system be affected? Time will tell.

CASE STUDY   BIOFUEL—ENERGY FROM VEGETATION

When most people consider the terms *vegetation* and *energy* they probably would think only of firewood for heat or industrial use or plant food that supplies us with the necessary carbohydrates and proteins to fuel our bodies. Some might think of fossil fuels, which resulted when the lush vegetation of ages past became covered with layers of earth or bodies of water and was converted through time into petroleum or coal. However, as we now know, these fossil fuels are dwindling in supply and may be gone by the turn of the century. Thus, it is somewhat reassuring to realize that our present-day vegetation is—and will continue to be—an invaluable energy-producing resource because it is renewable.

You might now ask, "How can this be? When you burn a log or eat a piece of corn, they are gone forever!" The answer is that we can cut down a tree or harvest the corn and then plant another tree or reseed the cornfield. Brazilian sugarcane producers have shown us that this is possible. After they have extracted the sugar from the cane, the stalks are burned to produce alcohol, which is then mixed with gasoline to produce *gasohol!* Sugarcane operators in Hawaii take the sugar, then use boiler/energy-producing machines to create electricity, which they then sell to the local power companies.

The potential for biofuel (sometimes referred to as "biomass") energy is tremendous and almost limitless because of the vast amount of agricultural land and forests. We can use rice hulls, straw from barley, wheat, and rice, cotton (extruded by the gins), walnut shells, orchard prunings, and unused waste from lumber mills.

However, with the increasing need for fuel,

Biofuel in action. This prototype "gasifier," developed by John Goss, Professor of Engineering, University of California, Davis, uses farm and forest residues to produce a methane gas flow that fires a boiler to produce electrical energy. It can supply equivalent energy at almost the same coast as natural gas and electricity today.

there is also the continued rise in the world's population. Because these humans need land, the available energy-producing crops will be replaced by energy-consuming humans.

REVIEW AND DISCUSSION

1. Why did people who resided in the world's desert regions differ in their attitude toward the gods and spirits of nature and the sky from people who lived in the colder, northern regions?
2. Explain the terms *epiphytes, thallophytes,* and *hydrophytes.*
3. What is a liana?
4. What is the prime difference between a shrub and a tree?
5. Explain the difference between *deciduous* and *evergreen.*
6. Explain the term *sclerophyllus.* How does it differ from the term *succulent?*
7. Why do some trees develop needle leaves?
8. What is the main difference in cultural and natural vegetation?
9. Explain the meaning of the term *plant succession.*
10. Explain climatic climax vegetation.
11. What is photosynthesis?
12. What is the effect of high air temperatures on plant growth?
13. What is permafrost?
14. What is meant by *timberline?*
15. Explain the meaning of the term *aspect,* with respect to sunlight.
16. Explain the term *xerophytic.*
17. What type of region would be most apt to have rain forests? Grasslands? Deserts?
18. Explain the differences between savannas, prairies, and steppes.
19. Explain the term *coniferous.*
20. What is the boundary between the tundra and the taiga?
21. How would the Third World's agricultural nations benefit from biofuel as a source of energy?
22. Do you believe there is a future in producing energy from vegetation?

NOTES

[1] S. R. Eyre (1963), *Vegetation and Soils: A World Picture* (Chicago, Ill.: Aldine), p. 4.
[2] Eyre, p. 108.
[3] Frits W. Went (1963), *The Plants* (New York: Time-Life Books), p. 82.
[4] Robert G. Kaiser (March 16, 1974), "Russian Plans to Open Vast Frontier," *Los Angeles Times.*

ADDITIONAL READING

Allen, D. L. (1967). *The Life of Prairies and Plains.* New York: McGraw-Hill.

Cain, S. A. (1971). *Foundations of Plant Geography.* New York: Hafner.

Dansereau, Pierre (1957). *Biogeography: An Ecological Perspective.* New York: Ronald Press.

Farb, Peter (1969). *The Forest.* New York: Time-Life Books.

Haden-Guest, Stephen, et al. (1956). *A World Geography of Forest Resources.* New York: Ronald Press.

Halacy, D. S., Jr. (1977). *Earth, Water, Wind, and Sun: Our Energy Alternatives.* New York: Harper & Row.

Leopold, A. S. (1961). *The Desert.* New York: Time-Life Books.

Polunin, Nicholas (1960). *Introduction to Plant Geography.* New York: McGraw-Hill.

U.S. Department of Agriculture (1949). "Trees," *The Yearbook of Agriculture.* Washington, D.C.: U.S. Government Printing Office.

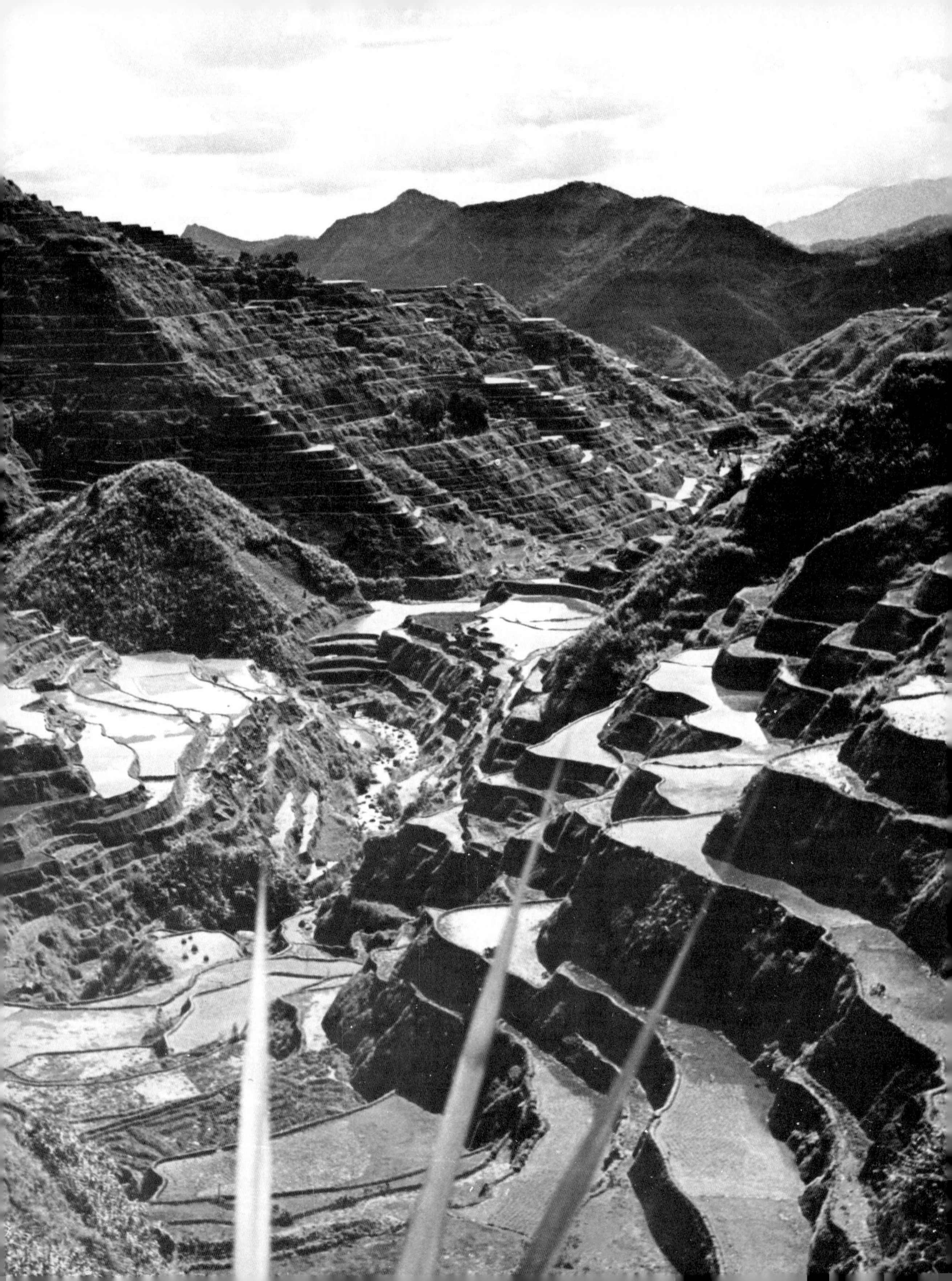

NINE

THE EVOLUTION OF SOILS

A dictionary might define soil *as something that is unclean; a filthy, defiling substance; something dirty. Or,* soil *might be defined as the upper layer of the earth composed of loose surface material. Soil scientists, however, are more apt to state that soil is not merely a part of the lithosphere, but a very important and complex segment of the ecosystem, the biosphere. Soil is the layer of the earth that supports humanity.*

The powerful forces of weathering and erosion operate to provide us with a physical environment that is both awe-inspiring and beautiful; they also produce the soils required for earth to maintain its vegetative cover. What is soil? It is a mixture of many substances: an almost endless variety of minerals, decomposing bodies of plants and animals, water, and air. When any one of these predominates, a particular type of soil results. Because conditions differ from place to place, soils differ. Even the soils in your neighborhood differ from one area to the next. To be sure, soils in the Amazon rain forest region differ from those of Australia's Great Sandy Desert, and soils from one section of Texas are unlike those from other sections.

Some soils extend downward to a depth of 2 or 3 feet (610 to 914 millimeters), while others are limited to a depth of only 2 or 3 inches (51 to 76 millimeters). Although some soils come into being almost instantly (as you will soon learn), most have evolved through several stages of development over hundreds or thousands of years.

FORMATION OF SOILS

Why do soils differ? *Pedologists* have shown that soil formation is extremely complicated, so generalizations or narrow considerations, such as descriptions of the amount of precipitation and sunlight or the type of vegetative cover, are inadequate in discussions of soil variation. To understand the earth's soils, we must consider, first, the important controls that create different structures, textures, and fertility; second, the main processes involved in the gradual development of soils; and, finally, the different methods of soil classification.

Soil-Forming Controls

Pedologists say that there are five major factors that control the formation of soils: climate, vegetation, time, parent material, and terrain.

Climate The dominant soil-forming control is climate because it is the source of sunlight and water. Precipitation in any of its forms can, like a double-edged sword, be beneficial or damaging. It can be beneficial when water from melting snow or falling rain penetrates the ground surface and gravity causes this water to move to lower levels. This **gravitational water** takes with it chemicals, organic nutrients (humus), and fine soil particles, which become lodged in crevices of the regolith. It can be detrimental if one of the layers becomes heavily compacted with fine particles and, therefore, impervious to further downward movement of water, resulting in the formation of a *hardpan*. If the area's precipitation is light and the rate of evaporation high, moisture may be drawn upward toward the surface by a movement called **capillary action.** Chemicals left behind may create a thick layer of deposits that is as impermeable as clay-filled hardpans. If calcium carbonate is the chemical compound, as is common in the American Southwest, the result is called *caliche*.

In areas of heavy precipitation, chemicals and soil nutrients may be taken away as the gravitational water moves down through the soil and unites with groundwater flow. This "robbery" of needed nutrients is termed **leaching,** and areas where this occurs are notoriously infertile. Farmers in such areas must constantly move their field of operations from place to place, clearing new lands in their search for fertile soils. If a hardpan develops in an area of heavy precipitation, the layers of soil above the compacted layer may become waterlogged as the air spaces become filled with water. Except for hydrophytes (water-loving plants), plants may develop root diseases and die. Because plants require oxygen to grow and perform their part in the formation of soils, the presence or absence of air is vital.

Temperature, whether consistently high or low, can encourage or discourage the prop-

Corkscrew Swamp, Florida Everglades. Waterlogged soils can support a thriving, luxuriant vegetation.

er evolution of a soil. High temperatures, if accompanied by sufficient precipitation, tend to increase the tempo of weathering activities. Bacterial activity speeds up, as does chemical decomposition; this results in the accelerated breakup of the regolith. On the other hand, low temperatures tend to slow down the formation of soils, although physical weathering processes may be increased.

Vegetation Soil development is greatly influenced by vegetation. First, the roots of shrubs, grasses, and trees extend down into the soil structure as they obtain nutrients and moisture. They aerate the soil (supply oxygen and nitrogen), and they draw moisture from it. Second, the vegetation life forms act as a protective cover to hold the forming soil in place; when they die, their bodies furnish humus to the soil. This is a continuing cycle. Vegetation requires humus so that it can grow and supply more humus.

Because conditions in grassland regions are favorable for humus formation and grasslands tend to retain humus near the surface, the grasslands of the world (for example, the steppes of central Asia and the midwestern lowlands and the Great Plains of the United States) have the richest soils. What kind of soil would evolve in a desert region with low precipitation and high evaporation? What kind of vegetation would develop and what kind of humus could it contribute? At first thought, we might erroneously conclude that such soils would provide only commercial minerals for mining enterprises. However, on-the-site research by pedologists has given us a different answer. The soils of arid desert regions such as Africa's Sahara and California's Imperial Valley are extremely rich in nutrients—the result of a different climate in ages past. The Sahara was once covered with lush vegetation, perhaps as recently as 10,000 years ago. Obviously, as the climate changed and the rains ceased falling and *desertification* ensued, the rich humus was left behind and is still there. Even though the present inhabitants of the Sahara are seriously endangered by the long drought of recent times, they might be able to rise above their terrible conditions if they could emulate the farmers of California's Imperial Valley who, with modern irrigation techniques, have turned this region into one of the most productive agricultural areas in the United States today.

Parent Material The initial bedrock, the **parent material,** that is the source of a soil contains chemicals and mineral combinations. If they are the right ones, fertile soils will evolve (assuming that other factors are favorable). Whether the soils have coarse or fine grains depends greatly on the original structure of the bedrock. Soil will evolve much faster from a sedimentary rock such as sandstone than it will from an igneous rock such as basaltic lava.

Time Another significant factor in the development of most soils is time. It may take thousands of years (under optimum conditions) for a few inches of soil to develop from a block of solid granite or a formation of lava. What about soils found along rivers or in the deltas that form at their mouths? When deposits are left by a river that has overflowed its banks and spread its **alluvium** (stream-deposited material) on the land, will the resulting soil be fertile and ready for use? It can be if the original source was composed of fertile topsoils. Why do you suppose the ancient Egyptians referred to the River Nile as the "Holy Mother?" For thousands of years this river carried alluvium from the mountains of Ethiopia and the Ruwenzori of the eastern Congo through the winding channel every spring. When flood stage was reached, the river overflowed its banks and deposited the rich, fertile alluvium on the adjacent lowlands on either side.

Terrain The word *terrain* means the surface of the earth, whether it is smooth and low-lying, with gentle inclines, or rugged, hilly, or mountainous, with steep slopes. Flat or moderately sloping terrain is more likely to evolve good soil (if other factors are favorable as well) than are steep slopes. Rugged, steeply sloping land allows water to run off too extensively and too rapidly, so lowlands that are surrounded by higher land may tend to become saturated with too much moisture (waterlogged) and form swampy or marshy areas where soil may have a difficult time evolving. Here, again, erosion is a factor. How can fertile soil evolve if it is always being depleted by violently running water or blowing winds? Often there is poor development of soil on mountain ridges because the newly evolving soil is continually being carried away.

Soil Horizons

So far in our discussion of soils we have referred to soil layers and soil structure. Pedologists, however, are more specific; they use terms such as **soil profile** and **soil horizons** when focusing on specific soils and the differences between them. Each soil has its own structure and profile, and each profile is composed of layers or horizons. Figure 9-1 is a generalized view of a soil profile, showing the major horizons of its structure. The various soils of the earth have unique profiles, and each may not match any other. They also have horizons that vary considerably in thickness. In some, the topmost layer (the A horizon) may be only 3 inches (76 millimeters) thick, while not far away, out on a grassland, the horizon may well be 3 feet (914 millimeters) thick.

The **A horizon** is the top layer; its upper surface is the surface of the earth. In many places you cannot see any soil at all because of the litter (plants, leaves, branches) covering the surface. If bacteria are present, they will convert this material into decomposed matter,

Influence of time on the development of agriculturally usable soils. (a) Relatively recent pahoehoe lava flow. (b) Commercially profitable sugarcane plantation, in Hawaii, on soil that evolved from an ancient lava flow.

Fertile soils along the River Nile, Egypt. (FAO Photo)

Well-developed soil profile. Profiles with distinct horizons are characteristic of grasslands.

FORMATION OF SOILS 255

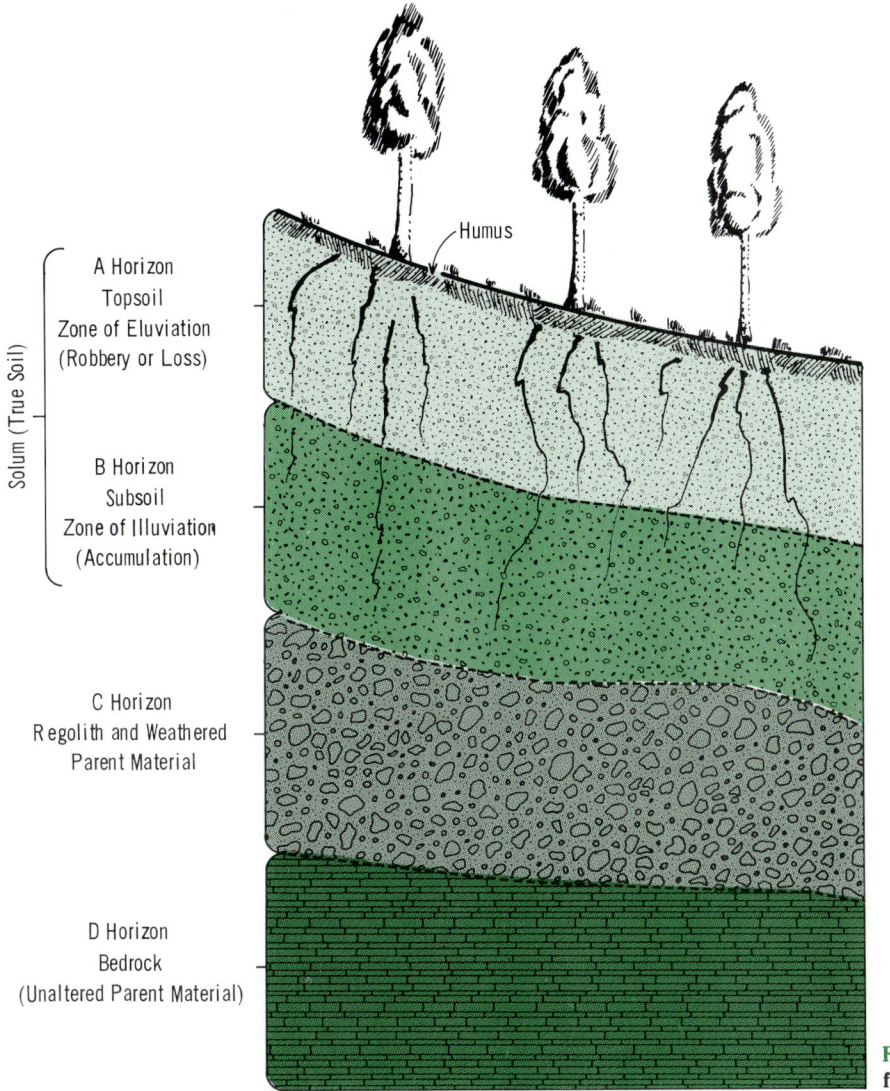

FIGURE 9-1 A soil profile (Not drawn to scale.)

Soil profile labels:
- Solum (True Soil)
  - A Horizon — Topsoil — Zone of Eluviation (Robbery or Loss)
  - B Horizon — Subsoil — Zone of Illuviation (Accumulation)
- C Horizon — Regolith and Weathered Parent Material
- D Horizon — Bedrock (Unaltered Parent Material)
- Humus

thereby supplying humus to the soil. If moisture is present, this decaying material will be carried into the earth, and if there is a sufficient quantity, the A horizon will be quite fertile. This is not always true for A horizons, however, as may be seen in the soils of rain forest areas. Here, the horizon may be 20 feet (6 meters) thick but nonetheless infertile because of excessive leaching. The average thickness of the world's A horizons is about 7 inches (178 millimeters). Pedologists call this horizon the **zone of eluviation** (to wash out), but farmers call it topsoil.

The **B horizon** is normally a more compacted layer than the A horizon because a heavier concentration of fine particles, such as silt or clay is brought down from above. This layer, therefore, is sometimes referred to as the **zone of accumulation.** Pedologists often use the term **zone of illuviation** (to move in). Because the air spaces have been filled with materials from the A horizon, water molecules have difficulty moving farther down through this horizon. In some regions, for one of two reasons, this horizon may actually be richer than the upper layer: First, farmers may have used the land too heavily and not too wisely, thus robbing the A horizon of its fertility; and, second, gravitational water may have caused excessive leaching of the A horizon. In order to replenish the upper level, farmers sometimes must plow quite deeply to "turn up" the subsoil.

The **C horizon** is composed of weathered material (regolith) that is separating from the bedrock and has not evolved far enough to be considered a part of the true soil, or **solum** (a term often used by pedologists to refer to the top two horizons as a kind of separate body: the soil mantle). The C horizon may contain moisture as well as chemicals and nutrients, but it is not considered an agricultural soil.

The **D horizon** is the unaltered parent material that has not yet evolved into a true soil horizon. It may be solid, even though it is fractured and jointed, or it may be fragmented in

Blurred soil horizon. Some soils have such indistinct horizons that it is difficult to construct a readable soil profile.

its topmost zone as it evolves and joins the regolith above it.

Horizon visibility can be quite clear, with the horizons having distinct boundaries; but sometimes it is extremely unclear, with blurred horizons and one zone seeming to blend into the next. Alluvial soils seldom have definite horizons because of the relatively short time it takes for them to evolve. Soils on mountain ridges and steep slopes may also have little or no A horizon because of heavy erosional action as the soil attempts to evolve.

Distinguishing Characteristics of Soil

A soil's color can aid in distinguishing its type. If there has been a heavy accumulation of humus on or near the surface, the A horizon will be dark, sometimes black. The famous steppes of the Ukraine are composed of dark soils, called *chernozem* (Russian for "black earth"). The dark color of these soils results from an abundance of humus in grassy areas. The color tends to lighten with increase in depth. The B horizon may be a medium tan and the C horizon considerably lighter.

In some instances the soil color can be attributed to the presence of chemicals in sufficient quantity to stain the soil. Soils in much of the state of Georgia are red because they contain iron oxides. In Arizona or New Mexico the soils tend to be almost white, or a very light gray, because of the abundance of calcium and gypsum. In areas of high precipitation the soils are light in color because both chemicals and organic matter have been leached out.

The texture of a soil is determined by the size of the particles of which it is composed. Pedologists consider soil to range in grain size from the smallest and finest particles, clay, then silt, and the largest granules, sand. If the particles are as large as gravel, the mixture is not true soil. The texture of a soil is very important because it influences the movement or retention of soil water. Sandy soils tend to lose moisture rapidly, and soils containing clay tend to resist absorption of water for some time (although in clay soils the air spaces eventually fill and the moisture will then remain). The best agricultural soils are somewhere in between, with a texture that allows both gravitational and capillary water to move freely through the horizons.

A soil's structure is determined by how the soil particles align themselves. If they form in thin, flat sheets, or "plates," we say the soil's structure is **platy.** If the soil is composed of minute granules with rounded shapes, the soil is **granular.** Uneven, irregular, sharp-edged grains are called **blocky.** And if the grains are aligned in straight lines, taking on the many-sided shape of a prism, the soil is termed **prismatic.**

Soil-Forming Processes

There are several soil-forming processes, but the three that are most important to humans are podzolization, laterization, and calcification. Each process operates differently, and each produces unique soils.

Podzolization The process called **podzolization** (from the Russian *podzol,* meaning "soil under foot") occurs in cool, humid regions under a vegetative cover that is primarily composed of coniferous (needle leaf) forests. Typically, the taiga forest regions of Canada and Siberia have podzolic soils. **Podzols** are acidic; silica is the prime mineral. Leaching removes most of the iron and aluminum oxides, and there is little humus in the A horizons. This results from the slow decomposition of the limited amount of plant and animal matter that does collect on the ground. There is little bacterial activity because of the cool climate, which lasts from eight to nine months each year. Because of the absence of humus and the pres-

Soil structure. (a) Platy. (b) Granular. (c) Blocky. (d) Prismatic.

ence of large amounts of silica, podzols are grayish to white in color. Soils forming during this process produce little in the way of agricultural goods other than timber. In addition, there are many areas within regions undergoing podzolization where the soils are found to be sticky, compacted, and, in some cases, very slushy. This occurs when there is inadequate drainage or moisture runoff that is caused by several factors, such as flat land or wide depressions with no outlet, or the presence of permanently frozen substrata that are too close to the surface. Thus, a clayish, boggy soil results. This subprocess is termed *gleization*.

Laterization The process of **laterization** (from the Latin *later,* "brick") produces soils called **latosols** (and, sometimes, *laterites*) and occurs in warm, humid climates, such as the humid subtropical and tropical rain forest areas. Leaching is common because of the heavy precipitation and results in the removal of the important mineral silica. The most widespread latosols are composed of iron and aluminum compounds (thus their name, *pedalfer,* from *ped,* "ground" or "soil," *al,* "aluminum," and *fer,* "iron"). Soil profiles in areas of laterization are very deep, and the A horizon may sometimes extend as much as 20 to 30 feet (6 to 9 meters) in depth. B horizons are difficult to distinguish as a separate zone. Groundwater movement is active because of constant replenishment from above.

Calcification The name **calcification** was derived from the dominant mineral combination in the process, calcium carbonate. Calcification occurs in climate regions that range from arid to semiarid to humid and are noted for their grass cover. The roots of the grasses reach far down into the B horizon and bring back calcium to the surface. Dying grass and good bacterial activity supply large quantities of humus, which becomes well mixed throughout the solum. The depth of these soils is not as great as the lateritic soils, but because of the semiarid conditions and lack of heavy rain, there is little leaching; therefore the soils retain their fertility and are more useful for agriculture and livestock ranching. Mixing carbon dioxide with the ever-present calcium produces calcium carbonate, which fosters lush grassland vegetation. The most famous of these areas are the fertile steppes of the Ukraine, the pampa of Argentina, and the Great Plains of the United States.

SOIL CLASSIFICATION

Pedologists have developed a number of ways to classify soils beyond the processes we have just discussed. The classic system, the 1938 USDA Classification, which evolved out of research by Russian and American pedologists during the late nineteenth and early twentieth centuries, divided the earth's soils into three major groups or zones: zonal, intrazonal, and azonal. The latest system, called the 7th Approximation and destined to become the standard, divides soils into ten basic groups. Because we find ourselves in a transition period, moving from the old standby to a new system, we will consider both of these soil classification systems briefly; and as we do, it would be wise for you to consult the maps of climate (Chapter 7), vegetation (Chapter 8), and landforms (Chapter 11).

The 7th Approximation*

The **7th Approximation** classification system was devised in 1960 by pedologists in the Soil Conservation Service of the U.S. Department of Agriculture. The ten soil orders, based on environmental controls (Fig. 9-2), are listed here along with a definition of the terms in parentheses. Each basic order is further subdivided into a number of suborders.

1 *Inceptisols (Beginning).* Soils with indistinct horizons resulting from a short evolutionary time, as in tundra areas, or those evolving from parent material that is resistant to change, as in mountainous and volcanic regions.

---

*For a thorough explanation of this new system, see U.S. Department of Agriculture, Soil Conservation Service (1960), *Soil Classification: A Comprehensive System: 7th Approximation* (Washington, D.C.: U.S. Government Printing Office); and J. G. Cruickshank (1972), *Soil Geography* (New York: Halsted Press; Wiley).

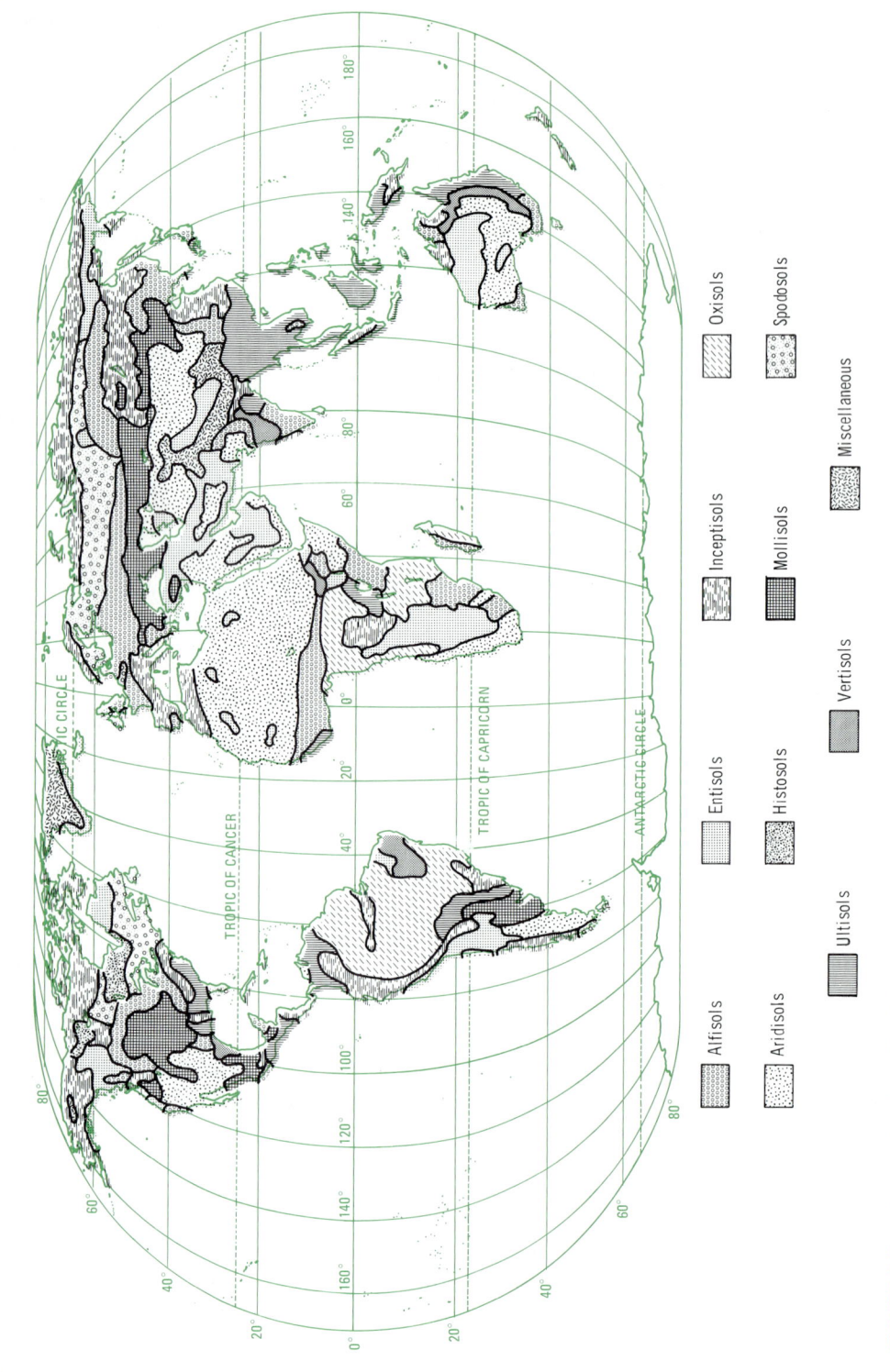

FIGURE 9-2 Soil classification: The 7th Approximation.

2 *Alfisols* (*Aluminum and Iron*). Mixed acid-alkaline soils found in subtropical to tropical climate areas in the main agricultural regions of the world.
3 *Spodosols* (*Ash*). Soils found in cool, humid climate regions; includes soils formed by podzolization.
4 *Ultisols* (*Ultimate*). Soils found in tropical wet and dry regions that are enriched with subsurface clay layers; considered to be good agricultural soils.
5 *Oxisols* (*Oxygen*). Tropical rain forest soils found in lowland areas; primarily lateritic in composition.
6 *Mollisols* (*Soft*). Grassland soils found in the prairie and steppe regions of the world, with covers of forest on the wetter margins and grasses on the drier areas; the chernozems are included in this order.
7 *Aridisols* (*Dry, Arid*). Desert soils that are deficient in both moisture and organic matter, so usually not used for agriculture.
8 *Entisols* (*New*). Includes new soils that have not had time to develop horizons or old soils that are thin and stony and are found in mountain regions, interior deserts, and some interior permafrost areas.
9 *Vertisols* (*Inverted*). Clay soils with a brittle and cracked surface.
10 *Histosols* (*Thin, Tissue*). Soils that form in swamps or bogs.

The 1938 USDA Classification*

This soil classification system (Fig. 9-3) may be rather ancient, but it is relatively easy to understand and has served pedologists well. It is based on the stages of development of the world's many soils. Well-developed soils are called *zonal*, immature soils are *azonal*, and intermediate soils are *intrazonal*. As we consider these classifications, we will also discuss some of the more important soil groups within them.

Zonal Soils The most highly developed of all soils, as well as the most widespread, are **zonal soils.** They include both fertile and infertile soils, depending on the climate region in which they are found and the type of vegetative cover that dominates. Following are subclassifications of zonal soils.

1 *Forest Soils.* These soils are represented by podzols and latosols. The podzols range from light grayish soils that develop under the coniferous forests of the higher latitudes (the taiga forests of Canada and Siberia) and some mountainous regions (southern Chile) to a more fertile, grayish-brown version that evolves under broadleaf deciduous forests on the North European lowlands (including much of Great Britain, France, and Germany) as well as the Ohio Valley region of the United States. The higher fertility of this podzol results from a lower rate of leaching. A red-yellow podzol develops in warmer, humid areas of the mid-latitudes (southeastern United States and southern Brazil). The latosols are tropical soils that are low in fertility because of excessive leaching and a low humus content. These reddish, reddish-brown, and yellow-brown soils are found in most tropical rain forest regions (Brazil, West Africa, South-East Asia and the East Indies, Malagasy, the Caribbean, and portions of Central America).

2 *Grassland Soils.* This group includes three important classes: the chernozem soils, the prairie soils, and the reddish prairie soils. The chernozems develop in warm to hot semiarid climates and are covered by grasses that provide excellent fertility, as evidenced by the pro-

---
*The 1938 USDA Classification has also been known by other names, such as Major Soil Orders and Great Soil Groups. This system was developed by American soil scientists using the results of a Russian pedology report written at the turn of the century that dealt with the close relationship of soils to the climate and the dominant vegetation of an area.

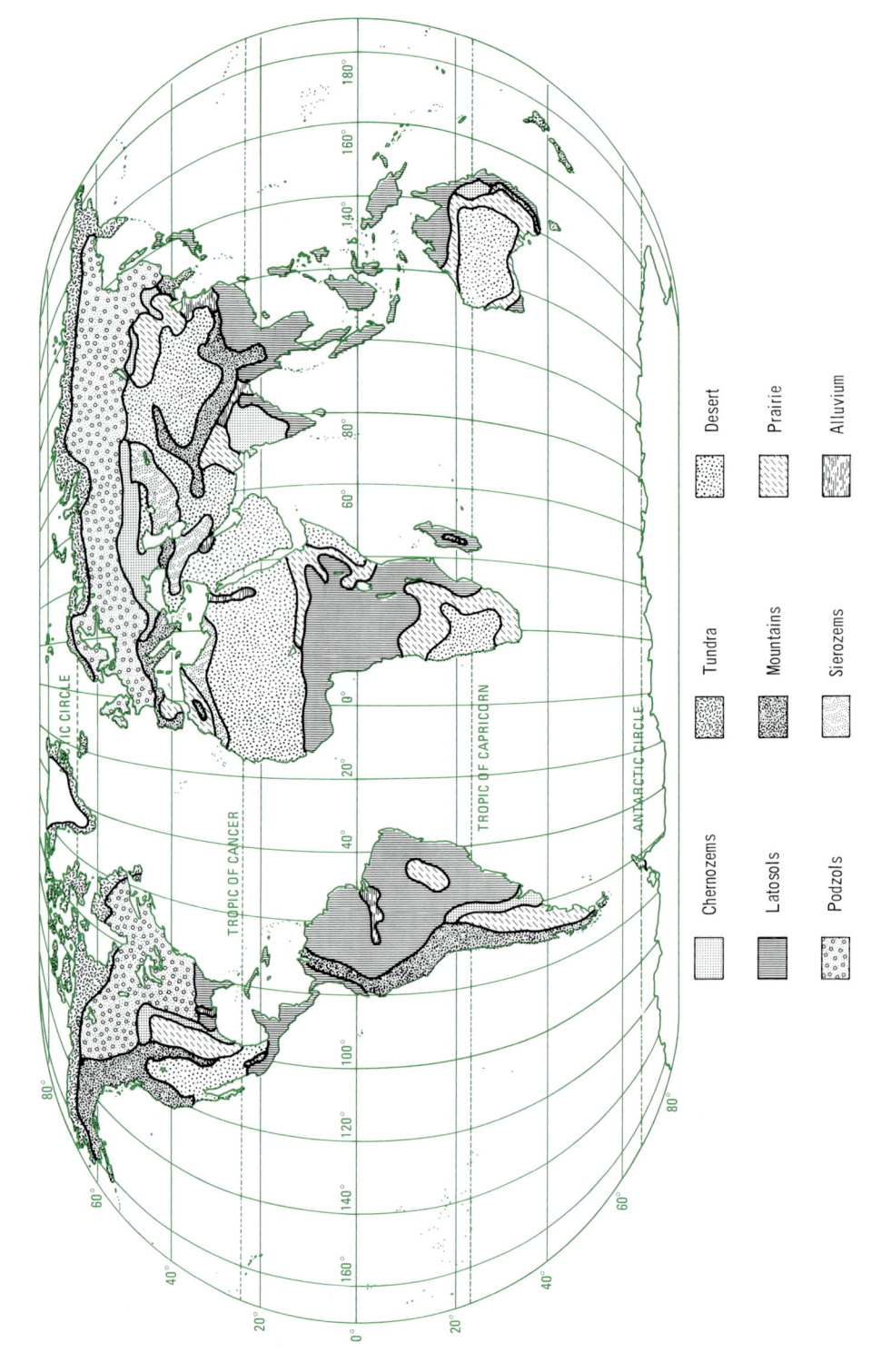

FIGURE 9-3 Soil classification by the 1938 USDA Classification System.

ductivity of the northern Great Plains of the United States and by the recent increase in agricultural activity throughout the steppe regions of Russia. Areas bordering the steppe regions produce the prairie soils, which are also extremely fertile and can be used for growing grains or grazing livestock. The reddish prairie soils develop around the borders of the tropical and subtropical climate regions and are typical of the tropical wet and dry climates: the veldt areas of East and South Africa, eastern and southern Australia, Argentina's pampa, northwestern India, Cambodia, northern Algeria, and the southern portion of the Great Plains of the United States.

3 *Desert Soils.* This group includes the chestnut-brown, gray desert, and red desert soils. Chestnut-brown soils develop in sections of the steppes where the precipitation is light and the cover is short grasses. Thus they are not considered to be very productive soils. Gray desert soils (also called *sierozem*) develop in cool, dry desert areas, such as Patagonia (Argentina), the Takla Makan and Gobi deserts of Tibet and Mongolia, the region east of the Caspian Sea, and parts of Wyoming and Colorado east of the Rockies. Red desert soils develop in hot, arid desert climates, such as in North Africa's Sahara, the Arabian Desert, the Thar Desert of India and Pakistan, Australia's Great Sandy, and the Sonoran of northern Mexico and the American Southwest.

4 *Tundra Soils.* These soils develop in the cold permafrost regions north of where trees will grow. They are found in Canada and Siberia in lands bordering the Arctic. The vegetative cover is composed of lichens, mosses, sedges, and dwarf willows, and the soils are not fully developed, except in a few places, because the development of horizons is limited by the permanently frozen subsurface. In some areas there is humus: but it reaches a depth of only about 9 inches (229 millimeters). Because of the long winters, short summers, and low annual range of temperature, commercial agriculture is impractical.

Typical azonal soil. Fertile alluvium deposited by a river over an original floodplain constitutes typical azonal soil.

Intrazonal Soils Less developed than the zonal soils, **intrazonal soils** show signs of immaturity. The parent material and the terrain exert control over these soils, as indicated by the three suborders.

1 *Hydromorphic Soils.* These are the waterlogged soils found in swamps and bogs, or they are found in meadows having better drainage, but also heavy clay deposits.

2 *Halomorphic Soils.* These soils are found in dry climates, especially where the rate of evaporation exceeds the precipitation. They include salty soils (*solonchak*) and alkaline soils (*solonetz*).

3 *Calcimorphic Soils.* These soils have a high lime content (calcium carbonate) and, because of their grassy covers, a high humus content. They include brown forest soils and rendzinas.

Azonal Soils The most immature (least developed) of the soils are the **azonal soils,** but they are not, however, necessarily the poorest in fertility. Although they may not have had sufficient time to develop definite horizons, some—alluvial soils—may be very fertile if their source region was high in mineral or organic content. Thin and stony soils found on mountain or hill slopes are called *lithosols;* pebbles and gravels that make up the soil of a rocky desert are called *regosols;* and sandy soils on sandy deserts are called *sand.*

REVIEW AND DISCUSSION

1. What is soil?
2. Why is climate considered the most important soil-forming control?
3. Explain the difference between gravitational and capillary water movement.
4. Explain the terms *caliche, leaching,* and *hardpan.*
5. What is humus?
6. Why is the parent material important in soil evolution?
7. Why is time important in soil development?
8. Why is alluvium sometimes considered fertile soil?
9. How is terrain important in the formation of soils?
10. What is a soil profile? Why is a consideration of soil horizons important?
11. Which horizon contains the most nutrients?
12. Explain the differences between *zone of eluviation* and *zone of illuviation.*
13. Why do soils vary in color?
14. Explain the difference between these soil structure types: platy, granular, blocky, and prismatic.
15. Explain these soil-forming processes: podzolization, laterization, and calcification.
16. Based on the brief explanations of the two major soil classification systems in use today, which do you prefer? Why?
17. What is the difference between podzols and latosols?
18. Why do soils develop differently under coniferous forests and grasslands?
19. Explain the difference among *hydromorphic, halomorphic,* and *calcimorphic.*
20. Why are some azonal soils important to humans?

ADDITIONAL READING

Baldwin, Mark, Charles Kellogg, and James Thorp, U.S. Department of Agriculture (1938). *Soils and Men: Yearbook of Agriculture, 1938.* Washington, D.C.: U.S. Government Printing Office.

Bunting, B. T. (1965). *The Geography of Soil.* Chicago: Aldine.

Cruickshank, J. G. (1972). *Soil Geography.* New York: David & Charles.

Kellogg, Charles (July 1950). "Soil," *Scientific American* (offprint no. 821).

Miller, C. E., L. M. Turk, and H. D. Foth (1959). *Fundamentals of Soil Science.* New York: Wiley.

U.S. Department of Agriculture (1957). *Soil: The Yearbook of Agriculture,* 1957. Washington, D.C.: U.S. Government Printing Office.

U.S. Department of Agriculture, Soil Conservation Service (1960). *Soil Classification: A Comprehensive System: 7th Approximation.* Washington, D.C.: U.S. Government Printing Office.

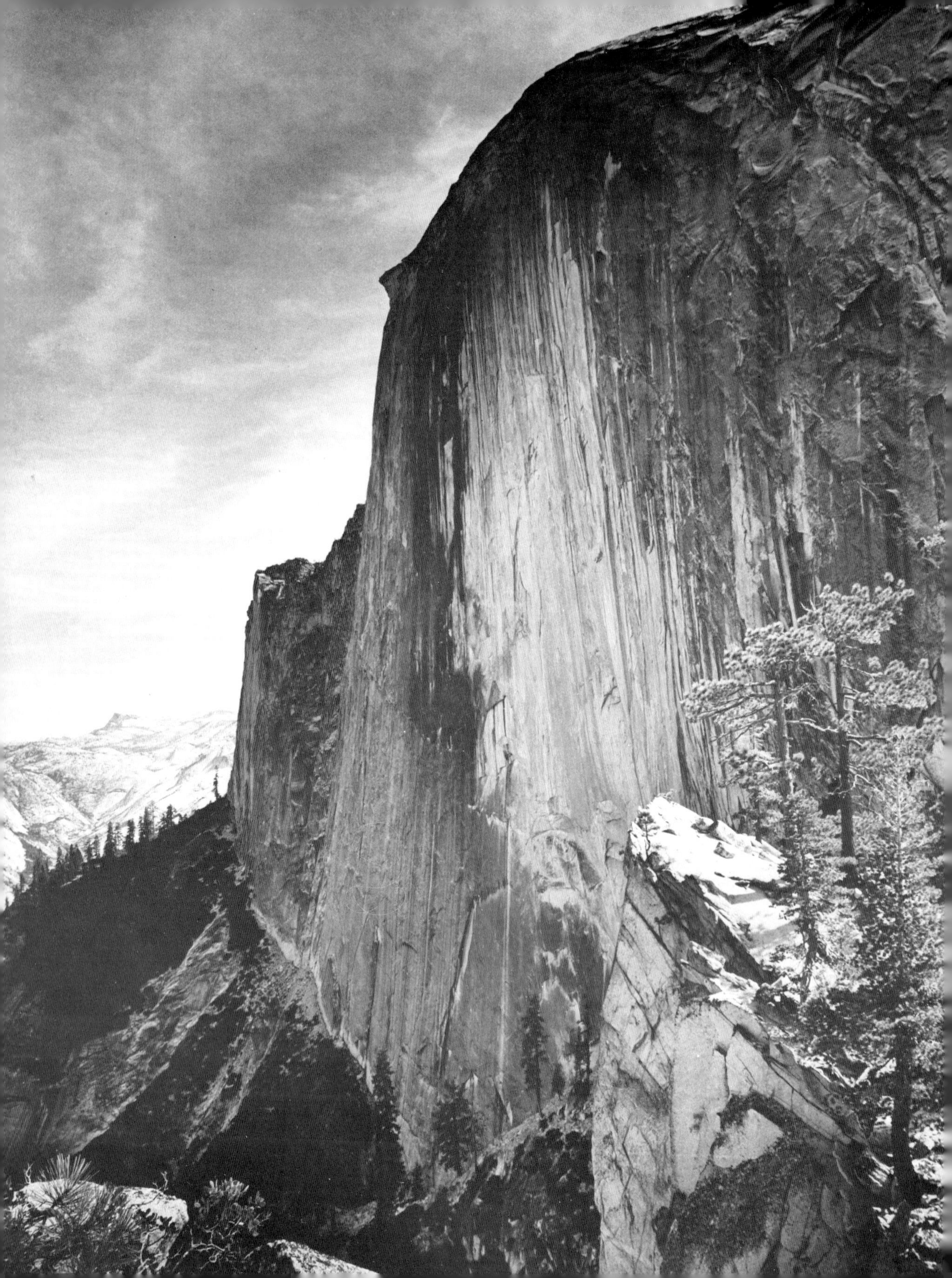

TEN
COMPOSITION OF THE EARTH

The whole valley, sparkling in the late sunlight, looks like a trim, polished, perfect existence. The dome Tissiack looks down the valley like the most living being of all the rocks and mountains; one would fancy that there were brains in that lofty brow. How grandly comes the gloaming over this pearly beauty!
JOHN MUIR

In 1869 John Muir, noted mountaineer and naturalist, spent his first winter in Yosemite Valley and was so overwhelmed by its spectacular beauty that he fell under the spell of the mountains and wilderness regions of the West. That he attained a feeling of oneness with the earth is evident in the words he wrote in 1890: "There is a love of wild Nature in everybody, an ancient motherlove ever showing itself whether recognized or no, and however covered by cares and duties."[1]

But Muir was not content with mere exhortations; he was a man of action, and he fought nobly to preserve our nation's forests and rugged places he held to be almost sacred. He founded the Sierra Club and was instrumental in persuading President Theodore Roosevelt to set aside for posterity the lands that now contain Sequoia and Yosemite national parks. He tried to convince all who would listen to him that conservation could not wait. Unfortunately, many would not heed his words, impelling him to write: "Most people are on the world, not in it—have no conscious sympathy or relationship to anything about them—undiffused, separate, and rigidly alone like marbles of polished stone, touching, but separate."[2]

For John Muir, ecology meant the preservation of the earth and the ecosystems interrelated with it, but he also believed that one could not function properly as an ecologist without reaching an understanding of the earth's structure and composition.

With this in mind, we will consider the earth's origin and the reasons it has evolved to its present state. We shall examine the earth's interior and exterior and learn about its basic mineral and rock composition. Then we shall consider the powerful forces that have elevated entire continents and great mountain ranges, as well as studying other forces that have operated to bring them down again. The earth's uneven and beautifully sculptured face did not get that way overnight. It took time—an almost inconceivable amount of time.

GEOLOGIC TIME

Since the beginning of our existence on the earth, there have been almost as many theories concerning earth's history as there were people. Because humans witnessed the alternating periods of day and night, the phases of the moon, and the cycle of the seasons, they thought only of days, weeks, months, years, and sometimes, in terms of millennia (thousands of years). Even though today we may say that we truly understand the length of earth's history, it is difficult to comprehend the vastness of cosmic time. Can you honestly say you can conceive of 1 billion years of elapsed time?

Geologists and geomorphologists have tried to fathom the riddle of the ages by measuring the thicknesses of different layers of rocks, the salinity of the seas, and the lowering of landmass surfaces. As you might guess, however, their conclusions do not always agree. A step forward came in 1904 with the discovery by England's Lord Rutherford that mineral elements disintegrated through time, each at a fixed rate of radioactive decay.[3] This meant that almost every substance lost radioactivity, and the determination of the various rates of decay of minerals, rocks, fossils, trees, and other bits of matter supplied geologists with a relatively accurate time clock. Geologists admit that radioactive dating is not absolutely accurate, but they do claim it is close enough to serve us until a better method is discovered.

The age of the earth is now estimated to be about 4.5 billion years. Because the earth did not come into existence exactly as it is now, geologists and geomorphologists have put their research findings on a chart that shows the evolutionary stages of its development.

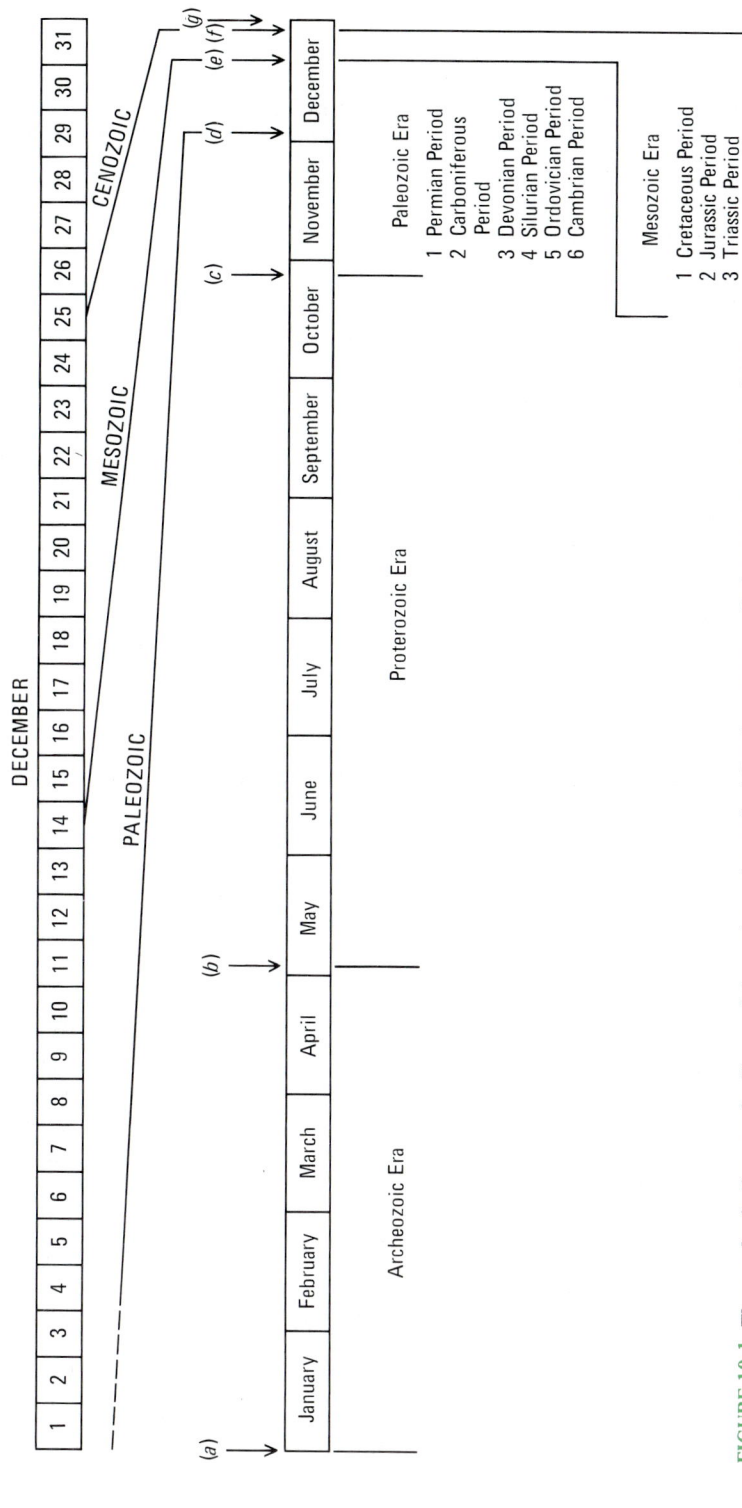

FIGURE 10-1 The geologic time scale The earth has existed for about 4.5 billion years, and geologists have divided earth history into intervals (eras, periods, and epochs) that are unique, each having special characteristics and features. To make geologic time more understandable, let us condense the earth's history into 1 calendar year, with the beginning set at January 1—(a). The formation of the earth's crust (through February), the creation of the oceans (March), and the release of moisture into the atmosphere (April) occurred during the 2 billion years of the Archaeozoic era (*archaeo*, "ancient"; *zoic*, "life"), which came to an end on May 3 with the appearance of the first life forms, single-celled algae and fungi (b). The Proterozoic era (*protero*, "fore," or "beginning") lasted for another 2 billion years and witnessed the rise of the supercontinent Pangaea and considerable mountain building. Toward the end of October, many-celled organisms evolved (clams, fish, and reptiles), and the Paleozoic era (*paleo*, "ancient") started (c). Pangaea began to break apart, forming the continents we now know. A number of intervals followed, Cambrian to Permian, each lasting millions of years. The first great forests appeared (d) about December 2, and a couple of weeks later, the Age of Dinosaurs (e) brought on the Mesozoic era (*meso*, "middle"). About 60 million years ago, December 25, the Cenozoic era (*ceno*, "recent") began with the first mammals (f). A week later, December 31, about 2 hours before midnight, human beings appeared (g). Many of the greatest human achievements have occurred during 1 second of geologic time (200 years). Imagine, in 1 second of time we have moved from the steam engine to space flight. What if we had an entire year?

GEOLOGIC TIME SCALE

ERA	PERIOD	EPOCH	SUCCESSION OF LIFE
CENOZOIC "RECENT LIFE"	QUATERNARY 0-1 MILLION YEARS	Recent / Pleistocene	
	TERTIARY 62 MILLION YEARS	Pliocene / Miocene / Oligocene / Eocene / Paleocene	
MESOZOIC "MIDDLE LIFE"	CRETACEOUS 72 MILLION YEARS		
	JURASSIC 46 MILLION YEARS		
	TRIASSIC 49 MILLION YEARS		
PALEOZOIC "ANCIENT LIFE"	PERMIAN 50 MILLION YEARS		
	CARBONIFEROUS — PENNSYLVANIAN 30 MILLION YEARS		
	CARBONIFEROUS — MISSISSIPPIAN 35 MILLION YEARS		
	DEVONIAN 60 MILLION YEARS		
	SILURIAN 20 MILLION YEARS		
	ORDOVICIAN 75 MILLION YEARS		
	CAMBRIAN 100 MILLION YEARS		
PRECAMBRIAN ERAS	PROTEROZOIC ERA		
	ARCHEOZOIC ERA		

APPROXIMATE AGE OF THE EARTH MORE THAN 4 BILLION 550 MILLION YEARS

This is termed the **geologic time scale** (Figs. 10-1, 10-2). There are five major divisions, termed **eras,** and a number of subdivisions, called **periods** and **epochs,** each with its own special characteristics and features. The oldest era is called the Archeozoic era; during this span of some 2 billion years, the earth's crust formed, moisture was released, the world ocean appeared, and the atmosphere came into being. Then followed the Proterozoic era, which lasted for about 2 billion years; during this time the first life forms evolved in the shallow waters of a continent (called Pangaea) surrounded by the seas. These two eras are often combined and referred to as the Precambrian era.

To many people, the most exciting events of earth's history occurred after the end of the Precambrian era, about 600 million years ago, because the life forms advanced from the single-celled algae and fungi to forms that we can visualize and understand: clams, fish, reptiles, grasses, shrubs, and trees. Physical events and the appearance of some dramatic new life form are the dividing factors, and the place where the event occurred or the fossil was found gave its name to the interval of geologic time. An example would be the Carboniferous period of the Paleozoic era. Forests covered vast areas of the earth through much of this period of time, and these forests have provided us with coal and other carbon products. Because these coal deposits were first studied in Pennsylvania and Mississippi, geologists in the United States gave the epochs those names.

Geologic time is important because it gives us a better understanding of the tremendous expanses of time that must occur before continents can be raised, mountains built or leveled, valleys carved, or nearly level plains created. As we move ahead with our investigation, refer occasionally to Fig. 10-1 so you will be able to maintain the proper perspective.

EARTH'S INTERIOR

You may have heard someone refer to the earth as terra firma (*terra,* "earth," and *firma,* "solid"), but is it really so solid? No, it is not. The earth's crust actually is in almost constant movement—as amply proved by the occurrence of many earthquakes and volcanic eruptions each year. Why must the earth experience these incessant motions? For the answer we must look below the surface and examine the earth's deep interior (Fig. 10-3), the mantle and the core.

The earth's diameter is estimated to be about 7927 miles (12,757 kilometers), and most of this dense mass exists in an intensely heated state. At the center is the **inner core,** a solid mass of nickle and iron about 1560 miles (2496 kilometers) in diameter with a temperature of around 7000°F (4000°C). Wrapped around the inner core is a liquid layer of the same metals, called the **outer core,** which has a temperature close to 6332°F (3500°C). It is from this core area that heat is constantly moving outward in what might be termed convectional streams, thus maintaining the temperature of the next region, the **mantle,** at a range of 6332°F to 3632°F (3500°C to 2000°C). The currents circulate throughout the mantle, tending to keep the earth and its crust in an unstable condition. If there were an abrupt increase in the temperature of a portion of the mantle, the heated region might expand and, in some cases, it might result in a lifting of the

---

FIGURE 10-2 Geologic Time Scale (Reprinted by permission from *Fossils: An Introduction to Prehistoric Life,* by William H. Matthews III, Barnes & Noble, Inc., New York.)

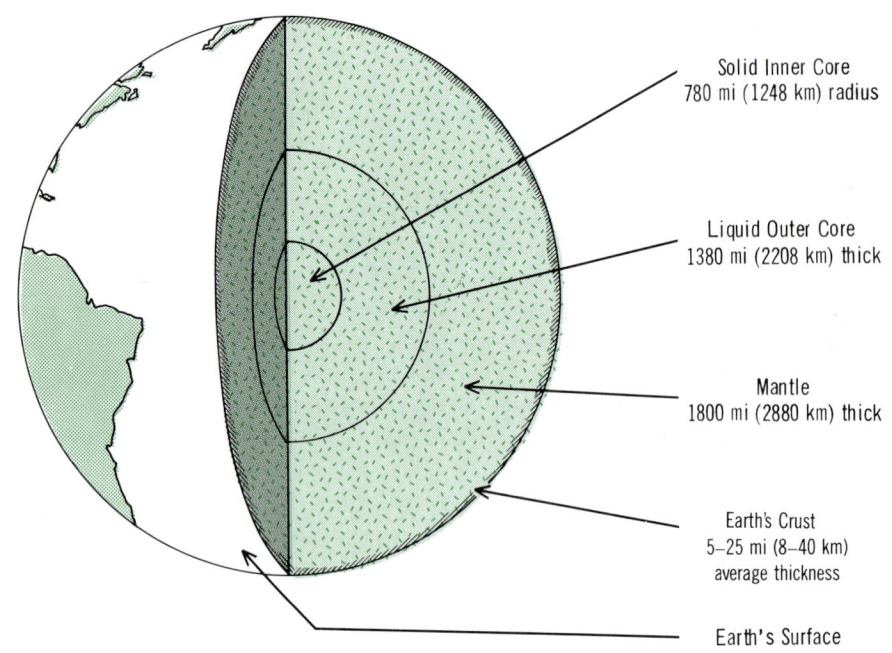

FIGURE 10-3 Earth's interior.

Solid Inner Core
780 mi (1248 km) radius

Liquid Outer Core
1380 mi (2208 km) thick

Mantle
1800 mi (2880 km) thick

Earth's Crust
5–25 mi (8–40 km)
average thickness

Earth's Surface

crust above. In other cases the upper mantle might liquefy (the rocks would melt) and volcanic activity would ensue. Sometimes portions of the earth's crust will settle where mantle rock has been displaced by the heat currents.

The upper reaches of the mantle are more pliable than the deeper regions, more responsive to the heat-circulation system within the earth. This upper zone is termed the **asthenosphere** (from the Greek *asthenes,* "weak"), and some earth scientists describe it as a flexible, almost plastic layer over which portions of the earth's crust can slide (as you will see in the next chapter when we discuss the theory of plate tectonics, continental drift, and seafloor spreading). The boundary between the asthenosphere and the crust is referred to as a **discontinuity** (distinct separation) and is called the **Moho**. This term is an abbreviation of the more correct term, *Mohoviĉiĉ discontinuity* (named after the Yugoslavian seismologist Andrija Mohoroviĉiĉ, who discovered it). The Moho varies in depth below the earth's surface; it is much closer to the earth's surface beneath the oceans than under the continents (Fig. 10-4). It is an uneven boundary, reflecting the weight of the masses above it.

As yet, no one has penetrated the Moho, although the scientific world is understandably enthusiastic about such a project; they want to know more about the mysterious mantle.* It is believed that most of the earth's internal movements begin within this region, which is 37 to 156 miles (60 to 250 kilometers) below the surface.

---

*For an interesting explanation of the earth's interior and its rather plastic mantle, see Don L. Anderson (July 1962), "The Plastic Layer of the Earth's Mantle," *Scientific American,* 209, pp. 52–59. Also available as offprint no. 855.

EARTH'S CRUST 273

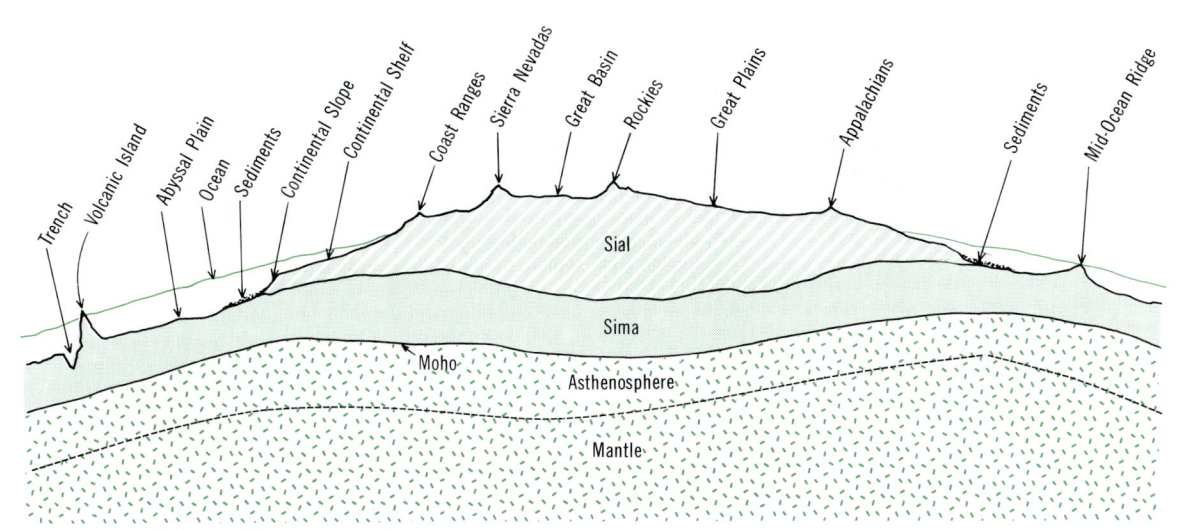

FIGURE 10-4 The earth's outer margin: crust and upper mantle This is an exaggerated (not to scale) cross section of the United States at approximately 38° north latitude.

EARTH'S CRUST

The ancient Greeks believed that the earth was composed of four elements: fire, water, air, and earth. When we think of the earth, which is our home, we think of four elements, too, but not the same ones. We call the air our atmosphere; the earth's waters, our hydrosphere; the life zone, our biosphere; and the earth mass itself, the lithosphere.

The term **lithosphere** is derived from the Greek *lithos*, "rocks," and *sphaeros*, "sphere"; thus we have a "rocky sphere." Some consider the lithosphere to include the entire earth, but most earth scientists use the term to define and delimit the region above the Moho. This is a region of less pliable rock materials made up of two distinct layers. The bottom layer rests directly on top of the asthenosphere and underlies the continental landmasses. It has been given a name based on its most common minerals, silicon and magnesium (thus we have **sima**), and it is composed of heavy, dense basaltic rocks. The upper layer is made up of lighter and less dense granitic rocks, and its most common minerals, silicon and aluminum, have earned it the name of **sial,** or the **sialic layer.**

The earth's crust varies considerably in thickness. Where the crust includes only the sima, as under the world's oceans, it is only about 5 miles (8 kilometers) thick, but where the sial and the sima are together, the thickness increases to about 25 miles (40 kilometers), as an average, and as much as 40 miles (64 kilometers) under great mountain ranges. The highest mountain, however, has an elevation amounting to about half the difference between the two averages. Thus, Mount Everest measures slightly over 5 miles (8 kilometers) in height. Both the sial and the sima are affected by heat currents operating within the mantle and by other forces operating there. We will continue this discussion of earth movements in Chapter 11.

EARTH MATERIALS

Many times during our discussion of the composition of the earth we have mentioned rocks. We have talked about mantle rock, the rocks of the earth's crust, heavy and dense basaltic rocks, and lighter, less dense granitic rocks. But what is a rock? Perhaps you can name some, such as granite, slate, limestone, and sandstone, but do you know of what they are made? How did they come into being? How many kinds of rocks are there? To arrive at some answers to these questions, we must start at the beginning. Rocks are composed of minerals. Some rocks have only one mineral; some have combinations of two or more.

Minerals

Minerals come in a variety of sizes and chemical combinations and have a variety of physical properties. **Mineralogists** have identified over 2000 different minerals, combinations of more than 100 elements that exist in and on the earth. A mineral can be made up of one element (such as gold or silver) or a combination of two or more (such as quartz, which contains silicon and oxygen). And what is an element? An **element** is matter that cannot be broken down or further subdivided by natural means. Of the many elements that are present in the earth's crust, the most abundant (Table 10-1) are silicon (28 percent) and oxygen (47 percent). This does not mean that the other elements are not significant. Some are scarce and cannot easily be discovered and mined, but they all are important because they make up most of the weight of the earth's crust.

Elements are made up of atoms, and each atom in a particular element shares in the physical and chemical characteristics of that element. We could go further into the mysteries of the atom and learn how it is made up of neutrons, protons, and electrons, but we will not. We must leave that to geophysicists and atom smashers. We can, however, tie it all up in a kind of equation:

Atoms form elements.

Elements form minerals.

Minerals form rocks.

Minerals can form in a number of ways: by crystallization from evaporation, when the moisture evaporates, leaving behind a mineral such as salt from seawater; by crystallization from a molten state, when once-fiery liquid rock, called **magma,** cools and allows minerals such as quartz or feldspar to form; and by condensation from a vapor, when, for example, sulfur is heated to a gaseous state by the intense heat within the vent of a volcano and, as it escapes, cools and condenses into beautiful yellow crystals around the mouth of the vent.

Some rocks can be made up of only one mineral (such as gold), but most rocks are a combination of two or more. Quartz, feldspar, and mica are the minerals that make up granite; if you pulverized a chunk of granite, the

TABLE 10-1 The Most Abundant Mineral-Forming Elements (Percentages Rounded Off) in the Lithosphere.

ELEMENT	CHEMICAL SYMBOL	PERCENTAGE (BY WEIGHT)
Oxygen	O	47 ⎫ 75
Silicon[a]	Si	28 ⎭
Aluminum	Al	8.1
Iron	Fe	5
Calcium	Ca	3.6
Sodium	Na	2.9
Potassium	K	2.6
Magnesium	Mg	2.1

[a]The most abundant mineral in the lithosphere is quartz, which is composed of silicon and oxygen. If nothing is added, quartz will look like pure, clear glass.

result would be tiny pieces of quartz, feldspar, and mica. Why? Because each mineral always retains its own, separate characteristics. When it is released from combination, it resumes its own identity.

Mineralogists have tried to simplify matters by considering the thousands of mineral combinations as "families." They group together all mineral combinations that have the same or similar basic composition (Table 10-2). But is that enough? Some geomorphologists want to have a more thorough understanding of the part of the earth they are studying, so they want to know the minerals contained in the rocks of the area. They examine the physical properties of the rocks by performing one or a number of tests. They may, for example, test for hardness, to discover what it takes to scratch a mineral. Talc is the softest of all the minerals and can be scratched with a fingernail. Diamonds are the most difficult to scratch, and in between diamonds and talc (in ascending order of hardness) are gypsum, calcite, fluorite, feldspar, quartz, topaz, and corundum. You can also test by examining the mineral's luster (whether it is shiny, like glass, or pearly, metallic, silky, fibrous, or dull), its cleavage (how it breaks up when hit by a hammer, whether it shatters or splits into sheets or fibers), or by visual inspection (how it looks to the naked eye).*

Rocks

Rocks, of course, come in a variety of sizes, shapes, and colors.† There are solid-appearing granite cliffs, accumulations of rock debris along a highway cut through a mountain or hill, rounded pebbles at the bottom of a stream, and black sand on a beach in Hawaii. **Petrologists** have neatly classified the rocks of the earth into three classes: igneous, metamorphic, and sedimentary.

Igneous Rocks

Igneous rocks are the result of volcanic activity. They are rocks "born of fire," rocks that were once in a molten state and hardened as they cooled. If molten rock (magma) works its way upward from the asthenosphere into the earth's crust and finds its

TABLE 10-2 **The Basic Mineral Combinations.**

FAMILY	COMPOSITION
Oxides	Among oxides, oxygen is the prime element, and it is usually united with another element. As an example, quartz is composed of silicon and oxygen.
Carbonates	Carbonates occur when oxygen and carbon are united. If calcium is added, calcium carbonate results, one form of which is limestone.
Sulfides	Sulfides are a combination of sulfur and oxygen. If another mineral, such as calcium, is added, a combination called gypsum would be the result.
Silicates	Silicon is the primary element in silicates. If oxygen is added, quartz results. If iron is added, along with magnesium, hornblende results.
Chlorides	Chlorine is the dominant element in chlorides. If sodium is added, sodium chloride (common table salt) results.

*For more detailed information on minerals and how to recognize them, see A. M. Bateman (1950), *Economic Mineral Deposits* (New York: Wiley); W. Ernst (1969), *Earth Minerals* (Englewood Cliffs, N.J.: Prentice-Hall); and B. Simpson (1966), *Rocks and Minerals* (New York: Pergamon Press).

†To be considered a true rock, an object does not have to be a solid chunk of granite, a pebble, or a boulder. If it is a solid mass, it is called a *consolidated* rock; if it is loose material, such as a handful of sand, it is termed *unconsolidated*.

Intrusive (plutonic) igneous rocks. (a) Devil's Marbles, round granite boulders in Australia's Northern Territory. (b) Columns of gabbro in the Hudson Palisades, New York. (c) Granite, an intrusive (plutonic) igneous rock. (d) Gabbro, also called diabase, an intrusive (plutonic) rock.

passage blocked, the magma must cool deep within the crust, and large-grained crystals will have time to form. Granite, gabbro, and diorite are examples of **intrusive** igneous rock. For igneous rocks that form underground, some petrologists prefer the term **plutonic** (from the ancient Greek god of the underworld, Pluto).

When magma escapes onto the outer surface of the earth in the form of a lava flow or in a violent blast of ashes or blobs of lava, the material must cool rapidly, and there will be insufficient time for large crystals to form. This type of igneous rock is termed **extrusive**, or *volcanic*.

Although color does have some value in identifying different kinds of igneous rocks, the grain or crystal structure is the most important difference. For example, the most common intrusive igneous rock is granite, which is large-grained; the most common extrusive rock is basalt, which is fine-grained. The Hawaiians have given interesting names to the lava flows of basalt on their islands. If the moving, or once-moving, mass is rough-textured and "blocky"—a jumbled pile of black rocks—they call it *aa* (pronounced *ah-ah*). If the flow is rather smooth-looking and ropelike, they call it *pahoehoe* (pronounced *pah-hoey-hoey*). Ashes blown from a volcano do not look like rock until they have had time to solidify and compact into a rock called *tuff*. Italy is a land of volcanic activity, and the ancient Romans constructed many buildings in Rome with bricks carved out of compacted tuff.

Sometimes blobs of magma are spouted from a volcano's crater. These red-hot globs of lava look like hand grenades or bombs, so they are called *volcanic bombs*. Sometimes lava splashes out of the crater, showering the vicinity with splotches of magma that cool so rapidly that there is no time for crystals to form. The result is a textureless black rock that looks like black glass. This is obsidian, called *volcanic glass*. Extrusive rocks that have been blown from a vent of a volcano are called **pyroclastic** ("broken up by fire") and **ejecta** (having been "ejected" from the volcano). Another kind of ejecta is a light-gray frothy-looking rock that is so filled with air spaces that it can float in water. This is *pumice*.

Sedimentary Rocks

Sedimentary rocks result from sediments composed of sand, silt, clay, or the skeletal remains of once-living organisms deposited by wind, stream, lake, ocean, or glacier. The sediments are deposited grain by grain on an open plain, a valley floor, a stream or lake bottom, or an ocean floor. After the grains have been chemically cemented together or compacted by the pressure of other layers above, they become known as sedimentary rocks. So common are these rocks that they make up about 75 percent of all the rocks on the earth's surface. They can be further classified as clastic, organic, and chemical.

Clastic sedimentary rocks are made up of rock fragments that range in size from boulders down through cobbles, pebbles, sand, silt, and clay. The type of clastic rock called *sorted* refers to *shales*, which are compacted flakes of clay or silt particles that easily split into thin, fine-grained layers, and *sandstone*, which is composed of grains of sand cemented together. Shales are the most abundant sedimentary rock, with sandstone second. The type known as unsorted refers to compacted mixtures of various-sized particles, with small boulders, pebbles, and sand or silt combined to form a *conglomerate*.

Organic sedimentary rocks are derived from the skeletal remains of once-living creatures. Examples include calcareous limestone, made up of shells and coral; coquina, made from shell fragments; reef limestone, made from the excretions and bodies of coral; and chalk, made from microorganisms that are invisible without magnification. Coal is another form of organic rock; it is derived from once-living trees and other plant forms that carbonized through stages, from bogs to peat to coal.

Chemical sedimentary rocks are the result of precipitation of dissolved chemicals through layers of rock or from water in which the chemicals were suspended. For example, table salt (sodium chloride) results from the evaporation of water from a salty liquid solution; after the water is gone, the salt remains. Another chemical rock that results from evaporation is gypsum, a common building material

Extrusive igneous rocks. (a) A photo of an aa lava flow north of Hatch, Utah. (b) A pahoehoe lava flow in the Craters of the Moon National Monument, Idaho. (c) Deep deposits of tuff in Big Bend National Park, Texas. Note the ejecta and volcanic bombs in the foreground. (d) A close-up of obsidian, pyroclastic ejecta also known as volcanic glass. (e) Pumice from Sunset Crater, Arizona. (f) Basalt.

(a)

(b)

(c)

(d)

Sedimentary rocks. (a) "Sandstone symphonies"—cross-bedding patterns of the sedimentary rock called sandstone, in Zion National Park, Utah. (b) Conglomerate rock in a road cut near Suffern, New York. (c) Petrified wood protruding from beds of sandstone in Arizona's Petrified Forest. (d) Coquina, a sedimentary rock made up of seashells; this outcrop is in the Fezzan Desert in Libya.

used in plaster or in drywall construction. Salt and gypsum are often called **evaporites.** While limestone can result from the skeletons of once-living creatures, it can also result when carbon dioxide is carried to the earth by rain and combines with calcium, forming a compound called calcium carbonate. This mineral is called calcite. When underground water carries the calcite in solution down to the roof of a cavern, stalactites and stalagmites (see Chapter 13, page 352) are formed. This type of chemical rock is called *travertine*. Another interesting form of chemical rock is *petrified wood*, which results when chemicals replace the original woody cells of a fallen tree.

Sedimentary rocks are especially interesting to geologists, geographers, and archaeologists as well as to rock-hounds, because only in these rocks do we find fossils and other features that give us insight into the history of our earth. The layers of rocks, called *beds*, or **strata,** can be read almost as though they were pages in a book by those who understand the "language." In some of these beds we find *ripple marks*, tiny "wavelets" that were pressed onto the surface of a sand dune or the floor of a lake or ocean, then covered up by new deposits, footprints of dinosaurs or birds, *geodes*, which are hollow rocks usually occurring in limestone formations that have beautiful crystals jutting from the inner shell and, of course, fossils of ancient creatures or plants.

Metamorphic Rocks **Metamorphic** rocks were either sedimentary, igneous, or metamorphic when formed. They then were either buried very deep within the earth's crust, where they were later subjected to great pressures, or they were subjected to extremely high temperatures from a nearby mass of magma. As a result, changes were produced and the rocks underwent metamorphosis (from the Greek *meta*, "changed," and *morphe*, "shape" or "form"). The original texture and minerals are altered, and the new rock has different physical and chemical characteristics. One example of a metamorphic rock is the industrial diamond, which was metamorphosed, by heat and pressure, from coal into pure carbon.

The transforming influence of heat is called **contact metamorphism,** which simply means that an igneous body of rock supplied sufficient heat to bring about changes in rocks nearby. Granite can become altered into a foliated (the appearance of plates or thin leaves) gneiss. Then there is **dynamic metamorphism,** metamorphism brought about by pressure. Great pressures constantly are being exerted on rock masses throughout the earth's crust, along vast breaks or fracture planes and through a variety of crustal movements. Here, we do not find radical chemical changes, and a realignment of the existing minerals into new crystal structures.

There are, of course, some alterations produced by the movement of chemically active solutions that dissolve existing chemicals, add new elements, and thus introduce a new structure with new crystallizations and shapes. For example, a rock called gabbro is transformed into another rock called serpentine by chemically active waters that have been superheated by intrusions of nearby magma.

The Rock Cycle

As noted earlier, there are a number of cycles that have an important bearing on planet earth. We have discussed the cycle of the seasons, the solar energy and heat balance, the hydrologic cycle, and the cycle of vegetation. We now will study the **rock cycle** (Fig. 10-5).

The rock cycle began billions of years ago as the earth cooled and the crust began to form, and it has endured throughout earth's history. When moisture was released by the cooling earth and the atmosphere allowed rain

Metamorphic rocks. (a) Banded metamorphic rock along the Hudson Highlands, New York: granite metamorphosed into gneiss and twisted into bands by the intrusions of a rock called pegmatite. (b) Slate from eastern Pennsylvania: a popular flooring material that results from the metamorphosis of shale. (c) Fractured and jointed blocks of marble (metamorphosed limestone) in a road cut near South Dorset, Vermont. (d) Quartzite, metamorphosed sandstone.

to fall, alternating periods of excessive and drying heat and heavy precipitation caused the newly forming igneous rocks to be weathered (a process we will consider further in Chapter 12) and broken down. Streams carried the smaller pieces down from the emerging elevations to lakes, oceans, and valley floors. Ashes, spewed forth by volcanic activity, settled on the land and in the waters. As time passed, these sediments were built up into layers, and the

FIGURE 10-5 The rock cycle.

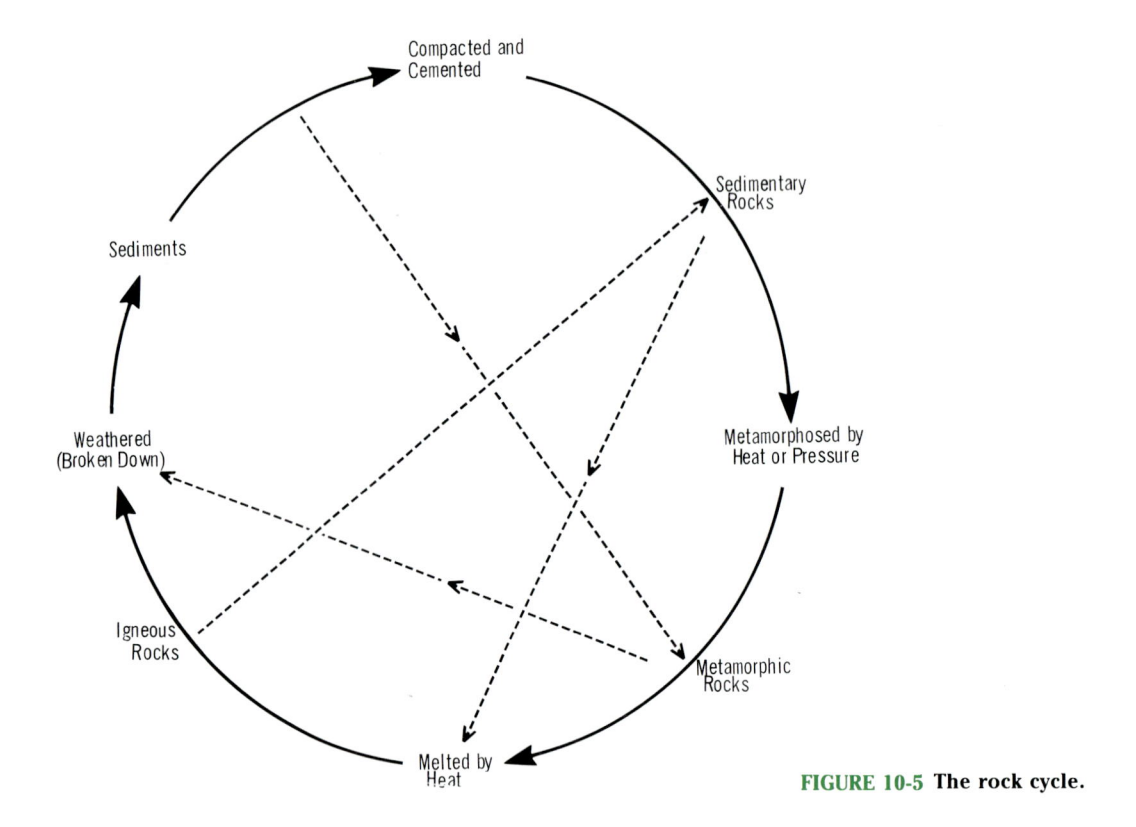

Fossils of a long-ago age.

sediments were compacted and cemented together by chemicals from the infiltrating waters. Sedimentary rocks were formed and, as more time passed, the tremendous weight of overlying beds created pressures that worked, along with the heat penetrating upward from the mantle, to change the mineral structures of the rocks. They became transformed into metamorphic rocks. Earth movements twisted the beds and folded them, and once again heat from the earth's interior caused the rocks to melt. Igneous rocks were created out of the old.

The cycle went on through the ages and, occasionally, there were interruptions in the cycle. Sedimentary rocks bypassed the perpetual round and were melted into magma without going through the stage of metamorphosis. Igneous rocks were subjected to intense heat and pressure and were transformed directly into metamorphic rocks. Metamorphic rocks were attacked by weathering agents and broken into fragments that eventually ended up as sedimentary rocks. There seems to be no end to this cycle. It still goes on.

CASE STUDY NUCLEAR ENERGY—FISSION OR FUSION?

Although the earth has supplied us with a tremendous variety of useful, indeed vital, minerals such as iron, copper, and tin, we have benefited mostly from the minerals that serve us as fossil fuels—coal, petroleum, and natural gas. We now know, however, that these energy resources are declining to the point of no return, and we must discover alternative energy sources that will be affordable and continuous and yet not cause environmental problems.

We have considered alternate energy sources such as geothermal, hydroelectric, tidal, and wind power, ocean thermal differences, and direct solar energy, but many scientists believe that nuclear (atomic) energy is the only positive answer to our future needs.

It all began about 2500 years ago, when the Greek philosopher Democritus came to the conclusion that any object, such as a rock, could only be broken down so far and that eventually you would wind up with a tiny particle that would be inseparable; that is, it could not be broken down further or subdivided. This indivisible particle he termed the *atomos*—the atom. Other scientists, including Galileo, Newton, Boyle, Lavoisier, Dalton, Avogadro, and Roentgen, gave considerable thought to the concept of the atom as time passed. Then, at the conclusion of the nineteenth century, Pierre and Marie Curie discovered that there was a spontaneous emission of radiant energy and gamma rays from the atomic center, or nucleus, of the uranium atom. This idea of nuclear fission was termed *radioactivity,* and the uranium atom was split into radium atoms. Albert Einstein next theorized that matter and energy are the same, only in two different shapes. As the atom splits, the mass becomes radioactive energy. Then, Ernest Rutherford found that the atom has a center, a nucleus, that is composed of particles called *protons,* which have a positive electrical charge. If negatively charged *electrons* are introduced into the atom's nucleus, they whirl about the nucleus, as the planets in our galaxy orbit about the sun. German scientists in the late 1930s were able to add particles that have no charge at all, or *neutrons,* to the heavy nucleus of the uranium atom, thus causing the atom to split into two pieces. This was the first occurrence of controlled *nuclear fission.*

*To learn more of the scientists who developed atomic theory, see the interesting book by Heinz Haber (1961), *The Walt Disney Story of Our Friend the Atom* (New York: Golden Press).

NUCLEAR FISSION

As one uranium atom splits, it triggers others to split; a chain reaction is under way, and massive amounts of gamma radiation are released, all in a fraction of a second. This can produce an enormous atomic blast (such as a hydrogen bomb), or it can be controlled in a *breeder reactor*. The reactor is an energy-producing plant where the atoms of uranium are splitting at a controlled rate because of the introduction of chemical elements (boron and cadmium) whose nuclei soak up the flying and expanding neutrons. If the rods of these elements are inserted into the enclosed uranium chamber, the reaction is slowed down; if the rods are pulled back, the reaction increases in intensity. The collision of these exploding atoms creates energy in the form of tremendous heat. If this heat is applied to water, it expands into steam, and this rushing steam can be directed at the blades of a rotary engine (turbine); this causes the blades to spin, creating electrical energy in much the same manner as the burning of coal, oil, or natural gas does. Unfortunately, however, the breeder reactor is far

TVA's first nuclear power plant—Browns Ferry. It is in operation on Wheeler Lake, 10 miles (16 kilometers) southwest of Athens, Alabama. It has a total generating capacity exceeding the output of all 30 Tennessee Valley Authority hydroelectric dams combined.

less efficient and, in addition, presents us with ecological problems of pollution.

NUCLEAR FUSION

When atoms of an element such as hydrogen, which has a lighter density, are pressed together, of fused, the protons in the nucleus overcome the electrical forces of repulsion, which have a tendency to keep them apart, and they collide with enormous force. This collision (fusion) is accompanied by intense energy and heat in the same way as our sun produces solar energy. To bring this about, a fusion reactor (sometimes called a *tokamak,* from the Soviet nuclear research) is used in which the thermonuclear reaction is excited by running an electrical current through a body of gaseous hydrogen. The current surrounds itself with a strong, bottle-shaped magnetic field that will not allow electrically charged nuclei to penetrate. As the temperature rises into the millions of degrees, the energized gas particles are held together and become fused.

Nuclear fusion seems to be the best approach to the use of atomic energy because the problems of environmental pollution are brought to a minimum and because the waters of the ocean can give us an almost limitless supply of the hydrogen. However, because the tokamaks cannot reach the necessary temperature range and sustain it long enough (1 second), it will probably take another 20 years or more before nuclear fusion becomes practical.

For additional information on nuclear energy, consult the following:

Anderson, William R., and Vernon Pizer (1966). *The Useful Atom.* Cleveland and New York: World Publishing.

Haber, Heinz (1961). *The Walt Disney Story of Our Friend the Atom.* New York: Golden Press.

Halacy, D. S., Jr. (1977). *Earth, Water, Wind, and Sun.* New York: Harper & Row.

Hammond, Allen L., William D. Metz, and Thomas H. Maugh (1973). *Energy and the Future.* Washington, D.C.: American Association for the Advancement of Science.

Patterson, Walter C. (1976). *Nuclear Power.* New York: Penguin Books.

Mann, Martin (1974). *Peacetime Uses of Atomic Energy.* 3rd rev. ed. New York: Thomas Y. Crowell.

Hunt, S. E. (1974). *Fission, Fusion and the Energy Crisis.* New York: Pergamon Press.

Redman, L. A. (1963). *Nuclear Energy.* New York: Oxford University Press.

Stoker, H. Stephen, Spencer L. Seager, and Robert L. Capener (1975). *Energy: From Source to Use.* Glenview, Ill.: Scott, Foresman.

REVIEW AND DISCUSSION

1. Describe the composition of the atmosphere, hydrosphere, biosphere, and lithosphere.
2. Make a sketch of the earth's interior and label the major zones. What are the distances involved?
3. What does the term *Moho* mean?
4. How did the geologic eras, periods, and epochs get their names?
5. Discuss the terms *Archeozoic, Proterozoic,* and *Paleozoic.* What types of life forms existed in each?
6. What is the mantle, and how does it differ from the earth's core?
7. What is the asthenosphere?
8. What is the difference between a mineral and a rock?
9. What are elements?
10. Which is the most abundant mineral?
11. Describe some of the different ways minerals are formed.
12. Discuss these mineral combinations: oxides, carbonates, sulfides, silicates, and chlorides.
13. What are some of the tests used to understand the different physical properties and structures of minerals?
14. What is magma? How does it differ from lava?
15. What is the difference between *aa* and *pahoehoe* lava?

16. Define the terms *pyroclastic, ejecta,* and *volcanic glass.*
17. How many kind of sedimentary rocks are there? Explain the differences.
18. What is the most abundant sedimentary rock?
19. What is the difference between the terms *organic* and *inorganic?*
20. What kind of rock is common table salt?
21. What is the difference between contact and dynamic metamorphism?
22. What terms are used to describe the layers of rocks?
23. What kind of rock is an industrial diamond?
24. When limestone is metamorphosed it becomes what kind of rock?
25. Discuss the rock cycle.
26. What is an *atomos?*
27. Explain the difference between nuclear fission and fusion.
28. Why is there mounting opposition to nuclear energy today?

NOTES

[1] John Muir (1938), *John of the Mountains: The Unpublished Journals of John Muir,* edited by Linnie Marsh Wolfe (Boston, Mass.: Houghton Mifflin), p. 315. With permission.

[2] Muir, p. 320. With permission.

[3] Edwin E. Larson and Peter W. Birkeland (1982), *Putnam's Geology,* 4th ed. (New York: Oxford University Press), p. 19.

ADDITIONAL READING

American Geological Institute (1984). *Dictionary of Geological Terms.* 3rd ed. Garden City, N.Y.: Doubleday.

Beiser, Arthur (1971). *The Earth.* New York: Time-Life Books.

Deer, William A. (1966). *Introduction to the Rock Forming Minerals.* New York: Wiley.

Dott, Robert H., Jr., and Roger L. Batten (1980). *Evolution of the Earth.* 3rd ed. New York: McGraw-Hill.

Ernst, W. G. (1969). *Earth Materials.* Englewood Cliffs, N.J.: Prentice-Hall.

Longwell, C. R., R. F. Flint, and J. E. Sanders (1969). *Physical Geology.* New York: Wiley.

McKenzie, G. D., and Russell O. Utgard, eds. (1975). *Man and His Physical Environment.* 2nd ed. Minneapolis: Burgess.

Shelton, John S. (1966). *Geology Illustrated.* San Francisco: Freeman.

Watkins, Joel S., Michael L. Bottino, and Marie Morisawa (1975). *Our Geological Environment.* Philadelphia: Saunders.

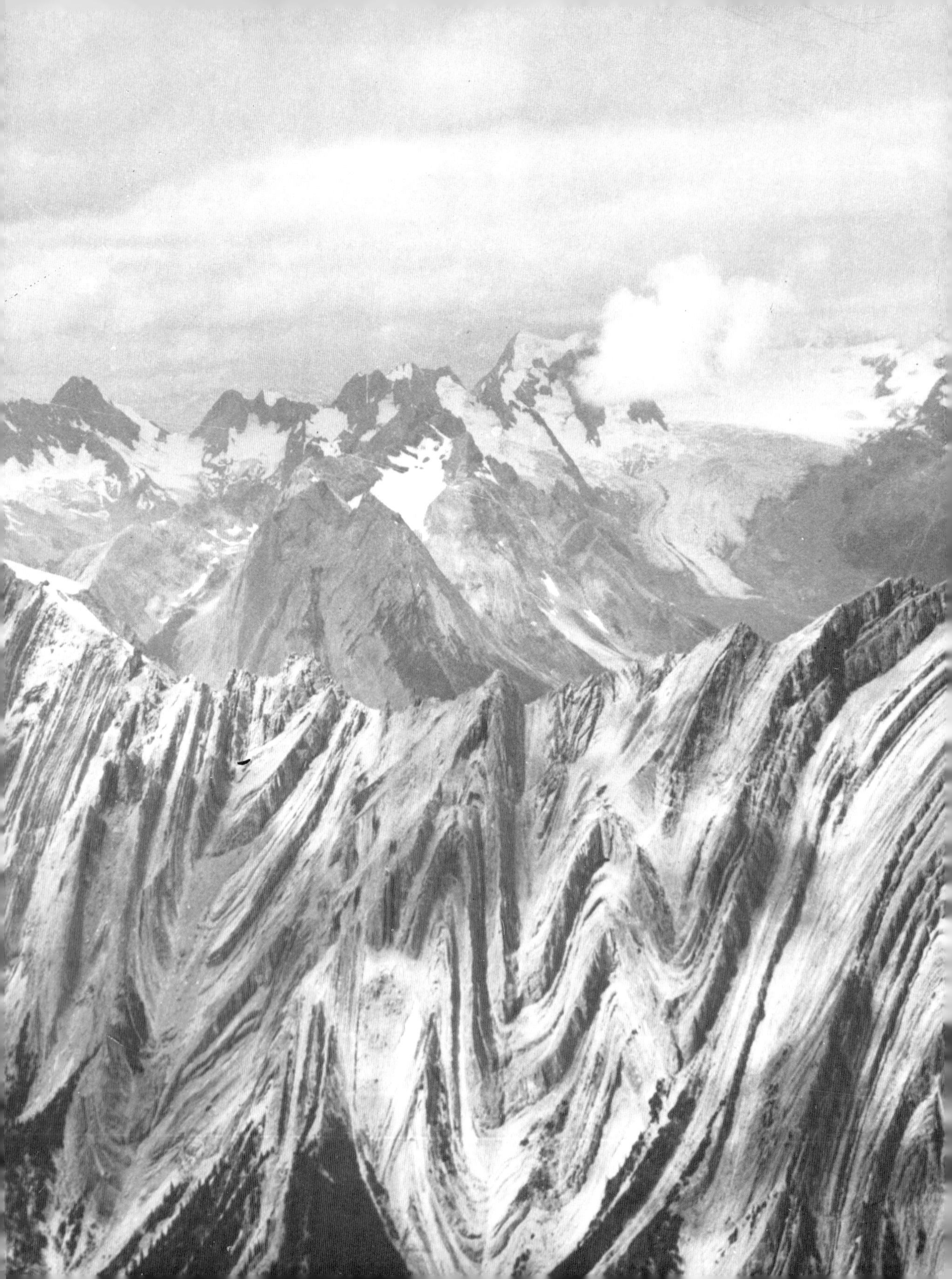

ELEVEN

LAND SURFACE AND SEA BOTTOM

When we dwell with mountains, see them face to face, every day, they seem as creatures with a sort of life—friends subject to moods, now talking, now taciturn, with whom we converse as man to man. They wear many spiritual robes, at times an aureole, something like the glory the old painters put around the heads of saints. Especially is this seen on lone mountains, like Shasta, or on great domes standing single and apart.
JOHN MUIR

We will now discuss the earth's surface features, the landforms that rose out of the mantle and were twisted, distorted, fractured, elevated and lowered, and shifted horizontally. We must keep in mind that mountains, valleys, hills, plateaus, and plains are not limited to the continents. Features present on the continents are also present on the ocean floors. If the waters of the oceans were to be drained away, the resulting landscape would appear much like that of the landmasses. Why is this? Why are there deep valleys, gaunt fractures, massive mountain ranges, and broad plains on the sea bottom? Are they caused by waves and currents? The answers to these questions can only be answered by examining the whole—the land surface and the sea bottom.

The same forces that moved the continents, elevated mountains, caused volcanoes to spew forth their fiery outpourings, and fractured the land are operating throughout the earth's crust, and they have done so since the beginning. Later in this chapter you will see how much alike the land surface and sea bottom really are, but for now let us concentrate on the internal, or tectonic, forces involved in the shaping of the earth's surface.

TECTONIC FORCES

Today, people tend to be scientific in their approach to the unknown. When the earth shivers and shakes, as it often does in Italy, Peru, central Asia, or Nicaragua, or when volcanoes erupt, as they do in Mexico, Hawaii, or off the coast of Japan, earth scientists speak of diastrophic forces, vulcanism, and plate tectonics. Earlier peoples, however, thought that a god or earth spirit was showing strength or displeasure at their behavior. The Japanese once called the "earth shaker" Nai-no-Kami (god of the restless earth) and built temples in his honor, hoping that by so doing they could placate his wrath.[1] The people of Nicaragua worshipped a goddess of volcanoes, Masaya, and threw human sacrifices into a crater to appease her.[2] The volcanic island of Lemnos in the Mediterranean gave rise to the cult of Haphaestus, a thunder god, god of the fires of the earth. He appears in Greek mythology as a beneficent god tending the earthly fires that allowed men to work with iron and other metals.[3] The Romans of later times adopted this god and called him Vulcan. He became their god of volcanoes and the home fires, and his name is now associated with the science of vulcanism, which studies this phenomenon.[4]

Such simplified answers to the question of the earth's restlessness may have been acceptable to those people of earlier times, but they have not satisfied earth scientists for several centuries. Many theories have been conceived by geophysicists and geologists trying to explain the tremendous forces and processes. Some disturbances and distortions within the earth's crust are easy to understand, but others are not. Cause and effect are obvious when a section of earth collapses over a mine shaft, but how can we conceptualize the unleashing of power immense enough to lift a great mountain range or to split an ocean basin?

Contraction
The most widely held theory of the nineteenth and first half of the twentieth centuries attributed the crumpling of the earth's surface to contraction, the cooling of the outer crust as the earth changed from a molten mass to a relatively solid body. As the earth released its moisture to the atmosphere and began shrinking, the process was equated to what happens to a plum that becomes transformed into a prune as it dries.

Proponents of this theory claim that the earth's interior is still cooling and reducing in size, so the earth's crust is constantly making adjustments to the new conditions. As the pressures mount in one place, a continent can be elevated and arched upward in its center or along one side and, in another area, a landmass can be compressed downward as its edges are rumpled into a series of mountains. A state of equilibrium can occasionally be attained, but it does not last because the cooling and shrinking are an ongoing process.

Convection

The theory of **convection** might also be referred to as expansion, distension, or swelling; it suggests that the earth is enlarging because of an increase in the temperature within the interior. As the heat generated by radioactive decay of minerals in the mantle rises to the surface, the materials of the mantle must expand, forcing continents to lift, tilt, or fracture. Not all earth scientists accept the idea that our planet is swelling, but many do agree that there are convectional currents circulating throughout the mantle. As heat is built up within the lower mantle, differences in temperatures of the upper and lower zones cause currents to flow through the more plastic layers within and near the asthenosphere. Rock materials in the simatic layer (sima) are affected by the intense heat below and are sometimes forced to move either vertically or horizontally. If the earth's crust were not so rigid and prone to break or fracture, the main result would be some volcanic activity or warping and folding of the layers above the moving masses of the asthenosphere.

Early proponents of the convection theory tended to limit its influence to the creation of mountains, the folding of rock strata, the uplift of landmasses, and the cause of earthquakes; modern earth scientists have shown that there is more to this theory than formerly suspected. This idea of constantly moving convectional currents leads us to another theory, that of plate tectonics, a theory that is rapidly gaining support by scientists all over the world.

Plate Tectonics

The term **tectonics** is defined as "the internal forces of the earth that cause crustal movements" and comes to us from the Greek *tectonicus*, "builder." The term **plate** means, of course, a "thin slab or piece of some material," but in this instance *plate* refers to the subdivision of the earth's crust into huge sections that are separated by sometimes violent fracture (or rift) zones. There are at least 10 or 12 major plates and a number of smaller ones. The plates extend down into the asthenosphere and are from 5 miles (8 kilometers) thick under the oceans up to 25 miles (40 kilometers) thick when measured through the landmasses. Figure 11-1 indicates the present positions of the most important plates, and you can see that each continent occupies its own plate. Later, when we discuss earthquakes and volcanoes, you will also see that the greatest numbers of each of these forces are found along the edges of these plates. This tells us something, does it not? There must be tremendous activity occurring in those areas, activity that has been going on for a long time. But when did it all begin? What is the existing evidence to prove that there are great plates and not merely powerful stress areas? Do these plates move and, if so, what was their original location? The answers can be found by examining the *theory of continental drift*.

The concept that the seafloor has been spreading and the continents have been drifting apart for millions of years is a relatively new theory, although the idea had been considered and rejected by a number of earth scientists through much of the nineteenth century. If you were to look at a globe and study the western coastline of Africa and the eastern coastline of South America, perhaps you might

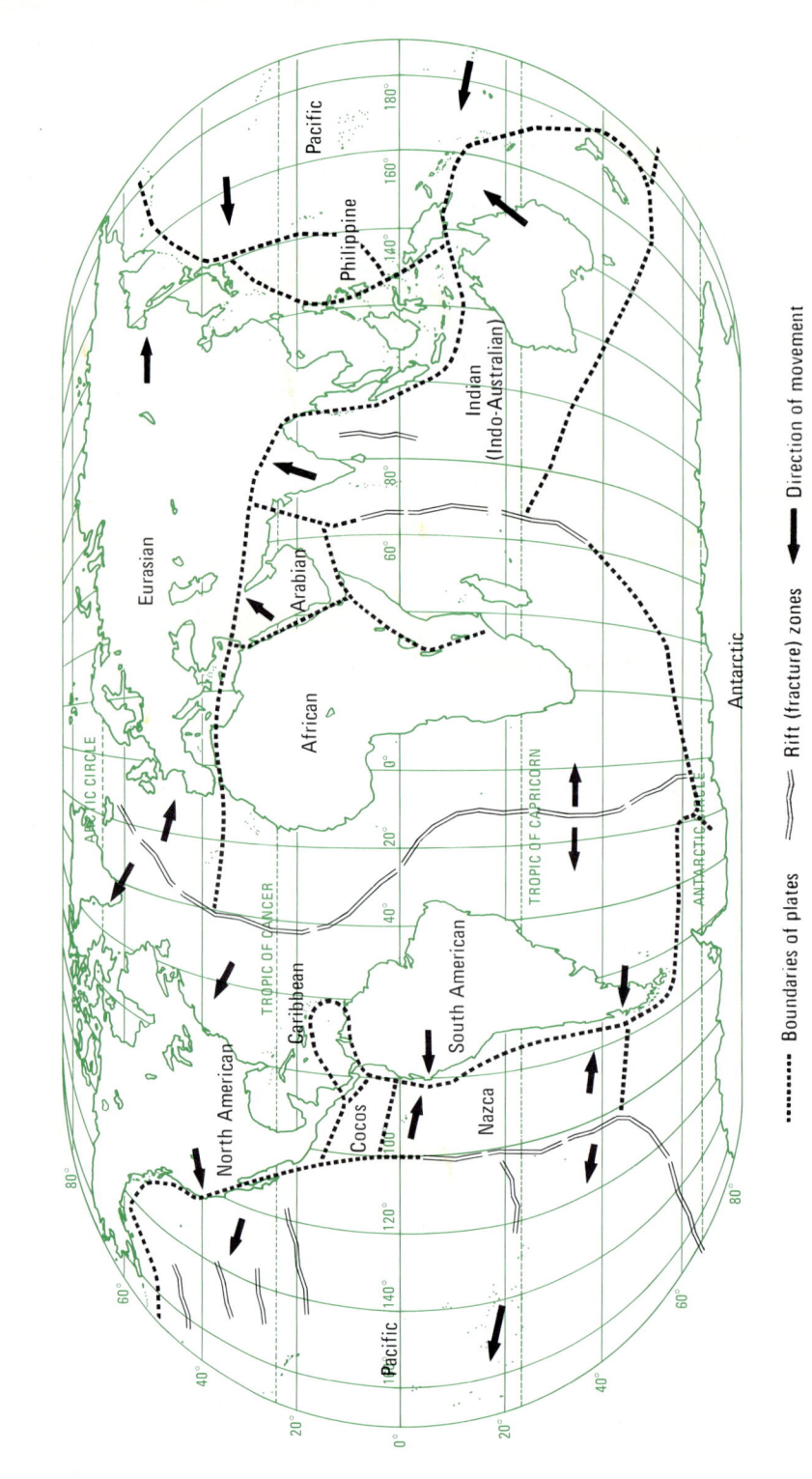

FIGURE 11-1 The earth's plates.

understand the reason for such a speculation. Do they not look as though they might have once been fitted together as *one* continent? Then, in 1912, a German meteorologist, Alfred Wegener, presented his views in an address to the Geological Association in Frankfurt, Germany.* He believed that there was once a single supercontinent, and he called it *Pangaea* (from the Greek *pan*, "all," and *gaea*, "earth"). In a sense, we might think of this single, huge continental landmass as the "parent" continent because from it developed the six continents that now exist.

Some earth scientists have grudgingly accepted the theory of continental drift, but others waged a series of stormy debates as they clung to the older theories. However, extensive geologic and geomorphic research on land and sea has turned up positive evidence in the structure of the rock strata of Accra, Ghana, and near the town of Sao Luis, northeast coastal Brazil. Radioactive dating determined that the ages of the layers of rock were the same on both sides of the Atlantic.† Continued research along the coastlines of Antarctica and southern Australia and of the southeastern United States and northwestern Africa has added further verification.

Pangaea is thought to have been surrounded by the world ocean about 2.5 billion years ago (near the end of the Archeozoic era), but its original position is not known. It is thought that Pangaea shifted around, breaking apart, coming back together, changing position periodically as it was pushed by the pressures exerted on the earth's crust from below. Field researchers have supplied a considerable amount of evidence to back up this claim. For example, scientists traveling through southern Africa, the Deccan Plateau of India, and parts of Australia, regions that are far from the present-day south polar ice cap and not believed to have been subjected to glaciation by an ice age, have found apparent proof that these lands were once positioned near the South Pole.

Another important discovery was made in the early 1960s, when it was learned that as rocks form, iron particles within them will line up like a compass needle and point toward one of the magnetic poles. Scientists now working in the field of **paleomagnetism** (study of the earth's ancient magnetic fields) found that lines of magnetism in ancient rocks of the different continents were not aligned toward the present-day magnetic poles. This could mean that the north and south magnetic poles either had traded positions or had wandered about over vast distances. By carefully measuring the magnetism of rocks from the ocean floors and the landmasses and feeding the data into computers, it was clearly shown that if the continents could be pushed together, the lines of magnetism would point fairly close to a common set of magnetic poles of ancient times. This does not imply that the magnetic poles do not move. Instead, it indicates that such movements are not extensive.

Physical oceanographers have long studied the ocean floor, and they could not account for the sediments on the sea bottom being less thick than they should have been considering the billions of years that materials from the landmasses had been washed or blown into the oceans. Instead of an expected depth of 10 miles (16 kilometers), the sediments average out to be about 0.5 mile (0.8 kilometer) thick. Specially designed oceanographic ships found

---

*To read a first-hand account of the research conducted by this pioneer in geophysics, see Alfred Wegener (1966), *The Origin of Continents and Oceans* (New York: Dover Publications). This is a translation of Wegener's original work, which was first published in 1924 and revised in 1929.

†Patrick M. Hurley (1968), "The Confirmation of Continental Drift," *Scientific American* (offprint no. 874).

that the floor of the oceans was composed of basaltic (igneous) rocks that formed the simatic layer of the earth's crust. In addition, the ocean floor was found to have extensive mountain ranges, high plateaus, and deep "valleys" (called **deeps**, or **trenches**). Enough proof was finally available to convince all but the most skeptical that our continents were once a part of Pangaea, the supercontinent on a superplate.

Based on the evidence, we can now say that Pangaea began its final breakup 180 to

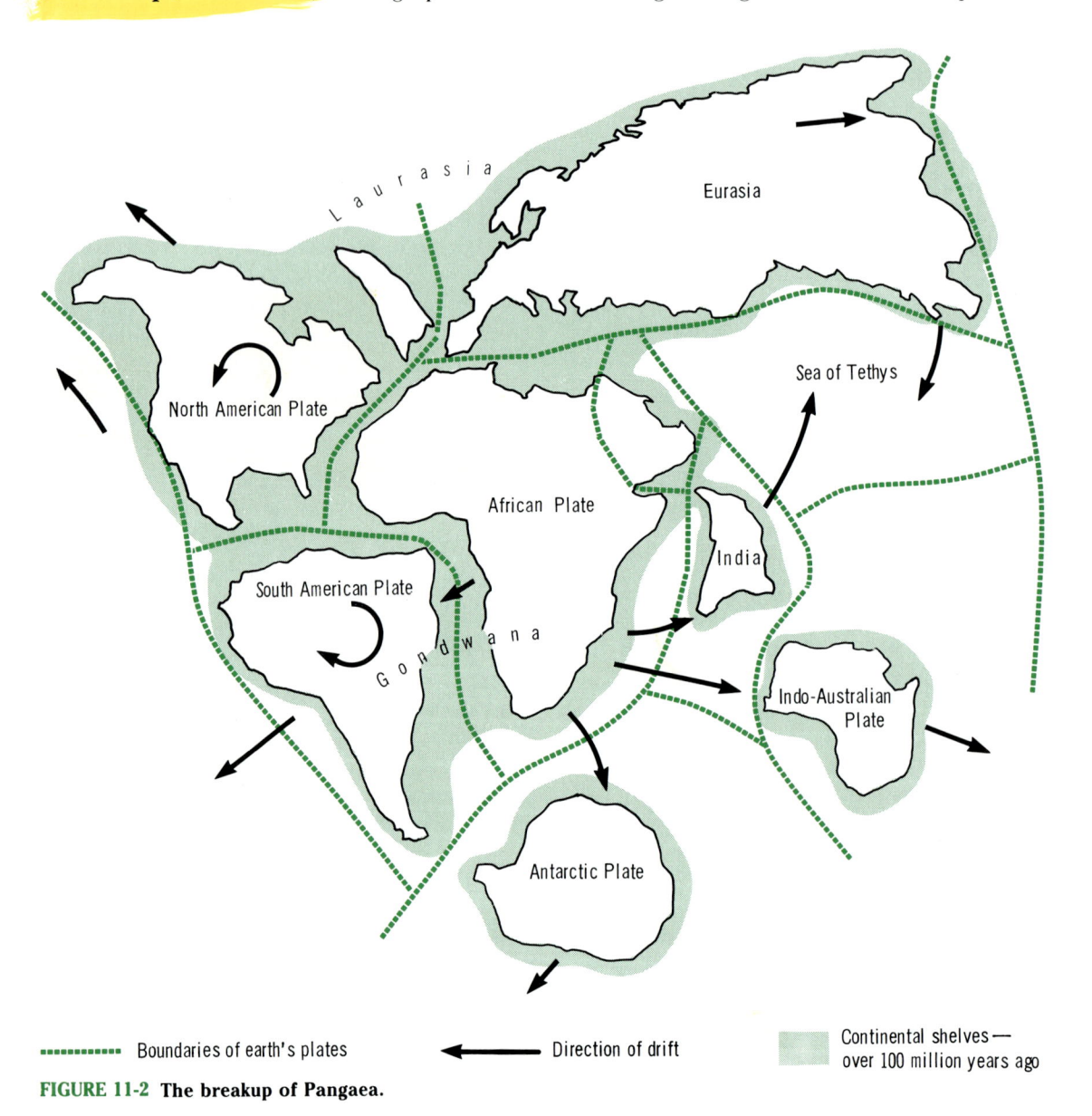

FIGURE 11-2 The breakup of Pangaea.

200 million years ago (Fig. 11-2). The first separation resulted when the northern section (Laurasia) broke free along a line that ran through the Mediterranean into what is now the Indian Ocean. The southern portion (Gondwana) began to split up, and one section, consisting of Antarctica, India, and Australia, swung away to the southeast, while South America began to drift westward. Africa appears to have moved slightly eastward, shifting somewhat in its north-south alignment. In time, Laurasia began to weaken under the irresistible pressure. North America began driving westward, and the North Atlantic Ocean came into being. Gradually, all of the continents began to move into their present positions.

It must not be assumed that a state of equilibrium has been reached. The prevalence of volcanic activity and earthquakes is convincing proof that continental drift and seafloor

California's San Andreas Fault.

spreading is continuing. One example can be seen by observing California's San Andreas Fault, which coincides with the boundary line between the American and Pacific plates. This is a violent fracture zone that is 600 miles (960 kilometers) long and extends about 20 miles (32 kilometers) down into the crust. Earthquakes occur regularly along this fault line, and it is estimated that the western side of the fault is moving northwesterly at a rate of about 1 inch (2.5 centimeters) per year. This means that someday (perhaps about 28 million years from now) the city of Los Angeles will pass San Francisco as it travels north!

Epeirogeny The movement of the earth's plates created the continents, and we use the term **epeirogeny** (from the Greek *epeiros,* "mainland," and *genesis,* "origin") when we talk of continent building. Such vast crustal movements usually take the form of widespread sinking, tilting, or uplifting of either an entire continent or a portion of it. A good example can be seen along the west coast of California, where the shoreline that had long been subjected to the work of ocean waves and currents was tilted upward so that wave-cut platforms (marine terraces) are now hundreds of feet above sea level. Later in this chapter we will examine additional examples of this force, called **diastrophism** (from the greek *diastrophe,* "distortion"), which faults, folds, and warps the land.

Orogeny Orogeny (from the Greek *oros,* "mountains") is another result of the moving plates; it refers to the process of mountain building. Figure 11-3 illustrates the uneven surface of the earth; a close examination will reveal the mountains and plateaus created by epeirogenic movements as well as by orogeny. If you compare this map with that of Fig.

Shoreline elevated by the power of epeirogeny.

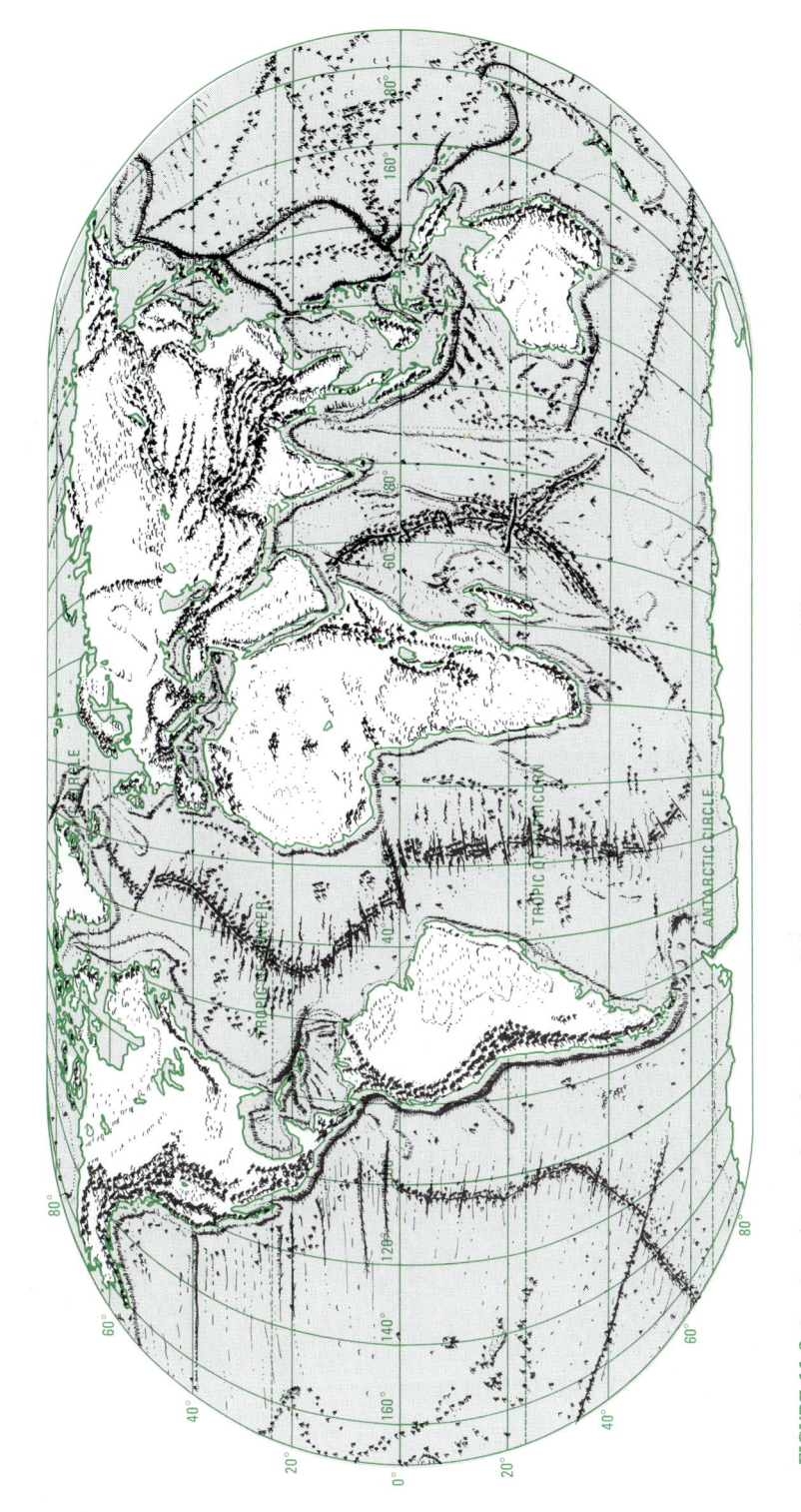

FIGURE 11-3 Land surface and sea bottom.

11-1, you will see that the mountains appear most often along or close to the edges of the plates.

Orogeny occurs when the lighter and less dense crustal combination of sial and sima is thrust against the heavier simatic crustal plate. It is almost like an irresistible force meeting an immovable object. Something has to give. Sometimes the moving plate will plunge downward at the boundary line, diving into the upper mantle; the resulting friction and increased heat causes some of the sima rocks to melt and thrust upward, creating volcanic mountains.

Sometimes a moving plate will buckle along or near its leading edge, and the upfolding of the sialic layer will create a mountain range such as the Andes of South America, the Rocky Mountains, Sierra Nevada, and coastal ranges of western North America, the Atlas Mountains of North Africa, and the Himalayas of Asia. In the latter case, the Himalayas were the result of the collision of the subcontinent India with the Asian landmass. India is believed to have been a part of a segment of Pangaea that included Australia and Antarctica, but it eventually broke loose and was pushed northward by the powerful forces of the moving plates through the ancient sea of Tethys, which is now the Indian Ocean. It must have taken millions of years to accomplish this task of elevating a former seafloor to heights of more than 25,000 feet (7500 meters). Mountain climbers have reported finding sedimentary deposits containing fossils of ancient sea life on top of Mount Everest!

The Himalayas. The Himalaya mountain system, which has resulted from the upward force of a moving crustal plate, consists of three main ranges. Amidst one, the Greater Himalayas, on the border of Nepal and Tibet (China), stands Mount Everest, 29,028 feet (8854 meters) above sea level.

Isotasy Another term often used by geomorphologists is *isostasy* (from Greek *iso*, "equal," and *stasis*, "to stand"). It implies a tendency toward equilibrium or balance between the lighter sialic (granitic) landmasses and the heavier basaltic ocean floor. If, for example, a heavy layer of sediments is deposited through time on the ocean floor, that portion of the crust will tend to sink, forcing the nearby landmass, which has lost materials by erosion, to rise a sufficient distance to maintain the balance between the two. If a great continental glacier were to move onto a landmass, the resulting increase in mass and weight might force the sialic landmass to be depressed. The sima of the ocean floor would then have to rise in compensation. When the glacier melted and disappeared, the landmass would be relieved of its burden and probably would return to its normal elevation. The ocean floor would gradually sink until all was in balance once again.

Diastrophism

We have been discussing large-scale crustal movements such as the drifting of the earth's plates, the raising of continents, and the building of mountains, all the result of the tectonic forces of diastrophism and vulcanism. Now we will focus on the smaller-scale actions of these forces and examine how they have altered the earth's structure.

Diastrophism implies crustal movement, and it can be defined as the distortion of the rock layers of the earth's crust by twisting (warping), folding (bending), and faulting (breaking) (Fig. 11-4). Irregularities created by these distortions are difficult to identify visually from the air or while driving across the land, unless you happen to be on a highway that cuts through a hill or mountain. The rupture or dislocation of rock layers is then laid out for you like a textbook diagram.

Geologists have taken core samples (by drilling down through many layers of rock) and are able to tell us what lies beneath our feet. The hills of Tennessee are surface features created by **folding.** This is the result of pressures exerted on a section of land from two sides. The central portion was compressed, and the rock layers began to shift. Some found release by moving up (creating an upfold, or **anticline**); some sank (creating a downfold, or **syncline**). If the horizontal (lateral) pressure were great enough, the layers of rock thrust upward and forward, as in the region of northwest Wyoming when the Grand Tetons were formed.

When the upper rock layers of the earth's crust are unable to withstand the pressures applied by the tectonic forces, one section may move past another as the rock layers fracture. This is called **faulting** (Fig. 11-5*a*). If the movement is primarily vertical and the strata on each side of the fault line are pulled apart by tensional force, one side may drop, or the other side may be lifted. This is called a *normal fault* (Fig. 11-5*b*). If the opposite occurs, if both sides are being pushed together by the force of compression, one side may be pushed up the dip slope of the fault plane, or the other side may be pressed downward, leaving a more abrupt escarpment. This is called a *reverse fault* (Fig. 11-5*c*). If the dislocation is primarily horizontal (sideways), it is called a **strike-slip** fault (Fig. 11-5*d*). This does not mean that such crustal movements are limited to either vertical or horizontal movement. In most cases, the movement will be a combination of both. For example, the San Andreas Fault is considered a "master fault" because there are a great number of smaller faults associated with it, such as the one that triggered the Coalinga earthquake on May 2, 1983, causing serious property damage to the small California area. Aftershocks continued on through the year. It is popularly thought of as a massive fracture

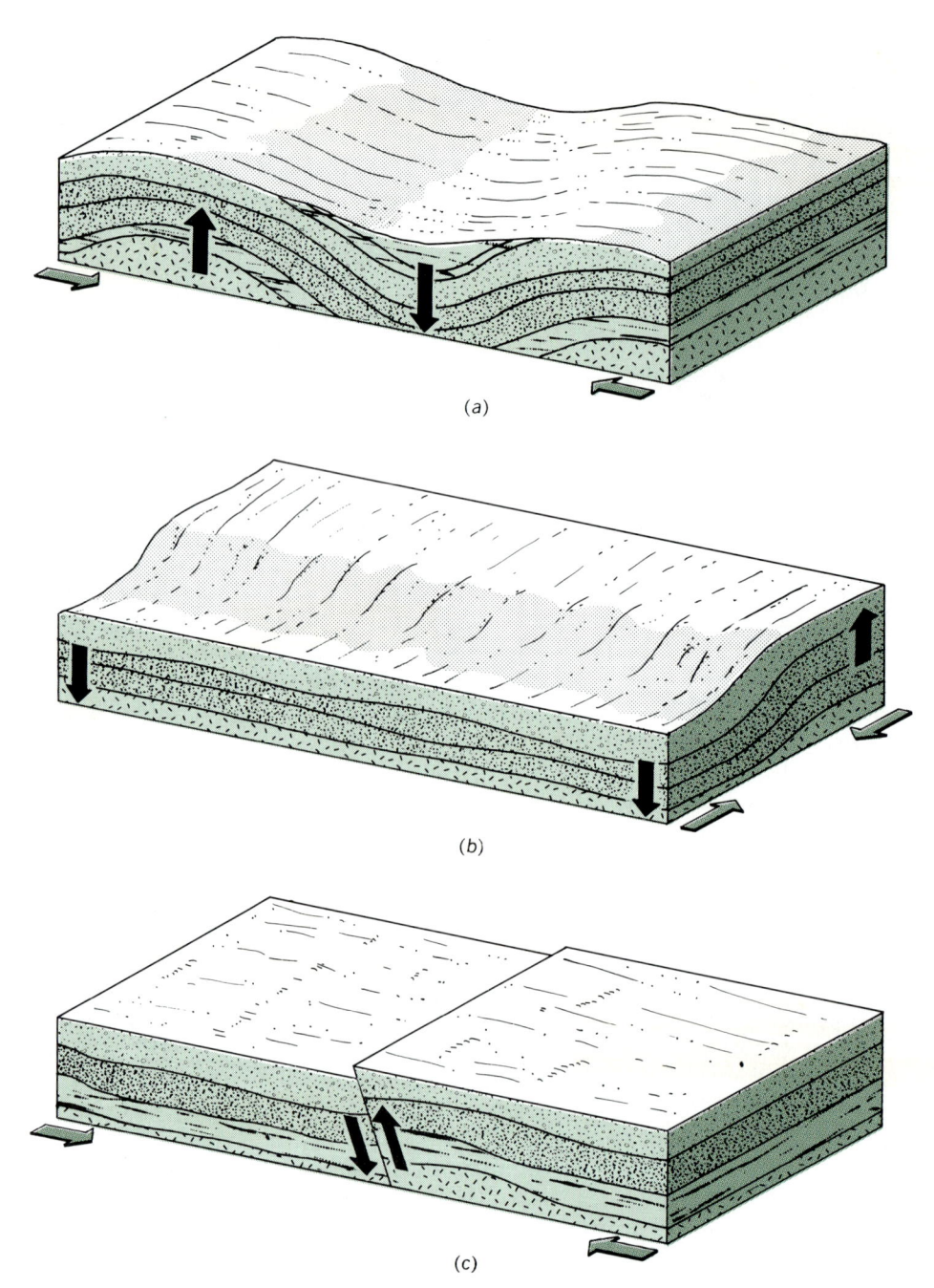

FIGURE 11-4 Forms of diastrophism (*a*) Folding: upfolds (anticlines) and downfolds (synclines). (*b*) Warping. (*c*) Faulting.

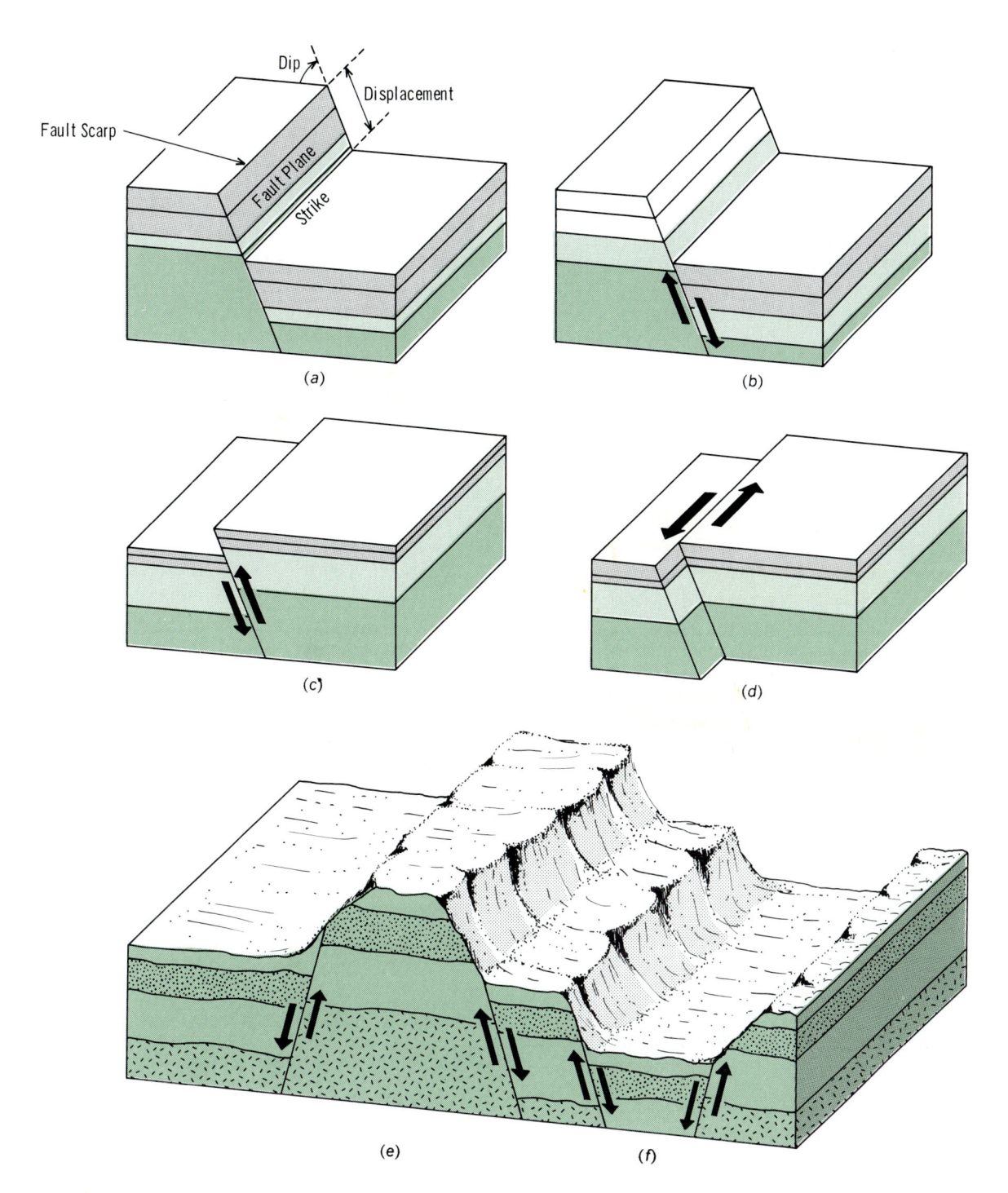

FIGURE 11-5 **Faults** (*a*) illustrates the terminology associated with faults; (*b*) shows a normal fault; (*c*) a reverse fault; (*d*) a strike-slip fault; (*e*) a fault-block uplift, also called a horst; and (*f*) a fault-block valley, called a graben.

that operates in a strictly horizontal direction, from the southeast to the northwest; however, the western side, which is a part of the Pacific plate, is rising. Measurements taken after the February 1971 earthquake, which resulted because of movement along this fault, showed that the western side had been uplifted approximately 4 feet (1.2 meters).* A similar event occurred on October 28, 1983, when a powerful earthquake struck the Pacific Northwest, centered near Challis, Idaho. The land split and the western side uplifted about 7 feet (2.1 meters).

Interesting landforms can be created by these crustal movements, such as hills or mountains created by vertical uplift or a central fault block (Fig. 11-5e) or by the tilting of one side of a crustal block. The landform created by this type of movement is called a horst (a German word meaning "ridge"), an example of which is the Sierra Nevadas of California. If the block between two fault systems drops because of the compression applied by tectonic forces, the result is a valley (Fig. 11-5f), called a graben (a German term meaning "trough"), or a rift valley. There are a number of excellent examples of this landform, such as California's Death Valley, the Rhine Graben, which extends from Basel, Switzerland, to Mainz, Germany, and the rift zone that stretches northward from Lake Nyasa in East Africa through Lake Tanganyika, the trough of the Red Sea, and the valley of the River Jordan in Israel.

When you think of the great age of our planet and observe evidence of past diastrophic forces such as elevated coastlines, great mountain ranges, and common sights such as fault and folds exposed in a highway cut, you may think that all of these things are from the past. When you consider how many millions of years these things have been happening, it is difficult to relate such processes to our own time. However, diastrophism is a continuing process, as the prevalence of earthquakes worldwide conclusively proves (Fig. 11-6).

Most earthquakes are noticed only by an instrument called a seismograph, which indicates the intensity of the earth's vibrations.* Some earthquakes, however, are quite violent and result in considerable damage to property and loss of human life. Severe earthquakes have occurred all over the world, as in Tashkent, Soviet Central Asia, in 1966, when 75,000 people were left homeless; Managua, Nicaragua, December 1972, 5000 killed, 20,000 injured, 250,000 left homeless; Fars, Iran, April 1972, 5000 killed; Lisbon, Portugal, in 1755, 50,000 killed; Yokohama and Tokyo, 1922, over 150,000 killed; Tangshan, China, July 1976, 655,000 killed; and San Francisco, California, 1906, more than 500 killed and much of the city destroyed.†

---

*Geologists have recently determined that another "master fault" exists in southeastern Missouri, near the confluence of the Mississippi and Ohio rivers, in New Madrid. This fault extends about 100 miles (160 kilometers) southwesterly into Tennessee, and although it is not as active as San Andreas, it does offer potential disaster threats to America's Midwest.

*The vibrations are of two basic types, a compression wave and a transverse, or shear, wave. Inasmuch as the compression waves travel faster through the earth, they arrive at a distant point ahead of the transverse waves, and thus are known as primary or P-waves. The transverse waves arriving later are referred to as secondary or S-waves. If you witness an earthquake, you will possibly note first a sharp thud, or blastlike shock, which marks the arrival of the P-wave; a few seconds later, a swaying or rolling motion may be felt, which marks the arrival of the S-waves. See note 5 at end of chapter.

†To learn more about the earth's devastating earthquakes, see the United States Department of the Interior, Geological Survey *Earthquake Information Bulletin*, published by the Government Printing Office, bimonthly; and "Earthquakes and the Earth's Interior," Chapter 18, in Larson, Edwin E., and Peter W. Birkeland (1982), *Putnam's Geology*, 4th ed. (New York: Oxford University Press).

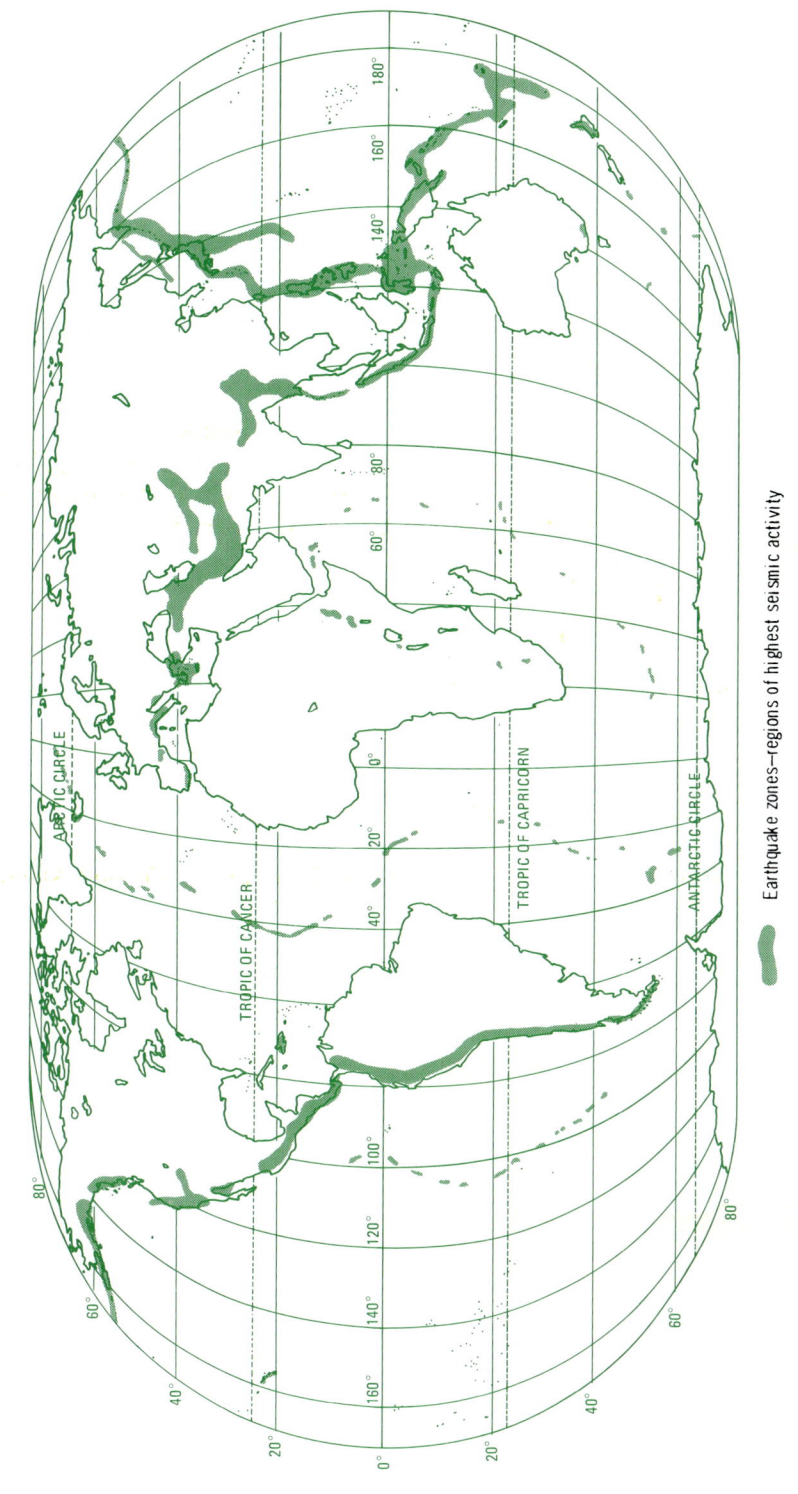

FIGURE 11-6 Earthquake zones of the world (Adapted from U.S. Department of Commerce, Earthquake Information Bulletin, 5, no. 2, 1971.)

One question always arises after every serious earthquake: Is there any way to predict or forecast an earthquake? Surely there must be some way to warn citizens in an active earthquake region that a seismic wave (from the Greek *seio*, "to shake") is on the way, as they do during the tornado season throughout the American Midwest. **Seismologists** have been working on this problem for a long time. They have produced instruments (seismographs) to locate the epicenters (central points) of earthquakes and scales (sets of conversion tables) to determine the magnitude of the quake (such as the **Richter scale,** which indicates the actual amount of energy involved, and the **Mercalli scale,** which describes the local effect or damage caused by an earthquake). They have not as yet, however, solved the problem of earthquake prediction.

The nature of the problem is complex. Not all earthquakes have the same origin. Some are primarily horizontal, sending great waves traveling through the earth's crust in a straight line and causing tall buildings to sway dangerously. Others have a rotary motion, and still others are short-lived, with a vertical thrust that makes the ground seem to jump up and down. Some occur within a landmass; others occur on the ocean floor, usually near the edges of the earth's plates.

Some seismologists believe that the **elastic-rebound theory** best explains how earthquakes occur. Rock layers beneath the earth's surface are subjected to intense pressures from many angles. First they bend, then they snap as the relieved sections try to return to their original position. Seismic waves are sent rolling through the earth's crust, and an earthquake results.

When an earthquake occurs in the floor of an ocean, the shaking of the sea bottom sets waves in motion that travel great distances. These seismic sea waves have inflicted great damage to coastal cities from Hawaii to Japan. They are popularly known as **tidal waves,** although the pull of the moon has nothing to do with their origin. The more proper term is **tsunami** (which comes to us from the Japanese, who have endured more than their share of seismic sea waves). Hilo, Hawaii, has suffered several tsunamis, the most notable one in 1960. The city received word that a tsunami was going to strike the waterfront, but despite radio warnings to move to high land, tourists ignored the advice and rushed down to the beaches to watch the "tidal waves" come in. Many were drowned as the 20- to 30-foot (6- to 9-meter) waves rolled into the low-lying city and poured into second-story windows of buildings.

The search for answers is continuing. A theory now more and more accepted suggests that before earthquakes strike, they always send advance notice in the form of gradually increasing shock waves. Proponents of this theory believe that a warning system of computer-connected instruments could be set up along active fault lines in earthquake regions. If, after a period of quiet, a tiny shock agitates the instrument and is followed by a growing stream of more intense shocks, the seismologist would know that an earthquake was in the offing and might be able to predict the time, place, and intensity of the quake.* Soviet and Japanese seismologists are also working on this problem and have indicated a willingness to share their discoveries with us.

Prediction is one thing; prevention is another. Some seismologists are trying to determine ways to eliminate earthquakes before they cause any damage. The idea is to lubricate

---

*To learn more about the possibility of earthquake prediction, see Christopher Scholz et al. (August 31, 1973), "Earthquake Prediction: A Physical Basis," *Science*, 181 (4102), pp. 803–809.

the fault line by injecting oil or water into the rift or fracture zone. Presumably, the fault blocks will then slide smoothly instead of jerking violently. People who reside in earthquake territory hope that these theories prove to be practical.

Vulcanism

Volcanic activity can be quite picturesque—if you are a tourist visiting the Hawaii Volcanoes National Park, where you can drive to within 100 yards (90 meters) of a steaming lava flow. You stand impressed and safe under the watchful eyes of a park ranger, knowing that the entire island was created out of a series of such flows over millions of years of time. But volcanic activity can be terrifying if you are forced from your home by a flaming wall of lava, as happened to the citizens of Heimaey, a volcanic island that is part of the Mid-Atlantic Ridge, just south of Iceland.

Earth's history records a number of extremely violent volcanic eruptions. In A.D. 79 Mount Vesuvius, in Italy, destroyed the city of Pompeii; in 1902 Mount Pelée, on the island of Martinique in the West Indies, leveled the town of Saint Pierre; in 1883 the volcanic island of Krakatoa, which is west of Java,

Eruption! The village of Vestmannaeyjar on the volcanic island of Heimaey is overwhelmed by the volcano Helgafel. Heimaey, part of the Mid-Atlantic Ridge, is south of Iceland.

306 LAND SURFACE AND SEA BOTTOM

Mt. St. Helen's blows her top on May 18, 1980.

erupted, sending a violent tsunami against nearby islands and causing the deaths of almost 40,000 people; and Mount Etna, on the island of Sicily, has erupted repeatedly (about 400 times since 475 B.C.), killing almost 15,000 people in 1169 and 20,000 in 1669.[6] In February 1943 volcanic activity erupted on a farm near the Mexican village of Parícutin. Not only was the farm lost, but the village as well, as the earth belched forth fire and steam, sending great clouds of volcanic ash into the sky. Where the farm once was, a 7451-foot (2273-meter) volcano now stands.

One must keep in mind, however, that the damage caused by volcanic eruptions is not limited to the immediate loss of lives or devastation of the nearby territory. Consider the effect of Mount St. Helen's eruption in the spring of 1980. The volcanic ash spewed into the sky and carried eastward by the prevailing westerlies created health problems for humans as well as vegetation across the United States. Also, we must consider the influence of the rise of the ashes into the upper level of the troposphere, where they combine with other forms of pollution to form a thicker layer of dense clouds that might greatly influence the Greenhouse Effect. We could be facing the advent of a new tropical age, or an ice age.

The causes of **vulcanism** are not known for certain. In our earlier discussion of the different theories regarding tectonic forces, the most likely to account for worldwide volcanic activity was the theory of plate tectonics, which included continental drift and seafloor spreading. Figure 11-7 shows the distribution of the world's volcanoes; it is obvious that most of them occur along the edges of the earth's plates. Because the Pacific Ocean is completely ringed with volcanoes, this collection is often termed the Pacific Ring of Fire.

Figure 11-8 presents several cross sections of the sea bottom. Figure 11-8a focuses on the concept of seafloor spreading and shows how the North Atlantic Ocean floor is separating along the fracture line called the Mid-Atlantic Ridge. As the American and Eurasian plates move apart, magma flows upward, creating volcanic mountains (islands) such as the Azores and the Canary Islands. Figure 11-8b is of the western edge of the North Pacific Ocean and illustrates how the Pacific plate plunges down into the mantle near the coastline of Asia, forming deep trenches (deeps) and allowing magma to rise and form volcanic islands such as the Marianas. Figure 11-8c shows how the movement of the Pacific plate over the mantle creates "hot spots" with magma rising through points of weakness to build volcanic islands such as the Hawaiian Islands and the Tuamotu Archipelago.

There are two basic types of volcanic activity: intrusive and extrusive. **Intrusive vulcanism** (Fig. 11-9) results when magma cannot find an outlet from the asthenosphere onto the earth's surface. If its way is blocked, the magma may ooze through cracks and fissures and send tiny fingers or broad sheets between the layers (strata) of sedimentary rocks. These slablike intrusions are called **sills**. If the magma rises through the beds of rocks, cutting across the layers, the result is a **dike**. The large chamber of magma deep beneath the surface is called a **batholith** (from the Greek *bathys*, "deep," and *lithos*, "rock"). If the chamber of magma lies close to the surface and forces the rock layers above to arch upward, creating a structural dome, the result is a **laccolith** (from the Greek *lakkos*, "pit"). The main difference between a laccolith and a batholith is size (one might say that a laccolith is a small batholith). Some batholiths, such as the Rocky Mountains, extend for great distances. In time, the overlying sedimentary rocks are worn away, expos-

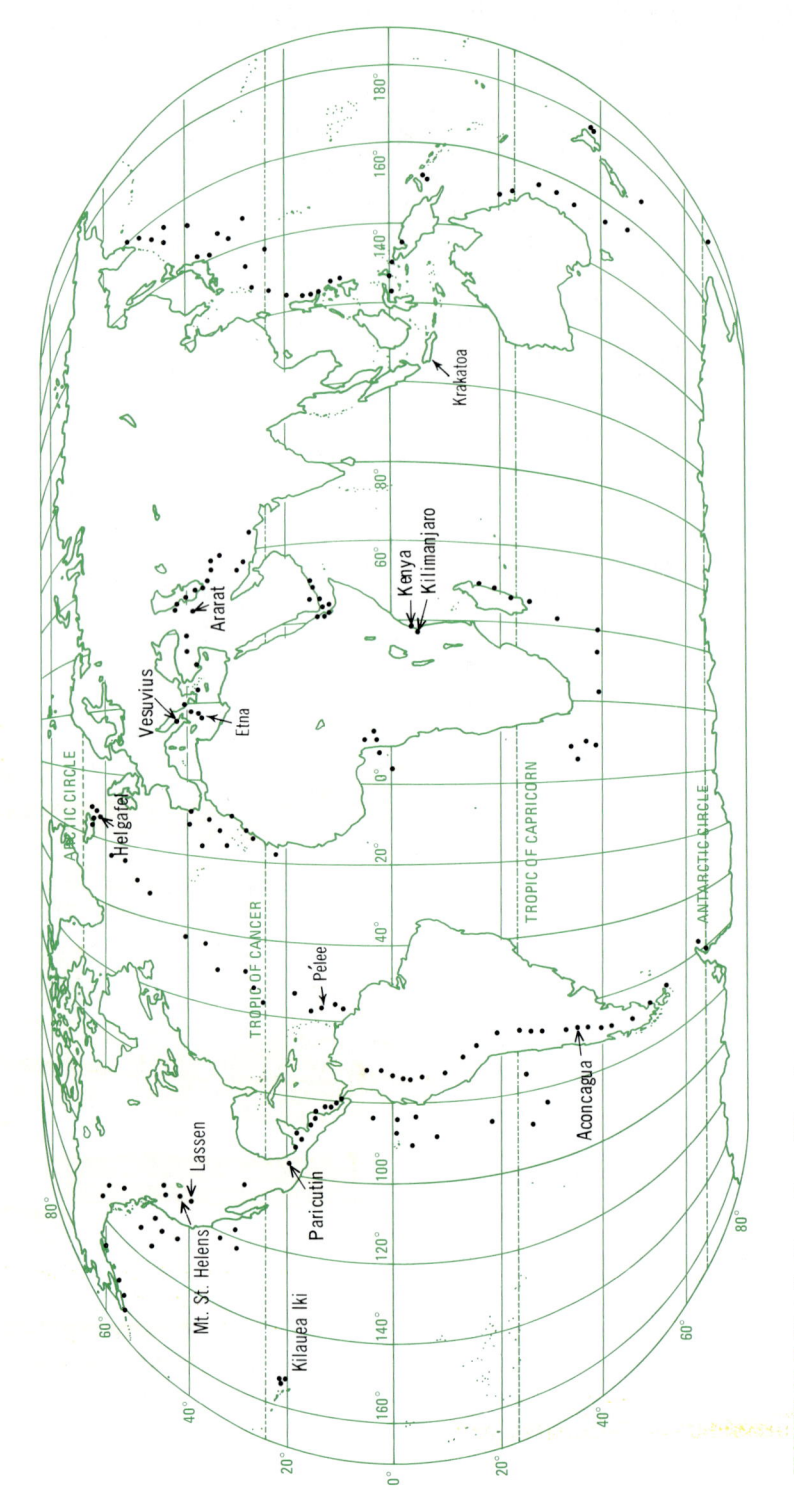

FIGURE 11-7 Distribution of world's volcanoes.

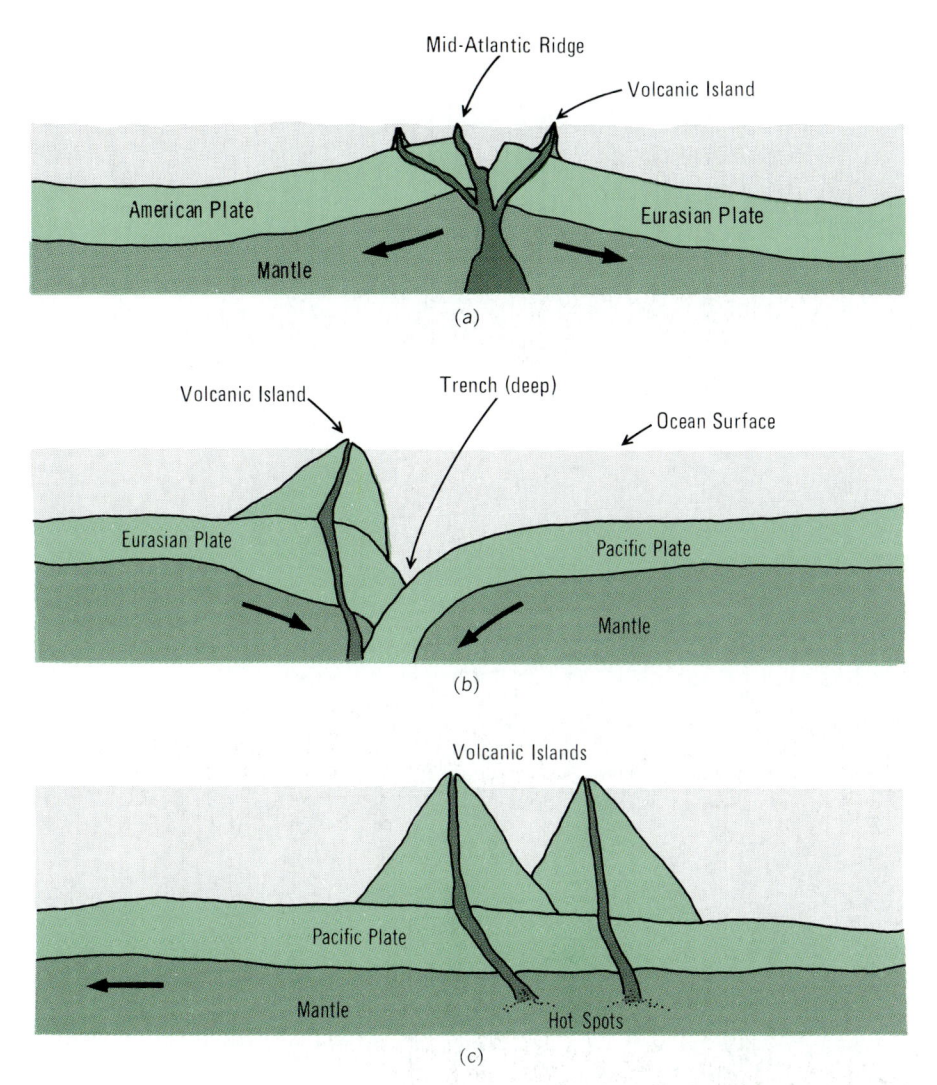

FIGURE 11-8 Vulcanism in the ocean floor.

ing the vast expanse of granitic rocks to view. Another exciting landform created by intrusive vulcanism is New Mexico's Ship Rock, an exposed network of dikes surrounding the pipe (or neck) of an ancient volcano.

Extrusive vulcanism occurs when magma issues forth from the neck of a volcano, either in liquid form as **lava,** flowing like a river, or as **ejecta,** with volcanic ashes or *pyroclastic* bombs being violently ejected. The landforms produced (Fig. 11-10) range from volcanic cones (Fig. 11-10a) to broad lava plateaus (Fig.

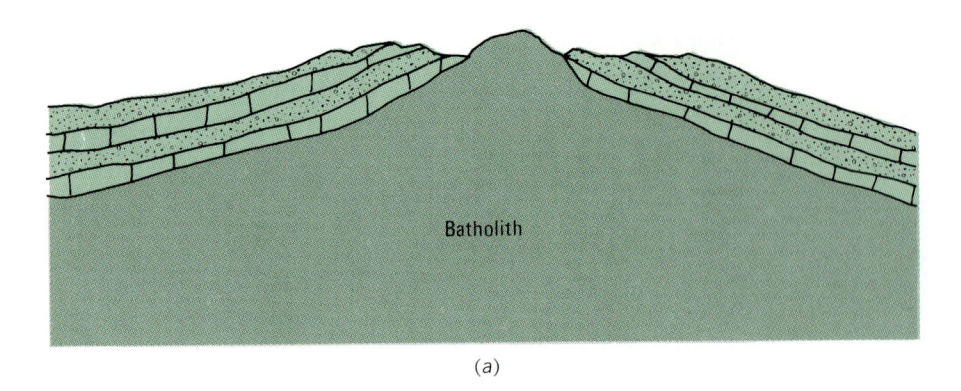

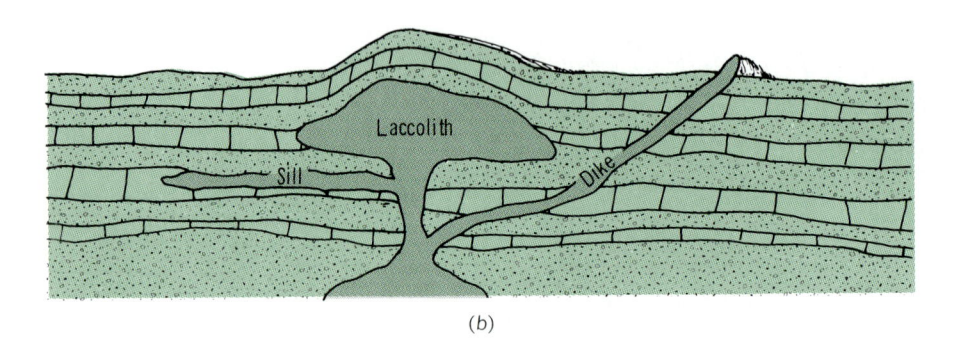

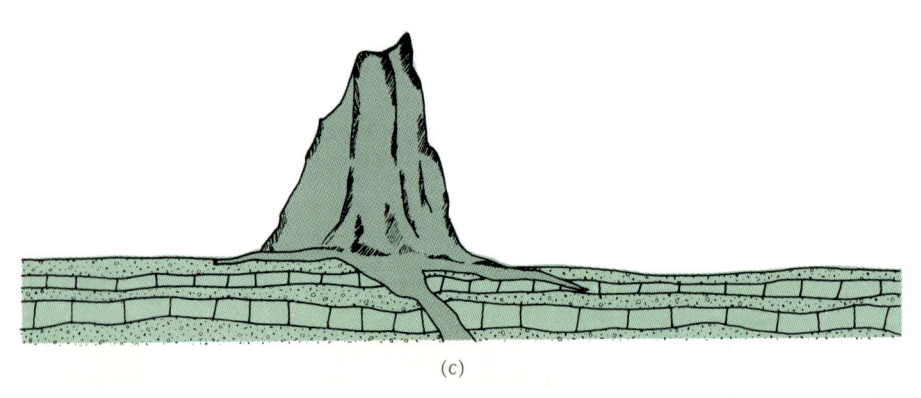

FIGURE 11-9 Intrusive vulcanism In (*a*) a drawing of exposed batholith, overlying layers, or strata, have been eroded away, exposing the grantic mass of the batholith. (*b*) shows smaller intrusions, including a laccolith, dike, and sill. The neck or vent of a volcano is exposed after the overlying basaltic lava and cinders have been eroded away (*c*).

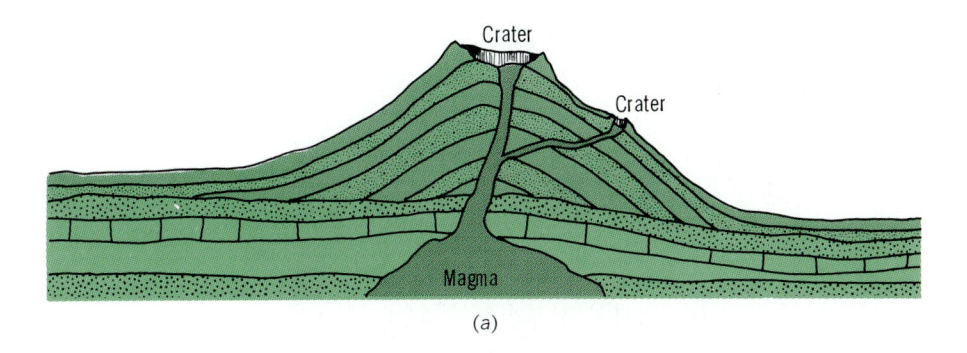

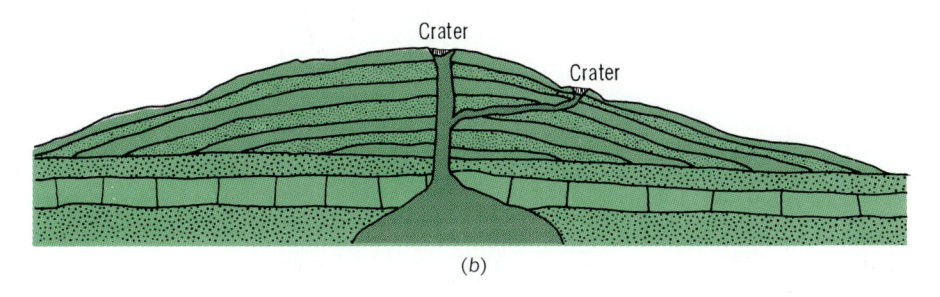

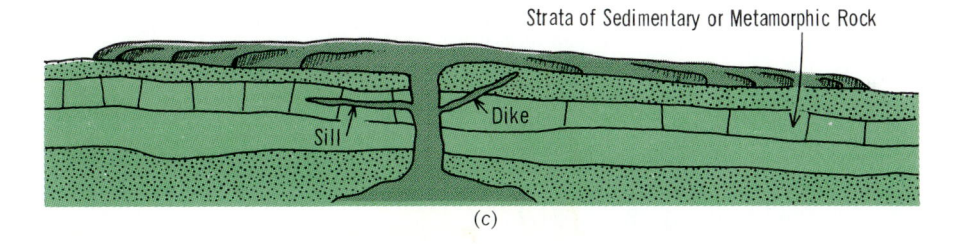

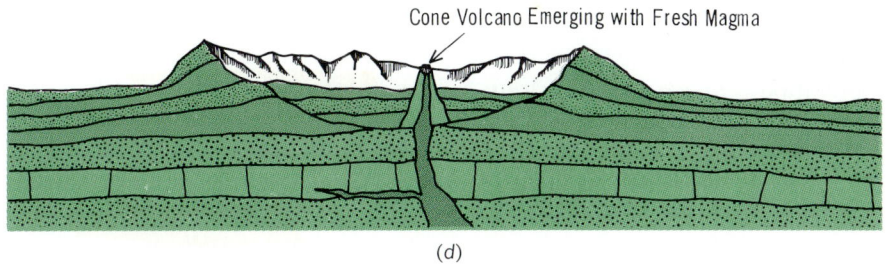

FIGURE 11-10 Extrusive vulcanism (*a*) Cinder cone, or conical volcano, composed of volcanic ash, cinders, or other forms of pyroclastic ejecta. (*b*) Shield, or dome volcano, also called composite or strato-volcano because it is made up of alternating layers of pyroclastic ejecta and lava flows. (*c*) Lava plateau, similar to a shield volcano, but with a very gentle slope and widespread coverage of many miles. (*d*) Caldera, an enlarged crater that results when the top of a volcano either blows its top off or collapses down into an empty magma chamber.

The Rocky Mountains: exposed batholith.

11-10c). The Columbia Plateau in Idaho covers a great area of the American Northwest, and one of the most interesting sections is Craters of the Moon National Monument. Astronauts who have returned from a landing on the moon say that this region is strikingly similar to the surface of the moon, complete with glassy black sand and numerous craters.

Volcanic eruptions can take many forms. If a developing hot spot happens to be situated near a large body of water, the magma may take in considerable moisture and thus become more fluidlike in consistency. When the magma is extruded from its chamber, it will flow smoothly out onto the surface. Gases in the magma tend to dissipate quickly, along with the water, in great clouds of steam. If the lava is fluid enough, the outpourings will be smooth-surfaced, much like the flow of chocolate syrup down the slope of a mound of ice cream. If this type of flow occurs repeatedly, the resulting volcanic mountain will be shaped like a shield or dome. A good example of a **shield volcano** is Mauna Loa, in Hawaii.

If the eruption comes in the form of a blast because its pipe (neck or vent) is blocked, pieces of rock and ashes will be flung high into the air. Subsequent similar eruptions will build a steep-sided **cinder-cone volcano,** such as Vesuvius. If there are alternating eruptions, some fluidlike, some pyroclastic, the volcano will have layers of ashes and ejecta, layers of lava, and the result is a **composite volcano.** Good examples for this type of volcano are Stromboli, in the Mediterranean, Fuji, in Japan, and Mount Lassen, in the United States.

When a volcano ceases to be active (that is, when there are no signs of activity over a long period of time), it is said to be dormant. We do not like to use the term *dead;* who knows when it may suddenly return to life and start erupting again? A good example is Mt. St. Helens in the state of Washington. It had been considered a dormant volcano, but since the spring of 1980 it has been shifting back and forth between dormancy and activity and no one can predict what the future holds for that region.

Other activities associated with vulcanism include geysers, hot springs, mud pots, and fumaroles. These result when water seeps down through cracks or fissures of the earth's crust and approaches close to the mantle or a chamber of magma. When the water is heated to extreme temperatures, it rises. Some of it is forced back out of the fissure as bubbling water; the remainder explodes upward as steam, spouting in great clouds above the earth. Geysers are the most popular attractions at Yellowstone National Park. Other countries noted for their geysers are Iceland, Japan, and New Zealand, all located on the edge of crustal plates and surrounded by oceans.

One final note about vulcanism. A United Press International news report from Hilo, Hawaii, dated March 18, 1973, tells of a circuit court judge's ruling that "any new land formed by a lava flow becomes the property of the owner of adjacent land." Thus ended a legal battle that resulted from a 1955 lava flow from Kilauea Iki.

Craters of the Moon National Monument, Idaho.

LANDFORM CLASSIFICATION

In order to assess the importance of the earth's many landforms, we should consider their individual characteristics. Figure 11-11 is one method of classifying the surface features of land and sea. We have divided the features into mountains, hills, plateaus, plains, and trenches (see again Fig. 11-3). Factors to be considered are local relief, elevation above or below sea level, and physical appearance.

Mountains and Hills

It is sometimes difficult to distinguish between mountain and hill landforms. Mountains are supposed to be rough and craggy, with a **local relief** (the difference between the highest elevations and the lowest in a given area) of at least 2000 feet (600 meters) and with an average elevation of 3000 to 4000 feet (900 to 1200 meters). Hills, on the other hand, are supposed to have a local relief of 500 to 1000 feet (150 to 300 meters), with an average elevation of less than 2000 feet (600 meters). In some regions, however, local inhabitants refer to hilly country in their locality as "mountains," even though the hills may barely top 2000 feet (600 meters). Some of the so-called mountains in the Los Angeles area, such as the Santa Monica Mountains, do not meet the specifications for height or ruggedness, and the elevations do not exceed 1800 feet (540 meters). Hills are usually smoothly rounded and can be formed by any of the tectonic forces (diastrophism or vulcanism) we have discussed. These are termed *structural hills*. In addition, they can be carved out of flat land by running water or glaciers. In Chapter 15 we will discuss how glaciers can also deposit small mounds of sediments that create a fertile, although hilly, landscape.

Plains and Plateaus

Both plains and plateaus are considered to be relatively flat lands, but plains have a local relief of less than 300 feet (90 meters), while plateaus have a local relief that may be greater. The prime difference between the two, however, is height above sea level. Plains are usually found at elevations of less than 500 feet (150 meters), while plateaus may reach higher altitudes, up to 10,000 feet (3000 meters). Some earth scientists refer to plateaus as "plains in the air," or "tablelands." There normally are two classifications of plains. The first is a **depositional** plain, which is made up of stream- or wind-deposited soils. Examples are delta (mouth of a river), flood (valley of a river), and **piedmont** (along the base of a mountain range) plains. The second classification is the **erosional** plain, where hilly or mountainous land has been worn away by stream, wind, or glacial actions. Plateaus are formed in the same ways as plains; thus their classifications focus on the activities that formed them. These classifications are **intermontane** (located between two mountain ranges), **piedmont, continental** (covering vast expanses, such as North Africa), and **dissected** (deeply cut by a powerful river, such as the Colorado Plateau).

Most of the great plateau regions of the world are to be found on the **continental shields** (Fig. 11-12). These are exposed bedrock areas, lands composed of granite and some metamorphic rocks that have been subjected to extreme glacial action. They were initially formed about 3 billion years ago and experienced tropical ages, with great forests and very warm climates. With the passage of millions of years, the climate of the world changed many times. Great ice ages occurred, sending vast sheets of ice (continental glaciers) moving equatorward from the poles. The ice scraped away any hilly country, leaving the granitic masses exposed. Every continent contains at least one of these shields as vast plains or plateaus, but time and the elements have treated them differently, and they are not all alike today. As climate varies with latitude and posi-

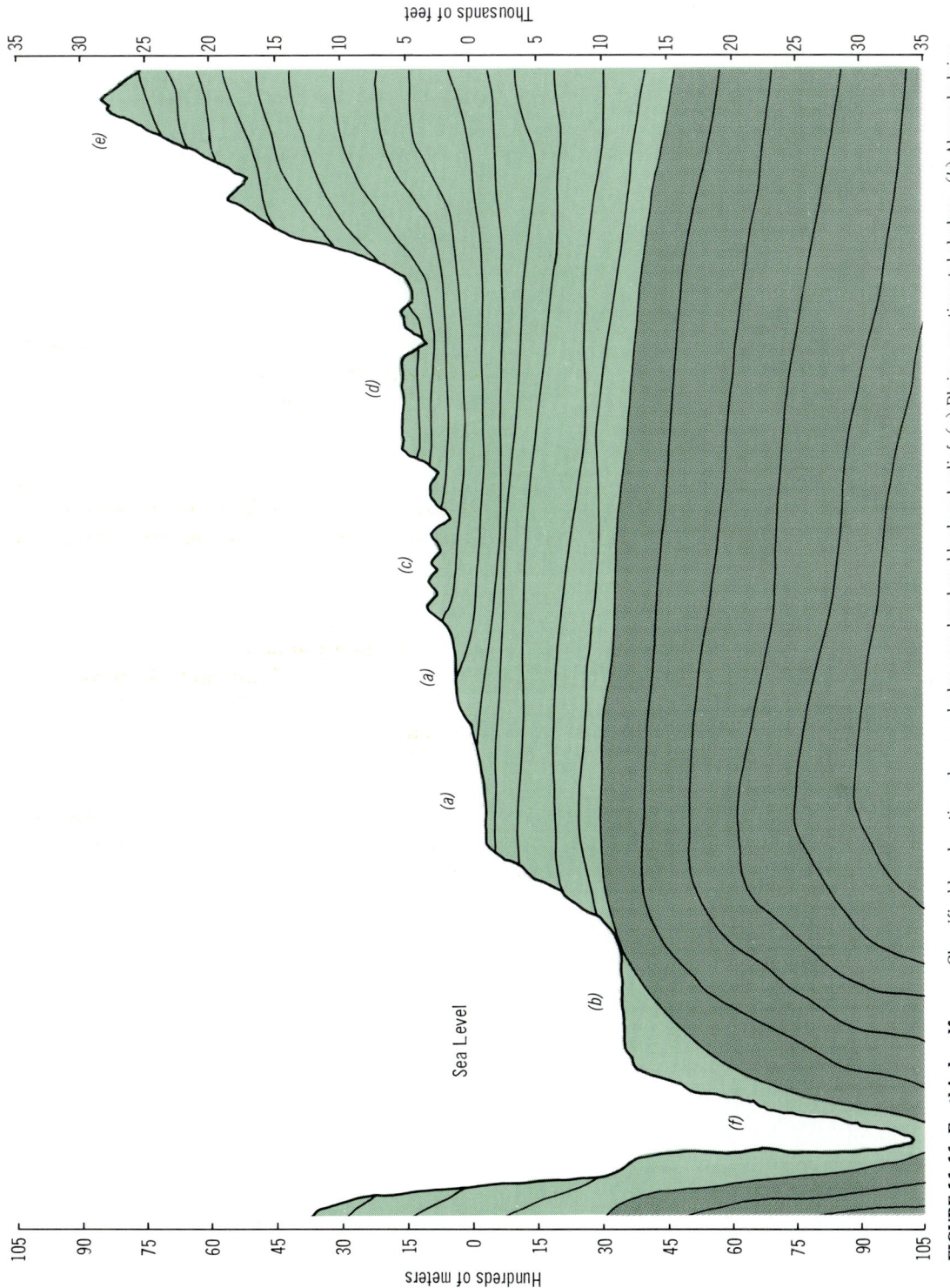

FIGURE 11-11 Earth's landforms Classified by elevation above or below sea level and by local relief. (*a*) Plains, continental shelves. (*b*) Abyssal plains. (*c*) Hills. (*d*) Plateaus. (*e*) Mountains. (*f*) Trenches (deeps).

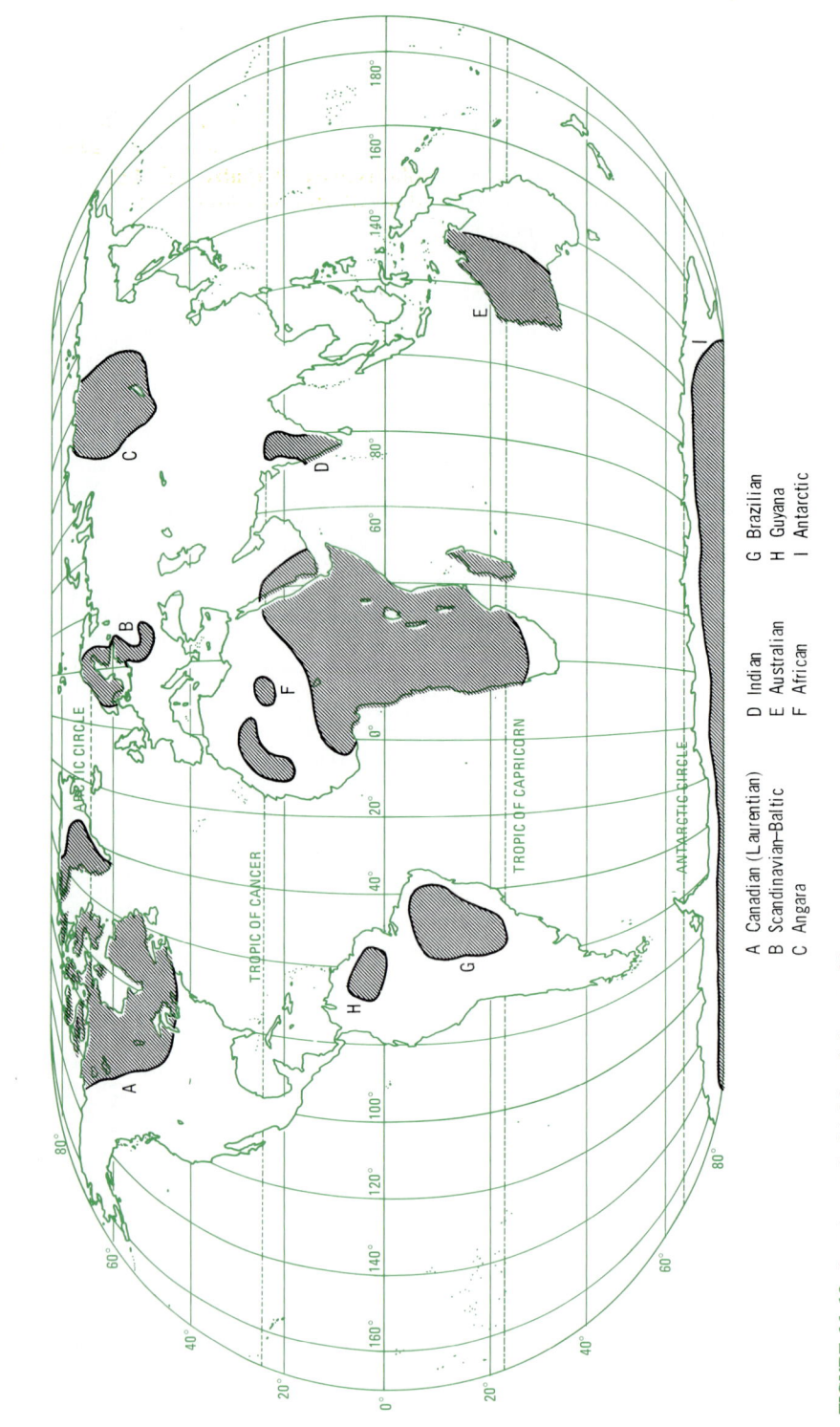

FIGURE 11-12 Continental shields of the world.

A Canadian (Laurentian) D Indian G Brazilian
B Scandinavian-Baltic E Australian H Guyana
C Angara F African I Antarctic

316

tion relative to the oceans and wind systems, the rock shields have not evolved at the same rate. The climate, soils, and vegetation of the Canadian and Angara shields cannot be compared to those of the Brazilian or African shields.

Another great plains region can be found on the floor of the ocean. These deep-lying plains are called **abyssal plains.** The outer borders of the continents, the edge of land that extends into the oceans, contains the continental shelves. We will consider these in greater

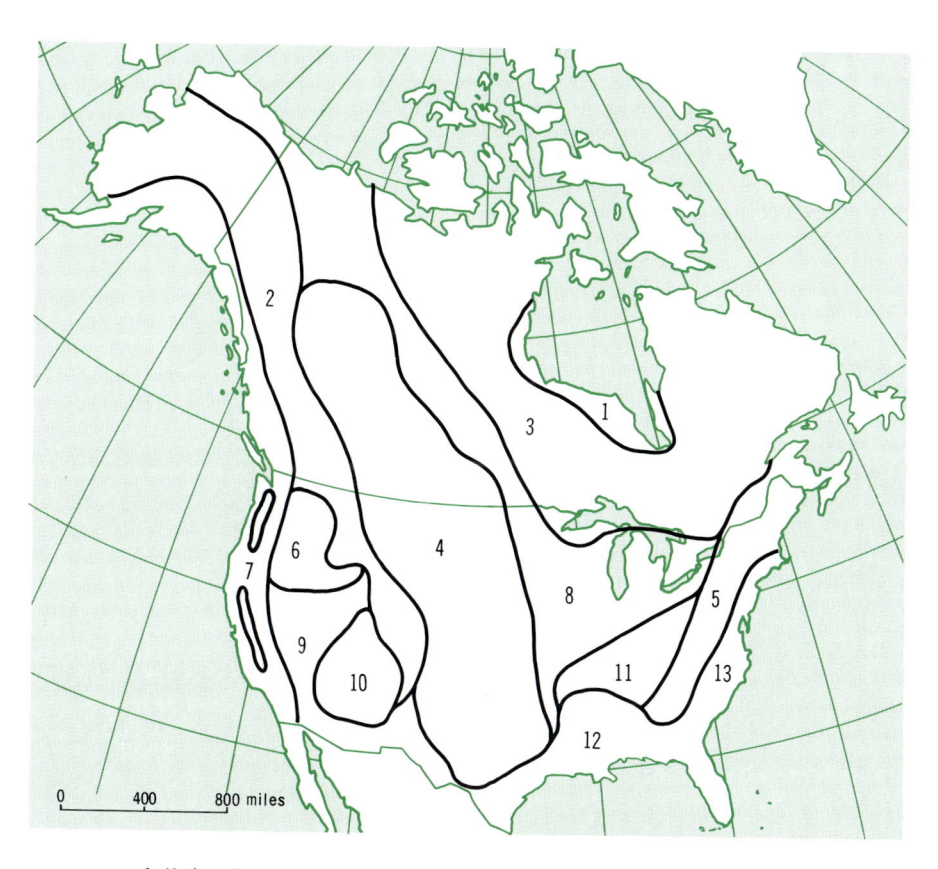

1 Hudson Bay Lowland
2 Rocky Mountains
3 Canadian Shield (Laurentian Upland)
4 Great Plains
5 Appalachian Uplands and Mountains
6 Columbia Plateau
7 Pacific (coastal) ranges
8 Central Plains (lowlands)
9 Great Basin (and range) Intermontane Plateau
10 Colorado Plateau
11 Interior uplands
12 Gulf coastal plain
13 Atlantic coastal plain

FIGURE 11-13 Landforms of the United States and Canada.

detail in Chapter 15, but for now let us state that these shelves are important to us because of the life forms that thrive in the shallow waters and the great petroleum deposits beneath the surface of the shelves.

Landforms of the United States and Canada

One of the most common ways devised by geomorphologists to explain and portray the landforms of the United States and Canada is the one shown in Fig. 11-13. This physiographic map shows the major landform features we have been discussing. You will note that some of the terms are not the same as those we have used. These different terms are employed to clarify further the differences between landforms that have similar characteristics. **Uplands,** for example, refers to hilly country or high plains areas that do not qualify as plateaus. **Lowlands** are low-lying lands that are fairly close to sea level in elevation and normally swampy or dotted with streams, rivers, and lakes. The term **basin and range** indicates a relatively flat plain or high plains region that is dotted with isolated mountains or ranges.

Mountain areas in Fig. 11-13 are indicated by the numbers *2, 5,* and *7;* hill country, *11;* plateaus, *6, 9,* and *10;* plains, *1, 3, 4, 8, 12,* and *13.*

REVIEW AND DISCUSSION

1. Explain the term *tectonic*. What are tectonic forces?
2. Explain the difference between the terms *contraction* and *convection*.
3. Explain the theory of continental drift. What was Pangaea?
4. How did a discovery made by scientists in the field of paleomagnetism help to make the theory of continental drift more acceptable?
5. What is the meaing of isostasy?
6. What does the Pacific plate have to do with California's San Andreas Fault?
7. Explain the difference between epeirogeny and orogeny.
8. Explain the difference between diastrophism and vulcanism.
9. Explain how folding occurs. What is the difference between folding and warping? What is the difference between folding and faulting?
10. What is an anticline? What is a syncline?
11. Explain the differences among these fault types: normal, reverse, strike-slip.
12. What is the difference between a horst and a graben?
13. Where would you find the greatest concentration of earthquake activity?
14. What is the difference between the Richter scale and the Mercalli scale?
15. What is the difference between a tidal wave and a tsunami?
16. What is meant by the term *Ring of Fire?*
17. Explain the difference between intrusive and extrusive vulcanism.
18. How does a sea bottom trench or deep come into being?
19. What is the difference between a laccolith and a batholith?
20. What is the difference between a sill and a dike?
21. Define the terms *ejecta, pyroclastic, lava,* and *magma*.
22. What is the difference between cinder-cone, shield, and composite volcanoes?
23. What does the term *local relief* mean with reference to landform classification?
24. What are the major differences between hills and mountains, plains and plateaus?
25. Explain these terms that refer to plateaus: *intermontane, piedmont, dissected,* and *lava*.
26. What is a continental shield?
27. What is the difference between abyssal plains and continental shelves?

NOTES

[1] Richard Aldington and Delano Ames, translators (1959), *Larousse Encyclopedia of Mythology*, (London: Paul Hamlyn), p. 423.

[2] Aldington and Ames, p. 447.

[3] Aldington and Ames, pp. 139–142.

[4] Aldington and Ames, p. 218.

[5] U.S. Geological Survey (1973), *San Andreas Fault*, publication no. 0-495-648 (Washington, D.C.: U.S. Government Printing Office), p. 6.

[6] Leon Bertin (1972), *The New Larousse Encyclopedia of the Earth* (New York: Crown Publishers), p. 163.

ADDITIONAL READING

Anderson, Don L. (November 1971). "The San Andreas Fault," *Scientific American* (offprint no. 896).

Beiser, Arthur (1971). *The Earth*. New York: Time-Life Books.

Bloom, A. L. (1969). *The Surface of the Earth*. Englewood Cliffs, N.J.: Prentice-Hall.

Butzer, Karl W. (1976). *Geomorphology from the Earth*. New York: Harper & Row.

Dalryample, E. S., and E. Jackson (May–June 1973). "Origin of the Hawaiian Islands," *American Scientist*, 61(3), pp. 294–308.

Dury, G. H. (1960). *The Face of the Earth*. Baltimore: Penguin Books.

Du Toit, A. L. (1937). *Our Wandering Continents. An Hypothesis of Continental Drifting*. Westport, CN: Greenwood Press.

Grove, Noel (July 1973). "A Village Fights for Its Life," *National Geographic*, 144(1), pp. 40–67.

Heirtzler, J. R. (December 1968). "Seafloor Spreading," *Scientific American* (offprint no. 875).

Herbert, Don, and Fulvio Bardossi (1968). *Kilauea: Case History of a Volcano*. New York: Harper & Row.

Hodgson, J. H. (1964). *Earthquakes and Earth Structure*. Englewood Cliffs, N.J.: Prentice-Hall.

Hurley, P. M. (April 1968). "The Confirmation of Continental Drift," *Scientific American* (offprint no. 874).

Iacopi, Robert (1971). *Earthquake Country*. Menlo Park, Calif.: Sunset-Lane Books.

Kay, Marshall (September 1955). "The Origin of the Continents," *Scientific American* (offprint no. 816).

Kennedy, George C. (December 1959). "The Origin of Continents, Mountain Ranges, and Ocean Basins," *American Scientist* 47(4), pp. 491–504.

Macdonald, G. A., and A. T. Abbott (1970). *Volcanoes in the Sea: The Geology of Hawaii*. Honolulu: University of Hawaii Press.

Matumoto, T., and Gary Latham (August 10, 1973). "Aftershocks and Intensity of the Managua Earthquake of 23 December 1972," *Science*, 181(4099), pp. 545–547.

Milne, L. J., and Margery Milne (1962). *The Mountains*. New York: Time-Life Books. See especially "Volcanoes," Chap. 3, and "Birth and Death of Mountains," Chap. 2.

Ogburn, Charlton, Jr. (1968). *The Forging of Our Continent*. New York: American Heritage.

Scholz, C., L. Sykes, and Y. Aggarwal (August 31, 1973). "Earthquake Prediction: A Physical Basis," *Science*, 181(4102), pp. 803–809.

Shelton, John S. (1966). *Geology Illustrated*. San Francisco: Freeman.

Snead, Rodman (1980). *World Atlas of Geomorphic Features.*, Krieger, Melbourne, FL.

Tarling, Don, and Maureen Tarling (1971). *Continental Drift: A Study of the Earth's Moving Surface*. Garden City, N.Y.: Doubleday.

Tazieff, Haroun (1961). *The Orion Book of Volcanoes*. New York: Orion Press.

Wegener, Alfred (1929, 1966). *The Origin of Continents and Oceans*. New York: Dover Publications.

Wilson, J. Tuzo (April 1963). "Continental Drift," *Scientific American* (offprint no. 868).

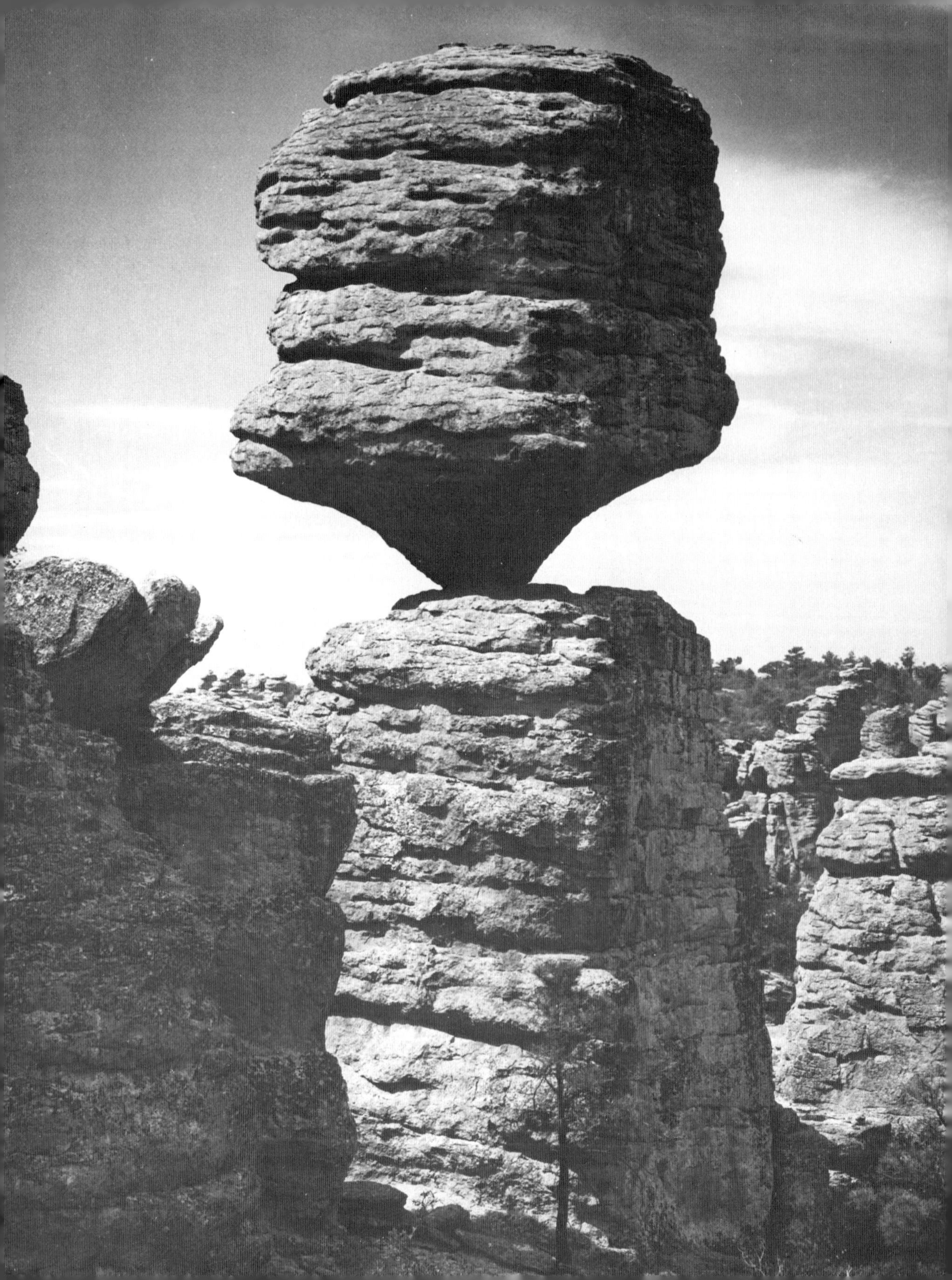

TWELVE
HOW THE SCULPTURING BEGINS

The earth must endure endless cycles:
  seasonal,
  hydrologic,
  vegetative,
  geographical.
Geographical cycle?
What goes up, must come down . . .
  Up . . . down . . . up . . . down . . .
There is no end. . . .

If the term *cycle* means a "round of events that recur again and again," how can there be a geographical cycle? Does this mean that the earth's surface is constantly being uplifted and then worn down? For an answer, consider the broad picture. The earth's tectonic forces operate to raise some portions of the crust and depress others; continents are set adrift; mountains are uplifted; plateau regions are elevated; and the ocean floors are spread apart. The story does not end there, because other external (outside) forces are working to level the land. These forces are called *gradational* because they operate to grade down or wear away elevated surfaces in gradual stages. As rapidly as an area is pushed up, the gradational forces work to bring it down. Earth materials are first broken up, removed from the higher places, transported to lower elevations, and deposited. In other words, the higher elevations are denuded (stripped of their cover), and the lower places aggraded (built up). When the lower place is uplifted, it will become denuded. This is, apparently, a continuous process, a true cycle. It is called, variously, the **cycle of denudation,** the **geomorphic cycle,** or the **geographical cycle.***

Our next step is to consider the agents working within the geographical cycle to sculpture the face of the earth, to level the land, and to allow the development of soils needed by humans. We will examine the erosional (loosening, wearing down, removal, and transportation) and depositional activities of the forces of gravity (mass wasting), running (surface) water, groundwater, glaciers, ocean waves and currents, and wind. First, however, we must learn how the bedrock, the parent material, is broken up and loosened, because weathering is the first step in this process.

WEATHERING

Have you ever wondered why the paint on your house peels and flakes away? Or have you looked at a car you bought five years ago and wondered why the paint looked different than when it was new? Or maybe you have stood in front of a towering granite cliff and noticed that the face of the cliff was slowly breaking down and was not as slick and shiny as you remembered it from when you camped in the area ten years ago.

All these things can be attributed to **weathering.** A number of factors are involved, such as the action of the weather—the amount of insolation received (based on the latitude, elevation, and aspect), the annual and daily ranges of temperature, and the amount and kind of precipitation. In addition, the activities of burrowing animals, bacteria, and the roots of plants are important weathering agents in breaking up rock materials. The process of weathering includes both chemical and mechanical (or physical) actions.

Mechanical Weathering

When rock masses are fractured and broken, we call the process **disintegration,** which means the rock falls and splits or crumbles into small pieces, grains, or particles. Disintegration is strictly a physical action; there is no chemical change within the rock structure. If you were to examine a chunk of granite and compare its chemical makeup with a handful of California beach sand, you would find that they have exactly the same mineral combination: feldspar,

---

*To learn more about this and opposing views, see William Warnitz and Peter Wolff (1971), *Breakthroughs in Geography* (New York: New American Library, Times Mirror), Ch. 7, Davis and Horton, "Theoretical Models for the Study of Landforms." Also, Kirk Bryan (Dec. 1940), "The Retreat of Slopes," *Annals of the Association of American Geographers,* 30(4), pp. 254–268.

mica, and quartz. The sand was granite before it was physically weathered. Rock layers within the earth's crust and upper mantle can be fractured (jointed) by tectonic forces, and if the overlying materials were removed by erosion, these once-buried masses would be subjected to further weathering.

Rocks can be fragmented by a number of processes. One such process is termed **frost action:** Water seeps into rock crevices and expands as it freezes, causing the rock to break apart along the fracture lines. Sometimes the moisture in the rock develops ice crystals as the temperature descends below freezing. The growth of these crystals exerts pressure on the rock mass from within, and the rock begins to break up. This process, termed **crystal growth,** also occurs in areas where the rate of evaporation is high and the heat draws moisture to the surface. This upward-moving moisture contains minerals it has dissolved out of the lower regions of the rock. On reaching the surface, the moisture rises through evaporation, leaving the chemicals (such as salt) behind, lodged in crevices of the rock. If the salt crystals increase in number and size, the pressure may force the rock to split apart. Frost action and the formation of ice crystals are common occurrences in the high- and middle-latitude regions, while the action of salt crystal growth is more prevalent in the lower latitudes and in dry regions.

Another physical weathering process involves the alternating of high and low temperatures, which causes rocks on or near the surface to expand and contract repeatedly, weakening the rock structure. This is called **thermal expansion,** or **thermal weathering.** For example, consider the effect of extremes of temperature on a large boulder. The outer portion, which is heated during the day, expands, while the interior, which receives little or no heat, does not expand. When night falls, the cooler air temperature chills the outer rock structure, causing contraction. The interior is not affected. The repeated expansion and contraction cause the rock to break up from the outside.

Another type of physical breakup is caused by plant roots working their way down into the joints and crevices of a rock layer. The roots expand as they grow, pushing against the sides of the crevice and forcing the rock to break apart. Tiny burrowing animals contribute to rock breakup by working their way through cracks in the rock, and their constant passage wears away the rock surfaces. This type of weathering is termed **biotic,** or **organic,** activity. It is the work of living organisms.

Rocks can also be worn down by the abrasive (sandpapering) action of running water and winds, the plucking (removal) of rock fragments by moving glaciers, and by a process called **unloading.** When a rock mass is buried under other layers, it is compressed by the resulting pressures. If the overlying rock layers are eroded away, the rock layer may expand and, as it does, fracture and break apart.

What about the human factor? Think of the effect of our technology on the rock structures: automobiles, motorcycles, trucks, bulldozers, and all kinds of earth-moving equipment that travels the roads and highways of the world. Do you think that the weight of these vehicles has any effect on rocks underneath? What of high-flying jet planes? If sonic booms can shatter windows, can they also fracture rocks? We might term this type of physical weathering cultural, or human-induced.

Chemical Weathering

When rock breakup is brought about because of alterations in the chemical composition of a rock, the result is termed **decomposition** (rot-

(a)

(b)

(c)

(d)

(e)

Weathering, erosion, and depositions: sculpturing agents. (a) The combined efforts of sun, wind, and rain have carved the fantastic columns of Bryce Canyon National Park, Utah. (b) Wind driven waves pound the rocky coastline of St. Andrews State Park, Florida. (c) Thundering waterfall along the Pigeon River (on the Minnesota-Ontario border) easily shatters and fractures highly resistant beds of rock. (d) Streams flowing over gently sloping ground in Wisconsin's cattle country carve deep gullies in unprotected land. (e) Silt transported by a hurricane overloaded this normally clear running stream, causing heavy deposition of the silt on both sides of the stream's valley.

ting). Sometimes chemicals are dissolved by gases or moisture and are carried away, leaving greater spaces between rock particles or grains. The weakened rock is left open to further decomposition and to disintegration.

If water enters a rock layer that contains carbon dioxide, a mild form of carbonic acid is formed that dissolves some existing chemicals within the rock, such as limestone (calcium carbonate). This process, **carbonation,** has created numerous underground caverns, such as Mammoth Caves, Kentucky, in limestone regions. If oxygen is present, rusting (**oxidation**) will occur.

When moisture creeps into rock cavities, it may unite with salts that are present to form an acid; the resultant chemical change causes the rock to break up. This action is called **hydrolysis.** If, as this occurs, the rock expands and increases the rate of alteration, the process is termed **hydration.** Hydrolysis changes a persistent mineral such as feldspar into clay, and hydration is associated with the oxidation of iron particles.

If a rock is weakened by the addition of moisture, which dissolves the chemicals that cement the rock grains or particles together, the rock will decompose as the binding cement is carried away. If some rocks are composed entirely of easily dissolved minerals, the water will bring rapid decomposition. This is similar to what would happen if you were to patch a broken ceramic vase with a water-soluble glue and then put the vase in a tub of water. The glue would quickly dissolve, and the vase would fall apart. This process is called **solution.**

The results of weathering. (a) Granular disintegration of boulders, Joshua Tree National Monument, California. (b) Joint-block separation of a quartzite cliff in Bennington, Vermont. (c) Spalling (exfoliation) of granite domes of Yosemite National Park, California. (d) Shattering of an ancient basaltic lava flow in Grand Coulee, Oregon.

(c)

(d)

The Results of Weathering

The basic rock structure of the earth is called *bedrock*. When weathering alters the bedrock by breaking it into smaller and smaller pieces, it eventually becomes **regolith** (from Greek *rhegos*, "blanket," and *lithos*, "rock"). This "blanket of rock" is the unconsolidated (loose) rock material that lies above the bedrock. Regolith in one place may not match that of another because many factors contribute to the eventual shapes of the rock fragments. Depending on the chemical composition, the granular structure, or the climate of an area, a number of different, but easily recognizable, shapes will result: small grains, curved shells, block- or bricklike pieces, or irregular chunks.

When rocks break up into irregular, sharp-edged, random-sized pieces because they do not have definite lines of weakness, cleavage, or fracture, we say the rocks have **shattered.** If the rock separates along previously induced joints, the result looks like huge steps leading upward; this is called **joint-block separation. Exfoliation** (from the Latin *exfoliatus*, "to strip of leaves") is the separation, through expansion, of curved shells or slabs from a solid rock mass that is uniform in structure. Excellent examples may be seen in the rounded domes of Yosemite National Park. Another term for this process is **spalling** (from the Old German *spellen*, "to split"). When a rock mass is composed of rocks with large crystalline structure, such as granite or sandstone, the rock may break up one grain after another; this process is called **granular disintegration.**

The processes of physical and chemical weathering are important, but the breakup of rock masses does not necessarily produce landforms. The actual sculpturing of the earth's surface is carried on by the erosional activities of running (surface) water, groundwater, glaciers, ocean waves and currents, winds, and the force of gravity. This latter erosional activity is often ignored, but anyone who has lived through a landslide or whose car has been struck by a rock falling from a cliff knows that the erosional agent termed **mass wasting** can be a dominant sculpturing agent.

MASS WASTING

The attraction of the earth on any mass is called gravity, and there are a number of ways that this important erosional activity induces the removal of weathered rock materials: soil creep, earthflows, mudflows, rockfalls, landslides, and slumps. The degree of the wasting away of a mass depends on the volume of the mass and its weight, the degree of slope, and the amount of moisture it contains.

The gravity movement called **soil creep** occurs in areas of moderate to gentle slopes where the rock material breaks up due to such weathering processes as frost action or thermal extremes. **Earthflows** are common in areas of heavy precipitation where waterlogged soils are common. If the slopes are relatively steep, the heavy soils may slide like a great sheet of mud over wide areas. In a sense, the face of the bedrock under the surface is like the metal slide in a playground.

An interesting form of earthflow occurs in the tundra regions of Canada and Siberia. In this vast region of permanently frozen ground (permafrost), the short summers allow only the top few inches of the frozen soil, or regolith, to thaw out. The frozen ground underneath acts as a slide, allowing the thick, souplike top layer to flow slowly down the slopes. This process is termed **solifluction** (from the Latin *sol*, "sun," and *fluctus*, "wave"). This thawing of the permafrost causes serious problems for anyone who is constructing a building. Without a layer of insulation, the heat radiating down from the floor of the house will thaw the uppermost layer of the permafrost and, if there is a slope, the house may slide. If the land is level, a portion of the house may sink down into the ooze. It is also difficult to maintain highways, even if they are paved, in permafrost areas because the roads buckle when the summer thaws begin.

Mudflows, which are similar to earthflows, are common in semiarid regions. They occur after violent thunderstorms or heavy rain showers have saturated the soil in the topmost layers. Pulled by gravity, the mud begins to flow down the slopes. If the rains continue, the flow of mud will become faster. It will sweep down slopes, pour over or knock down fences and retaining walls, enter through doors and windows of homes, and sweep out onto a valley or plain beyond. Mudflows can also result from excessive water seepage from glaciers high in the mountains.

Landslides are probably the most dramatic of all the various forms of mass wasting. In many cases there is no advance warning as a large rock mass composed of regolith and soils

Landslide. This photo gives an idea of the devastating effects of landslides on artificial features such as this railroad in Grand Valley, Colorado.

(a)

(b)

Mass wasting: transport of earth materials by the pull of gravity. (a) Massive earthflow (caused by water seepage high on the hillside) that covered U.S. 10 and the Cascade Irrigation Canal in Ellensburg, Washington. (b) Mudflow in Colorado that has been in existence for more than 100 years and flows 20 feet (6 meters) per year. (c) Gros Ventre rockslide near Kelly, Wyoming. (d) Solifluction in the Seward Peninsula region of Alaska; when spring arrives, the top few inches of the permafrost thaw, sending lobes of soil and rock sliding down the slopes. (e) Slump along Otter Cree, Vermont, with a vertical drop of about 30 feet (9 meters).

(c)

(d)

(e)

suddenly breaks loose and thunders down a steep slope, perhaps to inundate a highway, mountain cabin, or entire village. This occurs most often in high mountain areas such as the Alps, Rockies, and Andes. When a steep-faced cliff is undermined by erosional forces, the base may no longer be able to support the mass of rocks behind. Fractures occur as the base begins to move; then the entire mass slumps down, giving us the term **slumping.**

Another action, called **rockfall,** results when the earth movement is localized and the mass being wasted is composed of individual pieces or fragments of regolith that break away from the edge or face of a cliff and fall tumbling to the valley floor below. When the pieces of broken rock build up sufficiently at the base, a **talus slope** comes into existence. Sometimes the fragments construct a cone-shaped structure called a **talus cone;** a landform that can be very difficult to climb.

Mass wasting is only one of the many forms of the erosional forces that pick up weathered material and transport it to lower elevations, but they are all important because they exert significant influences on the development of soils. And what would we do if we had no soils?

REVIEW AND DISCUSSION

1. Explain the term *weathering*.
2. What are the differences between rock disintegration and decomposition?
3. Explain mechanical weathering.
4. What is meant by *frost action*?
5. What have temperature extremes to do with the breakup of rock?
6. Explain the term *crystal growth*.
7. Explain how small burrowing animals and microorganisms can contribute to weathering.
8. Can people be included in the weathering processes?
9. Explain the term *unloading*.
10. One of the processes of chemical weathering is *carbonation*. What does that term mean?
11. How does oxidation work to cause rock breakup?
12. Explain the difference between the terms *hydration* and *hydrolysis*.
13. Explain the term *solution* with regard to rock breakup.
14. What is regolith?
15. Explain the differences among shattering, joint-block separation, spalling, and granular disintegration.
16. What does *mass wasting* mean?
17. Explain soil creep.
18. What are the differences between earthflows and mudflows?
19. What is permafrost?
20. What is solifluction? Where is it most likely to occur?
21. Explain the difference between landslides and slumps.
22. What is a talus cone?

ADDITIONAL READING

Larson, Edwin E., and Peter W. Birkeland (1982). *Putnam's Geology.* 4th ed. (New York: Oxford University Press).

Reich, Parry (1950). *A Survey of Weathering Processes and Products.* Albuquerque: University of New Mexico Press.

Sharpe, C. F. (1938). *Landslides and Related Phenomena*. New York: Columbia University Press.

Shelton, John S. (1966). *Geology Illustrated*. San Francisco: Freeman.

Thornbury, William D. (1969). *Principles of Geomorphology*. 2d ed. New York: Wiley.

Thornbury, William D. (1965). *Regional Geomorphology of the United States*. New York: Wiley.

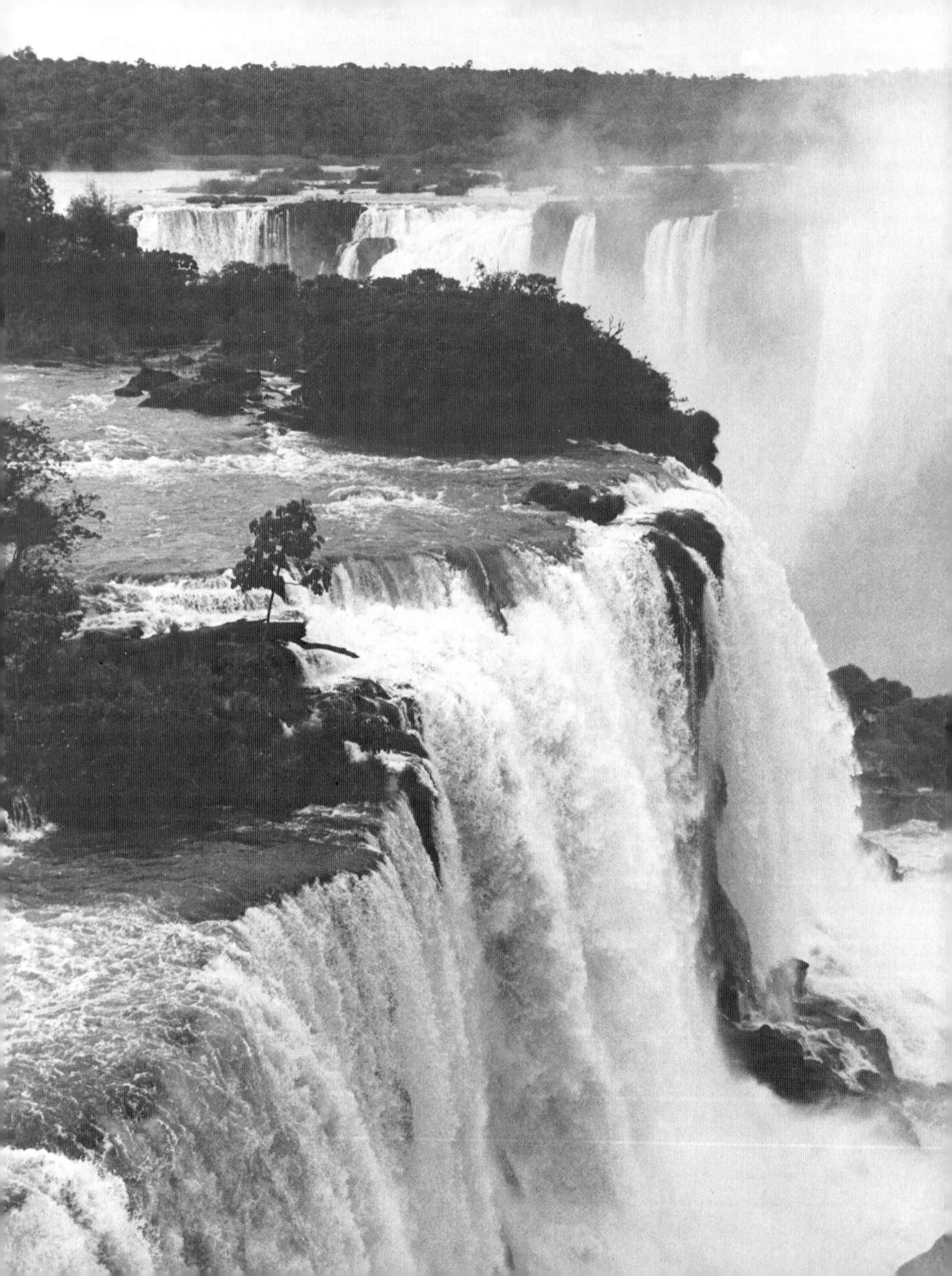

THIRTEEN
EMBOSSED BY RUNNING WATER

And see the rivers how they run
Through woods and meads, in shade and sun,
Sometimes swift, sometimes slow,
Wave succeeding wave, they go
A various journey to the deep,
Like human life to endless sleep!
GRONGAR HILL, JOHN DYER

The term *running water* can mean different things to different people: a dripping faucet, a flowing stream, or the movement of groundwater deep beneath the earth's surface. It can mean urgently needed moisture for food crops and natural vegetation, and it can mean life for humans. Hawaiians of bygone years kneeled along the banks of a stream and offered this prayer of thanksgiving.

> *One question I ask of you:*
> *Where flows the water of Ka-ne?*
> *Deep in the ground,*
> *In the rushing spring,*
> *In the ducts of Ka-ne and Loa,*
> *A wellspring of water to quaff!*
> *A water of magic power—*
> *The water of life!*
> *Life! O give us life!*[1]

The running waters of the earth, whether on the surface or underground, do more than supply us with the "water of life." They have efficiently embossed the surface and etched the deeper layers to provide us with magnificent landscapes of rugged highlands and fertile valleys. Running water is only one of the earth's sculpturing agents, but it is the most active.

Running waters are to be found everywhere—in every physiographic region and in all latitudes—and they are persistent in their endless quest to level the land. It would seem that their function is to counter the tectonic forces that operate to raise and wrinkle the earth's surface.

How does all this happen? Why are streams so active? What kinds of activities do they perform? How can we recognize the surface features created by running water and know immediately that they are not the result of the other erosional forces, such as blowing wind, grinding glaciers, or the heaving waters of the oceans? This chapter will provide some of the answers.

SCULPTURING PROCESSES OF RUNNING WATER

The sculpturing processes of running water—erosion, transportation, deposition—are important by-products of the hydrologic cycle. As part of the endless process, the water moves through the atmosphere from water bodies to the land, is precipitated, either soaks into the groundwater systems or runs off the surface toward some distant lake or ocean, and eventually returns to the atmosphere. As water moves along the earth's surface, it accomplishes a considerable share of the gradational activities of erosion.

Erosion

There are three methods of **erosion** (from the Latin *erodere*, "to gnaw out") by running water: splash erosion, sheetwash, and stream erosion.

Splash erosion is created by the impact of millions of raindrops falling on the earth's surface. As raindrops strike unprotected soil particles, the particles are dislodged from their original position and are sometimes thrown into the air. When they fall back to the ground, they are easily picked up by running water and transported elsewhere. Because water-retaining silts and clays, as well as bits of humus, are removed, the structure of the upper soil layer is altered. The amount of precipitation, the porosity of the surface layer, and the gradient have a lot to do with the effectiveness of splash erosion. Obviously, soft and weakly structured rock layers are broken up more rapidly than resistant rock such as granite, yet they all are modified in time. If a sturdy vegetation cover (or a mulch applied by humans) is present, splash erosion will be delayed for long periods, and sometimes actually prevented, thus saving agricultural soils for further use.

Sheetwash occurs after sufficient water has fallen to the ground, collected in pools or a thin sheet and, in response to the pull of gravity, begins to flow down a slope. The type of rock, the cover of vegetation, and the steepness of the gradient all influence the ability of sheetwash to perform erosional activity. If the soils are composed of fine, easily transported sediments, erosion will commence almost immediately. If the gradient is steep, gullying can transform a once-productive farming region into a wasteland. Farmers all over the world have tried to prevent gullying by using methods such as contour plowing and planting crops that tend to diminish the power and destructiveness of overland flow.

Stream erosion is accomplished by three processes, two of which (hydraulic action and corrasion) are physical; the other, corrosion, is chemical. As water flows down through its channel (streambed), it picks up small grains and pulls gravels and pebbles along with it, thus downcutting (eroding) the streambed by **hydraulic action.** As the grains, gravels, and pebbles are flung against the streambed, its banks and any large rocks in the way are reduced in size by the abrasive action of **corrasion** (from the Latin *corradere*, "to scrape together"). When the rushing waters of a stream pass over a hardrock base, then abruptly move across a section of soft, loose materials, a pothole is dug out and the flowing water becomes turbulent. **Corrosion** (chemical decomposition) is more apt to occur during movement of groundwater, especially in limestone areas and where carbonic acid is prevalent or in streambeds where the surface flow is intermittent and some of the chemical weathering processes can operate.

The velocity of a stream depends on three factors: steepness of gradient, volume of water, and the load. In a sense, these factors are closely tied together, like the links of a chain. If any one is altered, the entire chain is altered. A slow-moving stream, one that travels at less than 0.5 mile (0.8 kilometer) per hour, can carry only a light load of fine silts and sands, but if heavy rainfall upstream were to increase the water's speed to a velocity of 4 or 5 miles (5 or 8 kilometers) per hour, the stream could carry rocks with a diameter of about 3 to 6 inches (76 to 152 millimeters). If the velocity were increased to a speed of over 10 miles (16 kilometers) per hour, large boulders could be rolled along the streambed. Sometimes a stream erodes its bed down to such a gentle slope that the stream can no longer downcut and is said to have reached baselevel.

Baselevel (Fig. 13-1) is the lowest level to which a stream or river can lower itself: a plain at the mouth of a canyon, another stream or river, a lake, or the ocean. Since a stream cannot flow uphill, it can only degrade its bed until it joins the other body of water and its downward erosional power is ended. The time required for this process varies with the size of the stream. A heavy downpour over a weak rock area that has a steep slope will create what might almost be termed "instant valleys," deep, steep-sided gullies, and by storm's end the stream occupying the gully will have reached baselevel. On the other hand, a stream that must travel a great distance over gently sloping terrain will have to struggle for a long period of time to attain baselevel. Once baselevel is reached (for example, at the mouth of a stream valley), the stream will cease degrading and begin to reduce its entire course to baselevel.

The ultimate baselevel could be lowered only if the entire region were uplifted or if the lake, river, plain, or ocean into which the stream flowed were to have its surface level reduced. If either one of these things happened, the stream would immediately begin to reassert itself, degrading would commence again, and a new baselevel would ultimately be established. What if the terminal point were raised? If the surface of the river, lake, plain, or ocean should rise, the stream immediately

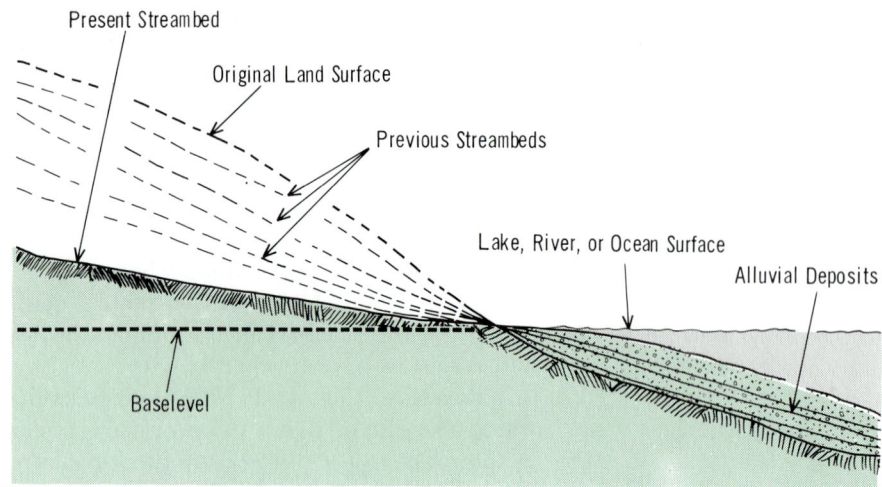

FIGURE 13-1 Baselevel.

would be forced to begin to fill in the old and lower bed with alluvial deposits and so establish a new and higher baselevel. The Colorado River has had its baselevel changed many times in its journey through the Grand Canyon, and the evidence is there for everyone to see: layers of marine sediments (left by oceans that once covered that area) resting atop shales deposited by river waters, which rest atop other marine sediments.

Graded streams are those that have attained a degree of balance with the important erosional factors of volume, slope, and load. A graded stream rarely degrades its bed vertically, and its erosional activity is primarily confined to lateral cutting of the sides of the valley. When the amount of water supplied combines with the slope just barely to transport the materials that have been eroded laterally, the stream is said to have reached *equilibrium*. This means that it has become a graded stream. However, if any of the three factors should be changed, the stream would be thrown off balance and would no longer be in a state of equilibrium. This could happen if a stream that had been eroding weak rock materials worked its way down to hard, resistant bedrock and thereby could move without being forced to carry a load. It could also happen if there were a climate change, with precipitation increasing the stream's volume, or if there were an increase in the stream's slope because of an uplift of the land.

Erosion Landforms

Interesting examples of the results of the erosional power of streams include waterfalls, incised meanders, terraces, and stream valleys. Stream valleys are perhaps the most important erosion landforms. We will consider them in more detail when we discuss the relationship of the geographical (or geomorphic) cycle to the development of stream valleys.

Waterfalls can result from a stream's passage over an escarpment, such as a recently uplifted surface, or they can be the result of the process of continued erosion of soft, loosely compacted beds of shale or sandstone that join along a boundary line with resistant rock (sometimes called "cap rock"). The es-

SCULPTURING PROCESSES OF RUNNING WATER 339

carpment or cliff is continually undercut, resulting in the fall of the overhanging portions and the subsequent receding of the cliff face. Niagara Falls is an excellent example of a "retreating" waterfall.

Incised meanders (Fig. 13-2) are the result of a gradual uplift of a region, which forces the stream to continue a vigorous downcutting in its previously determined meandering (wandering from side to side) pattern.

FIGURE 13-2 Incised meander of the Colorado River in the Grand Canyon.

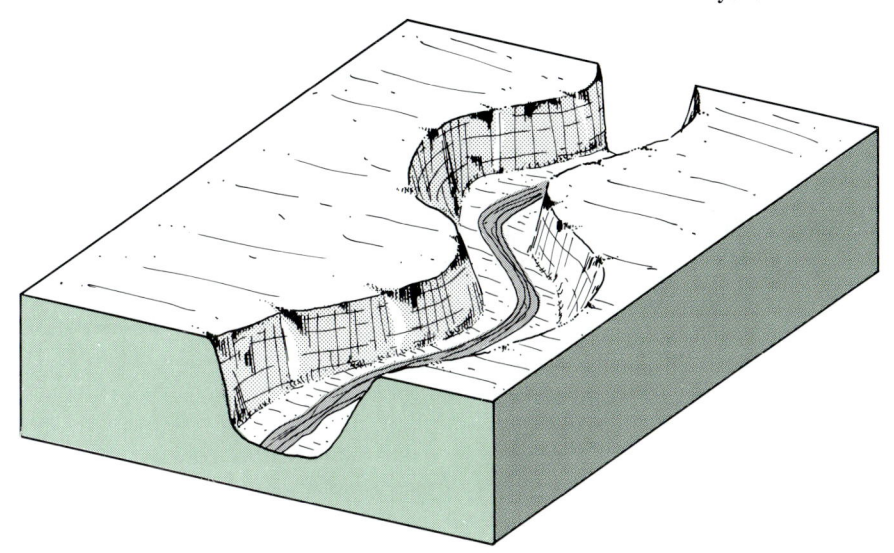

Stream-cut terraces are the result of a stream's wandering across its valley floor, eroding the sides as it moves. If the land were to be upraised, the stream would resume a vigorous downcutting and create a new, but narrower, valley floor. If this process continues, a series of the old valley floors will appear, much like shelves fastened to the valley walls.

Stream Transport

After rock materials have been weathered and eroded, the stream must transport the material to another place. The finest particles are carried in suspension, and the heavier grains or fragments are bounced or rolled along. The speed of the stream determines the amount of the load and what kind of material is carried, but all streams transport something. Even those that seem to be clear carry chemicals dissolved from the rocks. When you think of stream transport, you probably picture rivers such as the Mississippi, Amazon, or Nile because they carry tremendous loads of soil eroded from faraway highlands and seem to move sluggishly. Jet pilots flying from Africa to South America along the equator say that you can see the continent of South America long before you reach it; that is, you see the reddish, soil-laden outflow from the Amazon River about 100 miles (160 kilometers) before you can see Brazil.

Stream Deposition

We have been discussing the gradational (wearing down) force of erosion. Now let us consider the opposite force, deposition, which also works to level the land. Deposition does not tear or wear down, however, because this process is one of **aggradation,** a building up. One might say that erosion is destructive, and deposition is constructive. Fluvial (from the Latin *fluvialis*, "river") hydrologists tell us that sediments will be deposited out of a stream of moving water when the velocity of the stream is slowed to where it can no longer continue to transport material. This can happen when a stream flows out of a canyon onto a flat or a nearly level plain or as it wanders back and forth (meanders) in its valley, eroding one side and depositing on the other. A number of interesting landforms are created by this laying down of **alluvium** (stream-deposited sediments).

Alluvial fans (Fig. 13-3) occur when a stream rushes out of a steep canyon, and its velocity is slowed as it strikes a gently sloping area, forcing it to deposit its materials. The first materials to be sedimented are boulders, then smaller rocks, gravels, sands, and, finally, silt and clay particles. The initial shape is that of a cone, but as the stream continually blocks its own path with its deposits, it must seek new courses and the larger fan shape results. If there are a number of canyons fronting a broad valley or plain, the fans developing at the mouth of each canyon may coalesce (that is, come together), and the combination may stretch along the foot of the mountains for many miles. This alluvial landform is termed a **piedmont alluvial plain.**

Deltas form when a stream or river flows into a large lake or ocean, immediately slowing its movement forward (Fig. 13-4). The resulting buildup of deposits is almost identical to those of alluvial fans. Sometimes deltas form within the mouths of rivers, perhaps as part of a submerged coastline where the ocean has intruded into the river's valley. This delta is termed an **estuarine** (from the Latin *aestuarium*, "where the tide meets the coast") because the place of meeting is at the mouth of a river, an estuary. In time, a simple delta (Δ) shape may form as the sediments are pushed out into the lake or sea, but its future shape will be regulated by a number of factors. If the incoming stream is confronted by swift off-

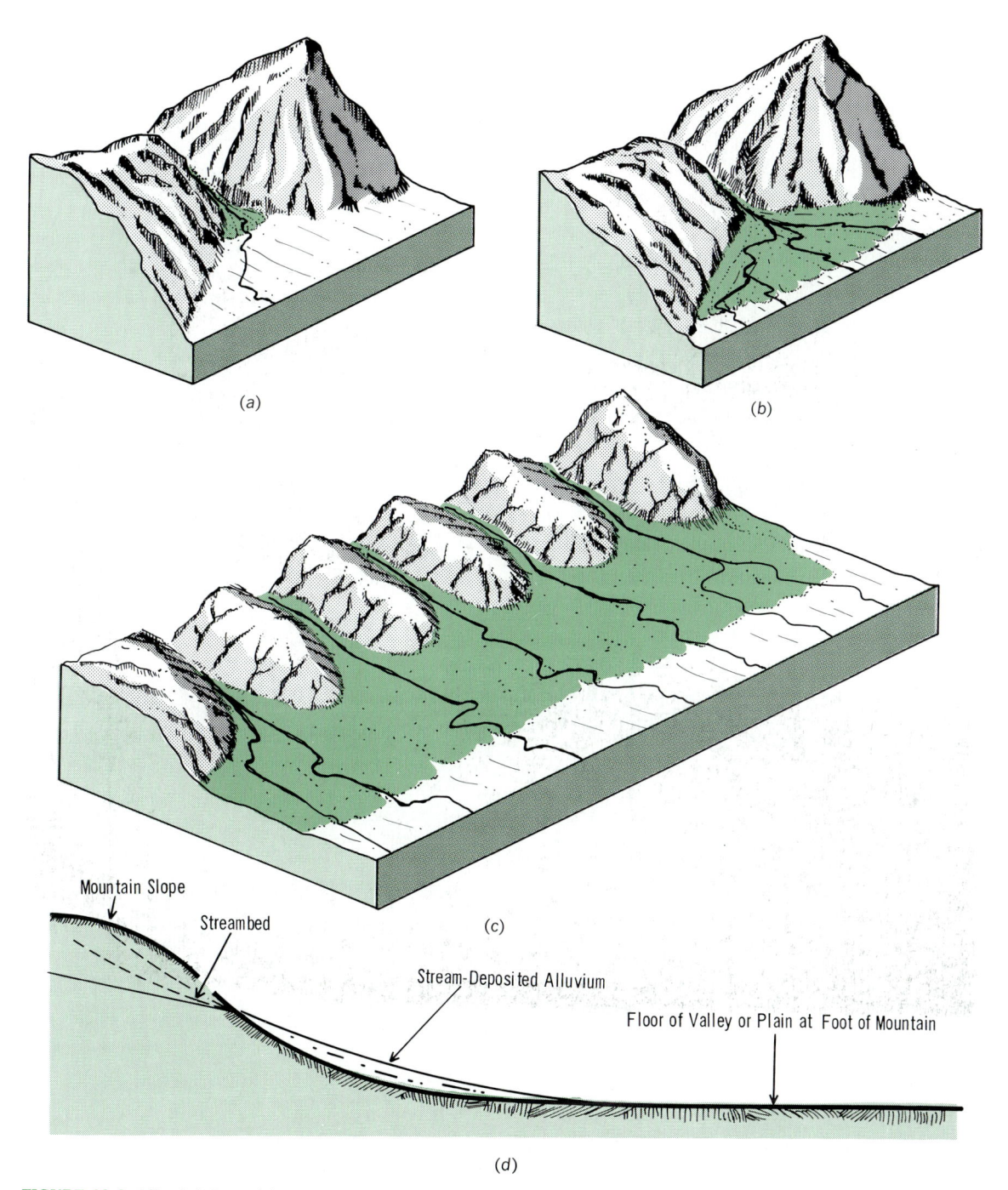

FIGURE 13-3 Alluvial fans (*a*) Alluvial cone. (*b*) Alluvial fan. (*c*) Piedmont alluvial plain. (*d*) Cross section of a typical alluvial fan.

341

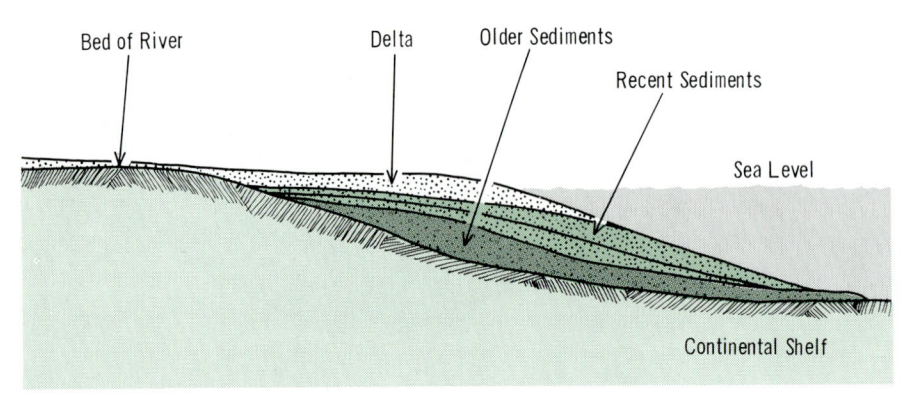

FIGURE 13-4 Formation of deltas.

shore currents, an **arcuate** delta will form; if the currents do not have a consistent direction but are constantly changing, a **cuspate** delta will form. If the waters offshore are calm and do not have the power to disturb the even flow of sediments, a **bird-foot** delta takes shape.

Some of these deltas are quite massive, extending many miles out to sea and complete with numerous *distributaries* (streams that branch off the main course). Some have heavy concentrations of population. A good example of the importance of deltas in the history of

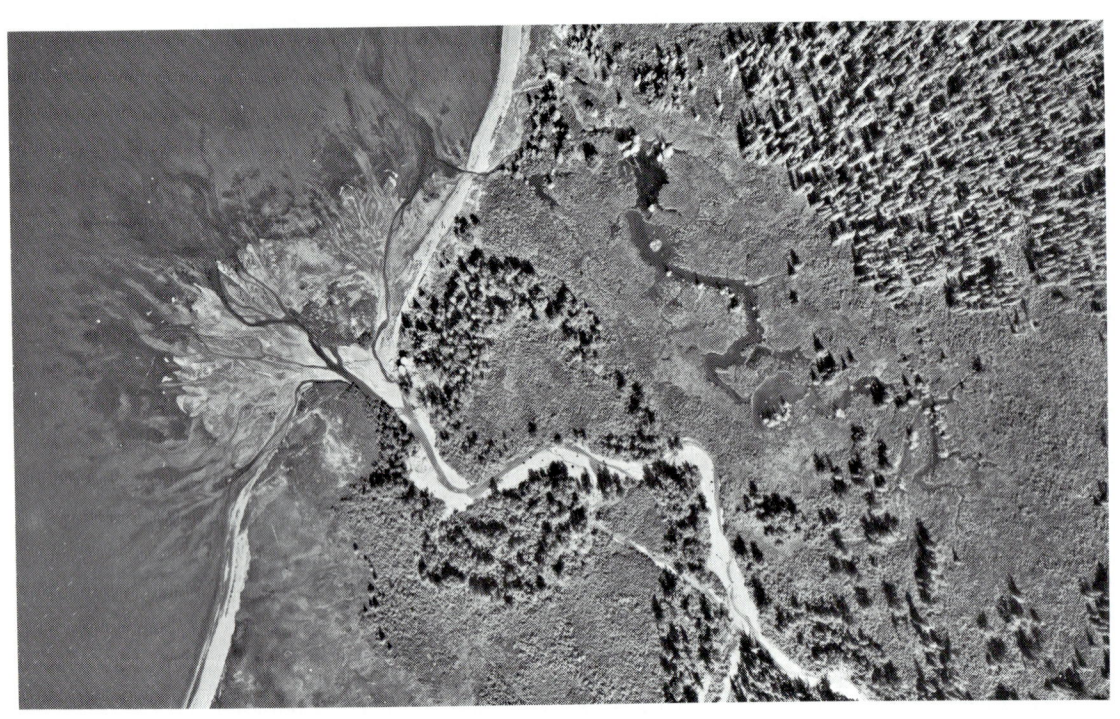

Arcuate delta, Chinitna Bay, Alaska.

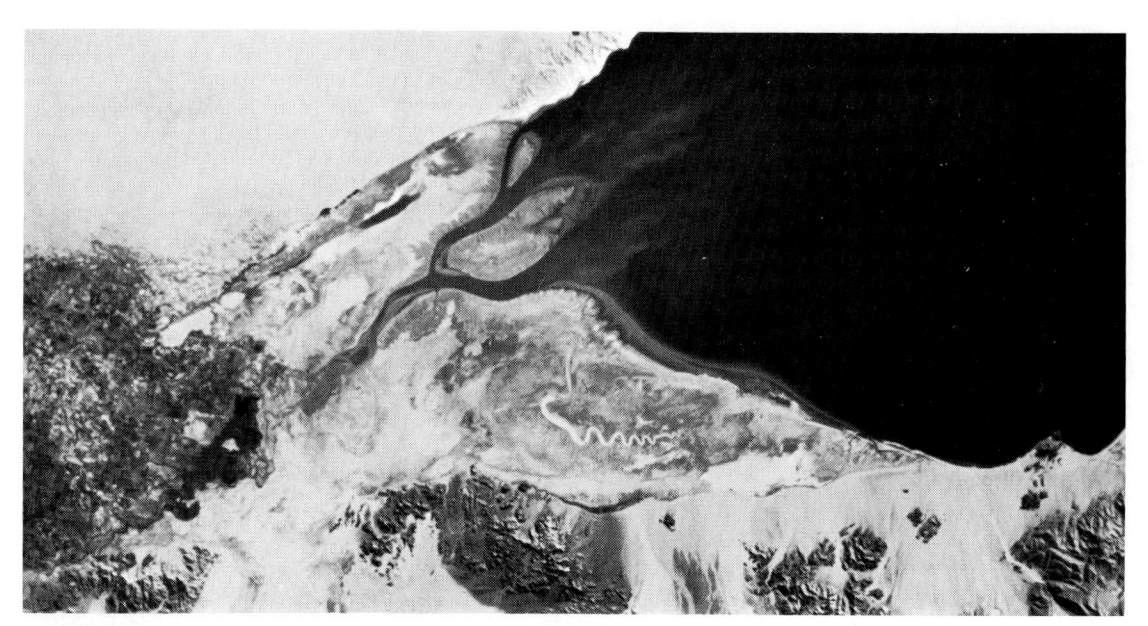

Estuarine delta The delta has formed at the mouth of the Colorado River where it empties into the Gulf of California.

Bird-foot delta of the Tongariro River, New Zealand.

The broad delta of the Nile River.

civilization is the Nile delta. The fertile soils of this area drew people to it, and the delta became Egypt's "cradle of civilization."

Floodplains (Fig. 13-5) are the result of a considerable widening of a valley by a meandering river. As the river flows across the land, it meanders, cutting away at the banks on one side and depositing sediments on the opposite side. In Fig. 13-5 the points of heaviest erosion occur at A, C, and E because the swiftest part of a meandering stream or river is on its outside edge. Deposition occurs at points B, D, and F because the inside portion of a river is the slowest, often forming swirls of backwater. Because this type of stream erodes laterally, there is little vertical, or downward, cutting, except during times of flood stage (when the volume of the river is tremendously increased). Generally, floodplains evolve through stages of development, widening through time and forming features such as natural levees and alluvial terraces.

When flood stage arrives, the height of the waters is increased by waters swollen by rain, which may have fallen in a distant area, until the river's banks can no longer contain the increase. Waters pour over the banks, carrying their load of sediments onto the valley floor on both sides of the river. As the speed of the rushing waters is checked, the heavier particles and then the lighter and finer ones are deposited. Thus a mound paralleling the river is

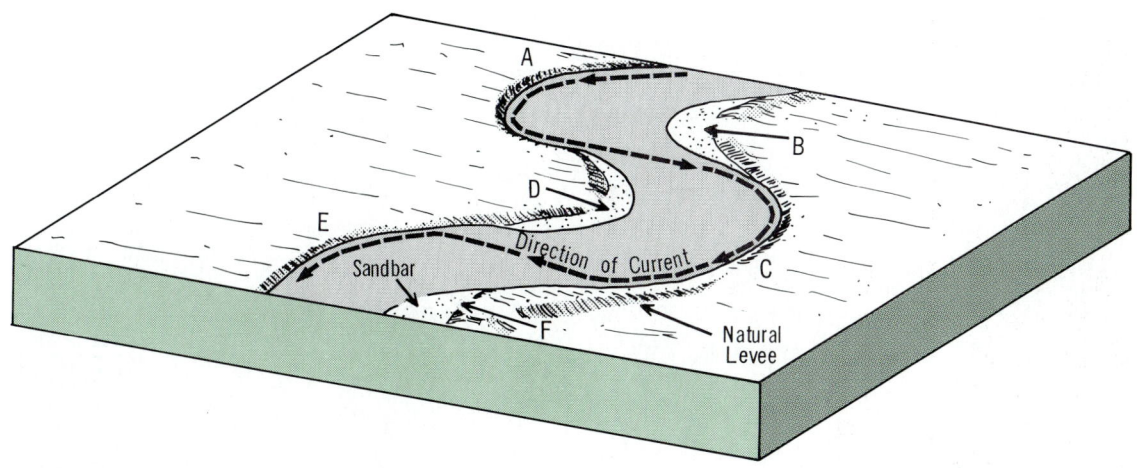

FIGURE 13-5 Development of a floodplain.

built up. This **natural levee** will contain the stream or river until the next high flood state. People combat this problem by building artificial levees (sandbags or piles of boulders) on top of the natural levees. Despite the dangers of residing in floodplain areas—the seasonal threat of flooding and the resulting loss of farm crops and human lives—these areas of the world seem to have the greatest concentrations of human population.

Alluvial terraces are composed of alluvium deposited by streams, arranged like the stream-cut terraces along the sides of stream valley. The main difference between these two different types of terraces is that the stream-cut terraces are usually "matched," that is, on each side of the valley the terraces are the same age, so they have the same elevation. Alluvial terraces, on the other hand, are not normally matched because the deposits on each side of the valley are laid down at different times and under different conditions of sediment load and water volume. As the land is uplifted, or as the stream downcuts, the terraces are exposed.

Oxbow lakes (Fig. 13-6) are to be found in relatively old floodplains that have had enough alluvium deposited through the years to raise the general level of the valley floor. The river no longer does much downcutting as it meanders sluggishly through the valley. When one of the meandering loops is bypassed or cut off from the main stream flow, an oxbow lake is formed. If enough water continues to seep into the lake, it will remain a lake; if not, the oxbow will slowly be transformed into a marshland and, eventually, into a grassy meadow or a farmer's field.

DEVELOPMENT OF STREAM VALLEYS

Valleys can be found almost everywhere, but they will not all have the same shape and size because of a number of controlling factors, such as the composition of the rock layers over which the stream must pass, the cover of vegetation that may hinder the movement of the running waters, the gradient of the slope, the regularity and amount of precipitation, the rate of evaporation, and time.

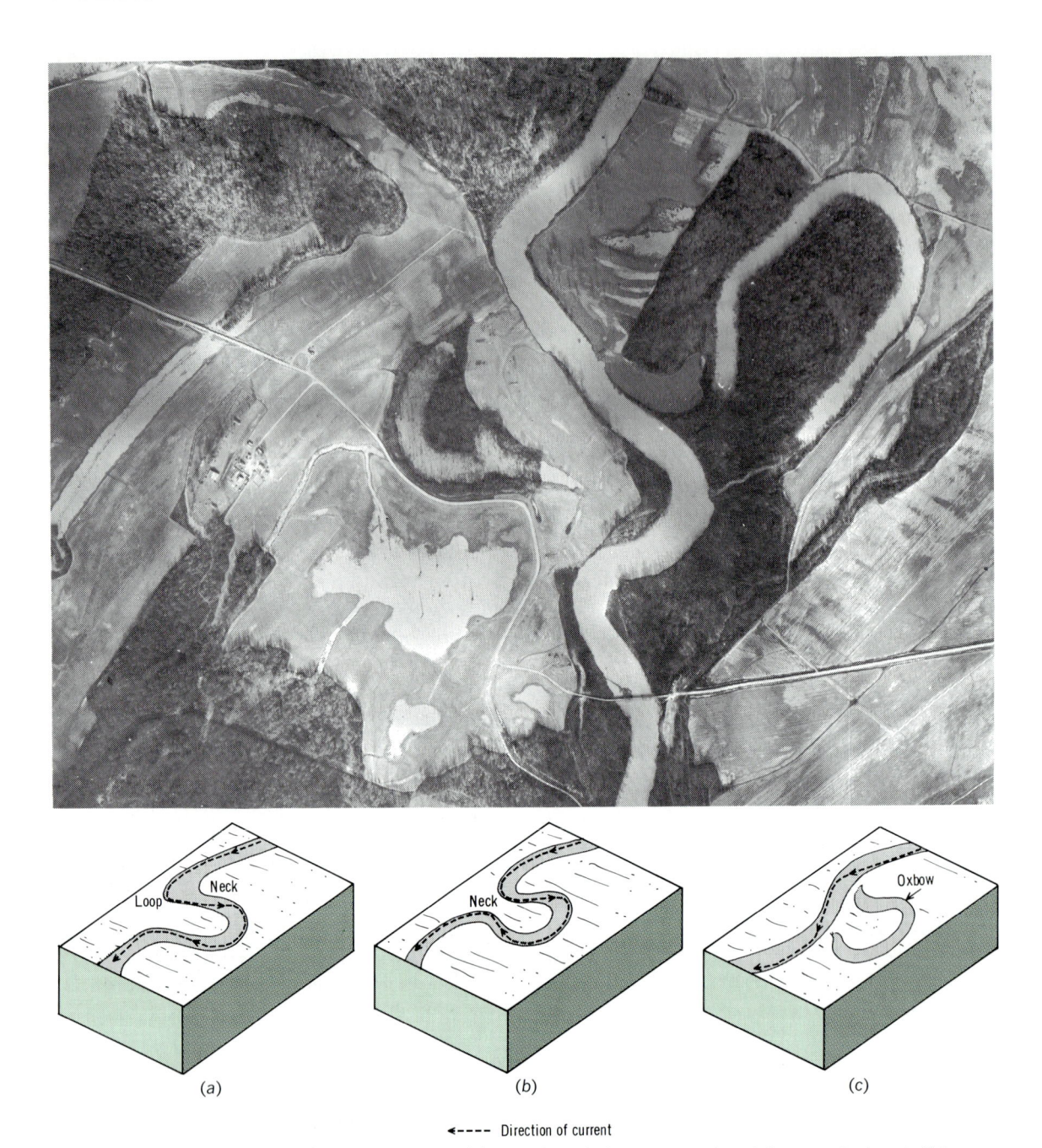

FIGURE 13-6 Development of an oxbow lake (*a*) A meandering river carves great loops on the valley floor. (*b*) The meander loops close in on each other. (*c*) The river cuts through the narrow neck and the meander is cut off, becoming a cutoff meander and, in time, an oxbow lake.

Stream-Valley Evolution

There has been considerable discussion during the twentieth century over the question of the evolution of stream valleys. Near the end of the nineteenth century a geologist-turned-geomorphologist, William Morris Davis, formulated a theory to explain a process he termed *the geographical cycle.**

After a long period of field study, Davis came to the conclusion that after the initial uplift of a landmass, running water would immediately go to work to bring it all down to a final, flat plain, the **peneplain.** He believed that someday all the mountains, highlands, plateaus, and hills would be gone, and only a flat plain would be left. Davis did suggest that a few rocky remnants might be left jutting up from the peneplain, resistant rock mounds he called *monadnocks* (after New Hampshire's Mount Monadnock). However, his general theory of the ultimate end of the cycle of erosion was avidly opposed. Earth scientists of today do not accept his theory because they say that the earth's surface does not reach a standstill. Diastrophism, vulcanism, and plate tectonics give ready proof, they claim, that the landmasses are continually rising at the same time they are being graded down. They say that individual streams may achieve baselevel, but only for a time, and broad areas will experience infinite crustal movements, thus preventing the attainment of a peneplain.

This does not mean, however, that we cannot apply Davis's concept of a geographical cycle to the evolution of stream valleys. Some valleys show their youthfulness by their rugged shapes, and others show their age by the flatness of their landscapes. The uplifting of landmasses and the subsequent grading down is a continuing process. Valleys, like mountains, work their way through a cycle of birth, life, and death, from youth to maturity to old age. After death (the peneplain) comes rebirth, and the wheel revolves.

Streams coursing down the slopes of young mountains with high elevations and steep gradients are strong and vigorous. The channels of these **youthful streams** are fairly straight, and their valleys are narrow and V-shaped. The land between two stream valleys, called *interfluves,* which means "between the streams," is high and broad. The general relief of the area is sharp and bold, not yet rounded by erosional forces. There is little in the way of lateral erosion because the steep slopes dis-

Youthful stream valley in the Black Canyon of the Gunnison River, Colorado.

---

*William Morris Davis (November 1899), "The Geographical Cycle," *The Geographical Journal,* 14(5), pp. 481–504; available in Bobbs-Merrill reprint no. G-48. For opposing views, see Kirk Bryan (December 1940), "The Retreat of Slopes," *Annals of the Association of American Geographers* 30(4), pp. 254–268; Bobbs-Merrill reprint no. G-28.

Mature stream valley of the Fremont River, Utah.

Old-age stream valley near Alsiak Bay, Alaska.

courage all but vertical cutting. As time passes, the *drainage area* (the valley between two ridges that catches the various forms of precipitation) will supply additional water in the form of tributaries that add to the widening of the valley.

If there are numerous obstacles in the stream's path, such as granitic dikes or other resistant rock masses, the valley may contain many waterfalls and lakes.

A **mature stream** is no longer vigorous in its vertical attack upon the valley floor; it has arrived at the state of equilibrium. The valley, instead of being steep-sided, would have been cut laterally by the stream for sufficient time to narrow the interfluves and widen the drainage area. It would be broad and flat, forming a well-developed floodplain with, perhaps, some small lakes, natural levees, and sandbars. In such a valley, tributaries are common and the drainage system is conducive to the creation of a meandering stream. There may even be a few oxbow lakes.

By the time a stream valley has reached the stage of old age, it has done just about all it can, unless some form of rejuvenation occurs. An **old-age stream** usually moves slowly, almost sluggishly, as it swings in great loops from one side of its wide floodplain to the other. It may be wide and shallow, smoothly functioning in a state of equilibrium, except during times of high water (flood stage), when it overflows its banks and natural levees to inundate the land beyond. Interfluves will be low in elevation and in local relief, smoothly rounded, narrow, and widely spaced. There will be marshes, oxbow lakes, stream-cut and alluvial terraces, numerous tributaries but a poorly developed drainage system, excellent farmland (if artificial levees can contain the seasonal rampages of the river), and a broad delta at the river's mouth.

Earth movements can cause disruptions at any point in the evolution of a stream valley. If there has been diastrophic uplift, as has occurred along the coast of Norway, an old-age stream will be rejuvenated and will begin carving out a youthful valley. If there has been a general subsidence of a large area, such as happened in the Mississippi River region, the floodplains will be widened and the river will flow even more sluggishly. In this case, a youthful stream valley might move directly into an old-age situation, by moving rapidly through the mature stage.

Climate and Stream-Valley Development

Stream valleys in humid (moderate-to-heavy precipitation) regions are quite different from those in dry or semiarid regions, primarily because of the vegetative cover and the amount of constancy of stream flow. Stream valleys in dry climates show the effect of thunderstorms. Sudden torrents are unleashed, attacking the loose-knit soils that, in such areas, have no vegetation to hold back the surging waters.*
Youthful valleys in humid areas are not as rugged as those of dry areas, and old-age valleys of dry areas are filled with alluvial deposits that seem to rise to the mountaintops. Humid areas are noted for their hills and low, rounded mountains, while dry areas are noted for their deeply dissected plateaus and mesas.

GROUNDWATER ETCHINGS

During our earlier discussion of the earth's waters we learned that groundwater supplies us with fresh water for drinking and for agriculture. But is that all we can say about groundwater? Does it serve us in other ways? The answer to this question is in the affirma-

---

*At this point you might reread the section on rainfall and desert vegetation (page 229), then look ahead to the discussion of the features of the desert landscape (page 408).

tive because groundwater performs in ways similar to surface water. It furnishes salts to the oceans and creates interesting and sometimes fantastic erosional and depositional landforms.

Erosion by Groundwater

When there is precipitation over a land composed of massive beds of limestone, water entering the ground is transformed into a powerful erosional force. Carbon dioxide in the air unites with the water to form carbonic acid, which literally eats its way through the limestone as it dissolves the calcium carbonate and carries it away in solution. Figure 13-7 shows the typical landforms created by groundwater that result in a landscape called **karst,** after the region of that name in Yugoslavia where the work of groundwater on limestone beds was first seriously studied. This, then, is another cycle of landform evolution. The state of youth is shown in Figs. 13-7a and 13-7b, where numerous surface streams sink into the ground and begin the work of dissolving the limestone. As the surface layers are attacked and dissolved, a number of depressions, called **sinkholes,** are formed. As time passes, some of these sinkholes increase in size and become deeper. These larger depressions are called **dolines** (Fig. 13-7c). After more time passes and erosion continues, the dolines become enlarged as surface streams run over their sides and disappear from view as they flow through underground channels to join the water table.

The state of maturity is reached when the karst area is dotted with large depressions,

Sinkholes on Karst Plateau of Yugoslavia North of Jajce.

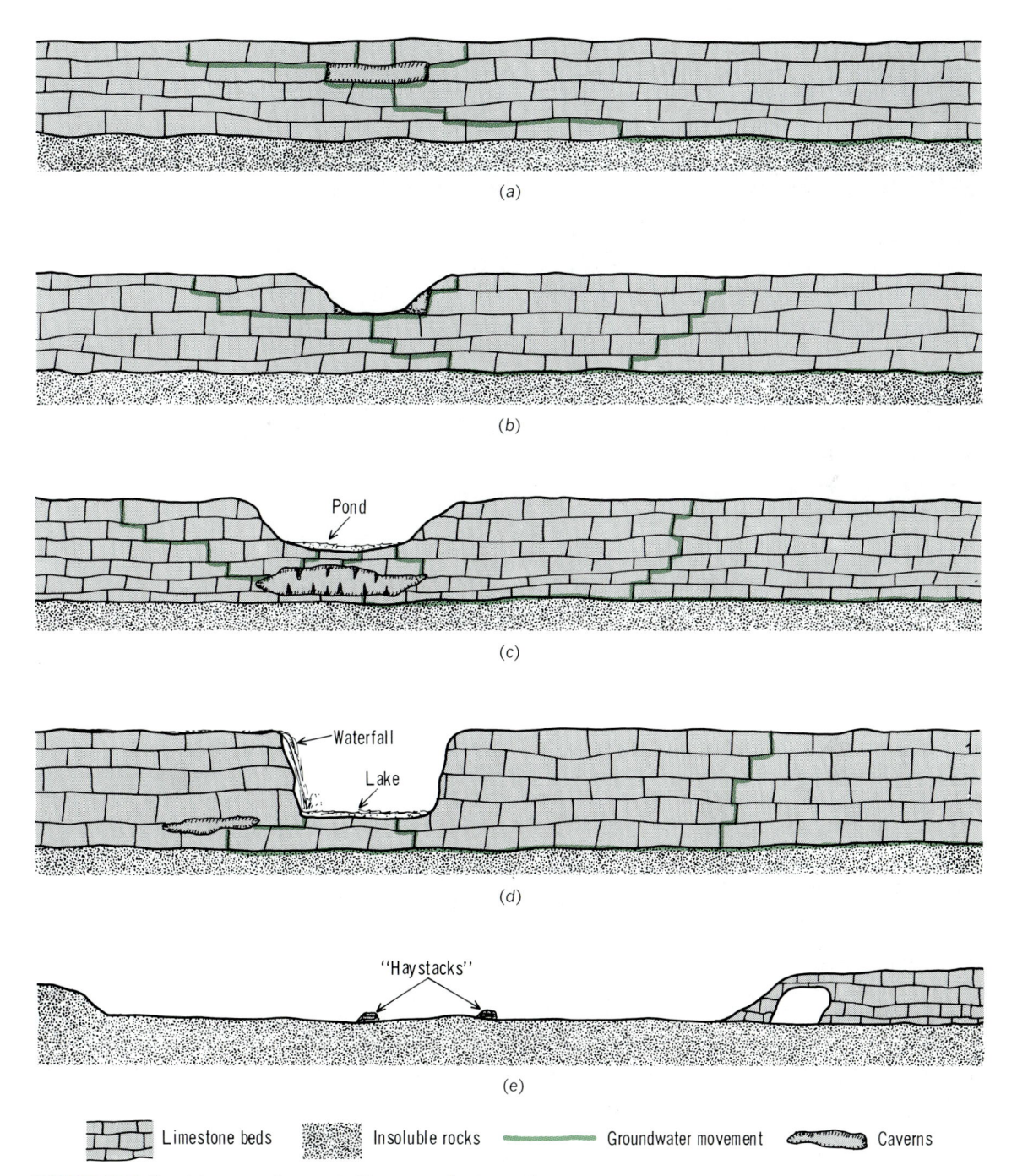

FIGURE 13-7 **Karst topography created by groundwater action.**

Limestone formations Stalactites and stalagmites that have formed in the Big Room, Temple of the Sun, in Carlsbad Caverns National Park, New Mexico.

which result from the collapse of the roofs or ceilings of underground caverns. These steep-sided sinkholes (Fig. 13-7d)—called *ponors* by the Yugoslavs, *swallow holes* by the English, and *cenotes* by the Spanish—have flat bottoms that may contain marshes, ponds, or lakes.

By the time the region has reached what might be termed old age, the sinkholes will have coalesced and the level of the land will be considerably lower than it was during its youth. Streams will once more flow across the insoluble rock surface that is now exposed, and the landscape of this *uvala* (or *polje*) will be dotted with the honeycombed remnants of limestone called *magotes* (or "haystacks"). There may even be a few natural bridges around the edges of the great valley (Fig. 13-7e).

Deposition by Groundwater

The most interesting depositional work done by groundwater can be seen only in underground caverns, such as those in the Mammoth Caves of Kentucky and Carlsbad Caverns of New Mexico. Here you will discover a fantasy land of weird and beautiful forms, created as the dissolved limestone from the overlying beds drips from the ceilings of the caves. As the moisture trickles down, deposits of calcium carbonate create fingerlike projections that hang from the ceilings (*stalactites*) and jut up from the floors (*stalagmites*). You may be interested to learn that one of our common building materials for bathroom walls and floors, travertine, comes from limestone caverns like these.

REVIEW AND DISCUSSION

1. Explain the difference between *corrasion* and *corrosion*.
2. What is splash erosion? What is sheetwash?
3. What three factors determine the velocity of a stream?
4. What is the difference between a stream-cut terrace and an alluvial terrace?
5. Define the terms *aggradation* and *gradation*.
6. Explain the structure and development of an alluvial fan and a delta.
7. What is a piedmont alluvial plain?
8. Describe the development of a floodplain.
9. Explain the terms *natural levee, oxbow,* and *meander*.
10. Explain baselevel.
11. What is a graded stream?
12. What is a peneplain? A monadnock?
13. What is an interfluve?
14. What is the effect of climate on stream-valley development?
15. Explain the difference between dolines, ponors, and uvalas.

NOTE

[1] N. B. Emerson (1909), *Unwritten Literature of Hawaii: The Sacred Sons of the Hula* (Washington, D.C.: U.S. Government Printing Office).

ADDITIONAL READING

Ellison, W. D. (November 1948). "Erosion by Raindrop," *Scientific American* (offprint no. 817).

King, Lester C. (July 1963). "Canons of Landscape Evolution," *Bulletin of the Geological Society of America*, 64, pp. 721–751. Also available as Bobbs-Merrill reprint no. G-116.

Mackin, J. Hoover (May 1948). "Concept of the Graded River," *Bulletin of the Geological Society of America*, 59, pp. 463–512.

Sweeting, Marjorie (1973). *Karst Landforms*. New York: Columbia University Press.

Thornbury, William D. (1969). *Principles of Geomorphology*. 2d ed. New York: Wiley.

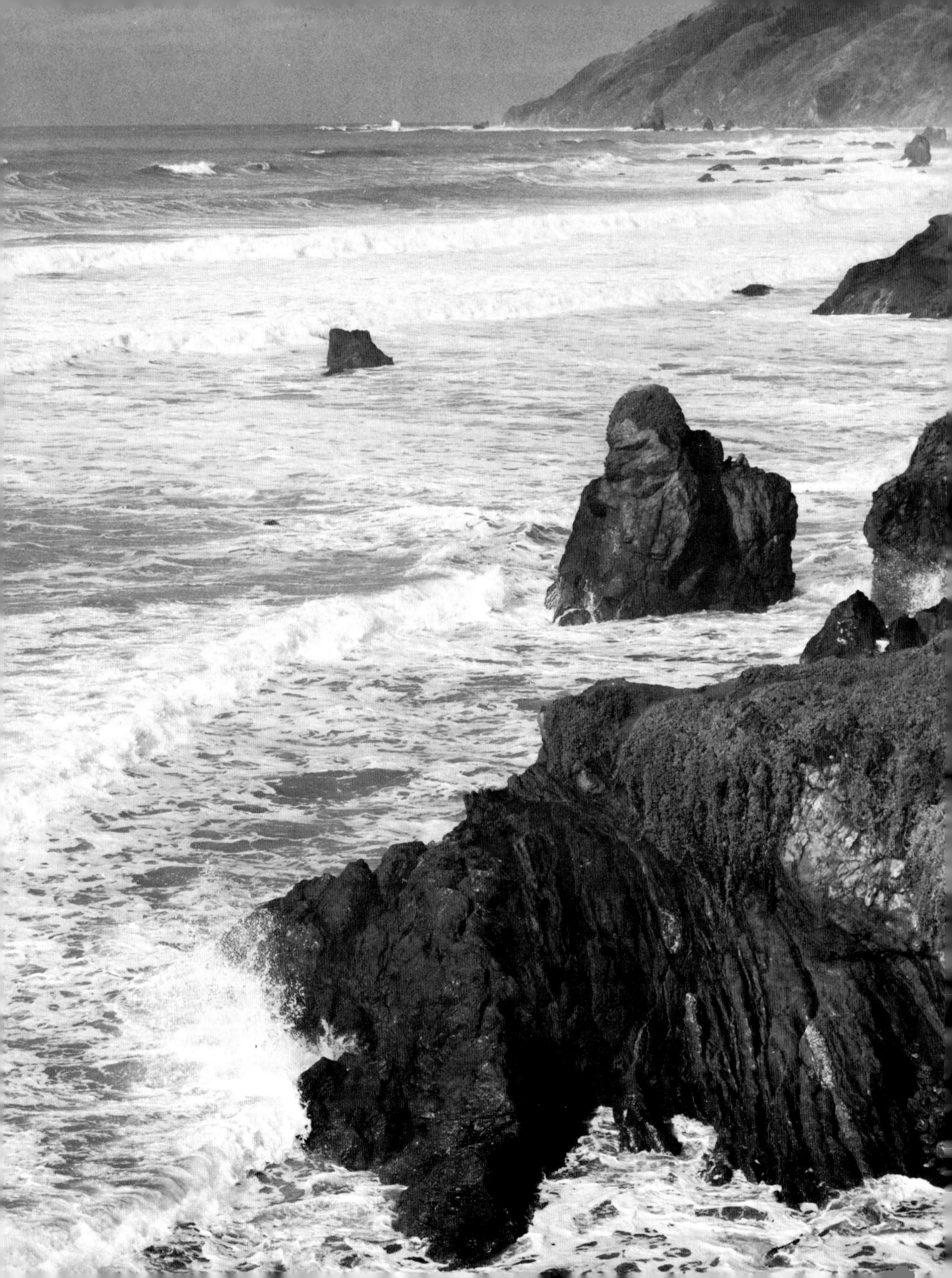

FOURTEEN

CARVED BY OCEANS

They that go down to the sea in ships, that do business in great waters; These see the works of the Lord, and his wonders of the deep. For he commandeth and raiseth the stormy wind, which lifteth up the waves thereof. They mount up to the heaven, they go down again to the depths: their soul is melted because of trouble. They reel to and fro, and stagger like a drunken man, and are at their wit's end.

PSALMS 107:23–27

Throughout the long history of human beings, many individuals have tried to fathom the mysteries of the "great waters," but it was not to understand the reasons behind the formation of the seas, the influence of the waters on their climates, or the sculpturing actions of the oceans on the lands. Because life was a daily struggle for survival for early peoples, the oceans were important as places that could be "farmed" with net and hook or used as "sea highways" to be traveled. Even though it was necessary for them to know of the rise and fall of tides and the movement of ocean currents, their knowledge and technology would not let them pursue the matter further.

The passage of time brought great changes. The study of the oceans was elevated to a respected place among the sciences by the work of men such as Benjamin Franklin, who charted the route of the North Atlantic's Gulf Stream; Charles Darwin, who traveled the oceans and, in 1840, wrote the epochal *Voyage of the Beagle;* and the American naval officer Matthew Fontaine Maury, who, in 1855, wrote the first oceanography text, *The Physical Geography of the Sea.* Modern oceanography became even more securely established when England's Royal Geographical Society sent the H. M. S. *Challenger* on a record-setting 3½-year voyage to conduct oceanographic research in the Atlantic, Indian, Antarctic, and Pacific oceans. The voyage ended in May 1876 after the ship journeyed almost 69,000 nautical miles.[1]

A century has passed since that memorable event. Oceanographic research is continuing as the nations of the world try to reach agreement on how they are to "use" the ocean waters and the sea bottom. It is important for us to understand what has been learned about these vital regions so that we can keep abreast of developments as news flows from the international conferences. We will begin expanding our knowledge of the oceans by examining that part of the ocean floor that is most accessible to us: the rim of the continental landmasses, the **continental shelf.**

CONTINENTAL SHELVES

The term *shelf* may be defined as a "flat ledge or horizontal surface usually fastened to a wall to serve as a support for objects of value." The definition is apt because the continental shelves do support objects of value to people, such as rich petroleum deposits and many varied forms of sea life. The continental shelves are a part of the land, an extension of the continents, forming a relatively flat but gently sloping margin of varying width. Figure 14-1 shows a typical cross section of the rim of the continents.

The term **shelf edge** marks the outer limit of the continental shelf and the beginning of the downward slope to the sea bottom beyond. This occurs, on the average, at a depth of about 600 feet (180 meters). This zone, from the surf line to the shelf edge, varies greatly in width, as you can see by studying a map detailing the world's continental shelves (Fig. 14-2). All countries lying in contact with the ocean do not have wide shelves. The width is near zero along some portions of the west coast of South America, but observe how wide they are around Great Britain and Ireland, northern Australia, and the East Indies. Some ocean-fronted countries have little or no sea life (fish, seaweed) or minerals to harvest, while others have shelves extending to 900 miles (1500 kilometers) that provide them with opportunities for bountiful harvests. Oceanographers estimate the average width of the world's continental shelves to be about 40 miles (64 kilometers).

CONTINENTAL SHELVES 357

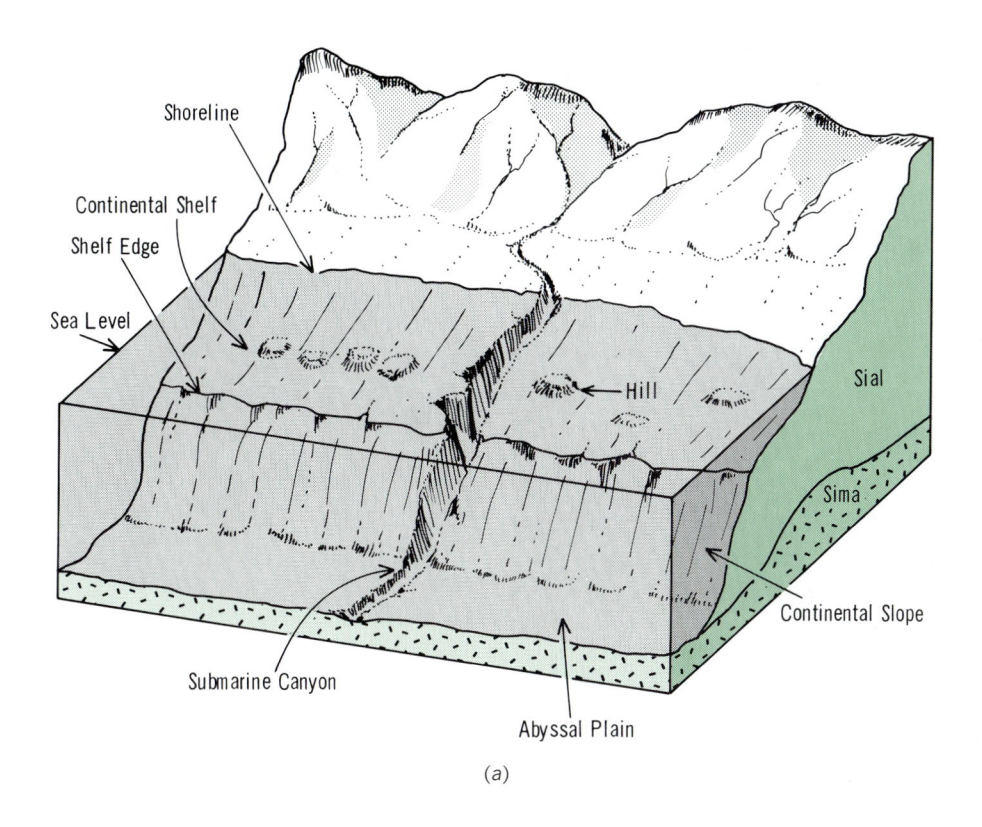

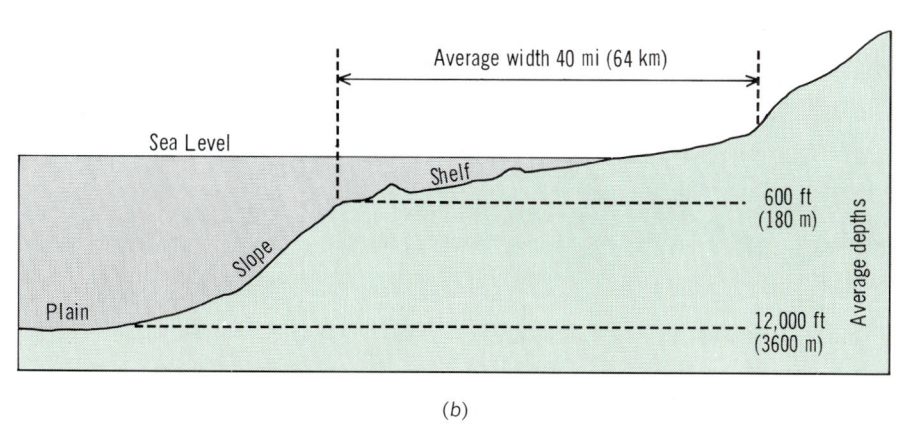

FIGURE 14-1 **Rim of the continents** (*a*) is a perspective view; (*b*) is a cross section. (Vertical scale greatly exaggerated.)

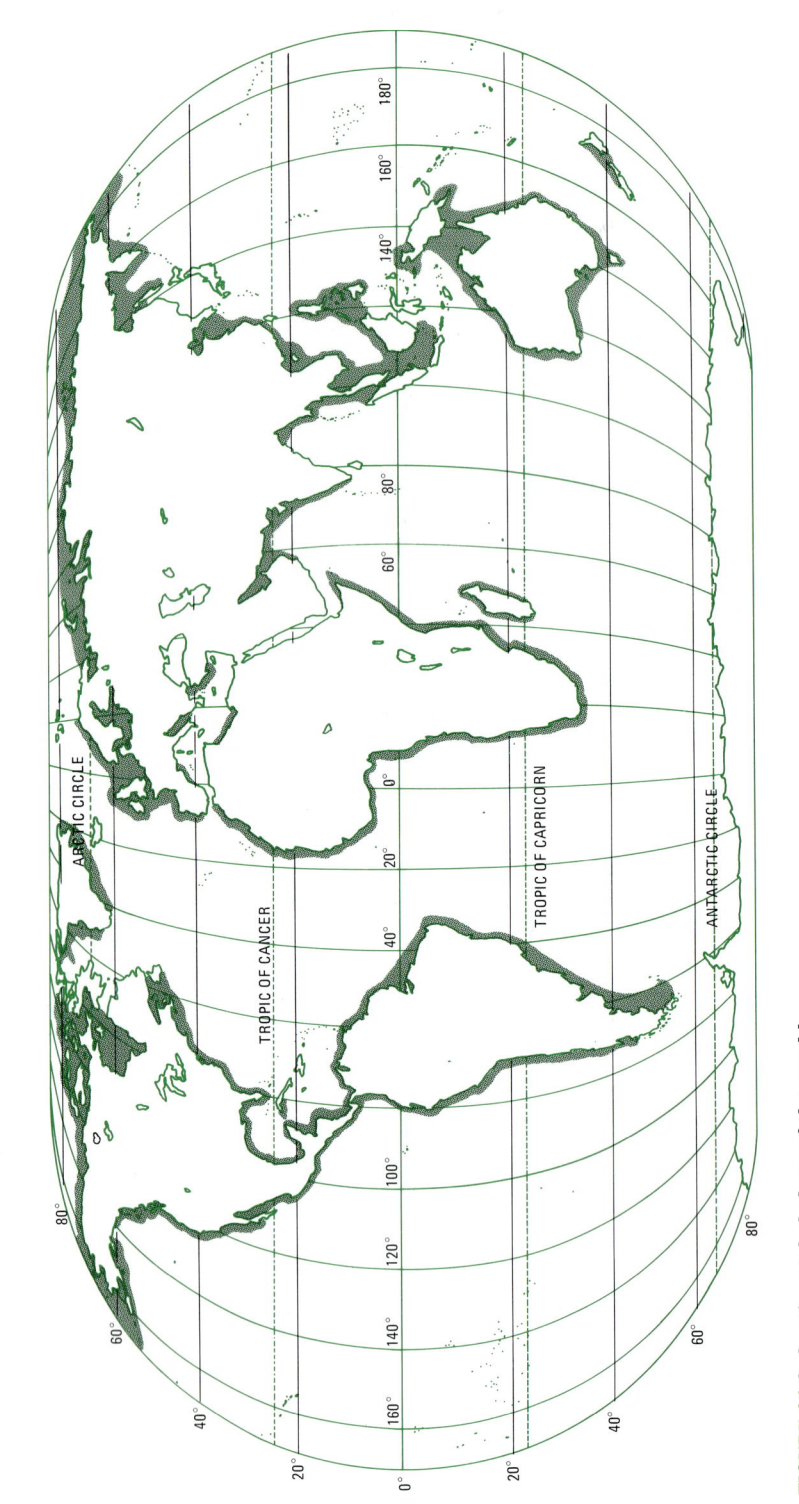

FIGURE 14-2 Continental shelves of the world.

Origin of the Continental Shelves

How did the continental shelves come into being? Geologists originally believed that they were the result of wave action attacking the granitic (sial) margins, wearing them down in some places and building them up in others. However, close examination of the structure of the shelves has shown that while the underlying beds are composed of granitic rocks, much of the covering material is made up of sedimentary rocks. This tells us that at some time in the past, perhaps at many times, the level of the oceans was far below what it is today. Earth materials were carried down and deposited on the shelves by streams, winds, and glaciers.

Geologists have proven that the earth experienced at least four great ice ages during relatively recent geologic time (about 2.5 million years). Within each ice age tremendous masses of ice formed at the poles and spread toward the equator. In the Northern Hemisphere the continental glaciers covered much of North America and Asia. From where did the water that formed the glaciers come? It is probable that moisture rising from the oceans was precipitated as snow over the polar regions, thus lowering the level of the seas. It has been estimated that sea level was 200 to 500 feet (60 to 150 meters) lower than its previous level during these periods. When the climate changed and the ice sheets melted, water was returned to the oceans, and the level returned to normal.

It is apparent that the continental shelves of the world were alternately exposed and submerged, with each interval lasting for many thousands of years.*

---

*For one of the best discourses on the subject of oceanography, see Francis P. Shepard (1973), *The Earth Beneath the Sea,* rev. ed. (New York: Atheneum).

Development and Structure of the Continental Shelves

The continental shelves are quite complex in their structure. Not all of them meet the same specifications. Some have gravels and coarse sand in areas close to the shoreline and fine clays and muds deposited farther out, near the edge of the drop-off point, the slope. Others have patches of gravels and clays arrayed at random across the width of the shelf. Why is there such a variation? A number of factors are responsible, such as the presence or absence of large streams or rivers, the type of rock on the land behind the shelf, whether glaciation occurred on or near the shelf, and the form and strength of the ocean currents operating offshore.

Rivers have been instrumental in building some stretches of continental shelves. As we have seen, large rivers can build extensive deltas at their mouths during the periods when the shoreline is submerged. When the sea level is lowered during an ice age, the rivers supply sediments to construct alluvial fans at their mouths. The shelves at the mouths of the Mississippi and Amazon rivers are much wider than shelves elsewhere along those coasts and are composed of fine soil particles, such as clay and some sand.

Glaciation in the vicinity, another force in the development of continental shelves, is evidenced by terrain that is not as flat or smooth as the average shelf. If a continental glacier actually moved across the shelf, it would leave its mark in the form of deep gorges and old river valleys that were scoured out by streams issuing from the glacier. The shoreline would look much like the rugged land surface, with many indentations engraved by the moving glacier. One would be likely to find sands, gravels, and rocks in the upper layer of the shelf, except where these deposits were covered by more recent deposition.

Topography can also be important, especially as a source from which earth materials are eroded and transported out onto the shelf. If the region behind the shelf is mountainous, the shelf might become dotted with rocky hills and its composition would be as varied as the composition of the mountains would allow. *Cobbles* (large, round, smooth stones), gravels and coarse sand would be in plentiful supply.

Ocean currents can be very influential in building up or wearing down a portion of a shelf. This is evidenced by the lack of wide shelves along the eastern coast of Florida, where the northward-moving Gulf Stream strikes; the east coast of Brazil, which undergoes constant attrition by the Brazil current; and the coast of Southwest Africa, which is washed by the Benguela current. If, however, the waters are slow-moving and relatively warm, the shells of dying sea life may build up thick deposits, such as those seen along Australia's north coast.

The Continental Slope

At the outer edge of a continental shelf is the shelf edge. This marks the beginning of the steep slope known as the **continental slope,** which drops off sharply down to the abyssal plain beyond. Many slopes are marked with deep submarine canyons, small mountains and hills, and a generally rough surface. The foot of a slope at its outer margin may be as much as 4500 to 12,000 feet (1350 to 3600 meters) below the ocean's surface. Geologists once believed that the irregular terrain of the slope was due to chemical decomposition of the materials of the slope by the ocean waters or by seismic wave action (*tsunamis*). Now, however, they believe that it results from (1) the scouring action of *turbidity currents,* fast-moving currents that are heavily laden with sediments, (2) erosional activities that occurred when the ocean was lowered during the ice ages, thus exposing the shelves and slopes, and (3) faulting along the continental coastlines. It is probable that all three processes should be considered. In addition, the fact that some of the slopes have very steep faces indicates that landslides are common along many of the shelf edges. Good examples of this can be seen in the slopes off Cuba and Ceylon, which have slope angles of 45°.

WAVE ACTION

In Chapter 6 (page 175) we studied the movements of ocean waters and learned why and how ocean currents and wind waves are generated. Now we will consider the effect of currents and waves on the landmasses of the world. Ocean currents are important to us as an influence on the climate and weather and as sculpturing agents.

How do currents compare to wind-produced waves as sculptured agents? The movement of ocean waters along and against a coast may be localized (that is, limited to specific areas), but if the current is fast enough, the depth of water shallow enough, and the wind-born waves powerful enough, the outer margin of the continents (or any landmass) will be greatly affected. If these conditions prevailed, erosional activity would be pronounced. If, on the other hand, the current moved slowly, the waters were deep, and the waves were gentle, depositional activity would result.

Wave Refraction and the Longshore Current

There are water movements that occur along a coastline independent of the major currents we considered earlier. These are currents born of waves. That is, waves rolling in toward a shoreline will follow the direction given to them by the wind that created them. As the waves advance toward the shoreline, the water behind them piles up in the shallower depths and the waves crest and break on the beach in lines parallel to the beach. Waves usually do not approach a shoreline that way, however. If the waters are deep, a wave advancing at an angle of about 45° would probably strike the beach at about that angle. If, however, the water depth were shallow across the shelf, the waves would tend to "bend around" as they advanced. Figure 14-3 illustrates what happens to such a wave advance. The circular wave motion, generated in deep water, gradually weakens as the wave travels a great distance. When it moves across the comparatively shallow waters of the shelf, the wave motion is greatly reduced and the wave begins to slow along its beach side, while the seaward extension continues at the original speed. The axis of the wave then swings around until the crest is almost parallel to the beachfront. This process is called **wave refraction**.

As the waters surge up (*swash*) the sloping beach, they are not advancing parallel to the beachfront, but when they are pulled back, they move at a right angle to the beach. This produces a zigzagging motion that moves both the water and sediments picked up along the beachfront. This down-the-shoreline movement is termed the **longshore current** and is responsible for the transportation of considerable quantities of sand and other sediments along the coast. If a submarine canyon were in the way, the sand would pour into the canyon and be lost to the beaches farther down the coast.*

*Encyclopedia Britannica Films has produced an excellent 16-millimeter movie, entitled *The Beach: A River of Sand*, on the effect of the longshore current on beaches. It tells the story of the rise and decline of beaches caused by robbery by submarine canyons and the cutting off of new sand supplies as dams are built across rivers that once carried sediments to the beaches.

A surf zone phenomenon closely related to the longshore current is the **riptide** (also known as *rip current*, or *feeder current*). Riptides are fairly common along beaches that have a relatively steep foreshore (beachfront). When the wave rushes back into the surf zone, it works its way out through a groove (or throat) gouged in the sandy floor to a position beyond the breakers. There, it swirls about in a circling motion at speeds of as much as 10 miles (16 kilometers) per hour. Strong swimmers often find it difficult to free themselves from the rapidly whirling waters.

Tidal Currents

It may be difficult to think of the rise and fall of water from high to low tide as a force strong enough to cause erosion or deposition, yet this does happen. An incoming tide, the change of low water to high water, can carry marine sediments onto a beach in appreciable amounts. The outgoing tide can reverse the process as it carries away previously deposited sediments. Such erosion and deposition is a minor force, and it takes considerable time for it to alter the shape of a continental shelf.

More pronounced changes occur in a river's estuary, such as Canada's Bay of Fundy (located on the continental shelf between Nova Scotia and New Brunswick) (see photos, p. 363). Here the tidal onrush of water raises the level about 50 feet (15 meters) at high tide, and the speed of the incoming tidal wave may approach 14 miles (22 kilometers) per hour. At low tide, the fishing boats lie in the mud, tied to a buoy that also lies in the mudflats. Then the new high tide comes in the form of a steep-fronted wall of water, and the boats are summarily floated, swung about to tug at their lines. The fishermen hurriedly take to the open sea, knowing that they had better be back at anchor before the next low tide.

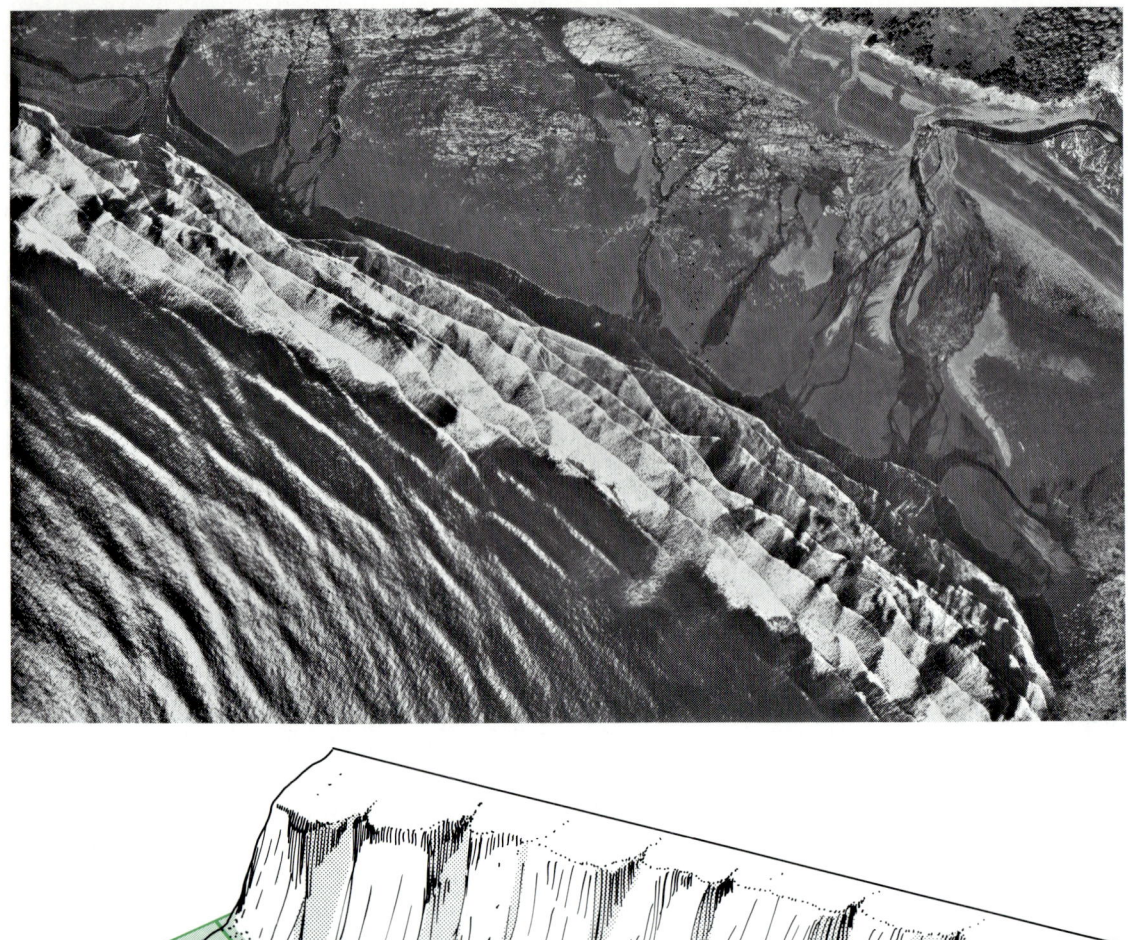

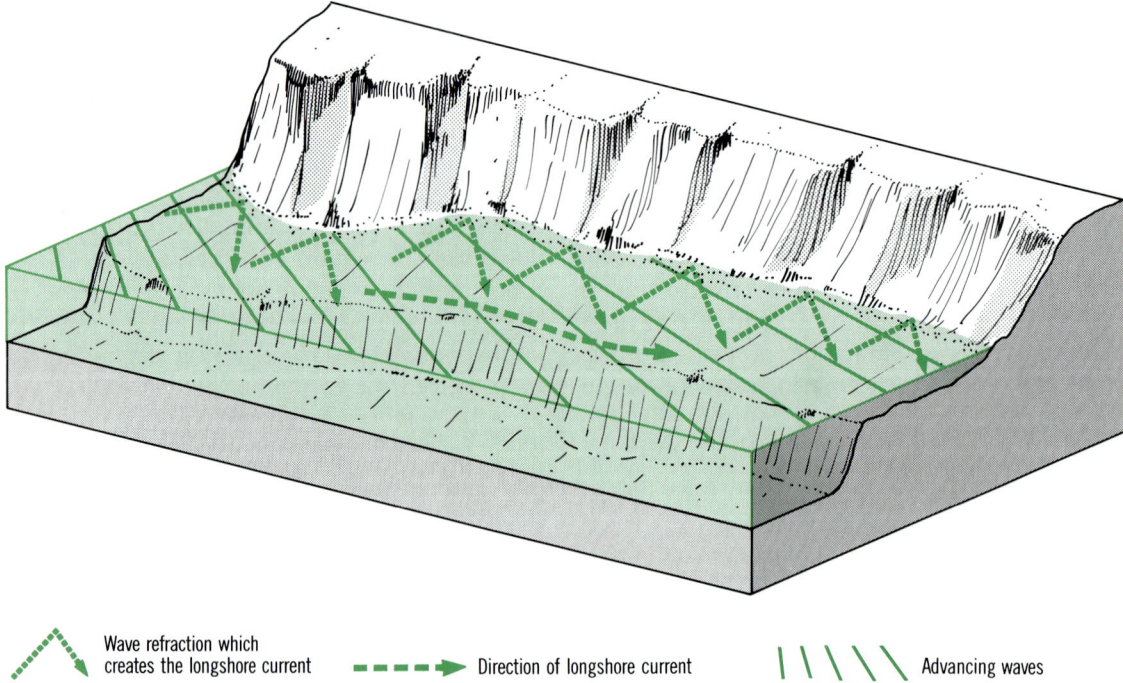

FIGURE 14-3 Wave refraction and the longshore current.

362

(a)

(b)

The Bay of Fundy, Nova Scotia There is a dramatic change from low tide (a) to high tide (b) as the ocean's waters rush through Minas Channel, pass Cape Split, and move up the river at Parrsboro.

Why are such tidal changes so extreme in estuaries? The reason is that the rising waters are compressed between narrow banks, forcing the water to move at a much faster speed. Somewhat similar is a phenomenon called a **tidal bore,** which results when a rising tide is blocked in an estuary by mounds of sands (sandbars). The water builds up pressure until it finally breaks through and races upstream.[2]

Some areas are noted for having exceptionally high tides, with serious damage resulting, such as on the continental shelf between England and the Netherlands. In February 1953 a very high tide invaded the Low Countries, submerging more than 6000 square miles (15,600 square kilometers) of land and drowning about 1800 people.[3] There are other areas where the tidal range is so low that it is scarcely noticeable, such as in the Mediterranean Sea, the Baltic Sea, and the Gulf of Mexico.

The Surf Zone

Although longshore currents, tidal bores, and tidal currents do perform some erosional work, it would seem that the major erosive force is produced by wind-created waves that endlessly attack the beaches of the world. This is accomplished by one or both of two erosion processes: hydraulic action and corrasion.

Hydraulic action refers to the actual pounding of a wave on an object in its way. An advancing wave can exert tremendous pressure as water piles up behind a cresting wave and then drives it forward like a sledgehammer. Robert C. Cowen tells of storm-tossed waves that struck the Tillamook Lighthouse on the East Coast of the United States and sent a 135-pound (60-kilogram) rock so high into the air that it came down through the roof of the 91-foot (27-meter) tall structure.[4]

Corrasion (another term for *abrasion*) refers to the sandpapering effect created when the advancing wave picks up sand or gravel and abrades the sea cliffs and other surface features of the shoreline and the shelf.

LANDFORMS OF WAVE EROSION

Figure 14-4 illustrates some of the more common landforms that result from wave erosion. All such forms were sculptured out of the original landmass or out of rock layers laid down at a later period of geologic time.

Sea terraces are the remnants of ancient shorelines that have been uplifted to their present height above sea level, and many of these exhibit landforms of wave erosion that occurred millions of years ago. The fronts, or faces, of these terraces are called **sea cliffs.** They offer proof of the erosive action of the waves. Sometimes heavy surf action grinds out a horizontal groove, called a **notch,** at the base of a sea cliff. If the notch is enlarged sufficiently so that the cliff front is no longer supported, it may fall, thus pushing back the face of the cliff. This is called the "retreat of sea cliffs."

Sometimes you will find a variety of landforms scattered along the front of an old, eroded sea cliff; **arches** (protruding masses of resistant rock that have had their undersides hollowed out by wave action), **stacks** (remnants of resistant rock after the bridge portion of an arch has collapsed), and **sea caves** (which result from wave action against easily eroded or fractured rock beds). The continued pounding of waves into the confined, hollowed-out space can break the rocks apart, thereby enlarging the sea cave. Eventually, the roofs of some of these caves are weakened by the wave action, and the overlying rock collapses into the cave, forming a canyonlike cleft. Some of the Hawaiian Islands have an interesting variation of a sea cave: a *lava tube* is the result of lava cooling

LANDFORMS OF WAVE EROSION 365

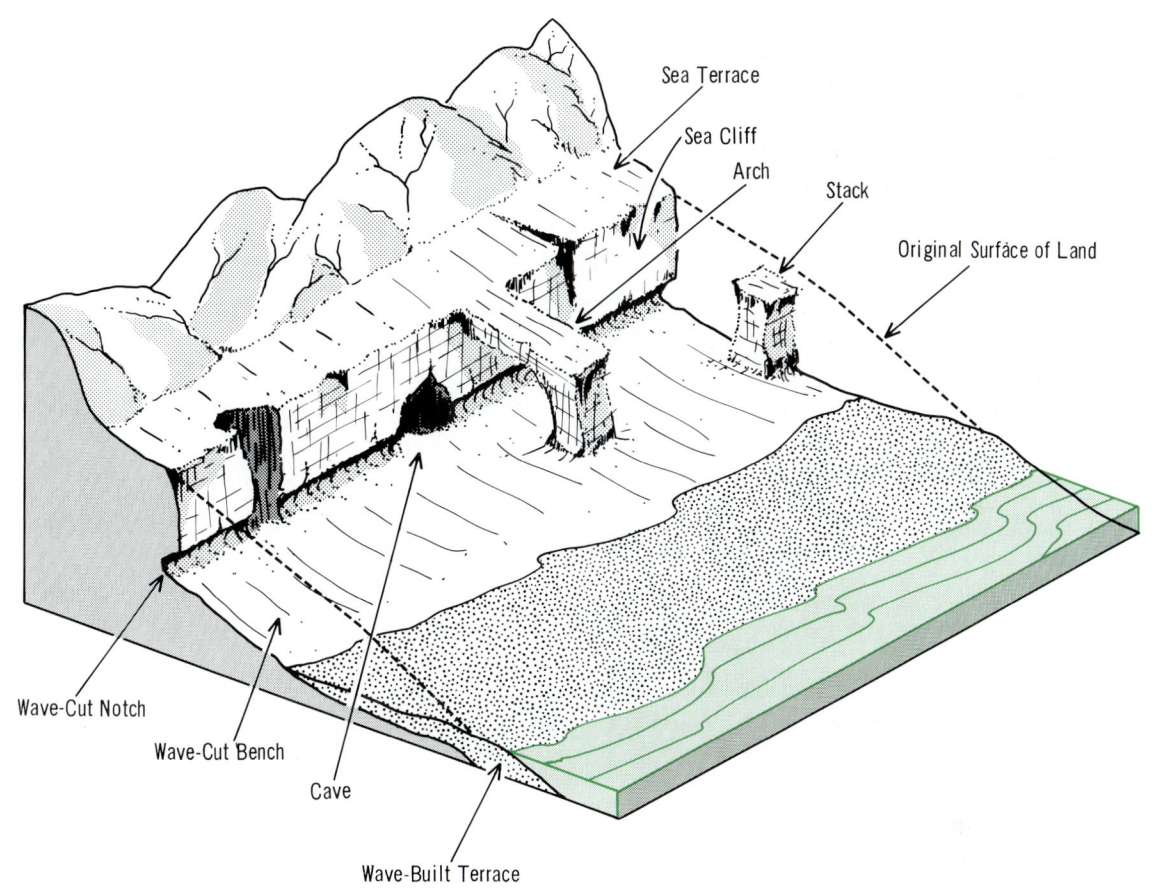

FIGURE 14-4 Landforms of wave erosion.

around a fast-flowing stream of lava, with the stream thereafter exiting from the surrounding hardened lava. When waves surge forward and enter the open end of the tube, they eventually break through somewhere along the line and spout into the air like a geyser. This gives them their name, *spouting horns*.

The beach fronting the sea cliff is often termed a **wave-cut bench**, or **abrasion platform**. Here, the layers of solid rock have been leveled and eroded by forward and backward action of the surf, leaving behind a relatively flat, benchlike surface that is usually covered with rock debris. It may be submerged during high-tide periods and fully exposed at low tide. Material taken from the abrasion platform is mixed with sands brought in from the shelf beyond to create a **wave-built terrace**.

Erosive power of ocean waves Shown in this photo of a portion of the coastline of France are a wavecut bench, sea cliffs, an arch, and an isolated stack.

LANDFORMS OF WAVE DEPOSITION

Figure 14-5 shows a typical beach scene, with landforms resulting from the deposition of sediments by wave action. Most of these forms are, of course, composed of sand.

Beaches can be produced by erosional or depositional processes, depending on any of several factors: the power of the surf, the structure of the underlying rock, the angle the shoreline presents to the incoming waves, and the level of the sea. The surface covering of any beach usually comes from somewhere else, brought down from inland areas by streams and rivers, and its composition would depend on the rocks from which the material was derived. Southern California beaches are light grayish, derived from weathered granite; the island of Hawaii has white, coral-derived sand on the east, black sand from weathered lava on the southeast, and green sand from weathered olivine on the northwest. Beaches along England's south coast are often referred to as *"shingle" beaches* because they are built of pebbles and cobbles that are flat, thus giving the beach the appearance of a shingled roof. The beach south of Ensenada, Baja California, is littered with large cobbles and small boulders that are laid out in parallel rows over an eroded granite base.

LANDFORMS OF WAVE DEPOSITION 367

Barriers include a number of landforms that always seem to have the shape of a bar; that is, their length is greater than their width. **Sandbar** is the most common term for a mound of sand located along a shoreline, but physical oceanographers have assigned these formations more representative names. An **offshore bar** is created by incoming waves that remove sand grains from the beach during high wave conditions (stormy seasons), drag them in the backwash to a position away from the shore, and deposit them in a pile that is parallel to the beachfront. Because the longshore current often works to extend these bars for a number of miles, they are sometimes called *longshore bars*. Some interesting examples of offshore bars that have become islands and serve as protective barriers to the coastline can be seen along the eastern coast of the United States and on the Texas coast of the Gulf of Mexico. Fire Island stretches for 27 miles (43 kilometers) off the coast of Long Island and has been described as a place that continually experiences "the ceaseless sound of booming surf, whistling wind-driven sand, and the crying of gulls."[5] It resulted from the interaction of the tides, ocean currents, and wind waves on the materials of the continental shelf. The island is perpetually altering its face and is, in fact, being driven in a westerly direction

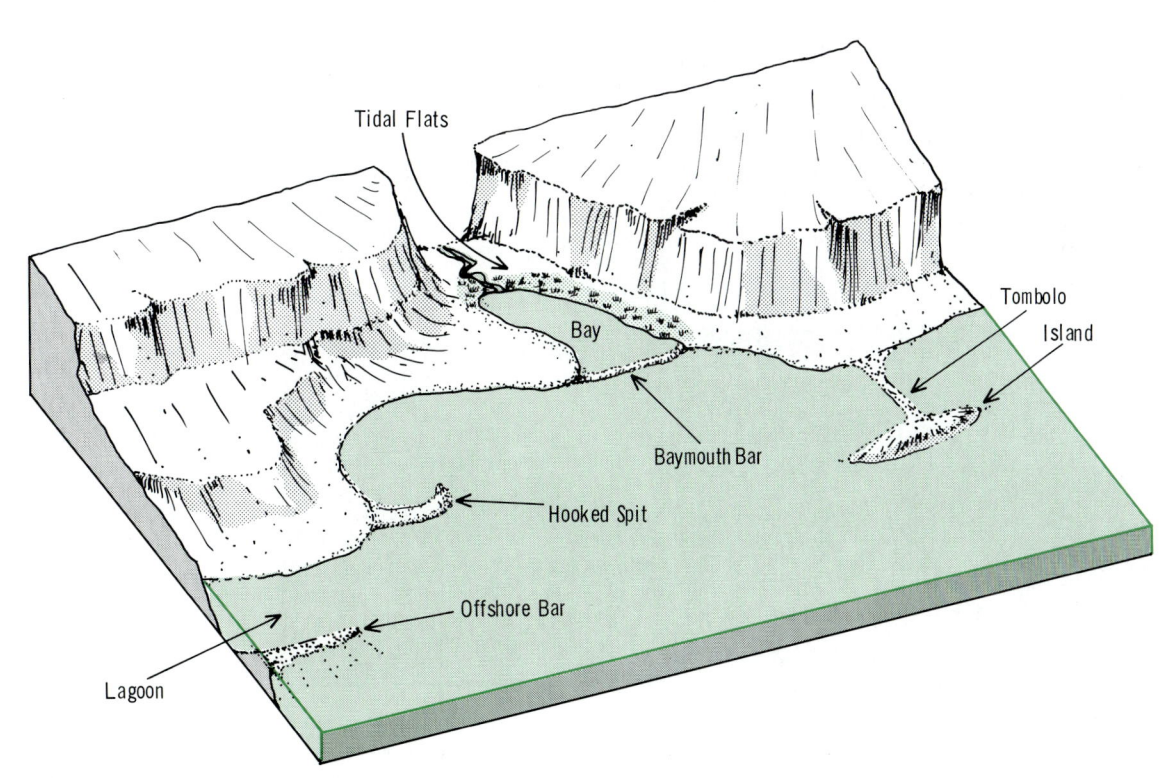

FIGURE 14-5 Landforms of wave deposition.

(a)

(b)

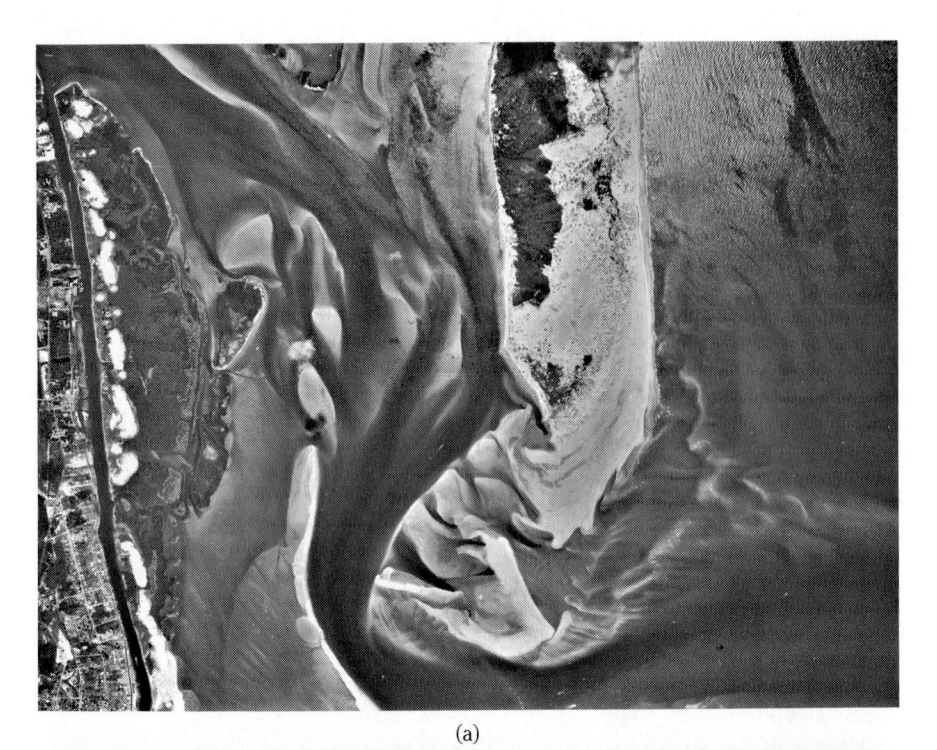

Barrier islands and sandbars (*a*) Offshore bars and long spits act as barriers to the beaches along North Carolina's coastline. This photo (taken at the end of summer) shows the numerous sandbars created by relatively gentle summer waves within the lagoon. (*b*) Hotel Row in Miami Beach, Florida, occupies a large offshore bar that might well be called a barrier island.

Spits The shoreline of Martha's Vineyard, Massachusetts, displays a variety of spits, including a hooked spit that has almost completely encircled its lagoon.

as its windward side is eroded and its leeward side is being added to by the elements.*

Padre Island in Texas, also a barrier composed of sand, is about 80 miles (128 kilometers) long. It is just one of several barrier islands that follow the coastline of Texas from Galveston around to Mexico. The barrier island that is the home of the city of Galveston is more properly termed a **baymouth bar** because it was created by longshore currents at the mouth of Galveston Bay. This bar comes very close to joining the southwesterly jutting Bolivar Peninsula. The body of water separating Galveston and its island from the mainland is a typical **lagoon**. In order to provide access by means other than boats, large causeways (elevated highways) were constructed across the waters.

Spits are long, narrow sandbars created by the depositional activities of a longshore current flowing past a cape or promontory that protrudes from the mainland. As the current curves around the end of the landmass, its velocity is slowed and deposition begins. One end is affixed to the land, the other points downcurrent. From above, the spit does look very much like the cooking tool used to hold meat over a fire. If the longshore current sweeps around a promontory or headland into a bay, the deposition activities will follow, forming a curving sandbar that is called a *hooked spit*.

---

*To learn more about the world's beaches, barrier islands, and other features bordering the oceans, especially such interesting places as Fire Island, New York, and Padre Island, Texas, see Robert and Seon Manley (1968), *Beaches: Their Lives, Legends, and Love* (Philadelphia: Chilton).

Tombolo is the name given to a ridge of sand that connects an offshore island to the mainland. When advancing wind waves and the longshore current are confronted by such a landmass, the sediment-laden waters must curve around the island, thus having their velocity slowed until the sediments can no longer be supported. Deposition begins along the beachfront directly behind the island and, in time, will reach out and become fastened to the island. The most famous tombolo is the one connecting the Rock of Gibraltar with the mainland of Spain.

SHORELINES

Shorelines differ greatly from place to place around the landmasses of the earth. There have been many attempts to classify them; they have usually resulted in more confusion than enlightenment.

Some shorelines have been exposed to the various erosional activities for vast periods of time. Mountains along a coast may have been worn down to low hills while streams and rivers were scouring out wide valleys and causing the formation of large alluvial fans. If this coastline were submerged by a sinking landmass or the rising elevation of the sea, the new coastline would be deeply indented with many estuaries and bays. It would be an irregular shoreline.

On the other hand, on a shoreline where deposition had been occurring all along the continental shelf, thus creating a fairly smooth and gently sloping plain, if the shoreline were upraised or the level of the sea were lowered, the result might be quite different. There would be hills and valleys, along with wave-cut benches and sea terraces, but they would be in the background. The forepart of the shoreline would look much like the piedmont alluvial plains seen along the base of great mountain ranges.

Complications occur, and it is sometimes difficult to distinguish a **shoreline of emergence** from one of submergence. For example, the shoreline along Monterey, California, is extremely rugged, deeply indented by clefts and bays, and the beaches are not all long and wide. It is a shoreline that has emerged, but it has all the characteristics of a **shoreline of submergence.** Why? Because this shoreline is situated on the eastern edge of the shifting Pacific plate, and the tremendous diastrophic forces operating there have crumpled the land as the coastal ranges were uplifted. Landslides crashed down from the newly emerging mountains and poured into the oceans, creating powerful turbidity currents that swept sand and other sediments before them as they gouged out submarine canyons and roughened the surface of the continental shelf.

Glaciated shorelines are not as confusing, because they are the result of processes that are still occurring. Along the coasts of Alaska and Norway the shoreline is extremely rugged, with an almost endless number of bays and deep valleys. A moving glacier does not have to conform to a baselevel; it can gouge as deeply as its weight and pressure will allow. A glacier moves down from the higher elevations, carving a valley as it goes. When the glacier confronts the ocean, it moves on out into the water throughout its span of existence, the front edge, or terminus, breaks off in pieces (bergs). When the glacier begins to melt faster than it can be resupplied from its source, it begins to recede back up through its valley, and the ocean follows. The glacial trough is said to be "drowned," and the result is an intrusion of the ocean into what looks like a common bay. This is called a *fjord* (fiord).

Coral shorelines are the result of the activities of living organisms, such as coral, algae, and shellfish. These organisms thrive in relatively warm waters of 68°F (20° C) or higher and make their home on the upper level of an

Glaciated coastline of Denmark Note the bedrock, which has been laid bare by advancing glaciers, and the fjord, now used as a harbor.

offshore bar, volcanic island, or sea mount. Coral growth invites other organisms to live in the growing reef, and as organisms die, leaving their shells to build up the formation, other organisms take over and the cycle continues.

Charles Darwin evolved a theory to explain why we find coral **fringing reefs** (coral growth and debris often found surrounding a submerged volcanic island) scattered throughout the South Pacific Ocean. Knowing that coral cannot survive in cold, deep waters, he surmised that the reefs must have had their beginning when the island was at or near sea level. When the islands subsided, the coral continued to grow and the reef maintained its position relative to the level of the sea.

There was considerable controversy over Darwin's theory, some geologists and oceanographers insisting that the coral grew when the sea level was lowered during the great ice ages. Time and research has proved most of Darwin's coral-reef theories to be correct.*

Barrier reefs are unlike fringing reefs in that they stretch for great distances along a coastline. The Great Barrier Reef off Australia's east coast extends for some 1200 miles (1920 kilometers) and is the most famous of all coral barrier reefs. Circular fringing reefs are

---

*To learn more about the "coral-reef controversy," see Robert C. Cowan (1969), *Frontiers of the Sea*, rev. ed. (Garden City, N.Y.: Doubleday), pp. 112–119; and Francis P. Shepard (1973), *The Earth Beneath the Sea* (New York: Atheneum), pp. 185–207.

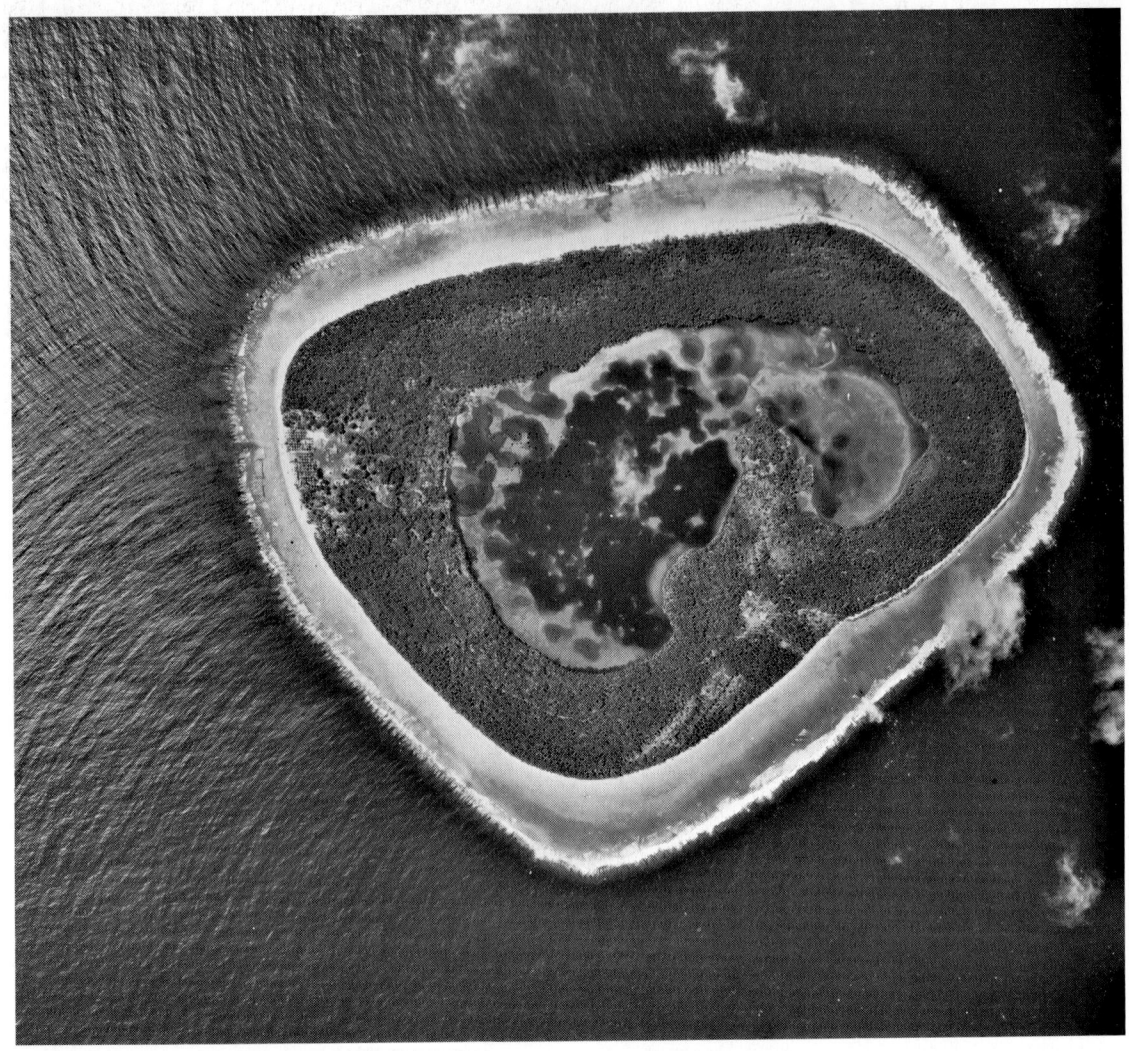

Swain's Island off the eastern coast of Australia: coral atoll and fringing reef.

called **atolls** if the mountain on which they are built is below sea level. The South Pacific is dotted with these atolls, which have a special kind of beauty: deep, clear blue waters in a calm lagoon, surrounded by weathered coral deposits where coconuts have taken root, providing an existence for humans who prefer life on an idyllic tropical isle.

Volcanic shorelines occur whenever lava flows have reached the surf zone. Here great black cliffs and terraces tower over the breaking white surf. In some places the beaches are of black sand, derived from the action of the surf on the lava; in other areas the shoreline has the appearance of a gigantic mass of coiled black rope.

Black sand beach, Kalapana, Hawaii.

REVIEW AND DISCUSSION

1. Define the term *continental shelf*. Where are the widest shelves to be found? How did they originate?
2. What is the relationship between the shelf edge and the continental slope?
3. Why are continental shelves important to people today?
4. What is the effect of turbidity currents in sculpturing the surface of the continental shelves and slopes?
5. What is the relationship between wave refraction and the creation of longshore currents?
6. What is a riptide?

7. Explain why tidal currents can be an effective erosional force.
8. Explain the difference between the terms *hydraulic action* and *corrasion* with reference to wave erosion.
9. Discuss the landforms of wave erosion, specifically sea cliffs, arches, stacks, and wave-cut benches.
10. What is a spouting horn?
11. How are beaches formed?
12. Describe the different barriers created by wave deposition. What is the difference between a baymouth bar and a spit? What is a tombolo?
13. Why is it difficult to classify shorelines of emergence and submergence?
14. Describe the formation of a glaciated shoreline.
15. What is the difference between a fringing reef and a barrier reef?
16. Describe a volcanic shoreline.

NOTES

[1] Robert C. Cowan (1969), *Frontiers of the Sea*, rev. ed. (Garden City, N.Y.: Doubleday), pp. 19–20.
[2] Leonard Engel (1961), *The Sea* (New York: Time-Life Books), p. 94.
[3] Leon Bertin (1972), *The New Larousse Encyclopedia of the Earth* (New York: Crown), p. 110.
[4] Cowan, p. 134.
[5] Robert and Seon Manley (1968), *Beaches: Their Lives, Legends, and Love* (Philadelphia: Chilton), p. 324.

ADDITIONAL READING

Flanagan, Dennis, ed. (1969). *The Ocean*. San Francisco: W. H. Freeman.

Bascom, W. (1980). *Waves and Beaches*. New York: Doubleday.

Carson, Rachel (1961). *The Sea Around Us*. New York: Oxford University Press.

Cromie, W. J. (1962). *Exploring the Secrets of the Sea*. Englewood Cliffs, N.J.: Prentice-Hall.

Darwin, Charles (1959). *The Voyage of the Beagle*. Abridged and edited by Millicent E. Selsam. New York: Harper & Brothers.

King, C. (1972). *Beaches and Coasts*. 2nd ed. New York: St. Martin's Press.

Maury, Matthew Fontaine (1963). *The Physical Geography of the Sea and Its Meteorology*. Edited by John Leighly. Cambridge, Mass.: Harvard University Press.

Pickard, G. L. (1979). *Descriptive Physical Oceanography*. 3rd ed. New York: Pergamon Press.

Vetter, Richard C., ed. (1973). *Oceanography: The Last Frontier*. New York: Basic Books.

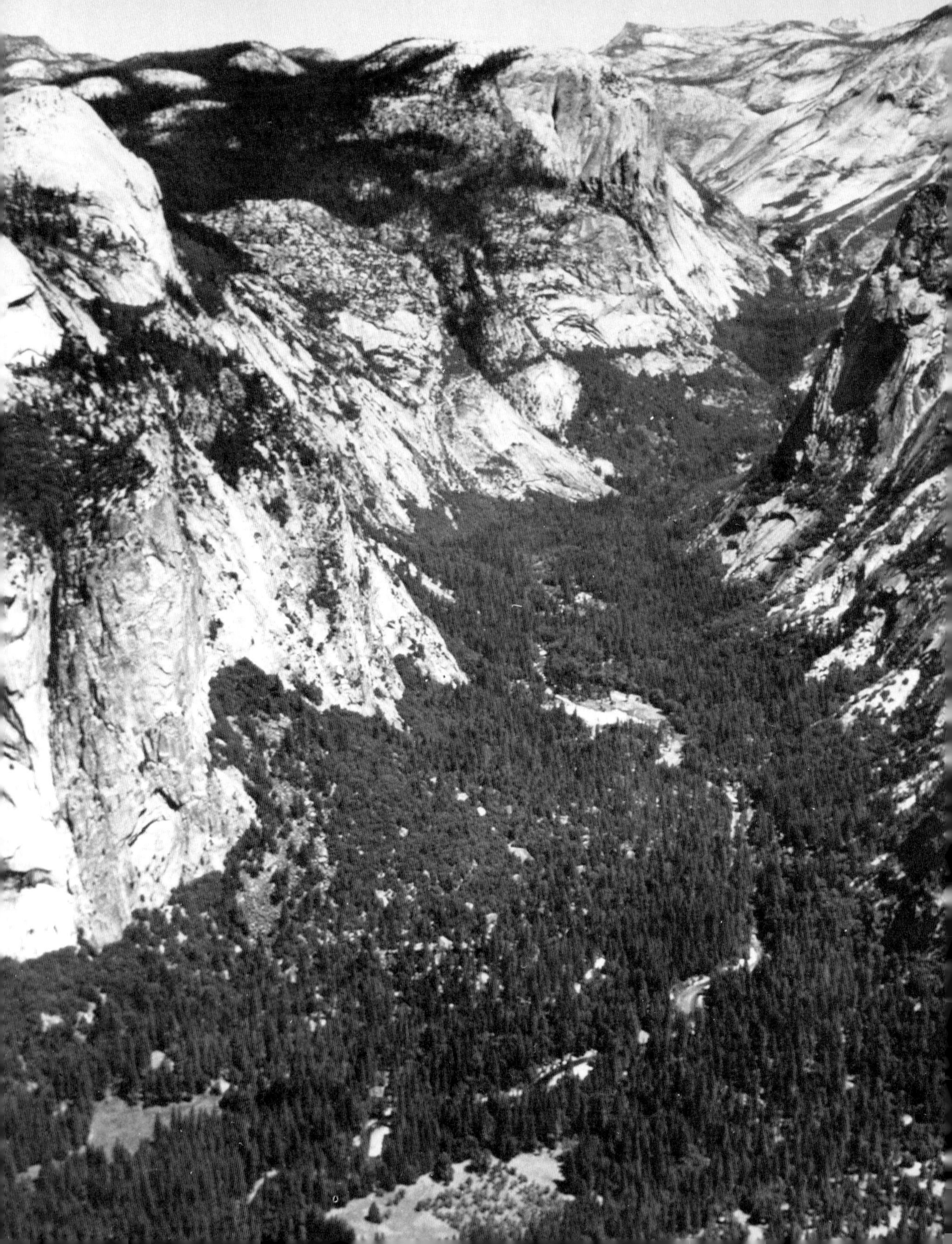

FIFTEEN
ENGRAVED BY GLACIERS

Glaciers mean different things to different people. To some, glaciers represent water: about 2 percent of the earth's total water supply and close to 10 percent of the fresh water. To others, glaciers represent an ever-present threat—for example, the citizens of isolated villages in Norway or Alaska, where a gleaming white wall of ice poises like the blade of a bulldozer before their homes. Yet, to still others, the results of glacial activity—such as in the magnificent valleys of Yosemite and Glacier national parks in the United States and Banff and Lake Louise in Canada—are sufficient evidence that some portions of the earth must endure the tortures of an icy hell before they can reign in all their majestic glory.

Curious individuals have long pondered the sometimes delicate, sometimes bold and sharply etched peaks, ridges, and valleys of the Alps, the Hindu Kush, or the Caucasus. Whose hands fashioned those engravings? And what of the huge boulders that litter the landscape from the foothills of the Alps northward to the Baltic Sea? Had playful or angry giants tossed them there? Had a great flood transported them from the north? Were they poured out of the earth's interior? And what of the towering granite cliffs that shone like polished metal under the noonday sun? Who or what made the deep scratches on high valley walls, scratches that were parallel to the valley floor?

The centuries rolled by, and although many were satisfied with the answers that evolved, others could not contain their sense of frustration. Then, in 1836, a young Swiss zoologist, Louis Agassiz, arrived on the scene.

Agassiz might well be called the father of modern glaciology; through his scientific reasoning, the answers to the preceding questions finally began to take form. He had studied the glacial activity of the high valleys of the Alps, and when he journeyed through the rolling landscape of the North European lowland, he noted that the huge boulders, along with the other earth materials and scattered rock debris, were remarkably similar to those found in the high alpine valleys. He evolved a theory that there once had been a great glacier covering the land, one that would make the glaciers of the Alps seem tiny by comparison. He theorized that the northern regions of Europe and Asia had suffered through a number of ice ages for a greater part of earth's history, and each time the ice sheets moved out, they scoured the land and transported rocks of all sizes, from sand grains to gigantic boulders.*

Agassiz's views generated considerable discussion and opposition, but his theories have prevailed. Glaciology (from the Latin *glacies*, "ice") moved from a purely hypothetical level to one of scientific analysis based on evidence.

Aside from gaining knowledge, have we benefited materially from a serious study of glaciers? The answer is an unqualified yes. For the proof, let us proceed with an investigation into glaciers, past and present.

THE FORMATION OF GLACIERS

When moisture is present, along with extremely low air temperatures, a number of different ice forms, such as frost, hail, sleet, permafrost, snow, and glaciers, are created.

Frost can occur almost anywhere in the world, whenever moisture-laden air close to the ground is chilled below the dew point. Hail and sleet are frozen water droplets that may have started to precipitate as raindrops, but they, like frost, are short-lived.

The permafrost regions bordering the Arctic Ocean are lands of underground ice, where the ground below the surface is permanently frozen. Although these regions offer a unique and sometimes exciting landscape, they are not glacier-producing areas today. They may receive up to 25 inches (635 millimeters) of precipitation annually, but most of that arrives during the short summer, when the top layer of the ground thaws. Glaciers are born from the accumulation of snow, but not all snowfields produce glaciers, because they may lack one or more of the formation controls.

---

*To learn more of Louis Agassiz, see Carrol L. Fenton and Mildred A. Fenton (1952), "Agassiz of the Ice Age," *Giants*

of Geology (Garden City, N.Y.: Doubleday), Chap. 10, pp. 111–123; Charles Ogburn, Jr. (1968), *The Forging of Our Continent* (New York: American Heritage), pp. 116–117; Ruth Moore (1971), "Agassiz: Ice!" *The Earth We Live On* (New York: Knopf), Chap. 7, pp. 132–156.

Formation Controls

The formation and growth of glaciers is determined by the operation of the hydrologic cycle. Glaciers can form only where the accumulation of snow remains throughout the year, which happens when the annual snowfall exceeds any losses resulting from evaporation or melting. Although the amount and type of precipitation are important, so are the temperatures of the air masses and the land over which they move. Some mountains in the Rockies may receive as much as 40 feet (12 meters) of snowfall during the winter months, but when summer finally comes to an end, only small patches of snow remain. Antarctica receives only about 6 inches (152 millimeters) of snowfall annually, yet scientists who participated in the International Geophysical Year (IGY) discovered that the ice sheet covering the land and extending out into the seas was actually growing.[1] Why do these regions differ so greatly? The answer lies in the differences in temperatures and the amount of precipitation.

Temperature, a very important snowfall control, is determined by altitude and latitude. Snowflakes or snow crystals will precipitate wherever the air temperature is cold enough for the water molecules to freeze as they condense. The snowflakes may melt if they fall through warmer air, or they may remain in a solid state until they strike the ground. There, if the temperatures are warm, they might revert to liquid form. Whether the fallen snow remains to become a part of the snowpack depends on the altitude of the snow line.

The term **snow line**, which varies with the latitude, means the lowest elevation at which a permanent snowfield or pack can perpetuate. The snow line is sometimes near the *tree line* which is the upper limit of tree growth. On Tanzania's Mount Kilimanjaro the elevation of the snow line is about 15,000 feet (4500 meters); in the Alps, 8500 feet (2550 meters); and along the northern borders of Alaska, close to sea level. This is not to imply that the snow line never varies in altitude from place to place along a specific parallel of latitude, because the snow line is influenced by a number of factors: the presence of dry or moist air masses; the prevailing wind systems, which may be cool or warm; the amount of insolation received; and the variations in daily, monthly, or annual temperatures. A storm moving out of the polar front may drop great quantities of snow on the Rockies, the upper Great Plains, and the lowlands of Iowa and Illinois. How long does the snow cover last in each of these areas? Obviously, a single storm, even a severe one, will not alter the level of the snow line permanently.

Snowfall is dependent on the air temperature, but it is also determined by the amount of moisture present when precipitation occurs. If a cold, dry air mass rises over a mountain range or a warm, dry air mass sweeps up from the subtropics toward the high latitudes, there will be no precipitation. However, if a moist air mass follows either of those paths, precipitation may result, and if the air temperatures are at or below freezing, sleet, hail, or snow will result.

Glacial Structure

Assuming proper freezing conditions at ground level and a plentiful supply of falling snow, the snowflakes or crystals will begin to accumulate in an ever-thickening mass. The greatest depths will occur in valleys, rather than on mountain ridges, simply because a heavy pack of snow would slide off a ridge. So, in high valleys the snowpack deepens from the head (the highest elevation) down to the snow line (Fig. 15-1). This is called the **zone of accumulation.** As the snow accumulates, the pressure on the bottommost layers increases. The crystals begin to break up, the snowflakes

Zone of accumulation In this view of Boston Glacier in Cascades National Park, northeast of Seattle, Washington, note the heavy compaction of snow (firn) in the zone of accumulation.

begin to melt under the pressure, and then they freeze. As this is repeated again and again, layers of ice crystals form; the impacted mass is called a **firn.**

When enough snow accumulates, gravity takes over and the newly formed glacier begins to move. When it descends below the snow line, the front edge, or foot, of the glacier begins to deteriorate as it loses moisture. If the underside of the icy mass moves across warmer ground, it begins to melt and thus releases streams of **meltwater.** If the upper edges encounter temperatures above freezing, surface thawing begins and moisture is lost through evaporation and melting. Sometimes the snow passes directly from a solid to a gaseous state without going through the process of liquifying. This is termed **sublimation.** This portion of the glacier is called the **zone of ablation** (from the Latin *ablatus*, "to carry away"), or the *zone of loss*. Further deterioration can occur if a glacier enters a body of water. Losses will result from wave action and the warmer temperatures of the water, and some sections of the glacier (for which the water does not provide enough support) will break off, to float away as icebergs or ice floes.

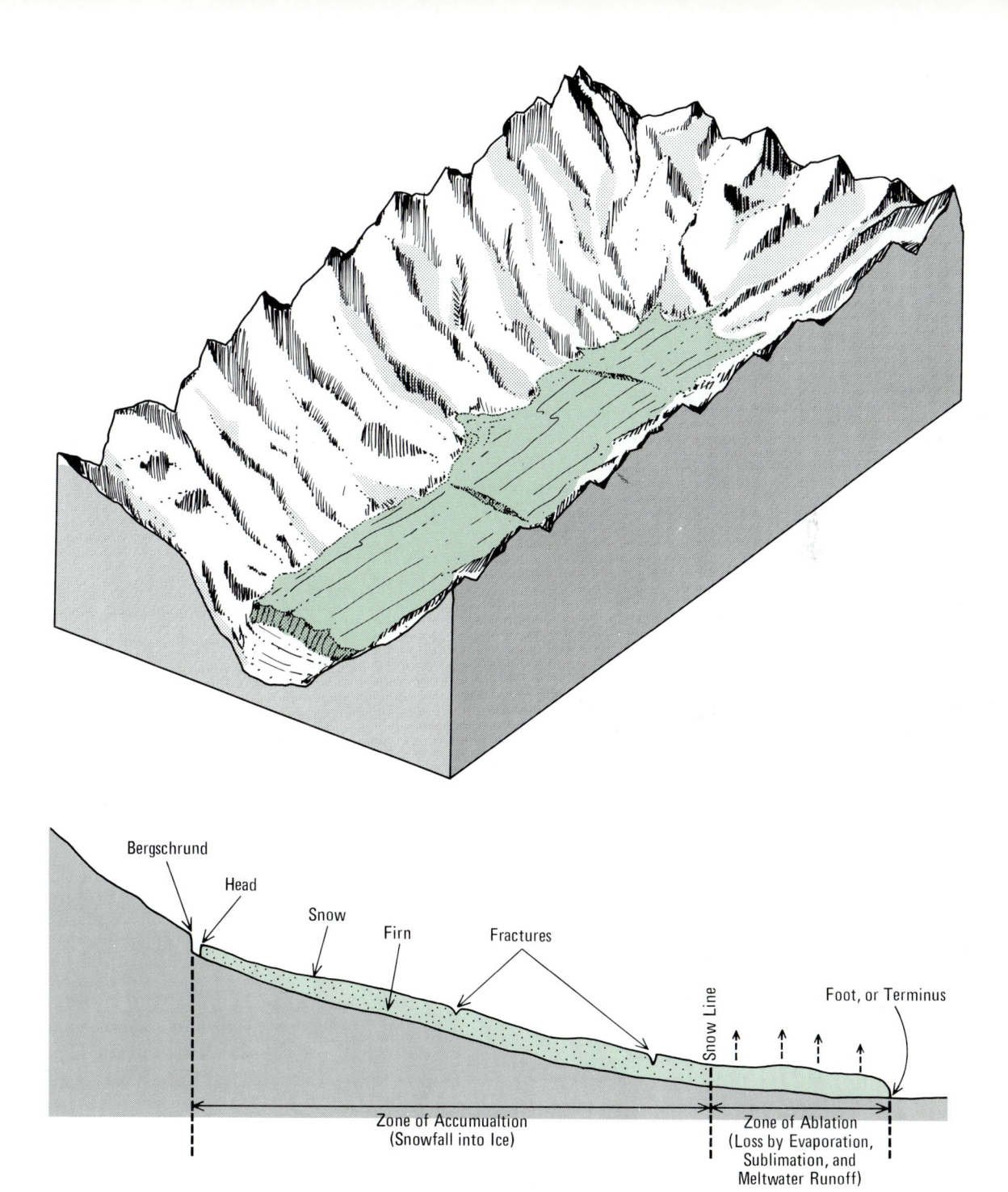

FIGURE 15-1 Formation of an alpine (mountain or valley) glacier.

Glacial Movements

As the firn continues to increase as it accumulates more snow, a newly forming glacier in a mountain valley may begin to move. Whether it does or not depends on a number of factors, such as the weight of the impacted snow, the steepness of the slope, and the terrain over which it must pass. Observers of glacial landscapes tell us that a mountain glacier must build to a depth of almost 66 feet (20 meters) before there can be any notable movement, and polar glaciers might remain stationary even when they have a depth of over 200 feet (60 meters).[2] Glacier movement begins with the lower portion stretching out, but the force is felt throughout the entire mass. Ice at the head of the glacier gradually pulls loose from the rock wall, and a crevasse (called a **bergschrund**) opens. The glacier oozes down the slope much like a river, although considerably slower and with less internal turbulence.

The speed of movement will vary, of course, depending on the preceding factors and on the time of the year, air and ground temperatures, and whether the pathway is straight or has many curves. Some glaciers speed up during the winter because of the increased weight added by new snowfalls; some speed up during the summer because friction is reduced by heavy losses in the zone of ablation. Some glaciers advance rapidly, moving at speeds of about 100 feet (30 meters) per day, which amounts to about 7 miles (11 kilometers) per year. Others seem to be entirely motionless. It is difficult to detect the motions of any glacier, but time-lapse photography enables us to "see" the flowing movements.* In this method, with the camera aimed at a section of the glacier, a foot of film, for example, might be exposed at noon each day. The filming might cover a period of three months, and when the film is processed and shown, the mass of ice, which seemed to be stationary, becomes a flowing "river" of ice. You see it flow over a ridge of bedrock, fracturing into crevasses as it goes, sending huge blocks of ice tumbling onto the glacier's surface.

As for movements within the glacier, the rock walls and floor of the valley tend to offer great resistance; thus the upper layers and the center of the glacier move more rapidly than the bottom or sides.

CLASSIFICATION OF GLACIERS

Because glaciers come in many different sizes and shapes and can be found at varying altitudes and latitudes, an uncomplicated classification system was devised that set up two major categories. Glaciers that form in mountainous regions in high valleys are called **alpine glaciers** (also called *mountain* or *valley glaciers*), and glaciers that cover vast expanses of land are called **continental glaciers.**

Alpine Glaciers

Alpine glaciers are the smallest glacial forms and originate in mountainous regions at generally high elevations. Their sculpturing efforts produce landscapes that are visually exciting, as anyone can attest who has driven down the Million Dollar Highway from Grand Junction to Durango, Colorado, stood on a hilltop in Alma-Ata, Central Asia, and gazed in awe up at the glaciers dotting the steep face of the Tien Shan, or hiked through the valleys near Banff and Lake Louise in western Canada.

How does such fantastic scenery come into being? When you think of the weight of the snow and ice pack confined in the narrow

---

*Encyclopedia Britannica Films has produced a remarkable film that uses the time-lapse photography technique to show the movement of a glacier down a valley. It is entitled *Evidence for the Ice Age*.

Zone of Ablation Streams of meltwater issue forth from the terminus of an Alpine glacier in the Wrangel Mountains of Alaska.

channel of a V-shaped, youthful stream valley, it becomes apparent that the pressures generated must be exceedingly powerful, especially against the rock walls of the valley sides. This pressure forces rocks that may already be jointed and fractured to break up into even smaller pieces. Moisture from the glacier seeps into the cracks and crevices, and if freezing occurs, the rocks will be plucked from the valley walls and floor as the glacier moves on. This process, called **plucking,** continues throughout the entire length of the glacier as long as temperatures are low enough to permit freezing. These jagged rocks, frozen to the glacier and sticking out like the spikes on golf shoes or the grains of sand on sandpaper, perform well as erosional agents as they scour and scrape the valley. This act of "sandpapering" is called **abrasion.** By these two processes great quantities of rock materials are forcibly removed from the glacial valley, eventually to be deposited in some distant place.

Erosional Landforms of Alpine Glaciation

Although a glacier may come into existence and then fade away because of the change in the region's climate, we can see the evidence of its former presence in the landforms left behind. Alpine glaciers perform both erosional and depositional activities, but Fig. 15-2 shows

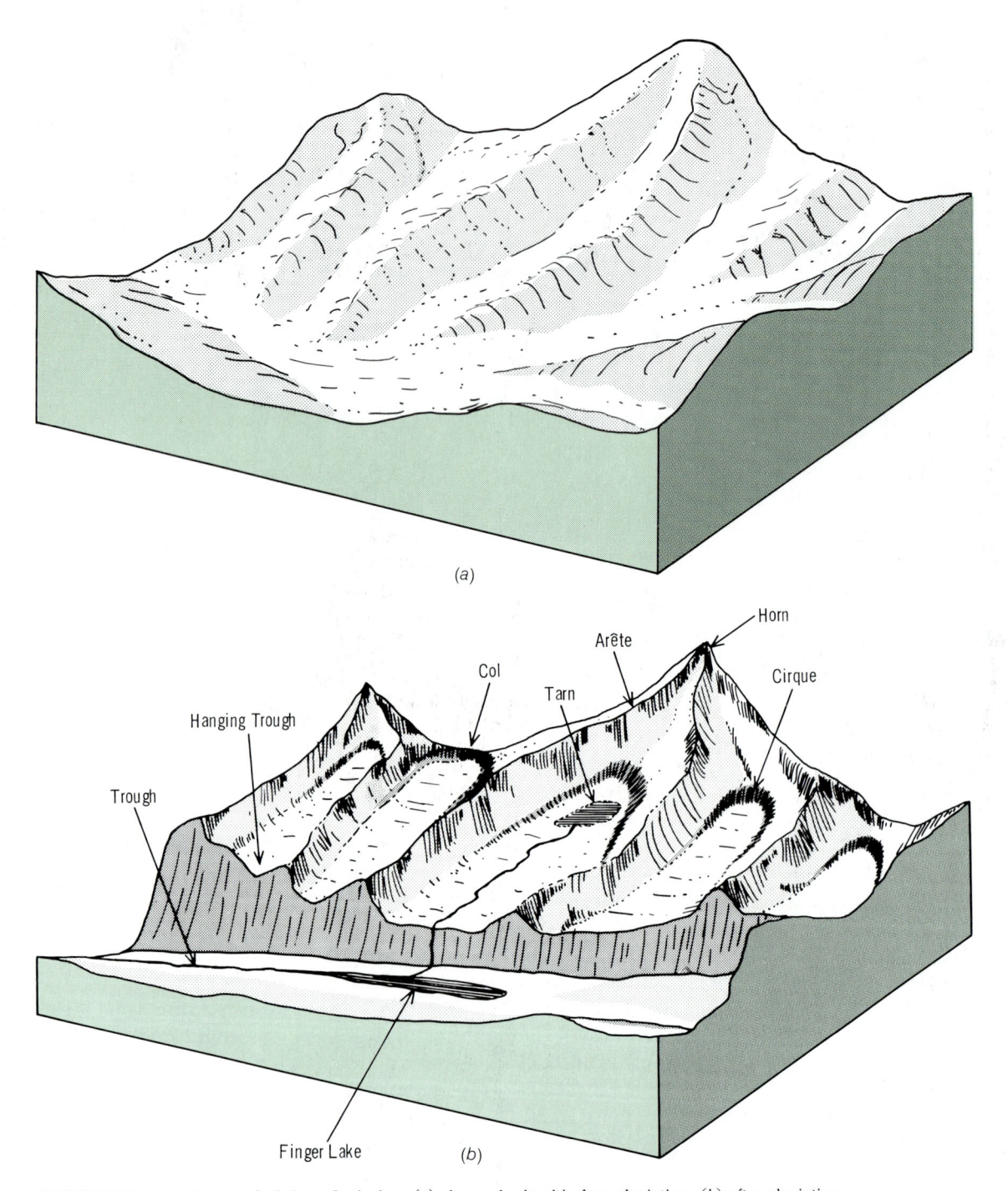

FIGURE 15-2 Landforms of alpine glaciation (*a*) shows the land before glaciation. (*b*) after glaciation.

that these high mountain glaciers are to be noted more for their erosional activities.

Erosional landforms have their beginnings when the firn develops within a valley at normally high elevations. The valley usually is a youthful stream valley that is narrow and V-shaped. When the glacier begins to move, plucking and abrading as it goes, the valley becomes rounded into a U-shaped profile. The head, or top, of the valley assumes an almost circular shape (much like the end of a bathtub), which earned it the name of **cirque** (from the French word meaning "circle"). When a glacier melts away, the only objects in the cirque may be a few scattered boulders deposited by the glacier and a shallow pool or lake (called a **tarn**).

As the glacial valley widens into a rounded **trough,** the erosional activities encroach on the valley sides, cutting away at the mountain spurs, or ridges, on either side. If two glaciers are operating in adjacent valleys, the ridge between will become sharply etched, almost knifelike. These are called **arêtes** (from the French word meaning "fishbone" or "sharp ridge"). If an arête is situated between two actively scouring and plucking glaciers, the two cirques may become joined at the top of the ridge, and a section of the arête will be cut away, leaving a **col** (from the Latin *collum*, "neck"), or mountain pass, through the ridge. When three or more glaciers are at work on a mountain peak, carving their cirques upward and creating arêtes on all sides, the resulting sharply chiseled peak is called a **horn** because it bears a striking resemblance to an animal's horn. In Switzerland the tallest horn in an area is called the *matterhorn* ("mother horn").

The rounded, U-shaped valley created by the passage of a glacier is called a **glacial**

The Tien Shan Mountains Glaciers and snowpack are the perpetual cover for these mountains which serve as a barrier between the Soviet Union and China. The Khan-Tengri peak has an elevation of 23,085 feet (6995 meters).

trough. If such a trough is formed in a valley leading to the shoreline and its base is below the level of the sea, the waters of the ocean will enter the trough as the face of the glacier melts and recedes. The resulting "drowned" glacial trough is called a **fjord** (*fiord*). The areas most noted for the number and picturesqueness of their fjords are the heavily indented coastlines of Alaska and Norway. Some of the Norwegian fjords reach inland as far as 110 miles (175

Switzerland's Matterhorn.

Sognefjord, Norway The longest fjord in the world reaches 110 miles (175 kilometers) into the heart of Norway.

kilometers).[3] Fjords predominate in these high-latitude regions (as well as along the coasts of Greenland, Scotland, and Chile, among others) because of the climate and the mountainous topography sculptured during the early ice ages.

If another glacier cuts across the foot of a glacial valley and erodes the land away, the tributary glacier will be left hanging high above the intruder's newly forming valley. The result is a **hanging trough**. A moving glacier often will scour out a long depression on the floor of its valley and, when the glacier has receded, the narrow, gouged-out cavity may fill with water issuing from the receding glacier; thus a **finger lake** is born.

What happens when a glacier encounters a ridge of resistant metamorphic or igneous rock? Does the glacier flow around the obstacle? Although a glacier flows somewhat like a river, the resemblance ends there. You may think of ice as being brittle and easily shattered, as not having the strength or resistance of a rock, but consider the power of a great mass of ice meeting an exposed volcanic dike. The glacier flows over it, abrading the *stoss* (from the German, meaning "thrusting") side as it climbs and plucking pieces off the lee side as it moves on (Fig. 15-3). Looking down on a glacial trough dotted with these rounded rock forms, you might think they resemble a herd of grazing sheep. This is why the Swiss named these erosional features **roches moutonnées,** which means "sheep rocks."

Glaciers sometimes move through areas of highly resistant bedrock, with the rocks plucked from places upstream scraping at the rock walls, making deep scratches or grooves called **striations.** If a mass of granite is polished smooth by the abrasive action of the ice,

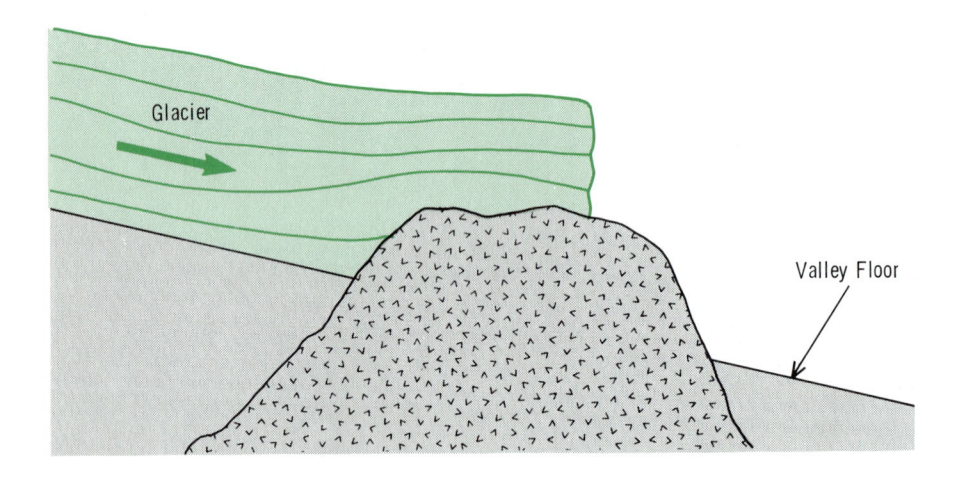

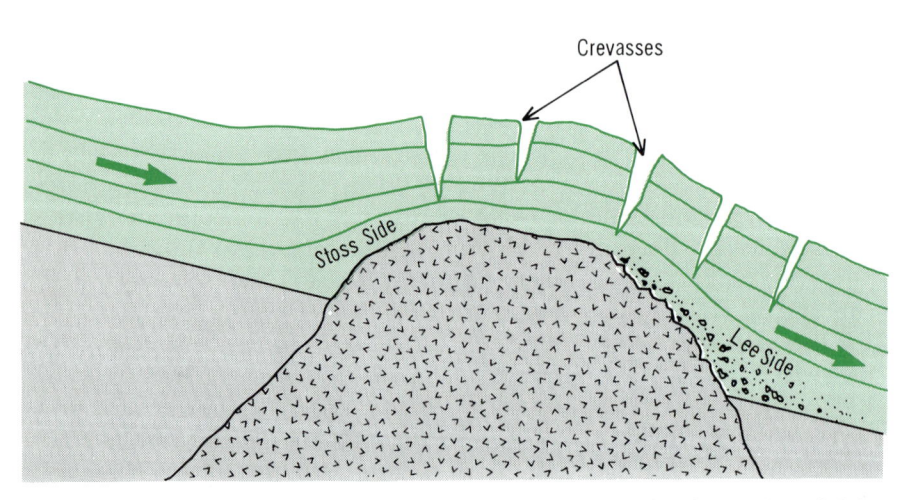

FIGURE 15-3 Glacial sculpturing A glacier moving across an extrusion of bedrock grinds it into a roche moutonnee ("sheep rock").

Glacial striations In this view, looking southeast towards the Hudson River and the George Washington Bridge, you can see the striations carved by an advancing continental glacier during an early ice age.

it will shine like a mirror; this is called **glacial polish.** These items may seem unimportant, but they do help us determine which processes created a particular landscape or landform.

Depositional Landforms of Alpine Glaciation Depositional landforms are created by advancing as well as receding or retreating glaciers. Glacial advance refers to the actual physical forward movement of a glacier as it surges down its slope. A receding glacier is not actually moving back up its slope. It only seems to do so. Glacial recession begins when a glacier no longer receives replenishment in its zone of accumulation to make up for the losses occurring throughout the zone of ablation. As pieces of the glacier break off or the ice melts, the **terminus** (front edge) of the glacier seems to retreat. If this process continues, the glacial trough eventually loses its glacier, and the only evidence of its existence is the erosional forms made by the moving glacier and the deposits left behind.

The area over which a glacier passes is strewn with earth materials, ranging in size from silt particles, sand grains, pebbles to large boulders (**erratics**). Such random, unsorted, and unstratified deposits are called **moraines** or *drift deposits*. The scattered debris found on the floor of a glacial trough forms a **ground moraine** or **till plain.***

---

*The term *till plain* is also used to refer to the ground moraine left by a receding continental glacier. The word

As a glacier moves down its valley, rock debris and fragments fall from the valley sides onto the edges of the glacier, resulting in morainal deposits called **lateral moraines.** When two glaciers extend beyond their valleys and merge in another valley, the joining of the lateral moraines produces a **medial moraine,** which shows as a dark stripe down the middle of the merged glacier.

Rock debris and soil materials carried by streams flowing over the surface of the glacier and deposited along the foot (or terminus), along with materials pushed by the glacier itself, form deposits called **terminal,** or *end,* **moraines.** If a receding glacier pauses long enough in any one location to reestablish itself (because of the increase in snow accumulation at its source), it forms a new terminal moraine before it begins melting away. One may discover a number of these **recessional moraines** in an empty glacial trough.

There are other depositional features associated with alpine glaciation, but because they are more commonly associated with the landforms of continental glaciation, we will consider them in that section.

Continental Glaciers

So far we have primarily considered the relatively small glaciers that perform their sculpturing and leveling activities in high mountain valleys. Now it is time to consider the vast ice sheets that we call continental glaciers. In ages past there were many ice sheets, but today we have only one that fits within this classification: the ice cover of the Antarctic continent. How do these great glaciers form? The answer to that question will be found by examining the conditions that created the continental glaciers of the geologic epoch known as the Pleistocene (from the Greek *pleistos,* "most," and *cene,* a revision of *kainos,* "new"; thus, "most new"), the most recent of the ice ages.

The Pleistocene Ice Age

The world's climate has changed at least four times during the past 1 million years, probably because incoming solar radiation was blocked off by an excessively heavy layer of clouds, allowing the temperature of the atmosphere to fall until the moisture precipitated over the polar regions fell as snow. (See Chapter 3, pages 67 to 74, and Chapter 6, pages 167 to 169). As snow continued to fall and the resulting snowpack became deep enough to form glaciers, great tongues (or lobes) of ice began to reach equatorward from the poles. Reaching out from the poles, time after time, they covered most of the northern portions of North America and Eurasia, advancing for thousands of years, then receding as the climate warmed and the ice melted, only to advance again as the climate turned colder.

Glaciologists and geologists have, however, been able to study the performance of the continental glaciers formed during the Pleistocene. Figure 15-4 shows the extent of this most recent ice age. Consider how dramatic the effects of this period of glaciation were on humans. With the level of the oceans dropping as moisture was trapped in the growing ice sheets, land bridges were exposed, enabling migrating people and animals to move from one landmass to another. Volcanic islands became mountains, standing above sloping plains that previously were part of the ocean floor. It is quite possible that the ancestors of today's Australian aborigines followed such an exposed pathway from Southeast Asia to the land "down under." It is also possible that when the glaciers finally melted and returned their wa-

---

till means "glacial deposits," which would apply to either alpine or continental glaciers, but the small glaciers do not produce the broad plains associated with the vast ice sheets of continental glaciation.

The convergence of glaciers Two glaciers in Greenland join and their lateral moraines merge into a quite visible medial moraine.

ters to the seas, the land bridge the aborigines followed became submerged, thus preventing return to their native land. The most accepted theory concerning the route taken by the ancestors of today's American Indian—the "first Americans"—is that a land bridge connected Alaska to Siberia during the Pleistocene.

All of this did not happen at one time. The Pleistocene is believed to have lasted about 1 million years, coming to an end about 10,000 years ago. During this span of time, the continental glaciers made four advances (stages) and retreats (interglacial periods). In North America (Fig. 15-5), the great glacier stretched from Alaska to Greenland and reached out to the south well into what is now the United States. The earliest advance (called the Nebraskan glacial period) reached to the present-day course of the Missouri River. After an interglacial period during which the ice melted

391

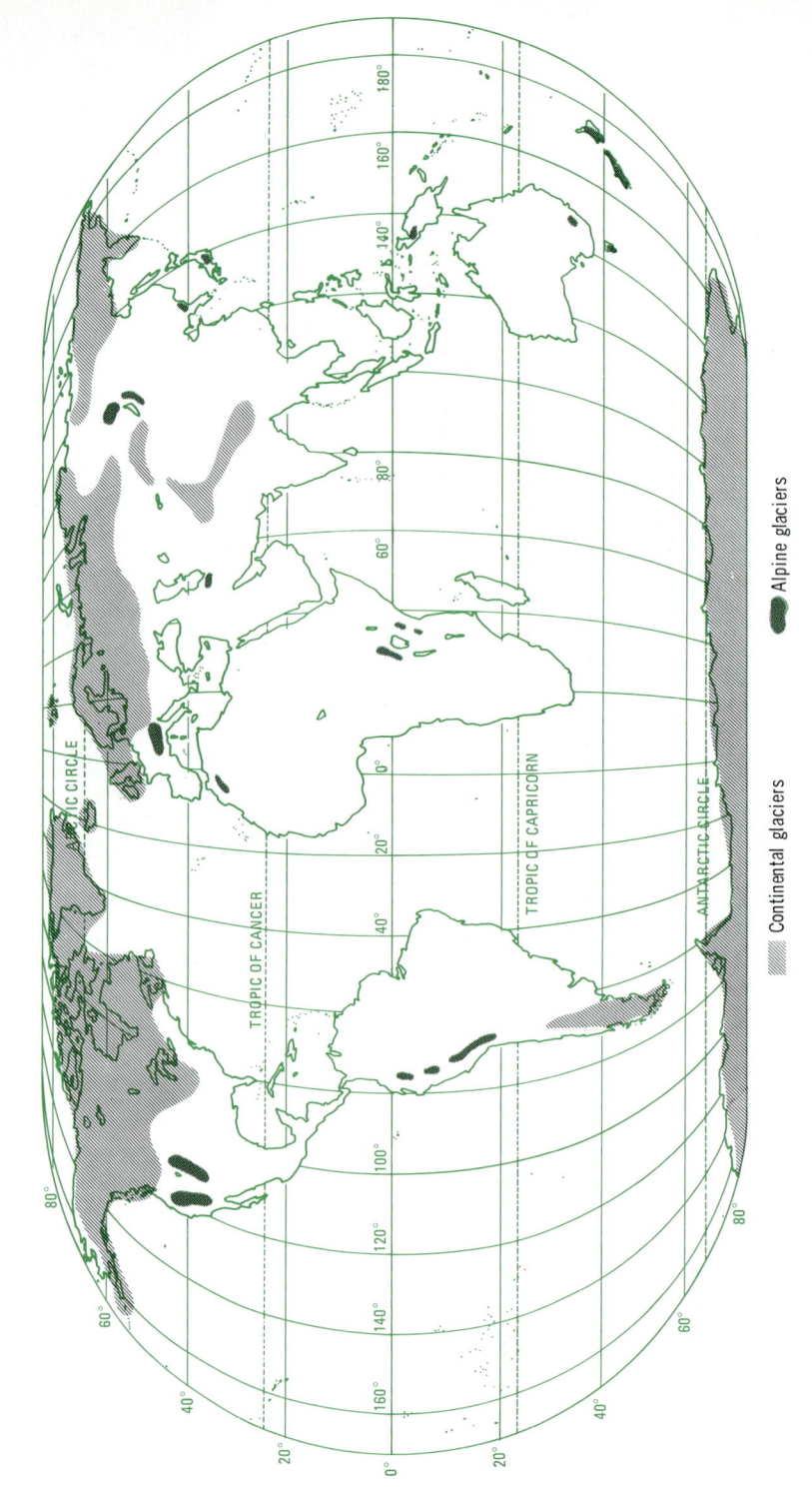

FIGURE 15-4 **Extent of Pleistocene glaciation.**

CLASSIFICATION OF GLACIERS 393

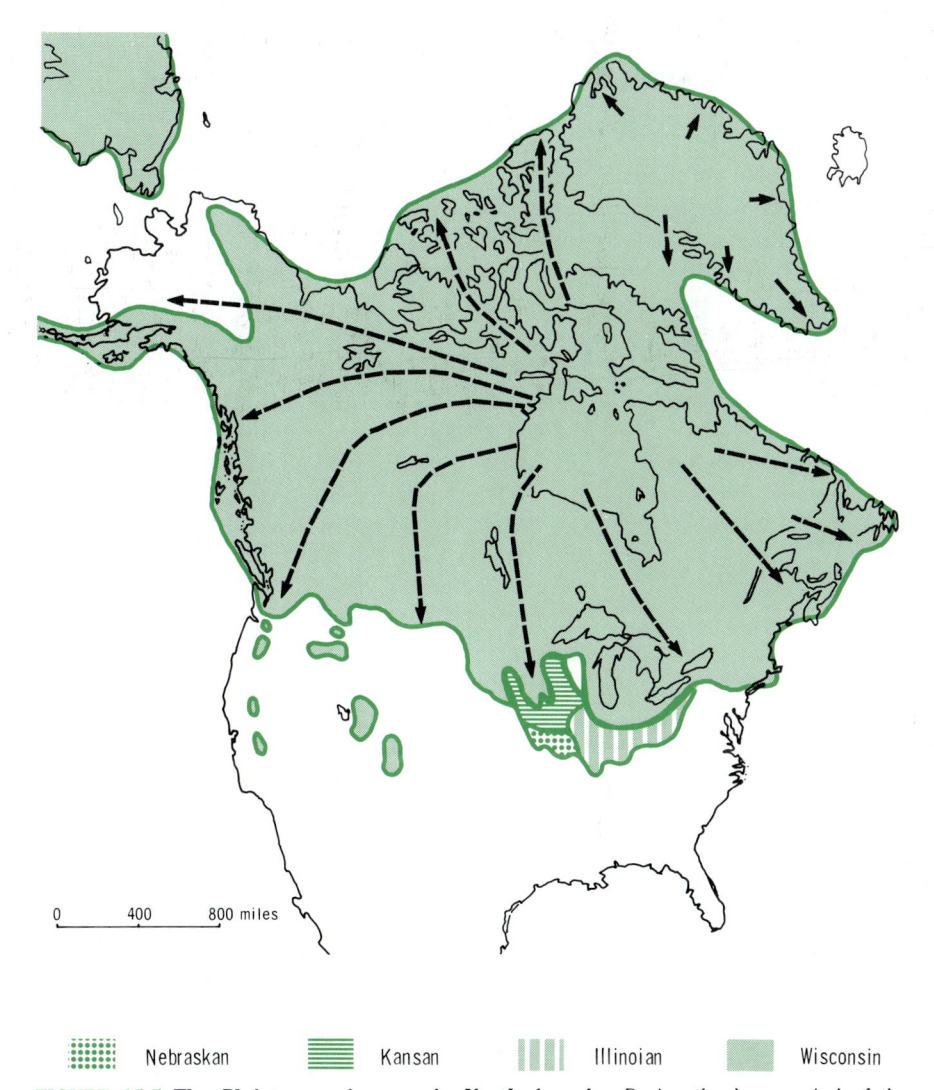

FIGURE 15-5 The Pleistocene ice age in North America During the long period of the Pleistocene, continental glaciers made four separate advances. The white area just west of Lake Michigan is called the "driftless area"; none of the advances ever crossed it.

Nebraskan   Kansan   Illinoian   Wisconsin

and the continental ice sheet was restricted to lands near the Arctic, a second advance (the Kansan glacial stage) entered Kansas, Missouri, and much of Illinois and Ohio. Another interglacial period elapsed; then the third advance (the Illinoian) moved south, making the farthest advance of all, reaching almost to the border of what is now Kentucky. The final stage is called the Wisconsin, and when it finally came to an end, the glacier receded, leaving only the evidence of its passing.

The continental glaciers of the Pleistocene

made a lasting imprint on the face of North America and Eurasia. They scraped away soils and laid bare the bedrock and shields of many areas. (Compare Fig. 15-5 with Fig. 11-12.) Streams that issued from the massive ice sheets deposited rich, silty alluvium over wide areas. Vast areas, previously uninhabitable (or at least not very hospitable) during the periods of glacial advance, were opened for human penetration. As the immense continental glacier receded, the climates of these northern lands improved, as did the resultant regrowth of vegetation. Then, as wild game moved northward, so did the human tribes who colonized the new territories that we now call Siberia, Scandinavia, Great Britain, and the New World.

Erosional Landforms of Continental Glaciation The erosion conducted by a moving alpine glacier is tremendous, but think of the erosion that was performed by a moving mass of ice not confined to a small valley, but spread over a vast land area and ranging from 5000 to 10,000 feet (1500 to 3000 meters) in thickness. Enough power, perhaps, to level hilly country and enough weight to depress the center of a continent? While it is sometimes difficult to detect the sculpturing effect of the great ice sheets, they did perform notably in both erosion and deposition.

Erosional landforms of continental glaciation can be found over wide areas, because a moving glacier is not confined to a narrow valley. The movement of such a vast ice sheet is slow, an outward flow from some central point in a radial pattern. This might be likened to what would happen if you were to pour a thick syrup onto a table top: the syrup would ooze slowly outward from the center. Operating from the pressure of the snowpack building up at its center and the pull of gravity, the continental glacier moves across the land, riding over and smoothing rock outcroppings, grinding down hills, scratching and gouging mountains, and filling in valleys with glacially transported materials. As alpine glaciers carve out depressions in valley floors and, in their passing, leave lakes behind, so do the continental glaciers. The main difference between the lakes produced is the size. North America's continental glaciers gouged out the land and created great valleys that are now occupied by the Great Lakes. The most easily identified erosional landform created by a continental glacier is the bedrock laid bare of its soil mantle, such as those found in many areas of Canada, especially along the coast of Labrador, and in parts of New England.

Depositional Landforms of Continental Glaciation Figure 15-6 illustrates depositional landforms of continental glaciation, which are more easily identifiable than the erosional features. Some would be invisible if the glacier were still present in the area, but once the ice has melted, its entire field of operation is exposed to full view. The area in front of the glacier's terminus has deposits that have been sorted and stratified (set down in layers) by glacial streams; we call this area the **outwash plain.** The area behind the terminal moraine, which was formerly beneath the glacier, is covered with deposits that are unsorted and unstratified and might be called rubble. This is the till plain. Sometimes we discover little holes in both of these areas; glaciologists tell us that they result from the melting of ice blocks that were covered with till deposits. When the ice melts completely, there is only a hole to show where it once was. These are called **kettles.**

Moraines are as common to continental glaciers as they are to alpine glaciers, only they are more prevalent and cover much greater

CLASSIFICATION OF GLACIERS 395

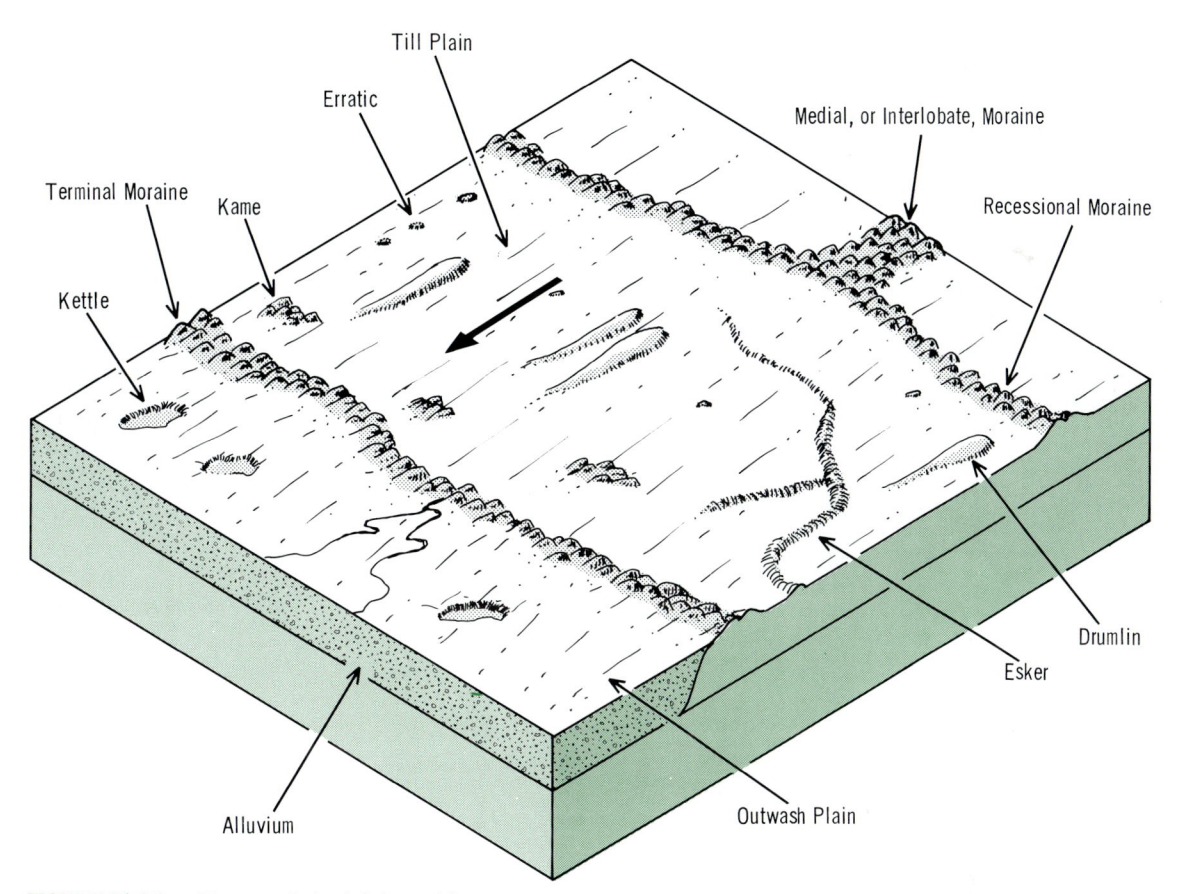

FIGURE 15-6 Landforms of glacial deposition.

distances. The same terms apply to the different types: lateral, terminal, ground, and recessional. The only exception is the medial moraine. Because a continental glacier does not operate as a single, small ice mass that can unite with another, but moves as broad fingerlike lobes, we call the moraine that results from the union of two of these lobes an **interlobate moraine**.

Streams flowing within and beneath a glacier create a variety of interesting landforms. A stream following a meandering path on or underneath a glacier's surface will deposit its load much as any surface stream does, but when the glacier has receded, the alluvial deposits laid down by the stream will be sinuously winding rounded mound that may be as much as 50 miles (80 kilometers) in length. These snakelike deposits of neatly stratified alluvium are called **eskers**.

Sometimes earth materials are deposited directly through the glacier, with the motion of the glacier giving the unsorted and unstratified mound a smoothly rounded shape.

Esker The esker shown in this aerial photo (a very good example, by the way) winds across the landscape of Boothia Peninsula, Canada's Northwest Territories.

Some of these little hills, called **drumlins,** have been found to be almost 1 mile (1.6 kilometers) in length and up to 50 yards (45 meters) thick. Conical hills, or mounds of sorted and stratified debris, deposited by a stream running beneath a glacier's surface are called **kames.** If these deposits are built up along the sides of a glacier, a *kame terrace* will form.

We have already discussed what is probably the most important contribution of glacial deposition to humans: the rich deposits set down by glacial streams and lakes over huge expanses of land that lay beyond the margins of the continental glacier.* Much of the stream-

---

*Aside from the Great Lakes, North America was given two other lakes of impressive size. One was Lake Bonneville, the ancestor of today's Great Salt Lake, which covered about 20,000 square miles (52,000 square kilometers) and had a depth of close to 1000 feet (300 meters), as compared to today's lake, which measures only 1700 square miles (4420 square kilometers) and is only 30 feet (9 meters) deep. The other was Lake Agassiz, named after

Antarctica and its vast continental glacier, viewed from the *Apollo 17* spacecraft.

deposited material is silt; in some places it is estimated to be more than 20 feet (6 meters) thick. When the winds pick up and transport this fine silt elsewhere, the resulting finely sifted material (called *loess*) is transformed into a fertile soil, such as that abounding in the Yellow River (Hwang Ho) region of northern China and in much of the United States below the southernmost limit of Pleistocene glaciation.

Present-day Continental Glaciers and Ice Caps
All that is left of the continental glaciers that covered so much of the earth is the island of Greenland and the continent of Antarctica.

the famous glaciologist-zoologist, which once covered much of North America's interior (Manitoba in Canada and Minnesota and North Dakota in the United States), an area of about 100,000 square miles (260,000 square kilometers). Canada's Lake Winnipeg and Lake Manitoba are the only remnants of this massive body of water.

Antarctica covers an area of 5,100,000 square miles (12,260,000 square kilometers), and its continental glacier covers about 98 percent of the land. In some places the ice sheet has a thickness of close to 13,000 feet (3900 meters). This considerable mass of ice presses down on portions of the continent until the actual surface of the land is now below sea level. If, however, the ice sheet were to melt away, the land undoubtedly would rise as the pressure was released. This region is said to be the coldest in all the world because monthly mean (average) temperatures range from 13°F (−10.6°C) in the summer to −80°F (−62°C) in the winter near the South Pole and are only slightly higher along the coastal regions. The world record for low temperature was set at the Soviet Union's Vostok Station in July, 1983: −128.6°F (−89°C).[4]

In many sections the ice sheet of Antarctica extends out beyond the edge of the continent in great shelves. The Ross ice shelf, for example, is about 1000 feet (300 meters) thick and extends almost 500 miles (800 kilometers) from land. When a section of an ice shelf is broken off and falls into the sea, an **ice floe,** a

Ice floe.

Icebergs at the terminus of the Portage Glacier, Alaska.

floating island of ice that may be as long as 50 miles (80 kilometers), is created. Can you imagine what sailors of ancient times thought when they first encountered such an iceberg as they rounded the southern tip of South America?

Greenland is the world's largest island. Its area is 840,000 square miles (2,184,000 square kilometers) and, of that, 670,000 square miles (1,742,000 square kilometers) is covered with an ice cap. It, too, supplies chunks of ice to the oceans, as great tongues of ice push out into the sea, break off (a process called **calving,** and become **icebergs** (ice mountains). Although they are not as big or as dramatic as Antarctica's ice floes, they represent a serious threat to the shipping lanes along the western North Atlantic Ocean. Some of these icebergs have been measured to be from 800 to 1000 feet (240 to 300 meters) thick, with most of the bulk hidden beneath the waves. An 800-foot (240-meter) iceberg would have only about 100 feet (30 meters) above the water.

REVIEW AND DISCUSSION

1. For what scientific theory is Louis Agassiz noted?
2. Explain why the permafrost regions near the Arctic do not produce glaciers today.
3. Explain the importance of air temperature to the formation of glaciers.
4. Explain the term *snow line*. Why does the snow line vary in altitude at different latitudes?
5. Explain the terms *zone of accumulation* and *zone of ablation*.
6. What is meltwater? What is a firn?
7. Explain the difference between the terms *evaporation* and *sublimation*.
8. Explain why some glaciers can move more rapidly than others.
9. What is the main difference between alpine and continental glaciers?
10. Define the terms *bergschrund, abrasion,* and *plucking*.
11. Discuss these terms, which apply to alpine glaciation and its erosional landforms: *tarn, cirque, trough, arête, horn, hanging trough, finger lake, col,* and *roches moutonnées*.
12. What causes striations and glacial polish?
13. As related to depositional landforms, what are an erratic, a moraine, and a till plain?
14. What is the difference between lateral, medial, terminal, recessional, and interlobate moraines?
15. What effect did the period of Pleistocene glaciation have on humans?
16. Describe the origin and shapes of the following depositional landforms: esker, drumlin, kame, and kettle.
17. What is the difference between icebergs and ice floes?

NOTES

[1] Willy Ley (1962), *The Poles* (New York: Time-Life Books), pp. 170–171.

[2] James Gilluly, A. Waters, and A. O. Woodford (1968), *Principles of Geology*, 3rd ed. (San Francisco: W. H. Freeman), p. 251.

[3] James L. Dyson (1962), *The World of Ice* (New York: Knopf), p. 5.

[4] U.S. Geological Survey (1969), *The Great Ice Age* (pamphlet 1969 0-357-128) (Washington, D.C.: U.S. Government Printing Office), p. 5.

ADDITIONAL READING

Dyson, James L. (1962). *The World of Ice*. New York: Knopf.

Field, W. O. (September 1955). "Glaciers," *Scientific American* (offprint no. 809).

Flint, R. F. (1971). *Glacian and Quaternary Geology*. New York: Wiley.

Ogburn, Charlton, Jr. (1968). *The Forging of Our Continent*. New York: American Heritage.

Shimer, John A. (1959). *This Sculptured Earth: The Landscape of America*. New York: Columbia University Press.

Vial, A. E. L. (1952). *Alpine Glaciers*. London: Batchworth Press.

Woodbury, David O. (1962). *The Great White Mantle*. New York: Viking.

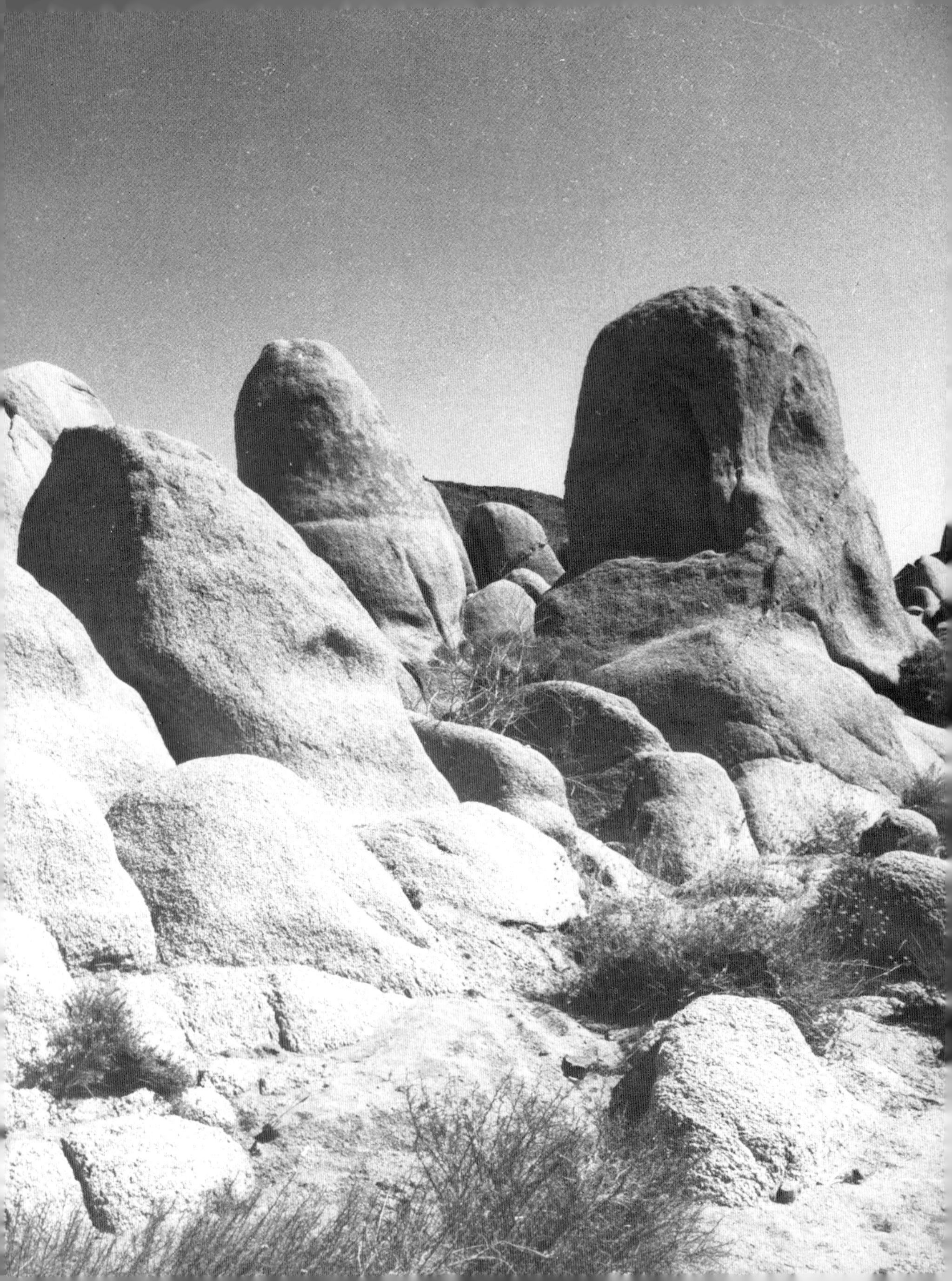

SIXTEEN
SCOURED BY THE WINDS

Whether you happen to be standing beside a tent in California's Death Valley, staring at an array of marching sand dunes or leaning against the driving force of a hot, dry wind on the outskirts of central Asia's Samarkand, marveling at the seemingly limitless height of a massive dust cloud that reaches to the horizons, you know you are witnessing something that has occurred again and again since time began. You have visible proof of the never-ending labors of the winds.

The Arunta, who live along the northern margins of Australia's Great Sandy Desert, have long been troubled by windstorms that race out of the desert in great whirling clouds of red sand to engulf their encampments and muddy their water holes. To them it is not a natural phenomenon because they believe the billowing whirlwinds are the abode of an evil spirit they call Koochee, the Red Demon. When one of these storms appears on the near horizon, the warriors pick up their large boomerangs and rush to the attack.

Should we consider the Arunta warriors merely "ignorant and superstitious savages" because they believe they can cut down the whirling cloud of red sand with their expertly thrown boomerangs? Perhaps the flat blades of their weapons do, somehow, interrupt the upward flow of spiraling wind, thus robbing the whirlwind of its power.

While it seems doubtful that the roaring column of red sand is inhabited by a demon, it is understandable that the warriors must try to protect themselves. Their land, the Australian outback, is not the most hospitable place; it is land of barrenness, red-pebbled plains, great stretches of shifting sands, and howling winds. If they believe the winds contain spirits, they

Winds: rivers of the lower atmosphere In this satellite view of part of southern California, including the Mojave Desert, a number of mountain ranges, and much of the county of Los Angeles (see the accompanying map), dust that has been roiled by the activities of recreational vehicles is picked up by the winds, resulting in "dust plumes" that have the appearance of flowing streams.

are not different from hundreds of other peoples living in different places around the world. Winds bring rains, winds bring storms, and winds bring drying conditions and drought. People can harness rivers, dam lakes, or position breakwaters to thwart the ravages of the sea, but except for rows of trees to serve as windbreaks or windmills to provide us with an alternate source of energy, no one has yet discovered a reliable way to control the winds.

No region on earth is spared the work of the winds. To the Yakuts, the nomads of northeastern Siberia, winds can mean a lowering of warm temperatures during their short summer or they can mean a chilling blast of icy air from the Arctic that will once again freeze

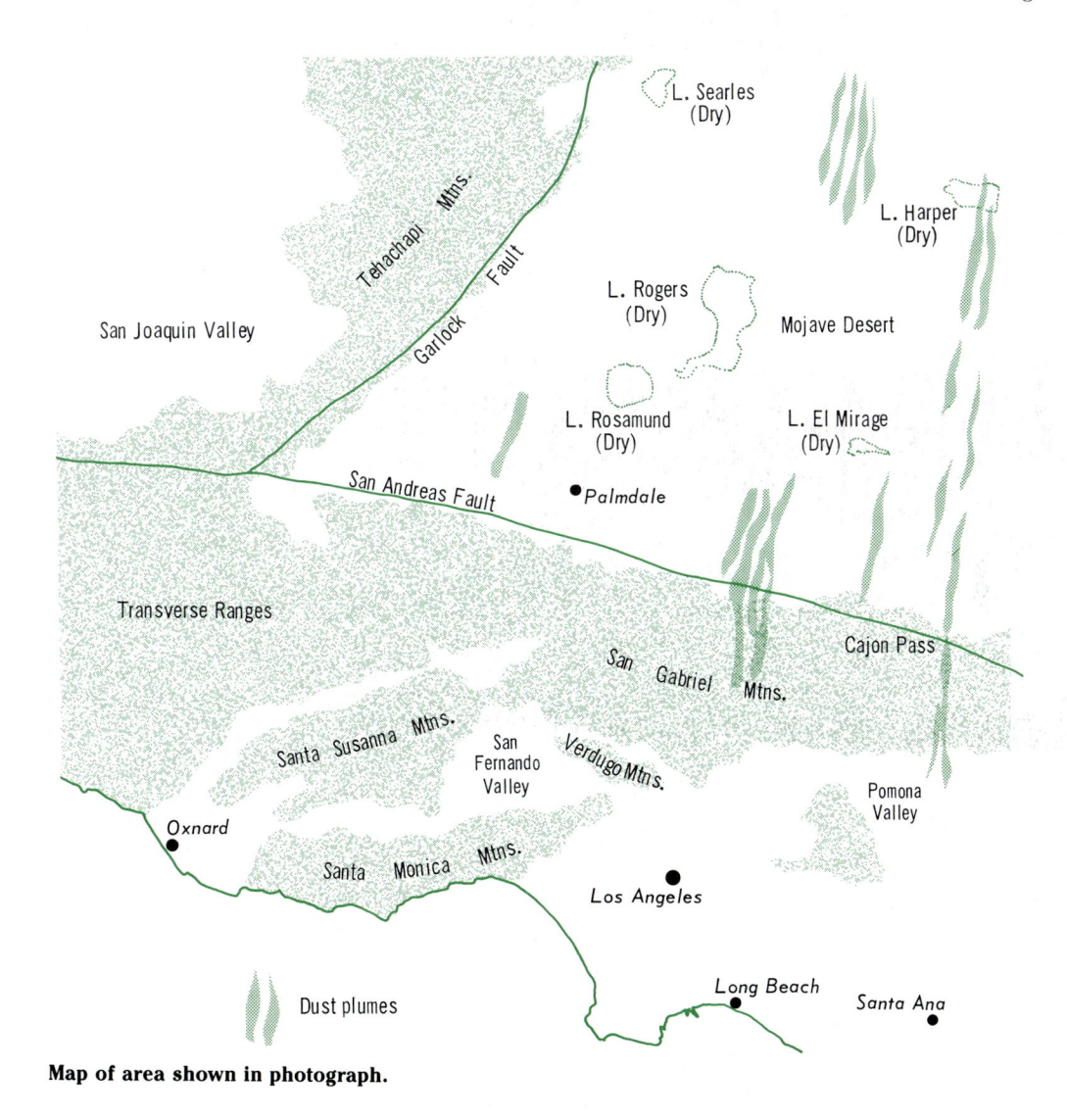

Map of area shown in photograph.

the partially thawed top layer of the permafrost. A century or so ago, the Finns sold wind to masters of becalmed ships. They provided a rope tied in three knots, and if they undid the first knot, a moderate wind sprang up; if they undid the second knot, half a gale blew; if they undid the third knot, a hurricane occurred.[1] Indians of South America flung fiery brands at a strong wind that threatened their homes and, in addition, tried to further frighten it with their screams.*

The winds never cease their restless whisperings or howlings as they prepare the land for the various weathering processes. They erode the land, transporting particles from one place to another, and they deposit their loads as they, too, along with the other earth-shaping forces, attempt to bring about a leveling of the land. One might infer that wind currents are somewhat like stream currents and therefore call them "rivers of the lower atmosphere."

WIND EROSION

Although winds operate in a manner similar to streams of running water, they are not confined between the walls of narrow valleys and so can operate freely over wide areas. The erosional processes of wind erosion are abrasion and deflation.

Abrasion

We used the term **abrasion** in our discussion of the sculpturing performed by running water, oceans, and glaciers, but the wind's action of wearing away a surface is more akin to sandblasting—sand grains being hurled against an object. The process of abrasion does not build landforms; it sculptures those already in existence. The various forces of weathering prepare landforms by breaking apart the rock structures, and then comes the wind, lifting the loosened particles, carrying them along, and throwing them violently against other objects in the path. Because of the weight of sand grains, most of the work of abrasion is accomplished within 3 feet (1 meter) of the ground If, however, the wind is powerful, moving at a high speed, it can carry a heavier load to higher elevations. Running water performed the major erosional work on the columns of Bryce Canyon National Park, but wind abrasion has softened the contours somewhat. Anyone who has been caught in a sandstorm while driving on the desert knows what windblown sand can do to the paint on an automobile. Wind abrasion can also shape large pebbles and rocks into one-, two-, or three-sided shiny fragments, called **ventifacts** (from the Latin *ventus*, "wind," and *factum*, "done"). Ventifacts are usually very small, with shiny, polished facets, or sides.

Deflation

The erosional process of **deflation** refers to the removal of sand or silt particles from an area that has been previously weathered, eroded, or deposited. Deflation occurs mostly in dry (arid or semiarid) climate regions that have little in the way of vegetation to hold onto the loosened particles of a weathered surface. It can also occur, however, in humid regions that have floodplains, old lakebeds, or the till or outwash plains of glaciated areas, and on elevated sea terraces or sea cliffs of an uplifted shoreline.

If the area is composed of loose sand particles, a depression may be created when the small particles are carried away. This hollowed-out space is called a **blowout**, or **deflation basin**. Blowouts vary in size and, although most are small, some attain lengths of up to 1

---

*For an interesting look at attempts throughout history to understand and control the workings of the atmosphere, see Sir James G. Frazer (1960), *The Golden Bough* (New York: Macmillan), especially Chapter 5, pp. 92–96, "The Magical Control of the Wind."

Desert pavement in Death Valley, California.

mile (1.6 kilometers). If an area also happens to have a great quantity of gravels and pebbles, the finer particles may be carried away, leaving the larger rock materials behind. This results in a stony, sandless desert floor that sometimes looks much like a paved country road. Geologists call this erosional landform *desert pavement,* but the Arabs of North Africa's Sahara Desert call it *reg,* and the Australians call it a *gibber plain.* Despite the movies and television shows that depict the deserts of the world as composed of majestically billowing sand dunes, stony deserts are the most common.

WIND TRANSPORTATION

As mentioned earlier, most wind-transported materials are carried along very close to the ground. This is because of the weight of the particles and the velocity of the wind (which is not usually great enough to transport materials heavier than sand or silt). Also, wind is more variable and turbulent than flowing water, so motion in a particular direction cannot be sustained by particles being carried along by the wind. Much of the wind's deflation and abrasion is a result of a process called **saltation** (Fig. 16-1). Briefly, saltation is the result of wind-driven sand grains bouncing along the ground, striking other sand particles, dislodging them, and causing them to bounce into the air or be pushed along slowly with the moving wind. The slow movement of sand grains downwind along the ground is called *surface creep.* If the particles are small enough, dust storms may result, and a considerable amount of soil may be lost to a marginal agricultural region. This happened to parts of Oklahoma and Kansas during the 1930s, when strong winds carried away soil that had been exposed by improper

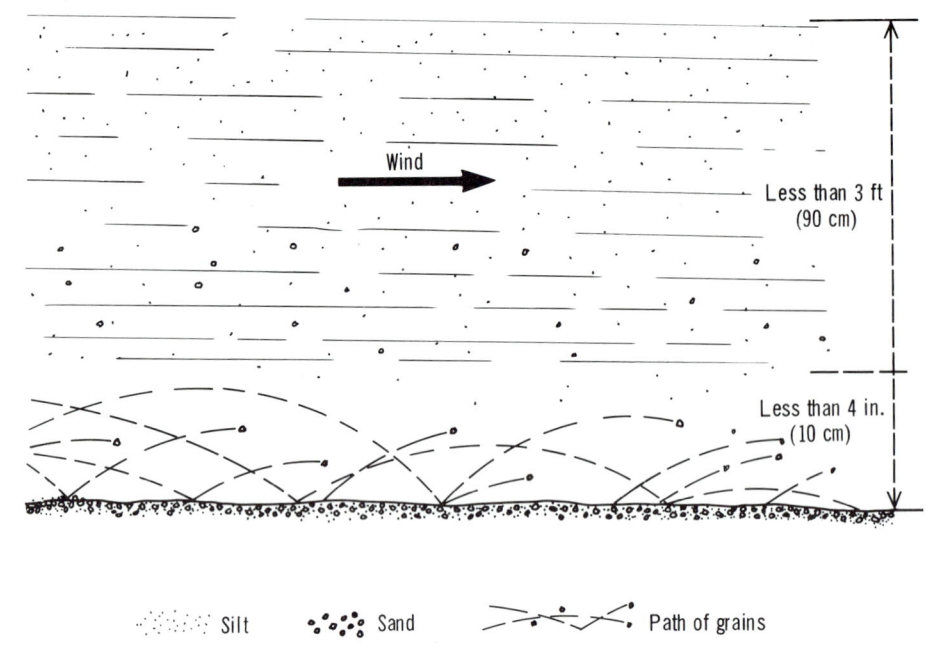

FIGURE 16-1 Saltation: the movement of sand grains by wind.

ranching and farming practices and created the Dust Bowl.

WIND DEPOSITION

Winds do perform some significant erosional activities, but they are more noted for their depositional accomplishments, such as those that result from the deposition of sand (sandy deserts and sand dunes) and silt (loess deposits).

Sandy Deserts

If you compare the map in Fig. 8-4, showing the world's deserts, to the map in Fig. 16-2, showing the location of the sandy desert areas, our earlier statement that stony or pebbly deserts are larger and more common than sandy deserts will be verified. Stony deserts do have their exciting features, but they cannot match the landforms to be found in the earth's sand regions. It might be noted that the most popular term used to denote a stony desert is the Arabic word *reg*, and for sandy desert the word is *erg*. It might also be noted that few deserts (if any) are either completely reg or erg. Most have features representing both. There usually are great stretches of reg, followed by large areas or pockets of erg. Blowouts and desert pavement, along with *wadis*, are the outstanding landforms of regs, while sand dunes and certain related features are the prominent landforms of ergs.

Sand dunes may be found in any area that has an extensive supply of sand, such as on the fringes of large deserts, along shorelines, in the floodplains of stream valleys in arid climates, or near areas that were once glaciated. The United States has a number of interesting sand-dune areas, such as the Indiana Dunes,

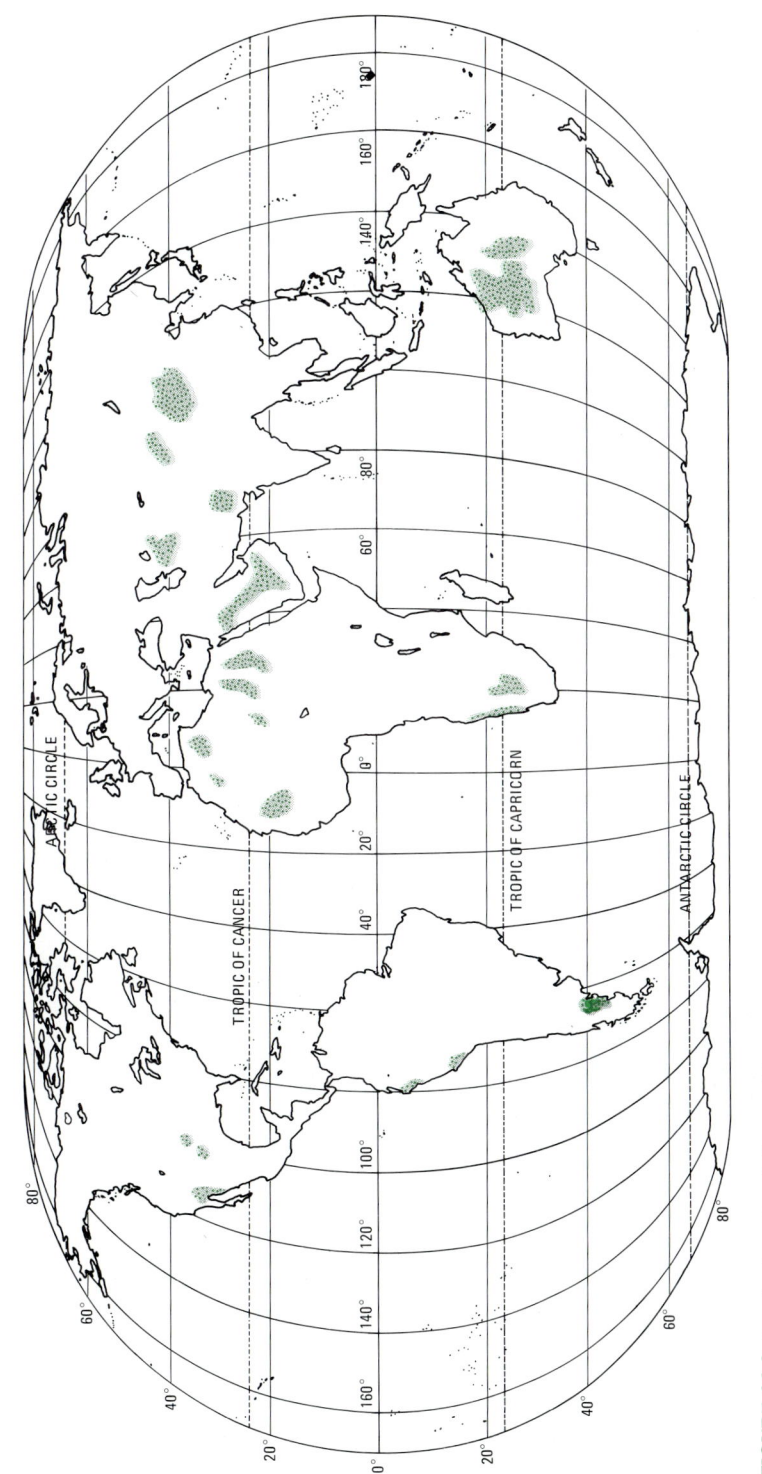

FIGURE 16-2 Sandy deserts of the world.

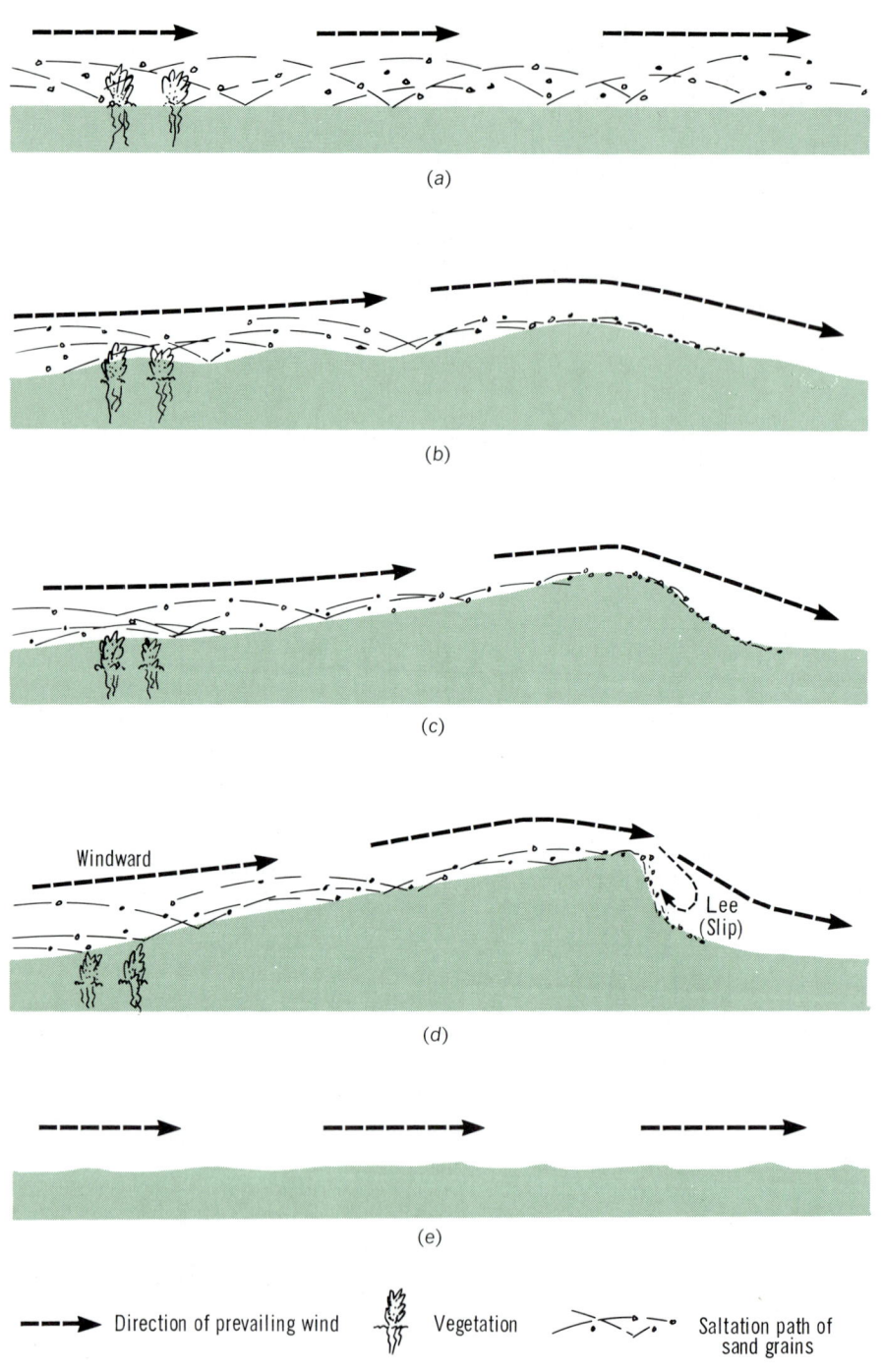

FIGURE 16-3 Evolution of a sand dune.

Checkerboard Mesa, a mass of cross-bedded Navajo sandstone, east of Zion National Park, Utah.

Little Sahara State Park in Oklahoma, Sleeping Bear State Park in Michigan, and the Great Sand Dunes National Monument in Colorado. Sand dunes occur in a variety of shapes, depending on the wind direction and velocity, the quantity and size of the sand grains, the amount and type of vegetation, and the topography of the land.

Figure 16-3 details the evolution of a sand dune. Wind drives grains of sand (by saltation) along a surface (Fig. 16-3a) until the velocity is slowed by some sort of obstruction (such as a depression in the ground, a small hill or rise, or, as in the illustration, some type of vegetation). The sand grains begin to pile up in the shadow (leeward side) of the shrub (Fig. 16-3b), forming a mound composed primarily of the heavier grains (because the lighter grains are transported further on by the wind). As the dune builds (Fig. 16-3c), the windward slope rises in a gentle incline (normally about 10°), and the lee side becomes quite steep (about 35°) as the heavy grains can no longer be pushed by the wind. They collect at the top of the dune; then, by the pull of gravity, they fall, or slip, down the lee (slip) side. With continuing action (Fig. 16-3d), the slip side sharpens and the dune advances slowly downwind as sand grains bounce up the windward slope or around the sides of the dune and move on. New layers of sand are added, giving the internal structure of the dune a stratified appearance. If you were to dig into an old dune, you would discover distinct strata, beds of different types of wind-deposited materials. If the wind becomes too strong or replenishment of sand losses is reduced, the dune will diminish in size and, eventually, be reduced to a thin, rippled surface (Fig. 16-3e).

The most famous of the forms taken by mounds of moving sand is the crescent-shaped dune called the **barchan.** The crescent shape is caused as sand grains race around the sides of the mound and form "horns" that extend downwind under the directing force of a wind that prevails in one direction and does not shift about.

When the supply of sand and the wind

Barchan sand dune The site of the dune area in which this well-formed barchan was found is about 20 miles (32 kilometers) east of El Paso, Texas.

Sand-dune march Constantly shifting sand dunes migrate through Great Sand Dunes National Monument, Colorado.

come from the same direction, perhaps at the mouth of a great valley, the sand will be strung out in a long ridge, parallel to the wind's direction, thus forming a **longitudinal dune.** Some of these may extend for 100 miles (160 kilometers) while attaining heights of 330 feet (100 meters).[2] An interesting combination of the longitudinal and barchan shapes occurs when the winds import fairly significant amounts of sand during a season and these deposits are reworked by winds coming from a different direction in a later period. The result is a string of barchans extending along the angle of the dominant wind like links in a chain. This form has been given the name **seif,** an Arabic term meaning "sword".

If there is a moderate supply of sand steady winds, and a sturdy vegetative cover, the mound of moving sand may find itself anchored on two sides by large clumps or shrubs that hold the horns while the main body of sand moves on to take a form that is the reverse of the barchan: the horns point upwind.

This is called the **U-shaped,** or **parabolic, sand dune.**

Where there is abundant sand and moderate winds, the sandy landscape looks much like the surface of an ocean; the "waves" seem to march across the land in an endless procession. The ridges of these dunes, called **transverse** dunes, are at right angles to the wind direction.

If the winds are not constant in their direction but come from all sides, shifting their directions throughout the day, the result is a **star** dune. Unlike the other dunes, star dunes remain stationary.

Other wind-deposited landforms include the **whaleback** and the **sand shadow.** Whalebacks are large, rounded mounds of sand that may extend for 50 miles (80 kilometers), their axes parallel to the prevailing winds. A sand shadow results when sand collects on the lee side of an object (house, boulder, or shrub) as the eddies of circling wind can no longer support their loads. This action is comparable to the deposits that result when a stream of running water is forced into a backwater eddy on the inside of a stream's meander.

The imprint of humanity on a sandy desert landscape The strange, footprintlike depressions are date-palm oases in Souf, Algeria (north of Touggourt in the Grand Erg Oriental). The farmers plant palm trees in areas where the water table is relatively close to the surface and erect small fences around the perimeter of each tiny grove. Winddriven sand (the arrow shows the prevailing direction) billows around the palm oases, building up along the sides. Once past the oases, the sand resumes standard shapes.

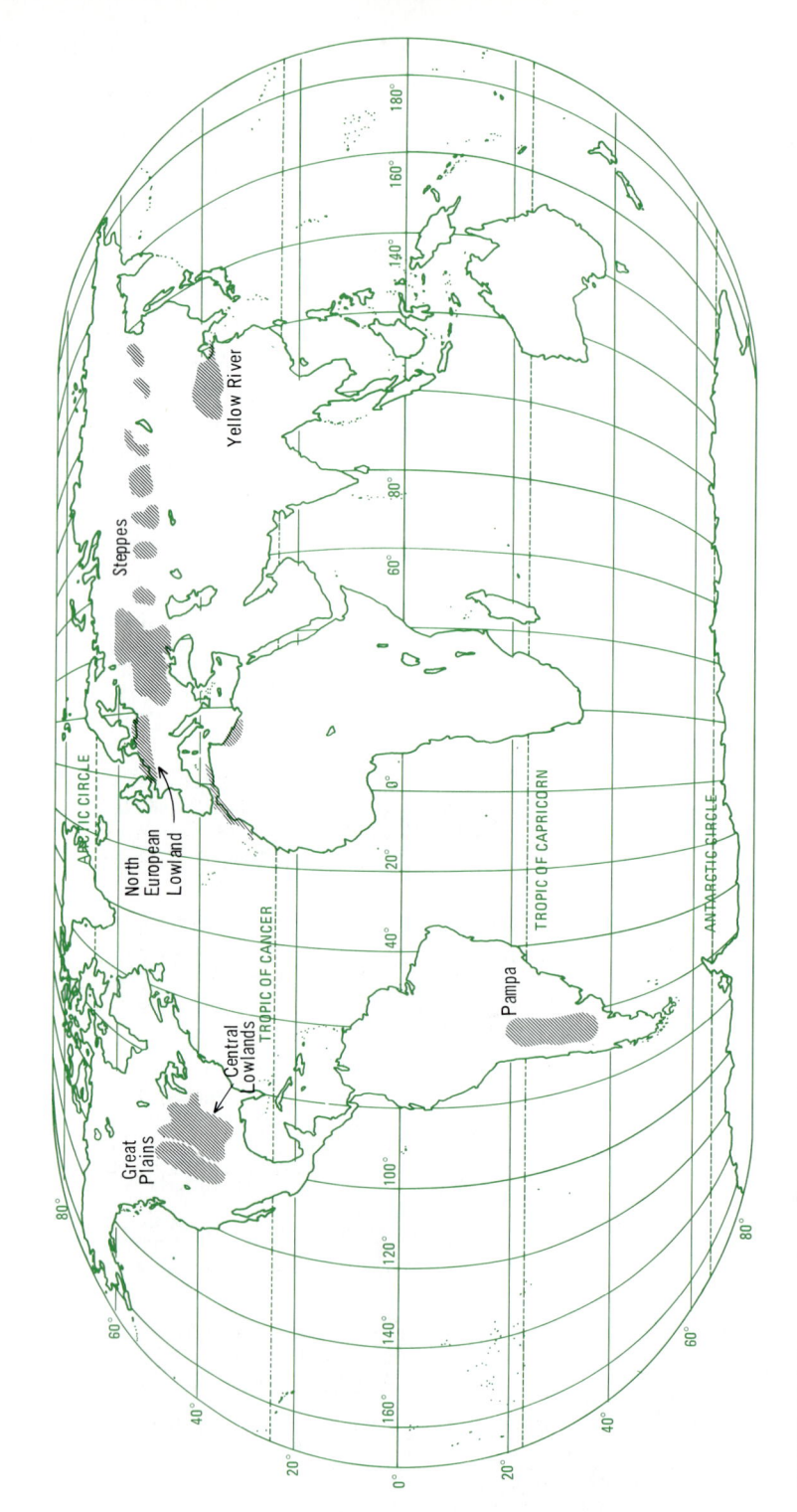

FIGURE 16-4 Loess regions of the world.

Loess Deposits

Loess (pronounced *lurse*) is wind-transported and deposited silt. It is usually composed of fine-grained, silty materials that are buff or grayish in color. (Figure 16-4 shows the location of the major loess deposits of the world; if you compare this map with the one showing the extent of Pleistocene glaciation (Fig. 15-4), it will become apparent that sediments deposited by streams issuing from the great continental glaciers or from large glacial lakes are the primary source of most of these deposits. Some of the world's most important agricultural areas are in regions where loess deposits are the deepest and enough time has passed to allow vegetation to add humus to what was potentially a fertile soil. The region of loess deposits through which China's Yellow River flows might be termed "China's breadbasket" because this rich, loamy soil helps to support that country's hundreds of millions of people. Argentina's pampa is famous for its cereals, grains, and cattle, as are the fertile steppes of central Asia and the Ukraine. Other productive loess regions are the northern European lowlands of Germany and Poland and, in the United States, the Great Plains, the central lowland, and the lower Mississippi River valley.

REVIEW AND DISCUSSION

1. Define *abrasion* and *deflation* in relation to wind erosion.
2. Why have winds sometimes been referred to as "rivers of the lower atmosphere"?
3. Describe a ventifact.
4. What is a blowout? What is a deflation basin?
5. What is the difference between *reg* and *erg* deserts?
6. What is a gibber plain?
7. What is saltation? How does this process differ from surface creep?
8. Why is most abrasion by the wind accomplished close to the ground?
9. What is a wadi?
10. Describe the evolution and construction of a sand dune.
11. What factors determine the size and shape of a sand dune?
12. Describe the appearance and form of construction of these sand dunes: barchan, longitudinal, seif, parabolic or U-shaped, and transverse.
13. What is the difference between a longitudinal dune and a whaleback?
14. How does a sand shadow form?
15. What is loess? Where are the most important loess regions of the world? What is the origin of loess?

NOTES

[1] Sir James George Frazer (1960), *The Golden Bough* (New York: Macmillan), p. 93.

[2] William D. Thornbury (1969), *Principles of Geomorphology*, 2d ed. (New York: Wiley), p. 296.

ADDITIONAL READING

Bagnold, R. A. (1971). *The Physics of Blown Sand and Desert Dunes*. New York: Methuen.

Bertin, Leon (1972). *The New Larousse Encyclopedia of the Earth*. Rev. ed. New York: Crown. See especially "Transport and Erosion by Wind," pp. 46–50, and "Eolian Deposits," pp. 51–53.

Gautier, E. F. (1970). *Sahara: The Great Desert*. New York: Octagon.

Leopold, A. S. (1961). *The Desert*. New York: Time-Life Books.

Lugn, A. L. (1962). *The Origin and Source of Loess*. Lincoln: University of Nebraska Press.

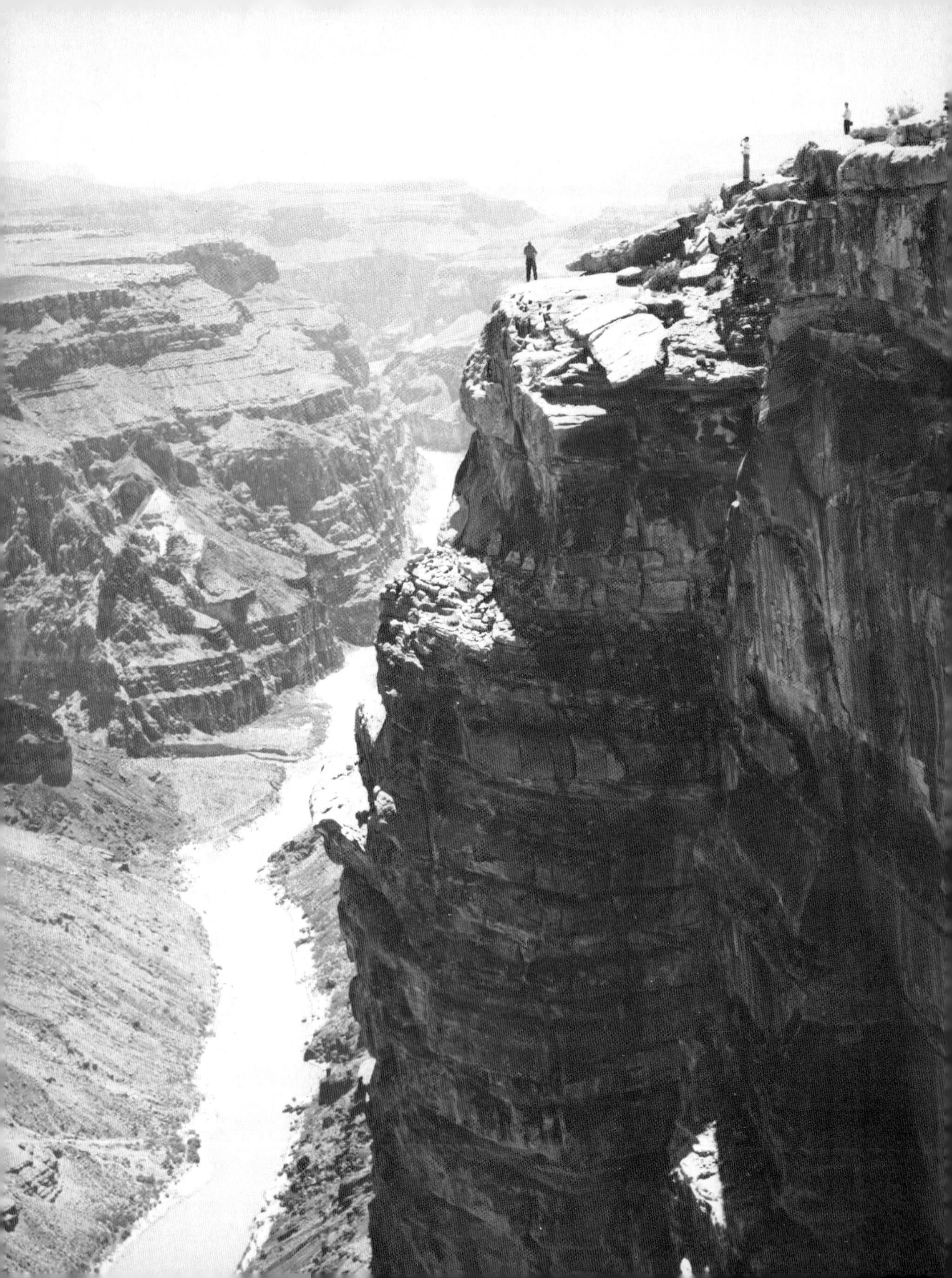

EPILOGUE

THE HUMAN ENVIRONMENT— TOMORROW?

*Two ways the rivers
Leap down to different seas, and as they roll
Grow deep and still, and their majestic presence
Becomes a benefaction to the towns
They visit.*

THE GOLDEN LEGEND, LONGFELLOW

As your voyage of discovery reaches its conclusion, hopefully you have gained a better comprehension of the massive and interrelated system we call our universe. You have a clearer understanding of the reasons why there are so many exciting differences in our physical environment. When you are on a vacation, you will find that you will be using the "scientific method" as you enjoy the view of glacially carved valleys in California's Sierra Nevadas or the beautiful sandbars and spits along the coast of Florida. When you are watching the news on television, you will know what the newscaster means when the terms *highs*, *lows*, and *cold fronts* are used. And if the newscaster describes the terrible damage to the coastline of Japan caused by a "tidal wave" resulting from a tremendous earthquake nearby, you will laugh, knowing that it was a tsunami, not a tidal wave. When you hear reports of earthquakes in Italy, Iran, Soviet Central Asia, or California, memories of plate tectonics and the collision of the earth's crustal plates will flash through your mind.

When the time comes, and you must consider moving to a new location, your decision now will not be based primarily on the cultural environment, because you will examine the differences in the climates as well as the possible physical risks. Suppose you were offered good employment opportunity in three different locations, such as Fairbanks, Alaska, Los Angeles, California, and Miami, Florida. Would you choose an environment where there are ten months of winter, with daytime temperatures averaging about 10°F (12.3°C), or a place where January temperatures average 72°F (22°C) but there are smog problems? Or would you choose a beachfront locale that is moderate in temperature but very high in humidity? Which would you prefer, residence in an earthquake zone or in the path of annual hurricanes?

You also are now more prepared to understand what is "behind" news of international importance, such as a major earthquake near Los Angeles, desertification in the Sahel, a monsoon that strikes Bangladesh, wheat crop losses in the Soviet Union, climate changes in the United States brought on by El Niño, or the dangers of acid rain. When the conversation turns to alternative energy sources, you will comprehend terms such as *solar collectors*, *geothermal*, *ocean thermal differences*, and *nuclear fusion*. When someone speaks of ecology, environmental protection, and the conservation of energy, you will be a participant, not merely an observer.

In Chapter 1 we quoted Marcus Aurelius; perhaps now, with the acceleration of your knowledge of the processes of nature and the interrelatedness of the earth system, you can understand his message.

We are standing on the threshold of new discoveries that may further increase our understanding of our environment and thus enable us to make decisions that will improve, instead of deteriorate, our quality of life and preserve, instead of destroy, our world resources.

Whether or not one plans to pursue a career in the field of geography, knowledge of the physical world—its history and its processes—becomes ever more important as our population and technological capabilities increase.

We have indicated that humans are both users and abusers of the earth. What can professional geographers do about the abuses? What opportunities and challenges are offered to those who would like to participate actively in the advancement of technology without destruction to the environment?[*] A number of

---

[*] See Salvatore J. Natoli (ed.) (1974), *Careers in Geography*, rev. ed. (Washington, D.C.: Association of American Geographers), for an interesting survey of the career possibilities in the field of geography.

directions can be taken. Some might aspire to solve the immediate problems confronting our neighborhoods, towns, or cities, while others might be more concerned with the long-range effects of human interference in the workings of nature.

Consider the problem of the shortage of petroleum. What should be done? Should we allow offshore drilling on the outer continental shelves? Should we encourage strip mining to reap close-to-the-surface coal deposits? Should we push for immediate development of the huge oil-shale deposits in the western regions of the United States? Should the Canadians strive to retrieve the oil reserves held in the tar sands along the Athabasca River north and east of Edmonton? Are there ways to tap these resources without upsetting the environment?

Perhaps geographers can do more to educate those who might despoil the land or sea, just as soil scientists have aided farmers in learning how to replenish their land by allowing it to lie fallow periodically. Miners or the oil drillers cannot replace the oil or the coal (or other minerals) they have taken from the land, but they can learn to restore its former contours and to replace its vegetation by reforestation, thus starting again its ecological cycle.

Geographers might become members of urban and rural planning and redevelopment commissions to encourage measures that would prevent, or at least minimize, pollution and deterioration. They might use their knowledge of climatology and physical geography to aid in deciding where to locate highways and railroads, factories and industrial developments, and new agricultural regions. As resource evaluators (or land-use ecologists, or community development engineers, or economic geographers), geographers might evaluate an area's possibilities with regard to soils, minerals, waters, forests, grasslands, agricultural crops, wildlife, fisheries, and recreation areas. As physical resource experts they might become concerned with the development of new forms of energy, such as solar, geothermal, and atomic energy. They might map the distribution of the resources of an area and aid in planning for the most beneficial development and utilization of those resources with the least detriment to the environment.

For those who plan to become teachers, the material presented here might advantageously be augmented with data on human or cultural geography to enable you to explain the origins and movements of people and ideas from one place to another throughout the history of civilization. Both teachers and students would then see more clearly why cultures and technologies have become so varied across the face of the earth.

Where can geographers find opportunities for putting their skills and knowledge into action? There are governmental agencies (federal, state, and local) that need well-trained geographers, specialists in the various fields. For example, consider the possibilities in the Department of the Interior's Bureau of Land Management, Bureau of Reclamation, Geological Survey, National Park Service, and Department of Forestry and in the Department of Commerce's Coast and Geodetic Survey (for cartographic work), the U.S. Weather Bureau, the Area Redevelopment Administration, and the Environmental Science Services Administration. There are also opportunities in the Soil Conservation Service of the Department of Agriculture.

Businesses need geographers to help them solve environmental problems, such as how to deal with sewage outfall or industrial wastes that pollute streams, lakes, and oceans or exhausts that pollute the air. Geographers can also aid businesses in developing markets at home and abroad, can help to establish transportation routes for materials and products.

For those who are interested in "emerging" countries, there are opportunities with the Agency for International Development (AID), which administers economic aid for foreign nations.

Technology has caused many of today's problems, but this same technology can also open doors to intelligent environmental planning. People who develop an awareness of the processes of our environment, whether or not they work as professional geographers, can contribute significantly to the reversal of the heretofore dominant trend to abuse instead of efficiently use the world's resources.

ADDITIONAL READING

Allsopp, Bruce (1972). *The Garden Earth: The Case for Ecological Morality.* New York: William Morrow.

Darling, F. Fraser, and John Milton, eds. (1966). *Further Environments of North America: Transformation of a Continent.* Garden City, N.Y.: Natural History Press.

Ehrlich, Paul R., and Anne H. Ehrlich (1970). *Population/Resources/Environment: Issues in Human Ecology.* San Francisco: W. H. Freeman.

Johnson, Cecil E., ed. (1972). *Eco-Crisis.* New York: Wiley.

Johnson, Warren A., and John Hardesty (1971). *Economic Growth vs. the Environment.* Belmont, Calif.: Wadsworth.

McKenzie, Garry D., and Russell O. Utgard, eds. (1972). *Man and His Physical Environment: Readings in Environmental Geology.* Minneapolis: Burgess.

Manners, Ian R., and Marvin W. Mikesell, eds. (1974). *Perspectives on Environment: Essays Requested by the Panel on Environmental Education, Commission on College Geography.* Publication no. 13. Washington, D.C.: Association of American Geographers.

Marsh, George Perkins (1965; originally published in 1864). *Man and Nature.* Edited by David Lowenthal for the John Harvard Library. Cambridge, Mass.: Belknap Press, Harvard University Press.

Odum, Howard T. (1971). *Environment, Power, & Society.* New York: Wiley.

Pursell, Carroll, ed. (1973). *From Conservation to Ecology: The Development of Environmental Concern.* New York: Thomas Y. Crowell.

Strahler, Arthur N., and Alan H. Strahler (1973). *Environmental Geoscience: Interaction between Natural Systems and Man.* Santa Barbara, Calif.: Hamilton.

APPENDIX A

TIME ZONES

When we discussed the earth's rotation and solar (sun) time in Chapter 3 (page 57), you may have gotten the idea that it is simple to figure out what time it is somewhere else in the world. Because we know that the earth rotates 15° every hour, it should be merely a problem of determining the number of degrees between the other location and your own. Then, knowing your own time, you should be able to add or subtract the difference to find out what time it is in the other location. For example, if a place is 90° west of your city, there would be six 15° units between that place and you. Therefore, the time at the other location would be 6 hours earlier in the day than in your city. However, this is not the case. The countries of the world do not set up their time zones according to 15° intervals. Instead, they choose their time zones to suit their particular interests and problems. Consider Fig. A-1, which shows today's world time zones, and you will immediately realize that the north-south lines separating one time zone from another do not follow meridians in a straight line from north to south. They vary tremendously. For example, check the variations at these places: the Galapagos Islands off Ecuador; Southeast Asia; India and Bhutan, which surrounds Bangladesh; England and France; and the People's Republic of China. You will note that the numbers +1, +2, and so on, are used to denote the number of hours east of Greenwich, England,

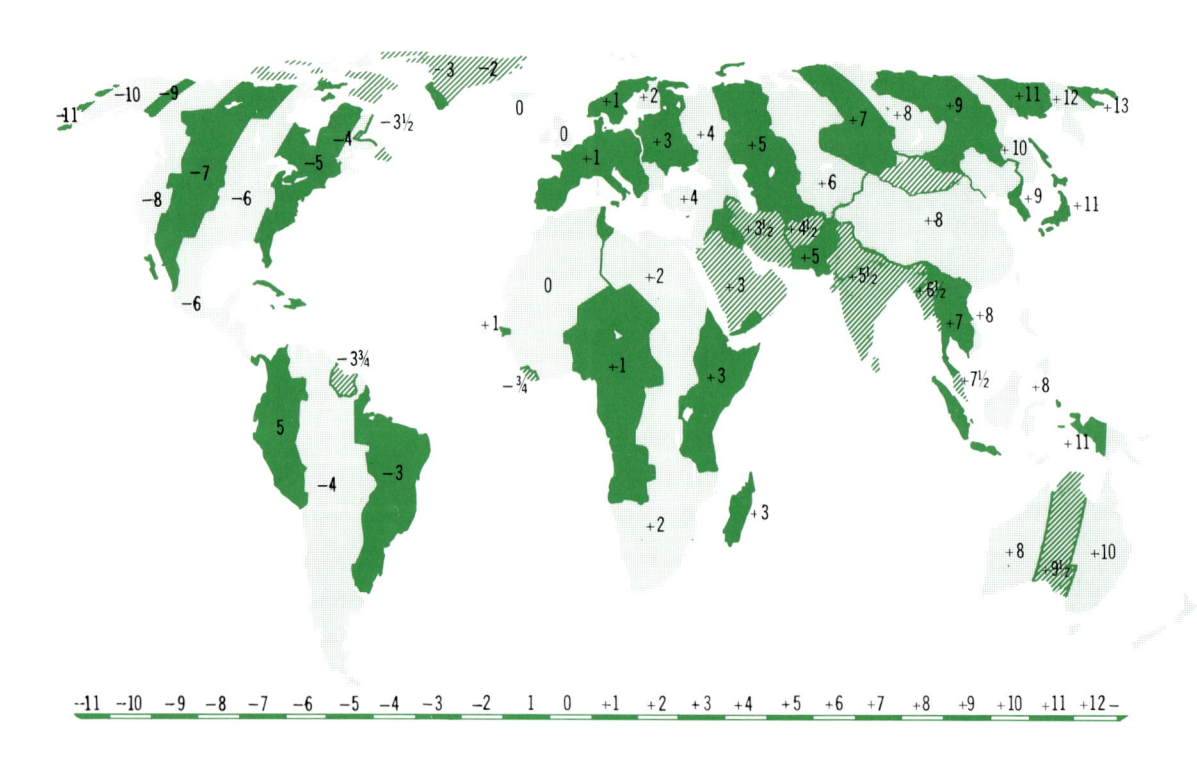

Areas which have not adopted a specific time zone, or where the time is more than a half-hour from neighboring time zones.

FIGURE A-1 World time zones.

422

and −1, −2, and so forth denote the hours west of Greenwich. Note also India's time is +5h 30 m, which means that this country is 5 hours and 30 minutes east of Greenwich. Liberia shows a time zone of −0 45 m, which tells us that Liberia is 45 minutes in time west of Greenwich—although that country sits astride the prime meridian.

For world travelers, whether businesspersons, college students, or tourists, the vagaries of "local time" can be confusing. People either must ignore their watch or constantly reset them as they flash across the earth from country to country. In other situations they might fly thousands of miles without changing their watches at all—as in China. If a group of people flew out of Shanghai at noon, heading westward from the Pacific Ocean coast, they would fly through 3 hours of solar time (45°), a distance of about 2500 miles (4000 kilometers),

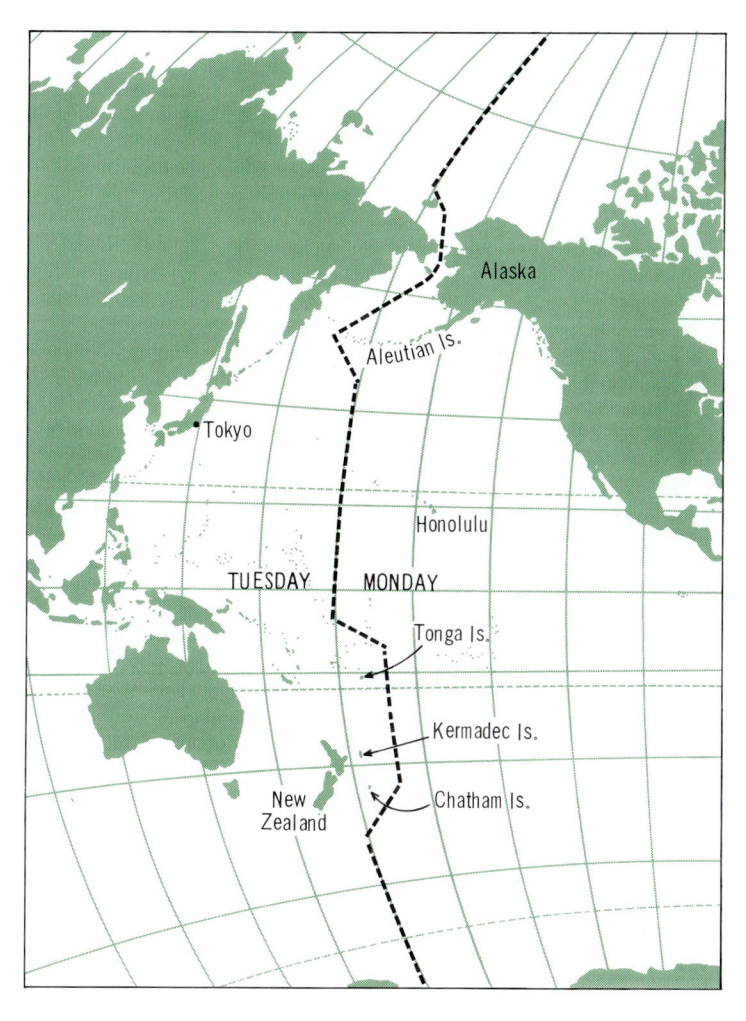

FIGURE A-2 The International Date Line (IDL) The IDL generally follows the 180° meridian (which is the other half of the great circle that includes the zero, or prime, meridian) and is a line of separation between east and west longitudinal positions. It is deflected occasionally from its straight north-south course to allow certain islands and other regions that are closely related to remain in the same time zone and in the same day.

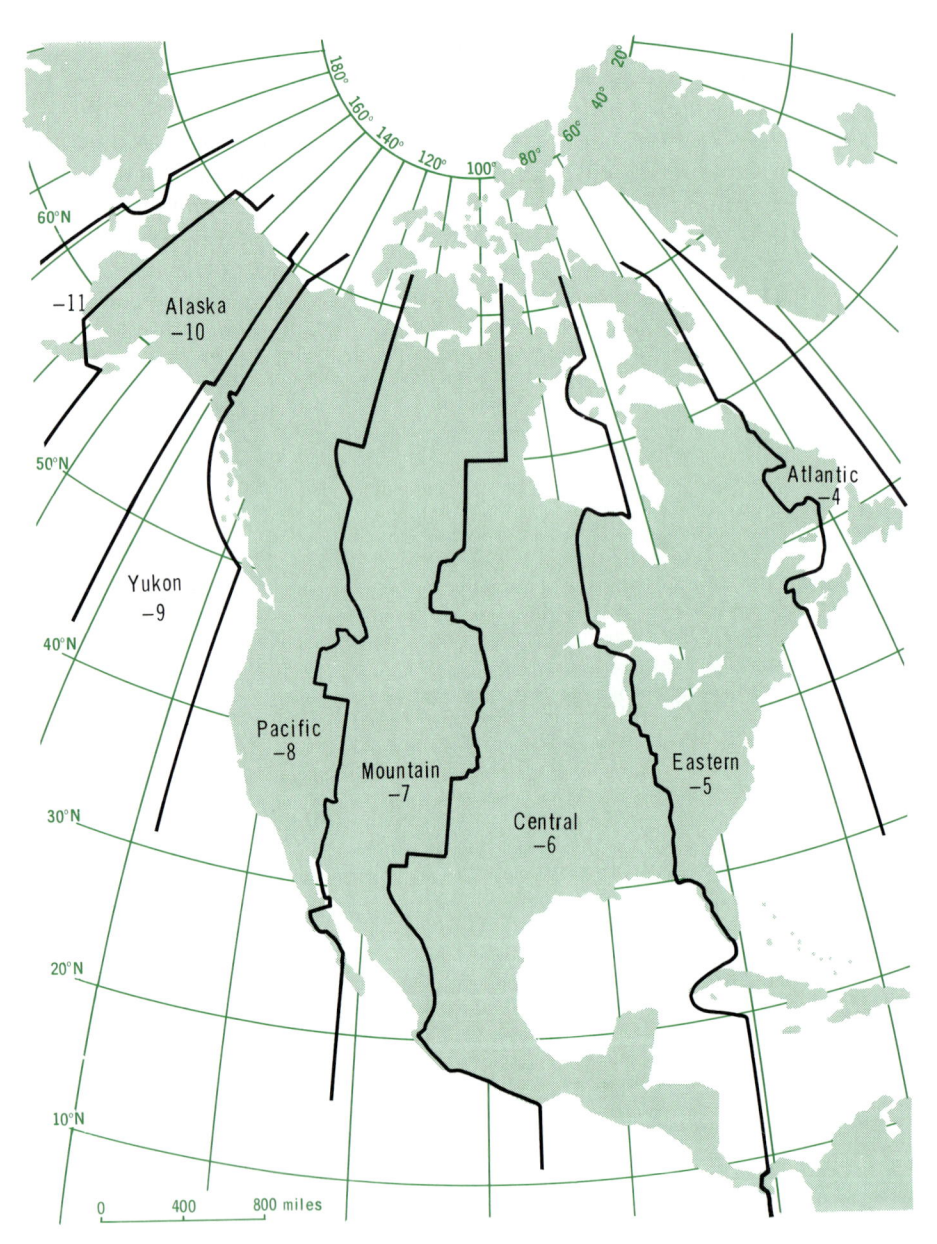

FIGURE A-3 Time zones: United States and Canada.

before departing China and entering the Soviet Union. They would not, however, alter the time on their watches—unless the plane flew at a speed of 833 miles (1337 kilometers) per hour; then, when they reached the border, they would set their watches back to noon.

The zero meridian is the internationally recognized starting point for world time. The prime meridian runs from the North Pole to the South Pole, passing through Greenwich, England. On the opposite side of the world is the 180° meridian, called the International Date Line (Fig. A-2). It is 12 hours distant from the zero meridian: When it is noon in Greenwich, it is midnight along the International Date Line (IDL). If you cross the IDL, traveling from the east toward the west, the day immediately changes. If you left Honolulu on Monday, the minute you crossed the IDL the day would become Tuesday. If it were Monday night at midnight, the day would become Tuesday night, midnight—and you would have lost an entire 24-hour day. If you were traveling from Tokyo to Seattle and it was Tuesday night, midnight, when you crossed the IDL, it would immediately become Monday night, midnight, and you would live Tuesday all over again. Note also that the IDL zigs back and forth, depending on the needs and wishes of the people residing in each region. See how the IDL bypasses the Aleutian Islands? This is to enable the islands to be in the same day as the state of Alaska. The same situation exists in the Southern Hemisphere, where people having ties with New Zealand want their area to be in the same time zone as that country.

All these problems are repeated within the United States and Canada (Fig. A-3). Instead of followng the meridians at 15° intervals, the boundaries of the various regions have been allowed to move east or west as the region's needs require. Consider the difference in latitudinal distance between Bismarck, North Dakota, and Montgomery, Alabama—14°29′; this amounts to almost 1 hour in time, yet both cities are in the same time zone. Both have the same time, although when it is noon (sun time) in Montgomery, it is about 11 A.M. in Bismarck.

APPENDIX B
TOPOGRAPHIC MAPS

Topographic maps, as all other maps, represent a portion of the earth's surface and can show natural or cultural features. They are laid out on a map projection, usually a modified polyconic projection, which keeps distortion to a minimum.*

However, a well-executed "topo map" is more informative than most other maps. Although any map can depict streams and rivers, mountains and valleys, highways and trails, residential areas, industrial sections, camping spots, farms and other agricultural areas, and swamps, beaches, and islands, it cannot do so quite as well as a topographic map.

Who creates the topographic map? In the United States, it is principally within the province of the U.S. Geological Survey (USGS). The USGS was established by an act of Congress in 1879 to consolidate a varied group of organizations that had been engaged in mapping and the study of the geology and topography of the land, especially public (not privately owned) lands. By 1882 the USGS had developed a general plan to standardize the projected series of topographic maps. Under this plan each map would cover a four-sided area, bounded by meridians of longitude and parallels of latitude; they were, therefore, called *quadrangle maps*.

There are several *series* of maps; a *map series* is defined as a "family of maps conforming generally to the same specifications or having some common unifying characteristics such as scale. Adjacent maps of the same quadrangle series can generally be combined to form a single large map."*

Specific steps must be followed in order to be certain of accuracy and usefulness, and these are control surveys, mapping procedures, and national standards.

Control surveys are undertaken to locate precise positions of latitude and longitude and height, or elevation. Horizontal control is needed to maintain the correct scale, position, and orientation of the map. The first step, therefore, is to make field surveys. These are conducted by surveyors and engineers who go out onto the land with accurate instruments to determine the geographic grid and to establish the elevation above sea level of the beginning point. When the "control point" is finally determined, a *bench mark* (or geodetic control point) is fastened to the ground by means of a metal tablet embedded in concrete, rock, or masonry. In the center of this circular piece of brass, the elevation above sea level is stamped.

Mapping procedures have changed considerably since early times. Although the survey teams still go out into the field, they no longer spend as much time there. Today we have aerial photography and photographs from reconnaissance satellites.† Costs to produce highly accurate topographic maps by these methods are about 10 percent of the cost of land-based teams. The process of converting information about elevation (to draw the

---

*A polyconic projection is one in which many cones intersect the surface of the globe; there is, in fact, a cone for each parallel of need. The meridians are ellipses, with the central meridian of the projection straight up and down. The parallels are not parallel, which would be a serious problem if a large area were to be represented. However, the U.S. uses only a small portion of the projection, so the defects are minimized.

*A booklet entitled *Topographic Maps* can be obtained from the Map Information Office, U.S. Geological Survey, Washington, D.C. 20242. Indexes showing published topographic maps in each state, Puerto Rico, and the Virgin Islands are available, free on request, from the USGS or from the Federal Center, Denver, Colorado 80225.

†A booklet published by the USGS Earth Resources Observations System, entitled *Studying the Earth from Space*, is available from the U.S. Government Printing Office, Washington, D.C. 20402.

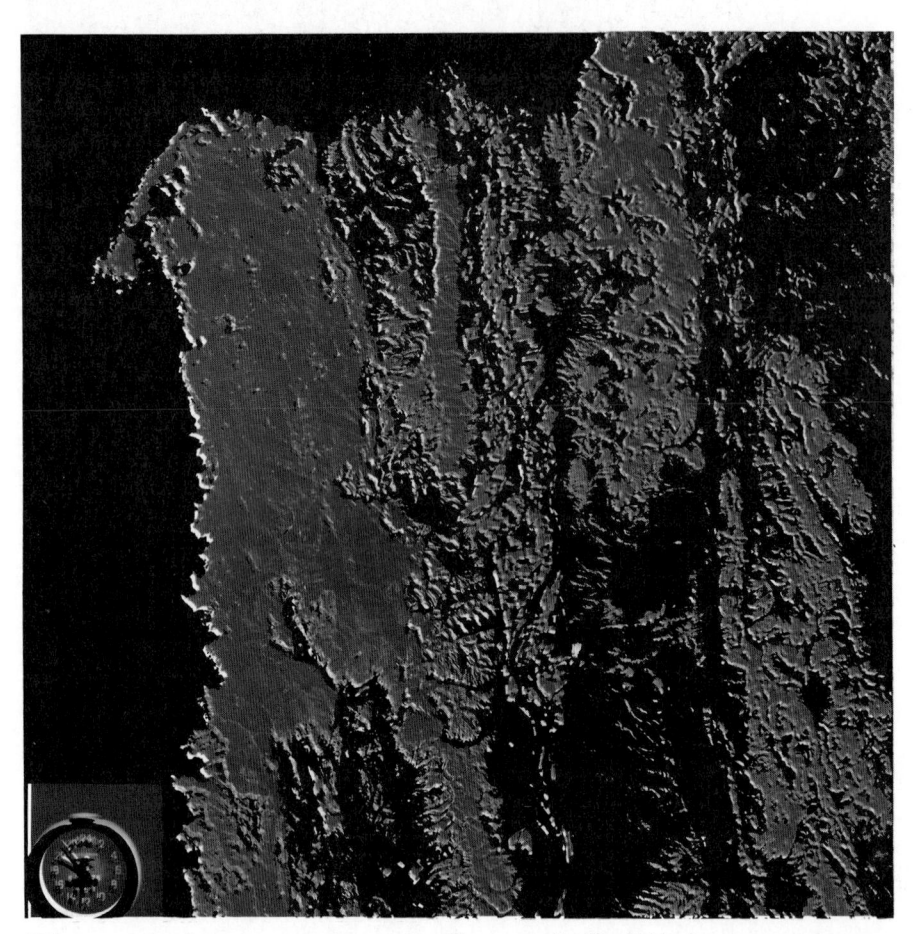

Airborne radar imagery: remote sensing The new scientific field of remote sensing uses electromagnetic equipment and cameras with infrared film. These techniques enable earth scientists to expand their knowledge of the earth far beyond that gained from the use of black and white or color aerial photographs or from study in the field. *Remote sensing* means, literally, "detecting the nature of an object without actually touching it." Remote-sensing equipment carried by high-flying aircraft and orbiting satellites can detect (sense) the differences in infrared reradiation emitted by soils, minerals, vegetation, waters, and man-made features. The images produced by the equipment provide information that cannot be known from closer observation, such as the existence of mineral deposits, the changing salinity of a body of water, the spread of disease in agricultural areas or regions of natural vegetation, and the presence of fractures or faults in the earth's crust. In this remote-sensing photograph of the Salinas River Valley south of Monterey Bay, California (taken from *Skylab 3* space station in earth orbit), the irrigated agricultural area is easily distinguishable, not only because of the highway and rectangular field pattern but also because of the color differences, enhanced by infrared photography.

contour lines) and natural and cultural features from aerial photography is called *photogrammetry*. Here, the cartographer uses stereoscopic viewers to study two photos that have overlapping sections, thus producing a three-dimensional representation of the earth's surface. Sometimes the land-survey party is "too close to the forest to see the trees," and a photograph from space provides startling information that was overlooked by those on land. When all the data are accumulated, the map is created. Additional field checking is undertaken, details are added, and, finally, place names and political boundaries are entered.

National standards of accuracy in both horizontal and vertical phases of each quadrangle are an important consideration, and each map that meets these standards carries a statement to that effect in the lower margin. For example, horizontal positions of at least 90 percent of the well-defined features must be correctly plotted within 1/50 inch on the published map, and elevations of 90 percent of the points tested vertically shall agree with the elevations interplotted from the contour lines within one-half the contour interval. This tolerance allowed for horizontal positions is equivalent to 40 feet (12 meters) on the ground for 1:24,000-scale maps and about 100 feet (30 meters) on the ground for 1:62,500-scale maps.

Map scales represent the relationship between a distance on the map to the actual distance of that part of the earth's surface represented by the map. Topographic map scales are usually presented as a graphic (bar) scale and as a representative fraction scale (Fig. B-1; see also Table B-1). For example, 1:24,000 means that one unit on the map equals 24,000 of the same units on the earth's surface. Therefore, 1 inch equals 24,000 inches, or 1 foot equals 24,000 feet, and so on. One can also convert the right-hand number to another unit form, such as converting the R.F. 1:24,000 to a verbal scale; for example, "1 inch equals 2000 feet," or "1 centimeter equals 240 meters." The decision of which scale should be used depends on the amount of detail needed. A map with a large scale is one that shows an extensive amount of detail and is especially useful for highly developed areas or rural areas where detailed information is needed for engineering planning or similar purposes.

Medium-scale maps (1:62,500) may be adequate for rural areas where less detailed planning is contemplated. Small-scale maps (1:250,000 and smaller) cover very large areas on a single sheet and are useful in the study of extensive projects or regional planning. As to which map series should be used by a camper, backpacker, or skier, it is obvious that it would be the largest scale map (Table B-1) available— the 7½-minute series, which covers about 49 to 70 square miles on one sheet.

The USGS is not the only agency producing and publishing maps for our use. Other federal agencies include the U.S. Army Map Service, Tennessee Valley Authority, Mississippi River Commission, and (for the person who enjoys ocean sailing). the U.S. Coast and Geodetic Survey, which publishes navigational charts of the waters off our coasts.

Topographic map symbols are used to explain the map. In a sense, the symbols comprise a "language" for us to read. Figure B-2 illustrates some of the more common symbols; it must be understood that symbols come in various styles, shapes, and colors and that each has its own definite meaning. Usually, the contour lines—the lines that connect points of equal elevation above sea level—are brown; streams and other water bodies are blue; natural vegetation is green; buildings are black; boundaries are designated by black dashes and dots; and so on.

Contours cannot be found on the earth;

LARGE SCALE
1:24,000 scale
1 in. equals 2000 ft.
Area shown is 1 square mile.

MEDIUM SCALE
1:62,500 scale
1 in. equals nearly 1 mile.
Area shown is 6 square miles.

SMALL SCALE
1:250,000 scale
1 in. equals nearly 4 mi.
Area shown is 95 square miles.

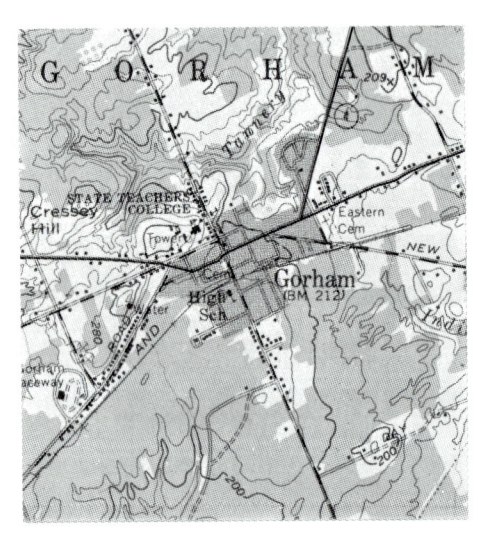

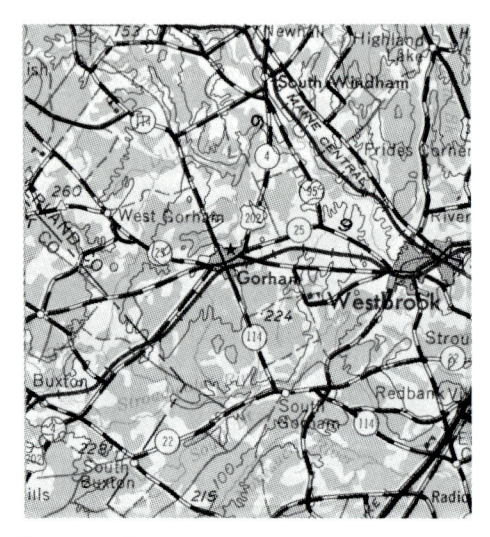

FIGURE B-1 Topographic map scales: large—medium—small.

TABLE B-1 National Topographic Map Series.

SERIES	SCALE	1 INCH REPRESENTS	STANDARD QUADRANGLE SIZE (LATITUDE-LONGITUDE)	QUADRANGLE AREA (SQUARE MILES)	PAPER SIZE E-W N-S WIDTH LENGTH (INCHES)
7½-minute	1:24,000	2000 feet	7½ × 7½ minute	49 to 70	22 × 27[a]
Puerto Rico 7½-minute	1:20,000	About 1667 feet	7½ × 7½ minute	71	29½ × 32½[a]
15-minute	1:62,500	Nearly 1 mile	15 × 15 minute	197 to 282	17 × 21[a]
Alaska 1:63,360	1:63,360	1 mile	15 × 20 to 36 minute	207 to 281	18 × 21[b]
U.S. 1:250,000	1:250,000	Nearly 4 miles	1° × 2°[c]	4580 to 8669	34 × 22[d]
U.S. 1:1,000,000	1:1,000,000	Nearly 16 miles	4° × 6°[c]	73,734 to 102,759	27 × 27

[a]South of latitude 31° 7½-minute sheets are 23 × 27 inches; 15-minute sheets are 18 × 21 inches.
[b]South of latitude 62° sheets are 17 × 21 inches.
[c]Maps of Alaska and Hawaii vary from these standards.
[d]North of latitude 42° sheets are 29 × 22 inches. Alaska sheets are 30 × 23 inches.

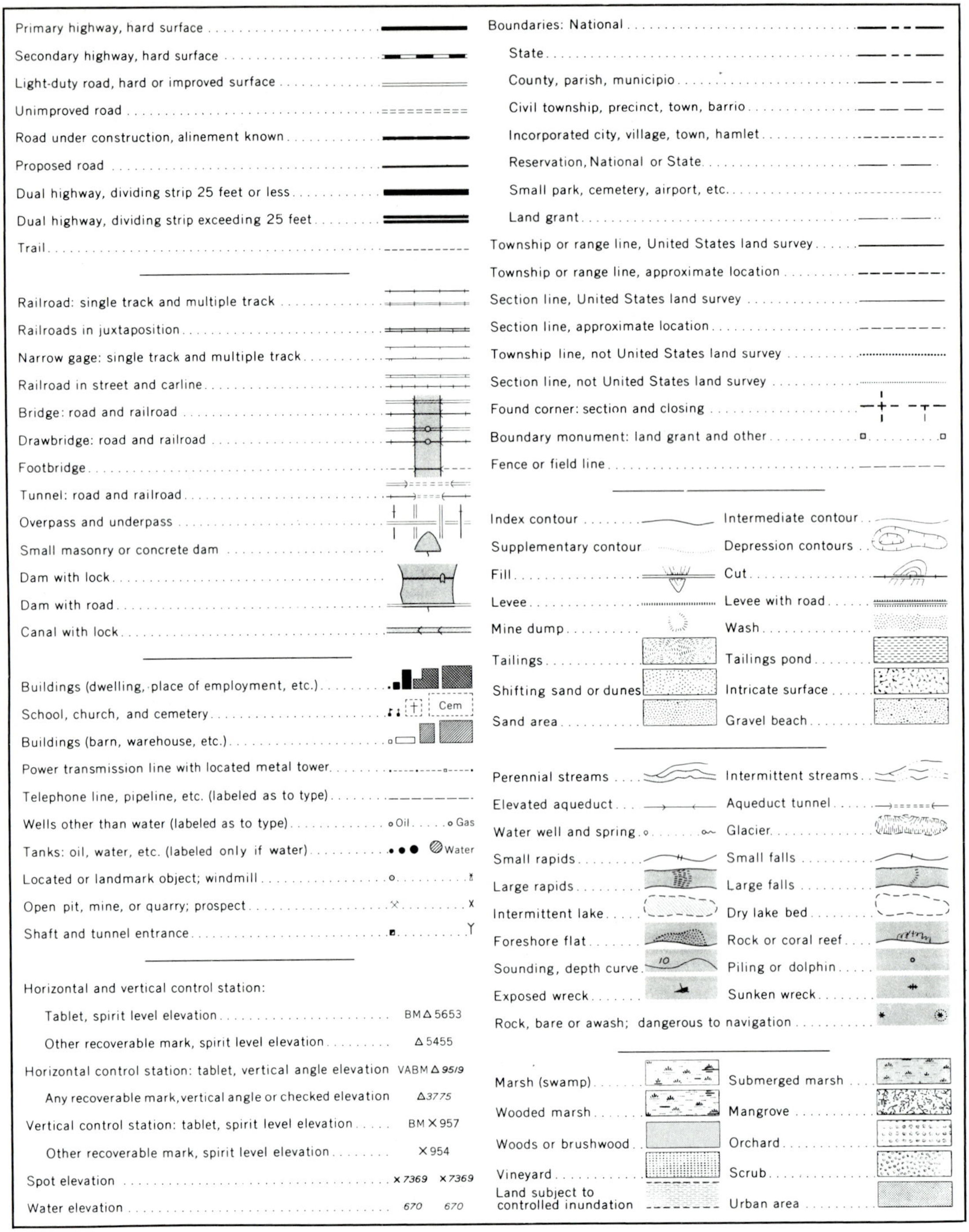

Primary highway, hard surface	━━━━━
Secondary highway, hard surface	━━━━━
Light-duty road, hard or improved surface	─────
Unimproved road	=====
Road under construction, alinement known	─────
Proposed road	─────
Dual highway, dividing strip 25 feet or less	━━━━━
Dual highway, dividing strip exceeding 25 feet	━━━━━
Trail	-----
Railroad: single track and multiple track	
Railroads in juxtaposition	
Narrow gage: single track and multiple track	
Railroad in street and carline	
Bridge: road and railroad	
Drawbridge: road and railroad	
Footbridge	
Tunnel: road and railroad	
Overpass and underpass	
Small masonry or concrete dam	
Dam with lock	
Dam with road	
Canal with lock	
Buildings (dwelling, place of employment, etc.)	
School, church, and cemetery	Cem
Buildings (barn, warehouse, etc.)	
Power transmission line with located metal tower	
Telephone line, pipeline, etc. (labeled as to type)	
Wells other than water (labeled as to type)	o Oil o Gas
Tanks: oil, water, etc. (labeled only if water)	● ● Water
Located or landmark object; windmill	o
Open pit, mine, or quarry; prospect	× X
Shaft and tunnel entrance	▪ Y
Horizontal and vertical control station:	
Tablet, spirit level elevation	BM △ 5653
Other recoverable mark, spirit level elevation	△ 5455
Horizontal control station: tablet, vertical angle elevation	VABM △ 95/9
Any recoverable mark, vertical angle or checked elevation	△ 3775
Vertical control station: tablet, spirit level elevation	BM × 957
Other recoverable mark, spirit level elevation	× 954
Spot elevation	× 7369 × 7369
Water elevation	670 670

Boundaries: National	
State	
County, parish, municipio	
Civil township, precinct, town, barrio	
Incorporated city, village, town, hamlet	
Reservation, National or State	
Small park, cemetery, airport, etc.	
Land grant	
Township or range line, United States land survey	
Township or range line, approximate location	
Section line, United States land survey	
Section line, approximate location	
Township line, not United States land survey	
Section line, not United States land survey	
Found corner: section and closing	+ T
Boundary monument: land grant and other	▫ ▫
Fence or field line	
Index contour	Intermediate contour
Supplementary contour	Depression contours
Fill	Cut
Levee	Levee with road
Mine dump	Wash
Tailings	Tailings pond
Shifting sand or dunes	Intricate surface
Sand area	Gravel beach
Perennial streams	Intermittent streams
Elevated aqueduct	Aqueduct tunnel
Water well and spring	Glacier
Small rapids	Small falls
Large rapids	Large falls
Intermittent lake	Dry lake bed
Foreshore flat	Rock or coral reef
Sounding, depth curve	Piling or dolphin
Exposed wreck	Sunken wreck
Rock, bare or awash; dangerous to navigation	
Marsh (swamp)	Submerged marsh
Wooded marsh	Mangrove
Woods or brushwood	Orchard
Vineyard	Scrub
Land subject to controlled inundation	Urban area

FIGURE B-2 Topographic map symbols.

mined by surveyors' measurements in the field or from photogrammetry. A certain specific interval (space) is established between the contours (Fig. B-3), such as 20 feet, and we can then tell whether the land is smooth and nearly level or rugged, with hills and valleys, by noting just how the contours are laid out. If the contours are located close together, it means a sharp or rapid difference, a rise or fall that is steep and abrupt. If the lines are widely separated, it means a gentle rise or decline. To determine the elevation of a point situated between two contours, you must interpolate, or approximate, the height of that point.

The margins of topographic maps contain much important information, such as the contour interval, the map's scale, meridians of longitude, parallels of latitude, amount of magnetic declination, geographic location of the mapped area, the year the map was constructed and if and when it was revised, the projection used, and the name of the quadrangle. Take a close look at Fig. B-4, which is a portion of the Bright Angel Quadrangle, Grand Canyon National Park, Arizona. Examine the information supplied in the margins and then study the map itself. What is the contour interval? Can you locate permanent and intermittent streams? Which are deep valleys and which are flat-topped mesas? What is the elevation of Pattie Butte?

Now compare this map with Figure B-5. The area encompassed by the map is exactly the same as that in Figure B-4, but there is a difference. This one has *relief shading* (some call it *plastic shading*) to make the buttes, mesas, canyons and valleys, and the river stand out more clearly. Which of these maps do you find easier to read and understand? The USGS has developed a new method of aerial photography to produce maps that are easy to understand. They are called *orthophotoquads* (Fig. B-6).

Using the topographic map is not difficult, especially after you have learned to read it and know how to use a compass to orient yourself, or to relate your actual location on the earth to that point on the map. Once this is done, you can maintain this relationship by lining up your compass with the geographic north and the magnetic north. Then you can take bearings (compass readings) on distant objects (Fig.

(b)

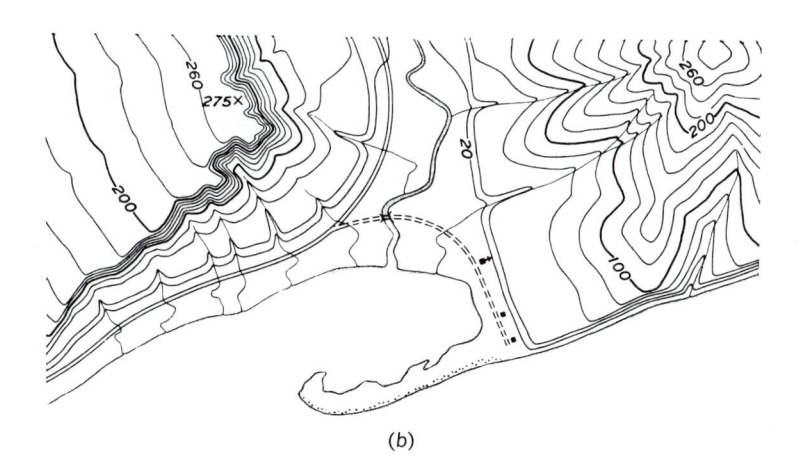

FIGURE B-3 Topographic maps Topo maps are drawn to represent a specific portion of earth space. This figure is a simplified topographic map of a hilly coastline complete with cliffs, streams, a highway, and a hooked spit. Contour lines are set at 20-foot (6-meter) intervals.

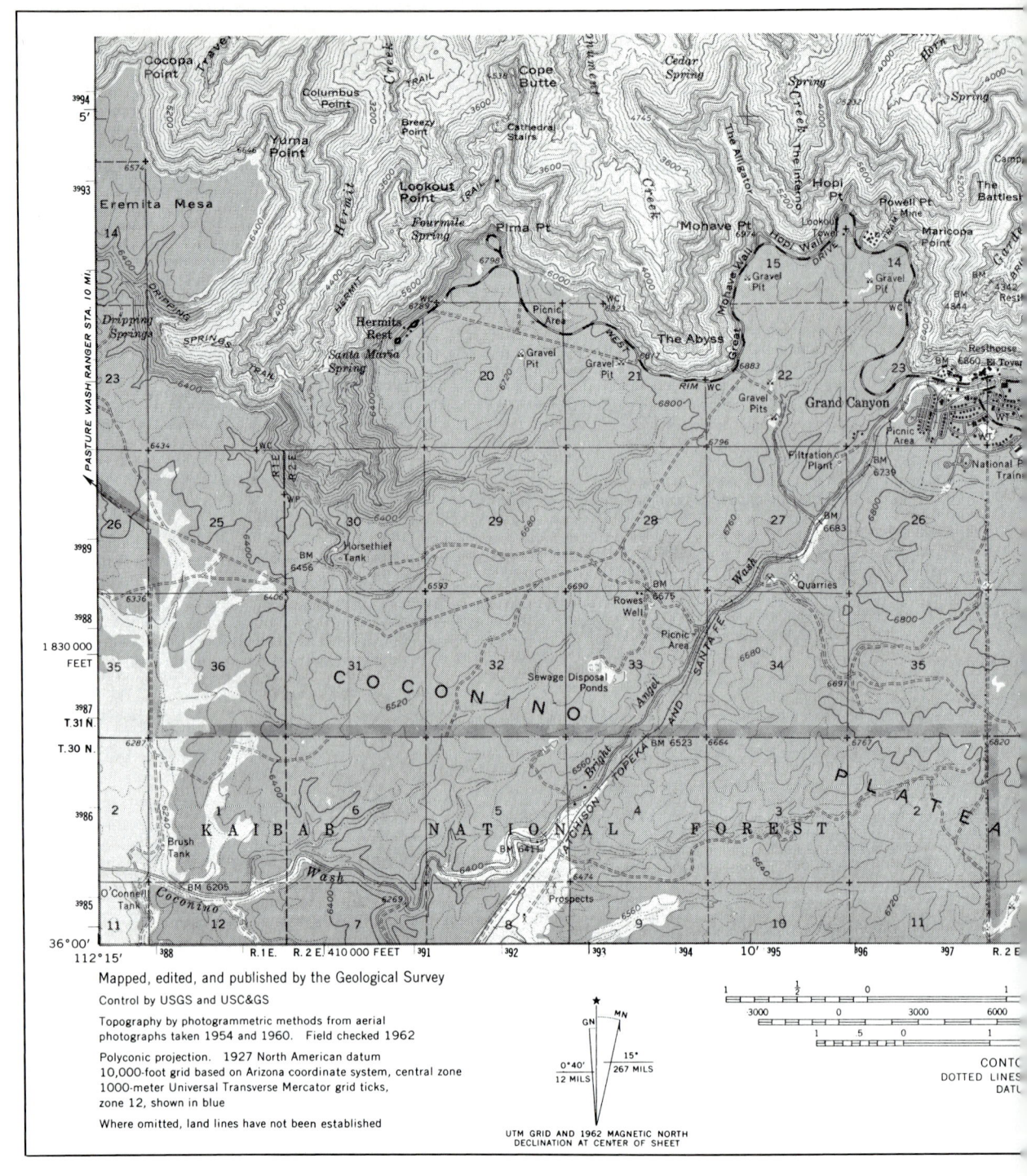

FIGURE B-4 Standard 15-minute series map: Bright Angel Quadrangle, Grand Canyon National Park, Arizona.

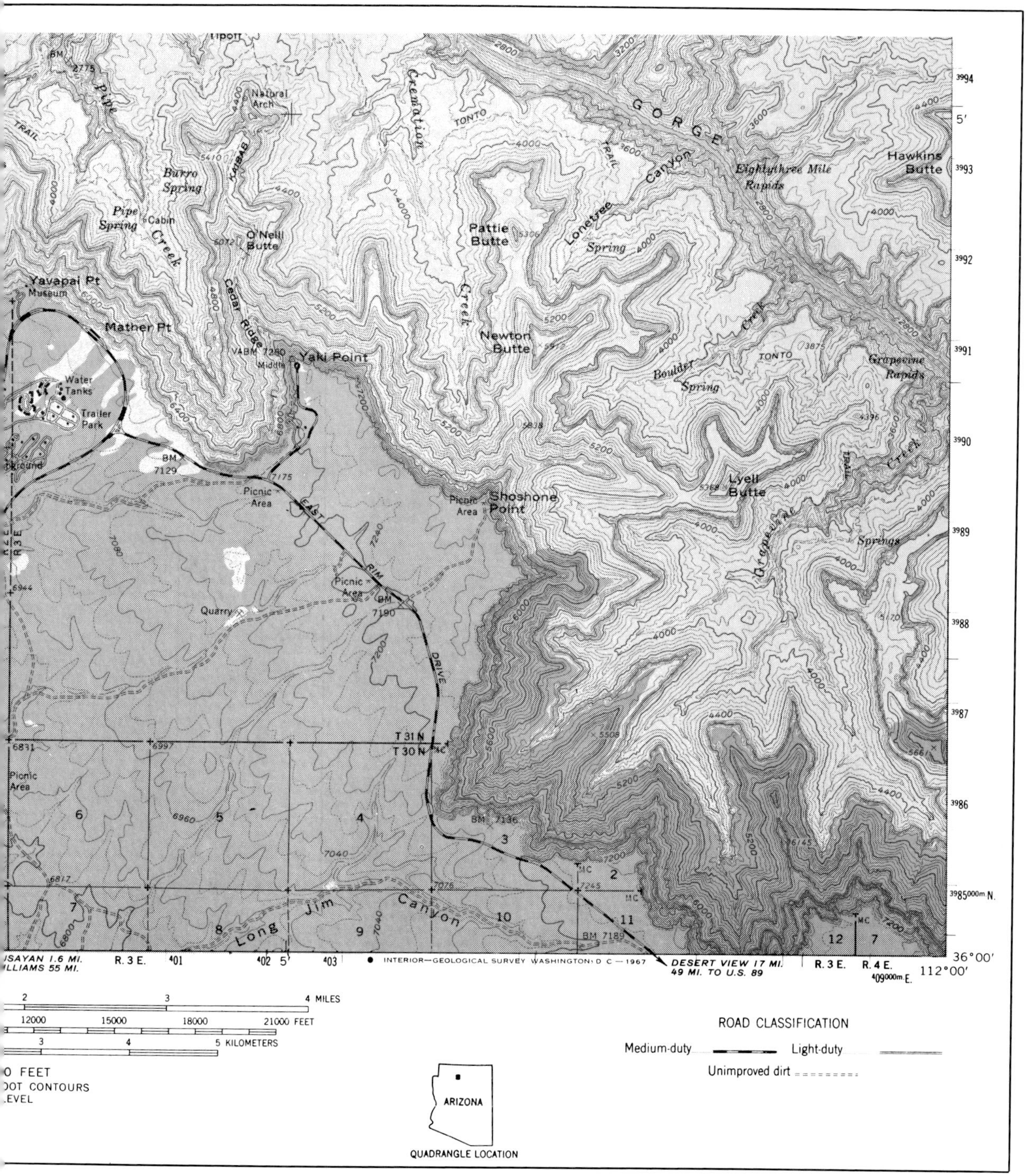
Natural Arch
Pipe Creek
Burro Spring
Pipe Spring
Cabin
O'Neill Butte
Cedar Ridge
Yavapai Pt
Museum
Mather Pt
Yaki Point
Water Tanks
Trailer Park
Cremation Creek
TONTO TRAIL
GORGE
Lonetree Canyon
Lonetree Spring
Eightythree Mile Rapids
Hawkins Butte
Pattie Butte
Newton Butte
Boulder Creek
Boulder Spring
Grapevine Rapids
Grapevine Creek
Springs
Lyell Butte
Shoshone Point
Picnic Area
EAST RIM DRIVE
Quarry
T 31 N
T 30 N
Long Jim Canyon
TUSAYAN 1.6 MI.
WILLIAMS 55 MI.
R. 3 E.
R. 4 E.
INTERIOR—GEOLOGICAL SURVEY WASHINGTON, D C — 1967
DESERT VIEW 17 MI.
49 MI. TO U.S. 89
36°00′
112°00′
4 MILES
21000 FEET
5 KILOMETERS
ROAD CLASSIFICATION
Medium-duty
Light-duty
Unimproved dirt
ARIZONA
QUADRANGLE LOCATION

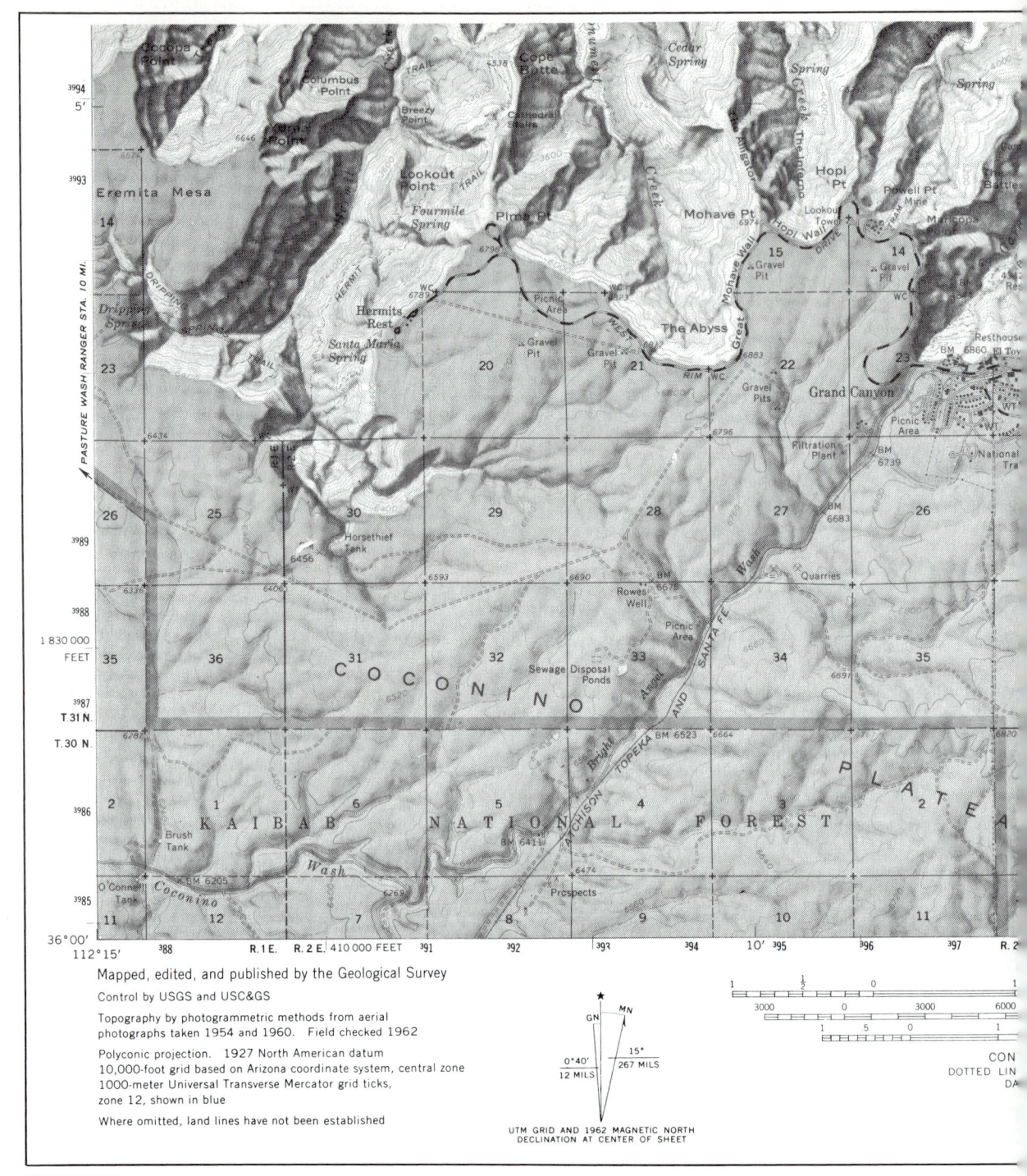

FIGURE B-5 **Relief-shaded map: Bright Angel Quadrangle, Grand Canyon National Park, Arizona** This map is of the same area shown in Figure B-4. The shading is added to accentuate the evenness of the terrain.

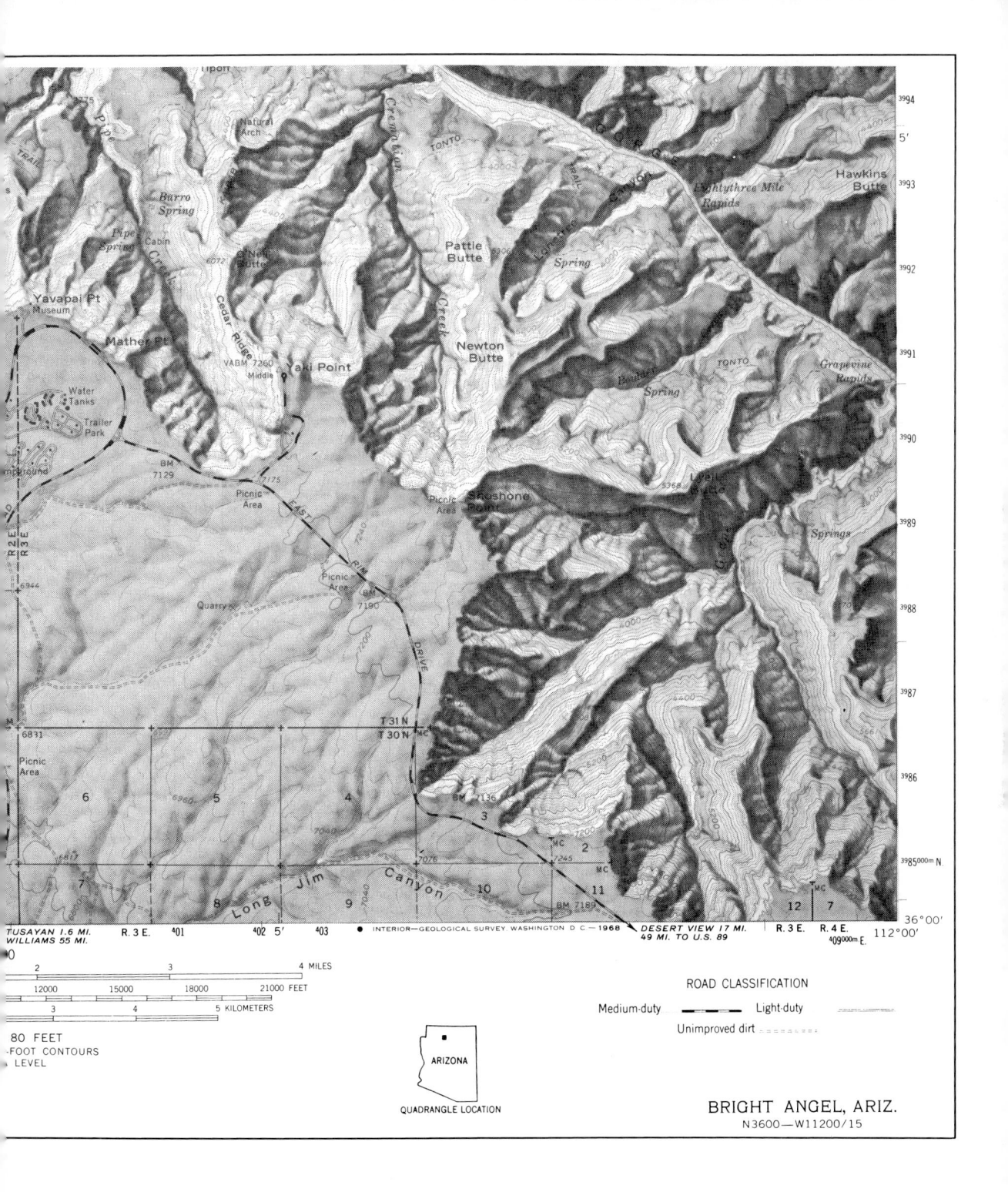

2 3 4 MILES

12000 15000 18000 21000 FEET

3 4 5 KILOMETERS

80 FEET
-FOOT CONTOURS
LEVEL

ROAD CLASSIFICATION

Medium-duty Light-duty

Unimproved dirt

ARIZONA

QUADRANGLE LOCATION

BRIGHT ANGEL, ARIZ.
N3600—W11200/15

(a)

(b)

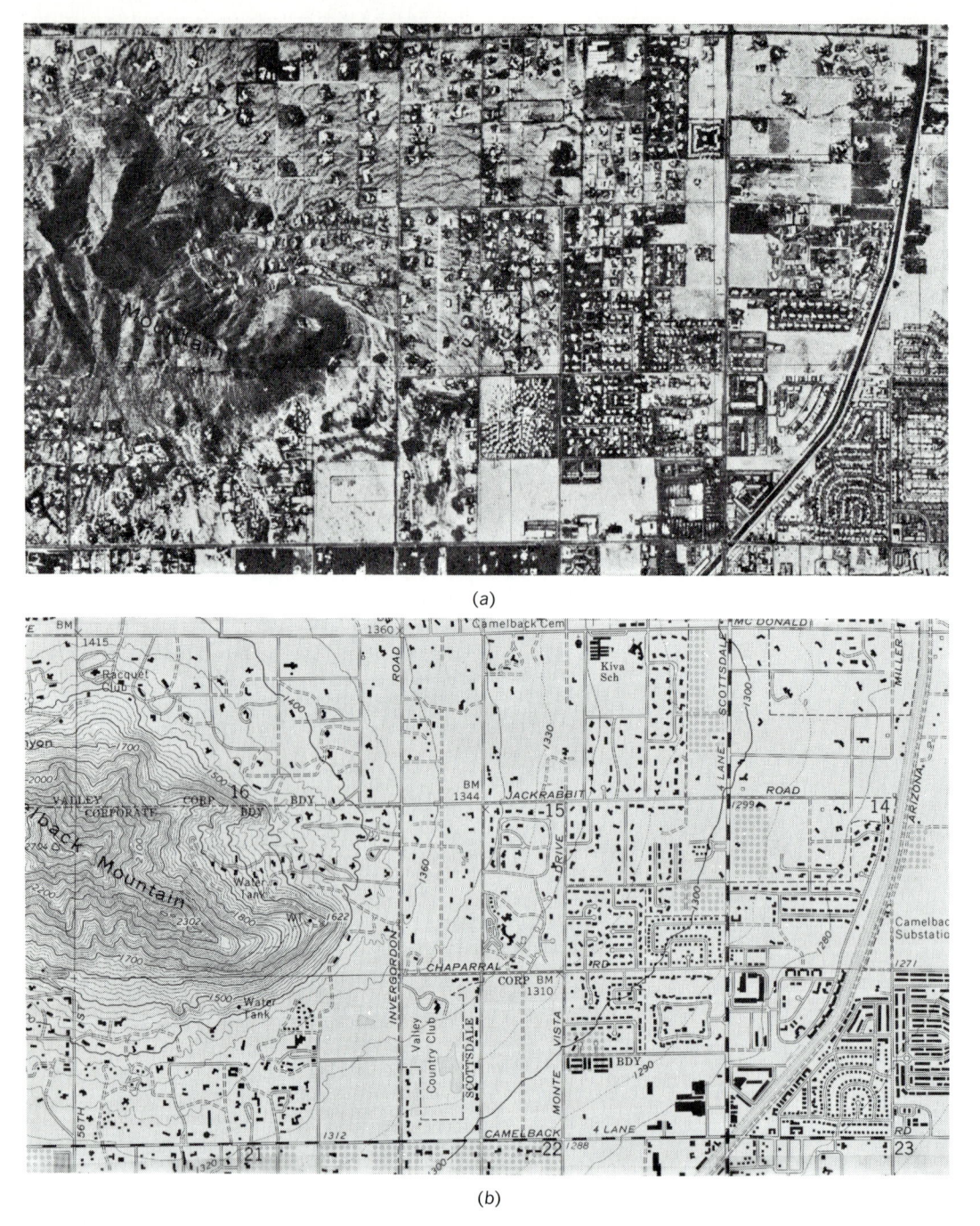

FIGURE B-6 Orthophotoquad—topographic map (a) is an orthophotoquad (orthophotomap) of a section of Paradise Valley, Arizona. The form of aerial photography illustrated by (a) is intended to depict the terrain and other standard topographic map features by color-enhanced photographic images. Compare with (b).

B-7), and you will always know your location. Geographers normally refer to a place's *azimuth** to denote its direction from the observer. Azimuth is a place's position with respect to (or relative to) a meridian, forming an angle that is expressed in degrees. A compass has a circle of 360°, and geographic north is 0°, due east is 090°, due south is 180°, and due west is 270°. When you are doing fieldwork or

*The term *azimuth* comes to us from the Arabic words *al* (the) and *zimut* (the way), thus, "the way" or "the direction" to travel.

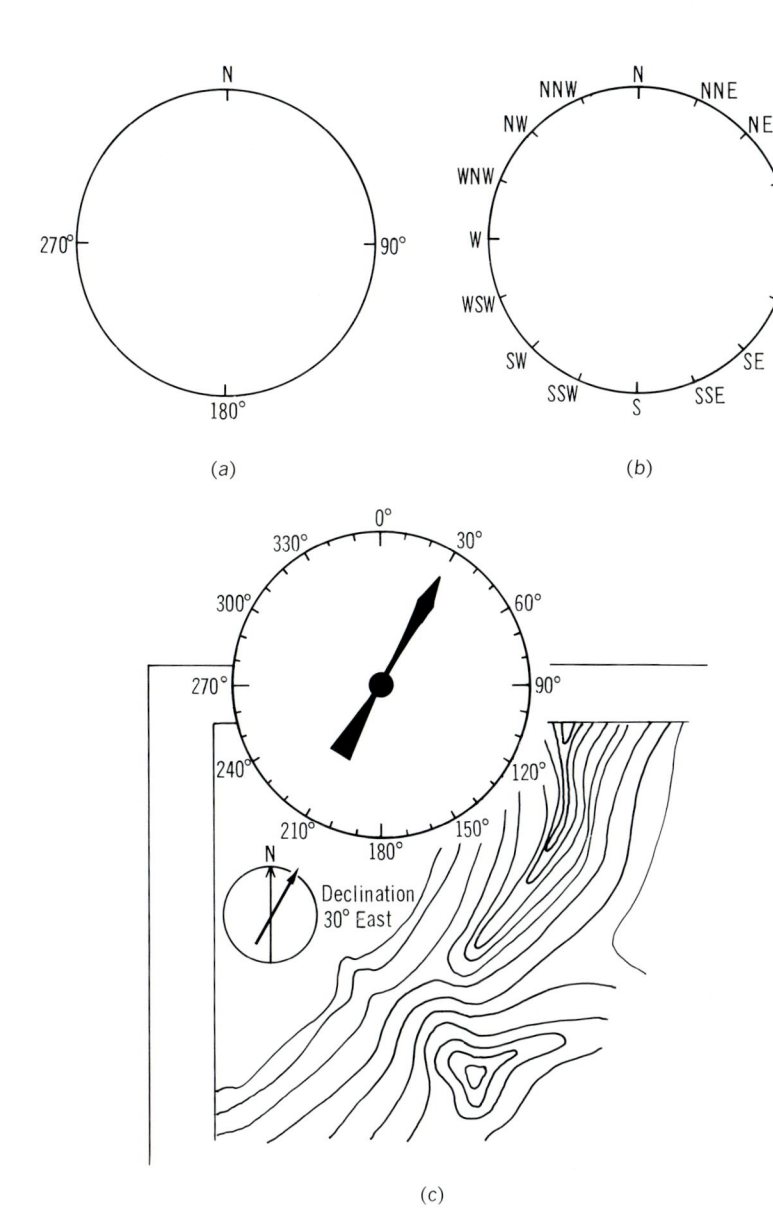

FIGURE B-7 **Direction finding with a compass** (*a*) shows the prime bearings of an object to true north. Due east is 90°, due south is 180°, and due west is 270°. (*b*) shows a mariner's compass card. Aboard a ship, the ship's course is given as "north by northeast" rather than "steer a course of 020°." (*c*) shows how to align a magnetic compass with a map's compass rose. Modern maps used for navigation charts or topographic maps also show the amount of magnetic variation, or declination (deviation from true north). For accuracy, geographers are apt to use a 360° circle, calling out bearings and azimuths as "azimuth 225°." Adding terminology such as "by north" or "azimuth 225° west of south" is unnecessary.

FIGURE B-8 Why maps must be revised The need for occasional revision of maps becomes obvious when one sees the changes wrought by people. (*a*) shows a portion of Pompano Beach, Florida, in 1946; (*b*) shows the same area in 1962. Note the inland spread of urbanization, the addition of more canals for marina use, and the major increase in the number of streets and highways.

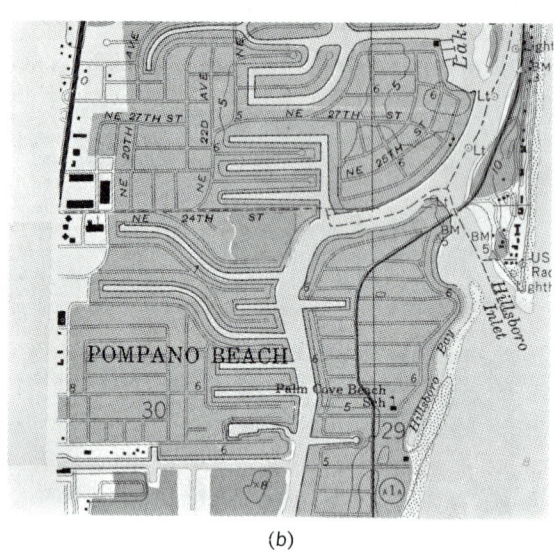

are out on a hike, you will use a magnetic compass to determine the azimuths of places or objects around you, but you have to keep in mind another fact. The needle of the magnetic compass does not point toward geographic north. Indeed, the needle can vary considerably from geographic north—perhaps as much as 15° to 30°. This is because the magnetized point of the needle aims for the magnetic North Pole. You must keep this variation, or declination, constantly in mind. Fortunately, your topographic map tells you what the variation is, so you can make the proper adjustment.

One final comment concerning topographic maps is that they are in need of frequent revision. The forces of nature, as well as the hand of humans, can bring about great changes in both the natural and the cultural landscape. Figure B-8 shows why the USGS believes maps must be revised. In this instance, a quiet and secluded oceanfront retreat became a relatively crowded residential community in less than 16 years. To be certain your map is up to date, check the date listed in the lower margin. After all, who knows what "they" may have done to your favorite vacation or camping area since you were last there.

ADDITIONAL READING

Estes, John E., and Leslie W. Senger (1974). *Remote Sensing: Techniques for Environmental Analysis.* New York: John Wiley & Sons.

Holz, Robert K., ed. (1973). *The Surveillant Science: Remote Sensing of the Environment.* Boston: Houghton Mifflin.

APPENDIX C
GROUND REFERENCING SYSTEMS

All map projections we have considered have used the geographic grid to pinpoint the location of landmasses (with the extreme exception of the cartograms) because latitude and longitude is the most efficient and accurate way of determining a place's location. But there are other methods. For example, do you recall reading a story as a child about a pirate who buried treasure he had taken from the ships along the Spanish Main? His map was crude, but the essential points were all there. "Walk away from the beach until you reach the large black rock; then walk 10 paces north, and turn to your left toward a gnarled tree. Walk 20 paces. Stop. Dig to a depth of 4 feet." For centuries this was the way people mapped the position of things, and this method was given a name: *metes and bounds.**

Confusion was the ultimate result of the migration of peoples to the New World: Spanish, French, and English, each with their own methods of determining landownership and boundary lines. For example, when Spanish and Mexican settlers migrated to California, they were given grants of land by the Spanish king and a description of the particular section of land being granted. However, the descrip-

*This term comes to us from the original Latin and the later Anglo-Saxon and means "to measure and fix the limits and the boundary" of a place without regard to any overall or public plan of survey system.

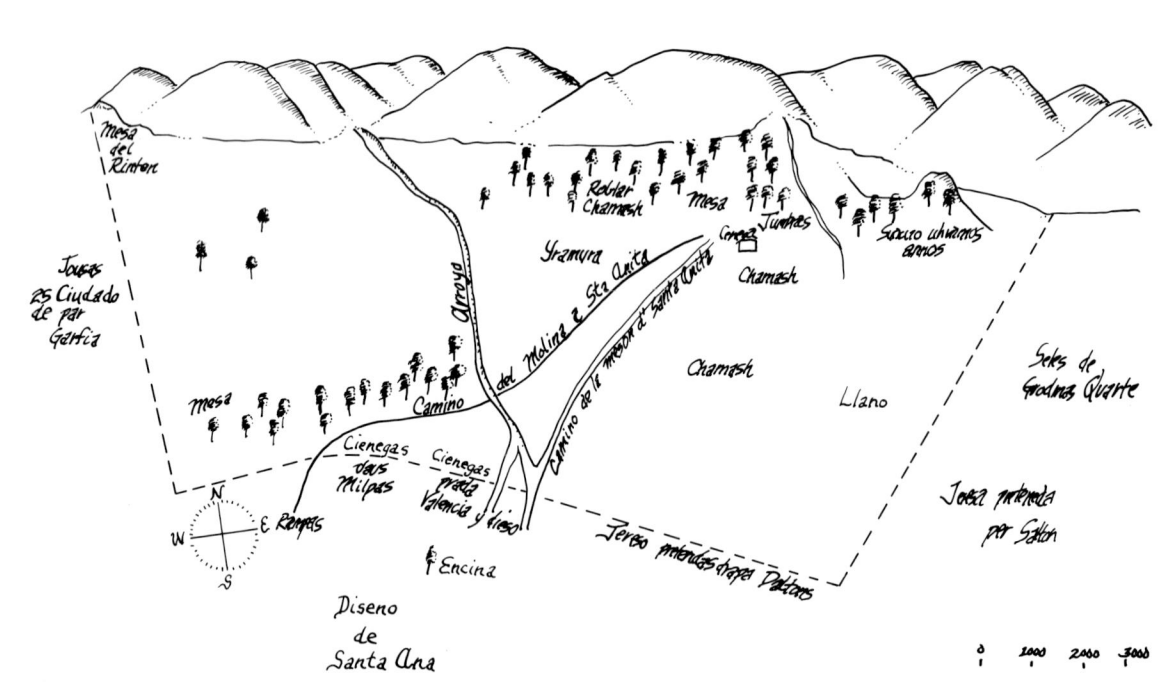

The early Spanish method of determining boundaries: the *diseño* system of early California. (Adapted from a sketch in W. W. Robinson [1953], *Panorama, A Picture History of Southern California* [Los Angeles: Title Guarantee & Trust Co.], p. 31.)

GROUND REFERENCING SYSTEMS 443

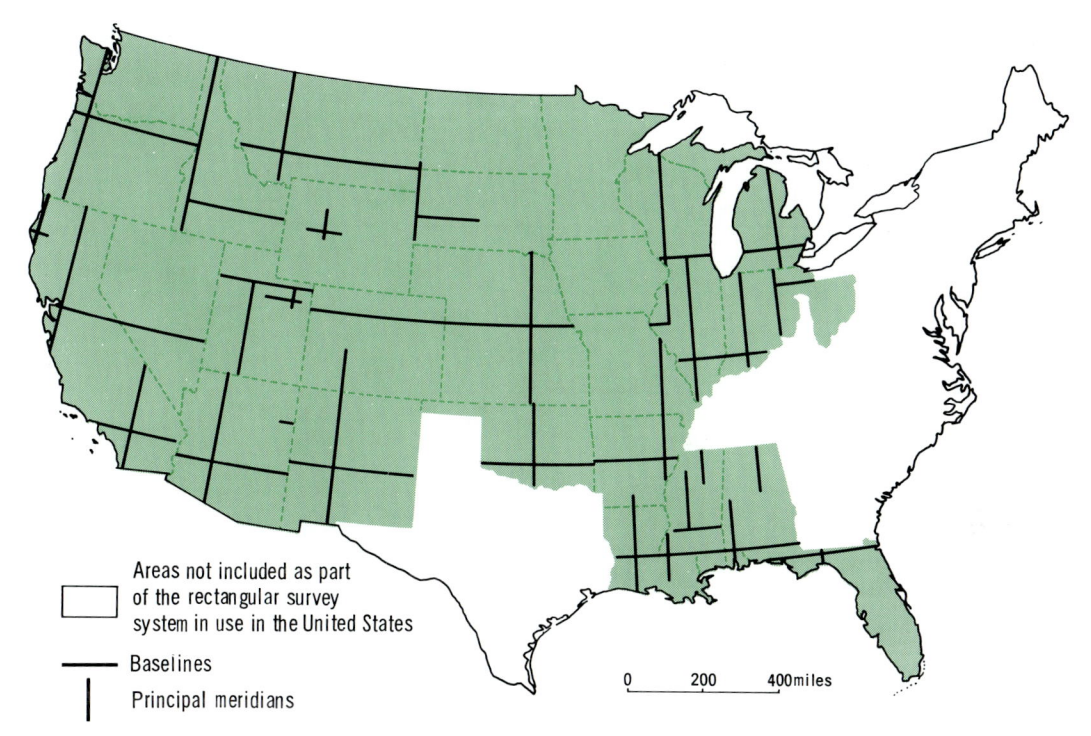

FIGURE C-1 Rectangular survey system in the United States [For further information on the land survey system used in the United States, see Francis J. Marschner, U.S. Department of Agriculture (1959), *Land Use and Its Patterns in the United States* (Agricultural Handbook no. 153), (Washington, D.C.: U.S. Government Printing Office).]

tion was not tied to any primary or fixed point. When California came under the jurisdiction of the U.S. government and its Public Land Survey System was instituted, arguments over the limits of each parcel of land erupted. A new grid system had evolved, with true compass directions and distance measurements more accurately given in feet instead of in steps or paces. Still later a more systematic method, the rectangular survey system, was developed. This method was primarily used for lands being settled in the Midwest and the western United States.

In this new land survey system each plot of land surveyed had definite boundaries, identified by its relationship to a selected meridian (called the *principal meridian*) and a selected parallel (called the *baseline*) (Fig. C-1). The rectangular shapes so arranged are called *townships* (which are 6 miles on each side), and these are subdivided into *sections* (which are 1 mile squares). Figure C-2 illustrates the way the township-and-range system is determined. As you can see, each township has its own designation, such as T4N, R3W, which means "township 4 north (or fourth township north of the baseline), range 3 west (or third range west of the principal meridian)." The meridians are always at right angles to the baselines and standard parallels. Each section can be further subdivided—depending on the land-use patterns of the people in that area.

444 GROUND REFERENCING SYSTEMS

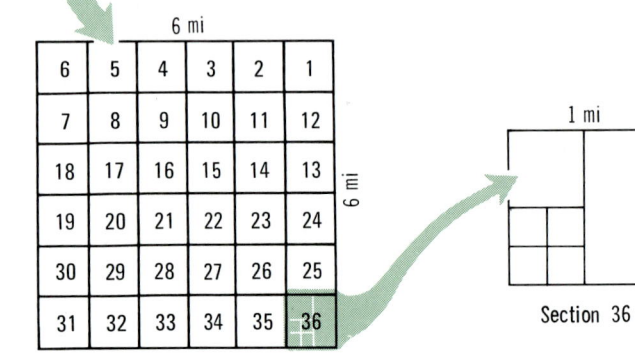

FIGURE C-2 Township-and-Range System.

APPENDIX D

THE METRIC SYSTEM—HOW TO USE IT

The Metric System—How to Use It

IF YOU KNOW	MULTIPLY BY	TO FIND
Temperature		
Fahrenheit (F)	5/9 (or 0.56), after subtracting 32	Celsius (C)
Example: 75°F	75 − 32 = 43	
	43 × 5/9 = 24	24°C
Celsius (C)	9/5 (or 1.8), then add 32	Fahrenheit (F)
Example: 30°C	30 × 9/5 = 54	
	54 + 32 = 86	86°F
Length		
Inches (in.)	25.4	Millimeters (mm)
Inches (in.)	2.54	Centimeters (cm)
Feet (ft)	30	Centimeters (cm)
Feet (ft)	0.31	Meters (m)
Yards (yd)	0.9	Meters (m)
Miles (mi)	1.6	Kilometers (km)
Millimeters (mm)	0.04	Inches (in.)
Centimeters (cm)	0.4	Inches (in.)
Meters (m)	3.3	Feet (ft)
Meters (m)	1.1	Yards (yd)
Kilometers (km)	0.6	Miles (mi)
Area		
Square inches (in.^{2})	6.5	Square centimeters (cm^2)
Square feet (ft^2)	0.09	Square meters (m^2)
Square yards (yd^2)	0.84	Square meters (m^2)
Square miles (mi^2)	2.6	Square kilometers (km^2)
Acres	0.41	Hectares (ha)
Square centimeters (cm^2)	0.16	Square inches (in^2)
Square meters (m^2)	1.2	Square yards (yd^2)
Square kilometers (km^2)	0.4	Square miles (mi^2)
Hectares (ha)	2.5	Acres
Volume		
Cubic inches (in.^{3})	16.39	Cubic centimeters (cm^3)
Cubic feet (ft^3)	0.28	Cubic meter (m^3)
Cubic yard (yd^3)	0.77	Cubic meter (m^3)
Cubic meters (m^3)	35	Cubic feet (ft^3)
Cubic meters (m^3)	1.3	Cubic yards (yd^3)
Cubic miles (mi^3)	4.1	Cubic kilometers (km^3)
Cubic kilometers (km^3)	0.25	Cubic miles (mi^3)
Mass (Weight)		
Ounces (oz)	28	Grams (g)
Pounds (lb)	0.45	Kilograms (kg)
Short tons	0.9	Tonnes (t)
Grams (g)	0.035	Ounces (oz)
Kilograms (kg)	2.2	Pounds (lb)
Tonnes (1000 kg) (t)	1.1	Short tons

PHOTO CREDITS

Chapter 1

Opener: Owen Franken/Stock Boston. Page 7: NASA, Florida News Bureau, Lick Observatory Photograph. Page 8: Photo Documentation Française, U.S. Air Force Photo, Tennessee Valley Authority, USDA-Soil Conservation Service, SDA photo, Royal Danish Ministry for Foreign Affairs, Geological Survey of Canada #1470, National Park Service, Italian Government Travel Office.

Chapter 2

Opener: Culver Pictures. Page 14: Musie De Turin. Page 17: NASA. Page 18: Norman J.W. Thrower, *Maps & Man: An Examination of Cartography in Relation to Culture and Civilization*, © 1972, p. 45. Reprinted by permission of Prentice-Hall, Inc., Englewood Cliffs, New Jersey. Page 19: NASA. Page 46–47: F.V. Thierfeldt Co. © F.V. Theirfeldt, Milwaukee, Wisconsin.

Chapter 3

Opener: NASA. Page 54: NASA. Page 58: Smithsonian Institution. Page 63: Courtesy of SAS Scandinavian Airlines. Page 66: Charles J. Ott/Photo Researchers. Page 80: © 1982 Joel W. Rogers/Earth Images, Alan Pitcharin/Grant Heilman, Jerome Wyckoff. Page 82: Ira Kirschenbaum/Stock Boston. Page 84: From "A Golden Thread" by Ken Butti and John Perlin. Page 85: Richard Choy/Peter Arnold.

Chapter 4

Opener: Richard Wood/The Picture Cube. Page 94: Carl Frank/Photo Researchers. Page 107: Russ Kinne/Photo Researchers, Runk-Schoenberger/Grant Heilman, Kahlseco. Page 118: NASA.

Chapter 5

Opener: Grant Heilman. Page 128: NOAA, Tennessee Valley Authority. Page 130: Grant Heilman. Page 135: Bonne Morrison. Page 140: NOAA. Page 143: David Petty/Photo Researchers, Grant Heilman. Page 145: NOAA.

Chapter 6

Opener: Grant Heilman. Page 154: Bonne Morrison. Page 159: Mike Mallaschi/Stock Boston. Page 160: Tom McHugh/Photo Researchers. Page 165: Hamilton Wright. Page 169: USGS. Page 179: Grant Heilman. Page 181: Ira Kirschenbaum/Stock Boston. Page 184: Michel Brigaud, French Embassy.

Chapter 7

Opener: Otto Angermayer/Photo Researchers. Page 200: Peter Menzel/Stock Boston. Page 202: G.R. Roberts. Page 203: Arabian American Oil Co. Page 205: Peter Menzel/Stock Boston. Page 207: Tennessee Valley Authority. Page 209: George Hunter. Page 211: Northway Survey Corp. Ltd. Page 212: Tia Schneider Denenberg/Photo Researchers. Page 214: Novosti Press Agency Photo. Page 216: Almasy. Page 218: Official U.S. Navy Photograph. Page 219: Official U.S. Navy Photograph.

Chapter 8

Opener: Jim Yoakum. Page 225: Photo Larousse, Photo Giraudon, National Museum Copenhagen. Page 228: Grant Heilman, Jerome Wyckoff, Carl Frank/Photo Researchers. Page 230: Jerome Wyckoff. Page 232: Grant Heilman, SAS Scandinavian Airlines. Page 233: U.S. Forest Service, National Park Service, Grant Heilman. Page 237: U.S. Forest Service. Page 239: Ken Jones. Page 240: Judith Currelly/National Film Board Photothèque. Page 242: Eugeny Shulepov/Sovfoto. Page 243: Steve McCutcheon. Page 246: Schneider/Department of Energy.

Chapter 9

Opener: W.H.O. Page 251: Florida News Bureau. Page 253: National Park Service. Page 253: USDA-Soil Conservation Service. Page 254: FAO, USDA. Page 256: USDA-Soil Conservation Service. Page 258: USDA, Jerome Wyckoff. Page 263: USDA.

Chapter 10

Opener: Photo Researchers. Page 276: Australian Information Service, Jerome Wyckoff. Page 278: Jerome Wyckoff, C. Milton/U.S. Geological Survey. Page 279: National Park Service, Jerome Wyckoff, G. H. Goudarzi/U.S. Geological Survey. Page 281: Jerome Wyckoff. Page 282: Courtesy of Utah Travel Council. Page 284: Tennessee Valley Authority.

Chapter 11

Opener: Geological Society of Canada, Ottawa. Page 295: R.E. Wallace/U.S. Geological Survey. Page 296: U.S. Department of the Interior, Bureau of Land Management. Page 298: Ira Kirschenbaum/Stock Boston. Page 305: Icelandic Photo & Press Service. Page 306: AP/Wide World Photos. Page 312: Grant Heilman. Page 313: Grant Heilman.

Chapter 12

Opener: Alan Pitcairn/Grant Heilman. Page 324: National Park Service, Florida News Bureau, Minnesota Department of Economic Development. Page 325: W.C. Alden/U.S. Geological Survey, USDA-Soil Conservation Service. Page 326: Grant Heilman, National Park Service. Page 327: Jerome Wyckoff. Page 329: The Bureau of Reclamation. Page 330: USDA-Soil Conservation Service, U.S. Department of the Interior Bureau of Land Management. Page 331: Jerome Wyckoff, P.S. Smith/U.S. Geological Survey, USDA-Soil Conservation Service.

Chapter 13

Opener: Peter Menzel/Stock Boston. Page 339: Grant Heilman. Page 342: National Ocean Survey U.S. Government Photograph. Page 343: NASA, National Publicity Studios. Page 344: NASA. Page 346: U.S. Army Corps of Engineers. Page 347: Jay Higgens/U.S. Forest Service. Page 348: USDA-Soil Conservation Service; Alsiak Bay, U.S. Department of The Interior, Bureau of Land Management. Page 350: Jerome Wyckoff. Page 352: National Park Service.

Chapter 14

Opener: Grant Heilman. Page 362: National Ocean Survey, U.S. Government Photograph. Page 363: Information Canada Phototèque. Page 366: Rapho Guillumette, Dominique Roger Pictures. Page 368: National Ocean Survey, U.S. Government Photograph, Owen Franken/Stock Boston. Page 369: National Ocean Survey, U.S. Government Photograph. Page 371: Royal Danish Ministry of Foreign Affairs. Page 372: National Ocean Survey, U.S. Government Photograph. Page 373: Boone Morrison.

Chapter 15

Opener: H.J. deBlij. Page 380: National Park Service. Page 383: Grant Heilman. Page 385: B. Bloshtein/Sovfotos. Page 386: Swiss National Tourist Office. Page 387: SAS Scandinavian Airlines. Page 389: Jerome Wyckoff. Page 391: Official U.S. Coast Guard Photo. Page 396: Information Canada Phototèque. Page 397: NASA. Page 398: Icelandic Photo & Press Service. Page 399: Grant Heilman.

Chapter 16

Opener: Jerome Wyckoff. Page 404: NASA and Geography Program University of California at Santa Barbara. Page 407: Jerome Wyckoff. Page 411: Jerome Wyckoff. Page 412: Jerome Wyckoff, Grant Heilman.

Appendix

Opener: National Park Service. Page 428: NASA.

INDEX AND GLOSSARY

Terms that are *italicized* refer to additional entries in this Glossary. Page numbers in **boldface** indicate maps, tables, charts, diagrams, or photographs. The letter *n* indicates a footnote.

Aa lava, 277, **278,** a rough, blocky, *extrusive lava* flow.

Ablation, 380, **381, 383,** the wasting away of a glacier by melting, *evaporation,* or *sublimation.*

Abrasion, the wearing away of rocks and other earth materials by the scraping motion of moving particles carried by streams (337), ocean waves (364), glaciers (383, **388**), and winds (406–408). See also: *corrasion.*

Abrasion platform, 365, **365,** the relatively flat area at the base of a *sea cliff,* also termed a *wave-cut bench.*

Absolute humidity, 126–127, **127,** see: *Humidity, absolute.*

Abyssal plains, **297, 315,** 317, the deep-lying flat areas on the floor of the ocean.

Acid rain, 231, sulfuric and nitric acids, which result from the burning of fossil fuels, that have a damaging affect on earth's vegetation.

Accumulation, zone of:

in glaciers, 379, **380, 381,** the upper portion of a glacier where falling snow accumulates;

in soils, **255**–256, the layer beneath the *topsoil* which gathers nutrients brought down by *gravitational* water movement from above.

Adiabatic processes, 75–78, **77,** changes of temperature within a gas or air mass that is ascending or descending to different elevations.

cooling, 76–78, **77,** the decline in the temperature of a gas or air mass due to the expansion of the molecules within the mass as it rises to higher elevations;

dry adiabatic, **75,** 78, the rate a moving air mass gains or loses temperature, if there is no *moisture* present to affect the change. This rate is 5.5°F (1°C) per 1000 ft (100m) of altitude increase or decrease;

heating, 71–76, **75,** the increase in the temperature of a gas or air mass due to the *compression* of the molecules, as the gas or air mass descends to lower elevations;

lapse rates, **75, 77,** 78, the rate of increase or decrease in the temperature of an air mass per 1000 ft (100m) of altitude increase or decrease;

normal lapse rate, 78, the rate of temperature change within a still, nonmoving air mass, which is 3.5°F (0.65°C) for every 1000 ft (100m) change in altitude;

wet adiabatic lapse rate, **77,** 78, the rate a moving air mass gains or loses temperature if the air mass is laden with moisture and *dew point* is reached, whereby *condensation* and *precipitation* may occur,

and the heat loss is lowered to a rate of 3.2°F (0.6°C) for every 1000 ft (100m) change in altitude.

Advection, 78, the horizontal movement of air.

Advection fog, 127, **128,** a fog layer created when a warm, moist air mass moves across a layer of colder air, land, or water, and the underside is chilled to the point where *dew point* is reached and *condensation* occurs.

Aeronomists, 90, scientists who study the upper levels of the earth's *atmosphere.*

Afghanets winds, 75, hot, drying winds that sweep down the *leeward* sides of mountains and are heated by *compression.* See: *Adiabatic heating*

Agassiz, Louis, 378.

Aggradation, 340, the building up of a stream bed or *delta* by the *deposition* of *sediments* carried by streams.

Agonic line, 39–41, **40,** the *isogon* on an *isogonic map* that connects all points of zero magnetic *declination* (*variation*) when the needle on a *magnetic compass* will point toward *true north,* and the *magnetic north.*

Agricultural meteorology, 220.

A horizon, 252–257, **255,** the upper layer of the true soil (*solum*) which receives the *humus* from the vegetation above. It is called the *topsoil.* It is also the *zone of loss* of minerals and other nutrients that are carried down into the B horizon by the movement of gravitational water.

Air: see *atmosphere.*

Air drainage, **75,** 76, the movement of cool air down a *slope.*

Air mass, 114–116, **115,** a relatively large body of air that is relatively uniform as to temperature and moisture content. The differing characteristics of the earth's variety of air masses are determined by the region (land or sea) and the *latitude* over which the air mass forms.

Air pollution, 79–**80.**

Air pressure, 101–112, **102, 104, 105, 106, 107, 108,** the weight of a column or mass of air. If the air is cooling and descending, it becomes heavier, thus creating a *high-pressure* area. If it is warm and rising, a *low-pressure* area is created. See also: *Cyclones; Anticyclones.*

Air temperature:

annual range in, **100,** 101, the difference between the average (mean) temperatures of the coldest and warmest months of the year;

449

Air temperature *(continued)*
    and climatic classification, 190–194, **195, 196;**
    composition of atmosphere by, 90–96, **93, 95;**
    decrease of, with change in altitude, 78, see: *Lapse rates.*
    diurnal cycle of, 108–110, the difference in air temperature between night and day;
    extremes of, 101, 201, 398;
    gradient in, 99–101, the rate of change in the temperature of air masses located close to one another. The gradient can be likened to a slope, which could be steep, thus creating a rapid change in temperatures. If the slope were gentle, the result would be a slow change;
    land and water influences on, **70**–71, 99–101, **100;**
    and ocean currents, **197, 198;**
    sensible, 101;
    and vegetation, 229, 245;
    world distribution of, 96–101, **97, 98, 100.**
Alaska, 196, **215,** 217, **218,** 331, 342, 393.
Albedo, 68, 217, the percentage of *insolation* (incoming *electromagnetic radiation* from the sun) that is reflected from the earth's surface and the atmosphere.
Algae and fungi, **232,** 234, 268, single-celled plants that absorb nutrients directly from the water in which they live.
Alluvial fans, 340, **341,** fan-shaped deposits of *alluvium* at the mouth of a stream valley.
Alluvial terraces, 345, relatively level and narrow plain with a steep frontal face bordering a river, that is composed of *alluvium.*
Alluvium, 252, **263,** 338, 340, **341,** stream-deposited *sediments.*
Alpine glaciers, 167–169, **168,** 382–390, **383, 384, 386,** glaciers that form in high mountain valleys. Sometimes referred to as mountain or valley glaciers. See: *Glaciation; Glaciers.*
Alpine vegetation, 243–245, plant communities that develop and evolve in very high elevations. See: *vertical zones of vegetation.*
Alps, 245, 378, **386.**
Altocumulus clouds, **128,** portions of cumulus clouds at very high altitudes.
Altostratus clouds, **128,** 129, a relatively thin layer of clouds at high altitudes.
Amazon River, **164,** 165, 359.
Anaximander, 7.
Ancient theories of the earth and its systems:
    Apaches, 15;
    Babylonians, 15, 152, 224, **225;**
    Central Asians, 15, 224;
    Chinese, 5, 155;
    Egyptians, **14,** 15, 163–**165;**
    Finns, 406;
    Germans, **225,** 226;
    Greeks, 5, 6, 15, 189, 290;
    Hawaiians, 277, 336;
    Hebrews, 152;
    Hindus, 7, 14;
    Japanese, 290;
    Mayans, 226;
    Mediterraneans, 290;
    Nicaraguans, 290;
    Romans, 290;
    Russians, 224, **225;**
    Scandinavians, **225,** 226;
    Tahitians, 15.
Andes Mountains, 94, 245, 298.
Anemometer, an instrument that measures the speed and, sometimes, direction of a wind.
Aneroid barometer, 102, **107,** an instrument that measures atmospheric *pressure* by its affect on a corrugated metal box.
Antarctic Circle, **24, 60,** 61, **64,** the *parallel* of *latitude* located at 66°30′ south of the *equator.*
Antarctica:
    climate of, 217, **219;**
    continent of, 167–169, **168, 169;**
    glaciers of, 167–**169,** 379, 390–399, **397.**
Antarctic Ocean, **170,** 171.
Anticlines, 299, **300,** upward folds of layers (*strata*) of rock, creating an archlike effect.
Anticyclones, 108, 137; *high-pressure* areas, where the air is descending vertically and spirals outward as it nears the surface of the earth. The direction is clockwise in the northern hemisphere and counterclockwise in the southern.
Aphelion, 54, **55,** earth's location in its *orbit* about the sun where the earth is at its greatest distance from the sun: 94.4 million miles (154 million kilometers) about July 4th.
Aquifer, 152–155, **153, 156,** a layer of rock that is both *porous* (allowing water to enter) and *permeable* (allowing water to move through it).
Arabian Desert, **198,** 201, **244, 409.**
Archeozoic Era, **269, 270,** 271. See: *Geologic Time.*
Arches, 364, **365, 366.**
Arctic Circle, **24, 60,** 61, **64,** the *parallel* of *latitude* located 66°30′ north of the *equator.*

Arctic climate, **215,** 217, **218.**
Arctic Ocean, 169–172, **170, 171.**
Arcuate delta, 342, **342.**
Arête, **384,** 385, a sharply defined ridge separating two glacial valleys, or *cirques.* See: *Glacial landforms.*
Arid Zones, **192, 193, 198,** 201, **203,** 210, 213, regions that are deficient in *precipitation* and have a very high rate of *evaporation.*
Aristotle, 6, 90, 189, 190.
Artesian wells, 155–158, wells dug into the ground that allow water flowing through an *aquifer,* which has a high *hydraulic gradient,* to reach the surface under its own pressure.
Aspect, of slopes, 230, **230,** 322, the direction a slope of a hill or mountain is facing in relation to the sun.
Asthenosphere, 272, **273,** 291, the upper region of the earth's *mantle,* a rather flexible layer of weak (pliable) rock that underlies the more rigid *lithosphere* which makes up the *earth's crust.*
Astronomy, 2, **7,** 54, the earth science dealing with the intimate interrelationship of the earth, moon, and sun.
Atacama Desert, 103, 176, **244,** a *piedmont plain* at the foot of the Andes Mountains in northern Chile and southern Peru. It is almost devoid of any plant life due to its position on the *lee* side of the Andes, thus not able to receive any of the moisture carried by the *southeast trade winds.* It is also greatly affected by the close proximity of the *upwelling* very cold waters of the *Humboldt (Peru) Current.*
Atlantic Ocean, 169–172, **170, 171.**
Atlas Mountains, 298.
Atmosphere, 2, 90–96, **91, 93, 95:**
  composition of, 90–96;
    by gases, 92–95, **93;**
    by temperatures, **95**–96;
  lower and outer, 90–92, **91;**
  pollution of, 79–**80;**
  pressure. See: *Air pressure.*
Atoll, coral, 408, **408.**
Atom, 283–285.
Aurelius, Marcus, 2, 418.
Auroras, 65, **66.**
Australia:
  aborigines, 404;
  artesian well, 155, **157;**
  deserts, 155, **157, 244,** 405, 407, 409;
  gibber plain, 407; Australian aborigine term for the Gibson Plain.
  Great Artesian Basin, 155–158, **157;**
  Great Sandy Desert, 155, **157, 244,** 405, 407, 409;
  Great Victoria Desert, **244;**
  "outback," 213, 404.
Axis, earth's, **20,** 57–64, **59, 60, 64,** the line passing through the center of the earth, from the North to the South Pole, around which the earth rotates.
Azimuth, 439n, **439,** the bearing of an object, measured in degrees clockwise from true north. Thus, an object east of your position would have an azimuth of 090°.
Azimuthal projection, see: *Map projections.*
Azonal soils, **263,** 264, youthful soils which do not have a well-developed *soil profile,* such as alluvium.
Azores High, 206.

Babylonian theories, of earth's origin, 15:
  of earth's vegetation, 224, **225;**
  of earth's waters, 152.
Bacteria in soil, 231, 251–252, 257.
Baikal, Lake, **164, 166,** 213.
Balance, heat, 66–67, **68.**
Balance, water, 124–**125,** 150, 245.
Balder, and the vegetative cycle, **225,** 226.
Barchan sand dunes, 411, **412,** crescent-shaped *sand dunes* that have their horns pointing downwind.
Barometer, 101, **107,** an instrument that measures the weight of a column or mass of air. See: *Aneroid* and *Mercury barometers.*
Bar, sand, 366–370, **367, 368, 369,** a ridgelike mound of sand or gravel deposited offshore that may be visible at low tide.
Barrier:
  islands, 367–370, **368, 369,** long and narrow offshore sandbars that are separated from the *shoreline* by a *lagoon;*
  reefs, 371, **372,** an offshore ridge composed of coral (skeletal remains of once-living organisms);
  sandbars, 366–370. See: *Bar, sand.*
Basalt, 273, **278,** a fine-grained *extrusive igneous rock* that is normally black or dark brown in color. See: *Aa lava; Ejecta; Obsidian; Pahoehoe.*
Baselevel, 337–338, **338,** the lowest level to which a stream can lower itself.
Baseline, 41, 443, **444,** the east-west line of reference used by the United States Bureau of Land Management in its *rectangular land survey system* to establish definite boundaries. The north-south line is called the *Principal Meridian.*
Basin-and-range topography, **317,** 318.
Batholith, 307, **310, 312,** (from the Greek *bathys,* "deep," and *lithos,* "rock.") A very large mass of *intrusive*

Batholith (*continued*)
igneous rock that lies deep beneath the earth's surface.
Baymouth sandbar, **367**–369, a sandbar located at the mouth of a river where it meets the ocean.
Bay of Fundy, 184, 361–364, **363**. See *Case Study: Energy from the Oceans*, 184.
Bayou, 345, **346, 348,** an *oxbow lake* created by a widely *meandering* stream.
Beaches, 366–370, **367.**
Bearings, compass, 39, 433–440, **439.**
Bedrock, 252, **255,** 322, 328, **388,** 394, the solid, unaltered rock composed of the original *parent material* that lies beneath the upper, evolving layers of soil.
Bench mark, 427, a control point established by the U.S. Coast and Geodetic Survey which marks a precise position of *latitude* and *longitude* and the height above sea level. It is usually marked on the ground by a metal tablet.
Benguela Current, **178, 216,** 360, a cold ocean current that rises as a spin-off of the *Antarctic Current* and moves along the western coastline of southern Africa. These cold waters offer little moisture to the *prevailing westerlies,* thus creating the barren Namib Desert.
Bergschrund, **381,** 382, a *crevasse* (fracture) formed at the uppermost portion (head) of a *glacier* where it pulls away from the rock wall.
Bermuda Triangle, 177.
B horizon, **255**–256, 259, the lower portion of the true soil (*solum*) is called the subsoil. It is the *zone of accumulation* of minerals and other nutrients leached down from the *A horizon* (*top soil*), which becomes the zone of loss.
Biofuel, 246, the conversion of vegetation into alternative forms of *energy,* such as burning waste material from sugarcane fields to produce alcohol, which, in turn, can be transformed into gasoline. Fossil fuels, such as coal and petroleum, are also biofuels, but they are not renewable. See *Case Study: Biofuel—Energy from Vegetation,* 246.
Biogeography: the scientific study of earth's life forms (flora and fauna) as to their patterns of distribution and their interrelationships.
Biological activity, 231, the interaction of living organisms on the evolution of a region's vegetation.
Biosphere, 224, from the Greek *bio,* "life," meaning the "life sphere," which includes the upper margin of the earth's crust, the waters, and the *troposphere.*
Biotic complex, 224, 268, see: *Ecosystem.*
Bird-foot delta, 342, **343,** the *delta* formed by the accumulation of *alluvial sediments* at the mouth of a river and extend like fingers into the open waters beyond.
Bjerknes, Vilhelm, 137, the founder of the *Polar Front Theory.*
Black-earth region, U.S.S.R., 213, **214,** 242, 251, 257.
Block mountains, 299–303, **301,** see: *Horst.*
Blowout, 406, a depression often found in dry climatic regions and deserts, where loose particles of sand are carried away by the wind. Also termed a *deflation basin.*
Bogs, 229–230, **233,** flat bottom lands that are unable to drain off excess *precipitation,* and become marshy and swampy as they fill with peat moss.
Bore, tidal, 361–364, **363,** a sudden, rapid onrush of an ocean wave as it enters an *estuary* as a high *tidal change* occurs. The common and improper use of the term *tidal wave* stems from this phenomenon.
Boreal:
climates, 213, **215, 216,** located in subpolar regions of northern Canada and the Soviet Union;
forests, **238, 240,** the needle-leaf forests termed the *taiga.*
Braided stream pattern, **348**–349, created when a number of streams in a wide valley become interconnected and interlaced as they divide from the parent stream and rejoin at intervals along the course of the streambed.
Brazil Current, **178, 198,** 360, a warm current that is a southerly spin-off from the south equatorial current and sweeps along the coast of Brazil.
Breeder reactor, 283–285, a device in a nuclear power plant that transforms uranium 235 into plutonium 239, thus creating electrical energy. See: *Case Study: Nuclear Energy—Fission or Fusion?*, 283.
Breeze, land and sea, 109–**110.**
Bright Angel Quadrangle, **434**–**437.**
Broadleaf summergreen deciduous forests, **238,** 240.
Broadleaf vegetation, 235, **236.**
Bryce Canyon National Park, **324,** 406.
Budget, heat, 66–71, **68, 69.** See: *Heat budget.*
Budget, water, 124–**125,** 150, 245. See: *Water balance.*

Calcification of soils, 259, the soil-forming process in semiarid climate regions in which calcium carbonate (lime), a fertile nutrient, is accumulated in the *solum.*
Caldera, **311,** a large depression formed at the top of a volcano when the top is either blown off during an eruption, or it collapses downward.
Caliche, 250, 264, a less fertile *calcimorphic soil* that develops in desert regions.
California Current, 129, **178, 198,** the cold ocean cur-

rent that sweeps along the western coast of North America.

Calimorphic soils, 264, *intrazonal soils,* developed in dry-land regions through *calcification.*

Calving (of glaciers), 399, the breaking off of pieces of a *glacier* that extends out into a body of water, creating *icebergs.*

Campos, 199, **241,** 242, the *savanna grassland* region in south-central Brazil.

Canada, 382, **393,** 394, 396n, **396.**

Canadian Shield, 314, **316,** 317.

Cancer, Tropic of, **24, 60,** 61, **64,** the *parallel* of *latitude* circling the earth at 23°30′ north of the *equator.* It is the northern limit of direct rays from the sun.

Capillary action, **153,** 250, the upward movement of moisture within a soil due to tension created between water molecules in their space of confinement.

Capillary fringe, **153,** 154, the *interface,* or contact surface, between the region of soil moisture above the *water table,* and the region below, the *zone of saturation.*

Capricorn, Tropic of, **24, 60,** 61, **64,** the *parallel* of *latitude* circling the earth at 23°30′ south of the *equator.* It is the southern limit of direct rays from the sun.

Carbohydrates, 246, organic compounds composed of oxygen, hydrogen, and carbon (sugars, starches), formed by green vegetation.

Carbonates, **275,** one of the basic *mineral* combinations, composed of calcium, carbon, and oxygen, which form rocks such as *limestone.*

Carbonation, 326, 349–352, **350, 351, 352,** the process by which *limestone* (calcium carbonate) is dissolved as *carbon dioxide* enters a rock layer by water infiltration, thus creating carbonic acid. See: *Carlsbad Caverns; Chemical weathering.*

Carbon dioxide, 92, a heavy, colorless gas formed as organic materials proceed through the process of combustion and *decomposition.* It is absorbed by vegetation from the air through the process of *photosynthesis.*

Carbonic acid, 326, 350.

Carboniferous Period (of Geologic Time), **269, 270,** 271.

Carlsbad Caverns, **352,** underground caverns in New Mexico, created by the dissolving of *limestone* beds of rock by the infiltration of *carbonic acid* through *groundwater* movement.

Cartograms, 44–46, **48–49,** maps that are deliberately distorted in order to indicate a specific geographic point of view.

Cartography, 3, 27–49, 427–440, the making of maps, charts, and diagrams.

Cascades National Park, **380,** a mountain range near Seattle, Washington.

Caspian Sea, 167, the largest saltwater lake in the world. It is supplied by the Volga and Ural rivers.

Caverns, **351, 352,** See: *Karst topography.*

Celsius, 52n, the most commonly used system for measuring temperature is based on the freezing point of water at sea level at 0° Celsius, and the boiling point is 100°C. Also known as *centigrade.* See *Case Study: Thinking Metric,* p. 9, and Appendix D, p. 446.

Cenotes, **351,** 352, See: *Karst topography.*

Cenozoic Era (of Geologic Time), **269, 270.**

Centigrade, 52n, see: *Celsius.*

Challenger, H.M.S., 356.

Chaparral, 206, **238, 239,** the natural *climax vegetation* of southern California and other areas within the *West Coast Dry Subtropical* (Mediterranean) *Climate* regions. Plant geographers include it within the vegetation known as Mediterranean Woodland and Scrub. See: *Maquis.*

Chemical weathering, 323–326, the breakup, *decomposition,* or dissolution of rocks by the alteration of their chemical composition by gases or moisture, or by a combination of both. See: *Weathering.*

Chernozem, 257, 261–263, **262,** extremely rich and fertile soils that are high in the accumulation of *humus* and are very dark in color.

Cherrapunji, India, 199.

Cheyenne, Wyoming, 213.

Chinese theories of earth's systems:
  Chuang-Tse, 5;
  Confucius, 5, 155.

Chinook winds, **75,** 75n, 213, the Native American term for the hot, descending, drying winds that sweep eastward down the slopes of the Rocky Mountains. See: *Adiabatic heating, compression.*

Chlorides, 172, **275,** a compound of chlorine with another *element,* and if the other element were sodium, the result would be sodium chloride, table salt.

Chlorophyl, 227n, the green-colored pigment in plants that causes *photosynthesis.*

C horizon, 252–257, **255,** the loose rock material (*regolith*) that has separated from the bedrock but has not evolved far enough to be considered a part of the *solum.*

Chronometer, 25, a very precise and accurate timepiece, or clock.

Cinder cone, **311,** 312, a cone-shaped *volcano* shaped by the accumulation of volcanic *ejecta* (ashes and other *pyroclastic* material).

454 INDEX AND GLOSSARY

Circle of Illumination, **60,** 61, the dividing line, which is a *great circle,* between the light of day and the darkness of night on the earth, moon, or any rotating planet. Also known as *terminator.*

Circles, great and small, 21–22, 44, are circles (lines) drawn on the surface of a sphere. Great circles always divide a sphere into two equal parts. Small circles do not. *Meridians of longitude* are halves of great circles. All *parallels of latitude,* except the *equator,* are small circles.

Cirque, **384,** 385, a steep-sided, crescent-shaped mountain valley carved by an *Alpine* (mountain, valley) *glacier.*

Cirrus clouds, **118, 128,** 129, thin strings of clouds at very high altitudes.

Cirrostratus clouds, 129, a high layer of thin, wispy clouds.

Climate, 189ff, the average weather conditions of a place over a long period of time:
  and humanity, 90, 92, 194, 217–220, 223;
  and soils, 164–166, **195–196,** 231, 250–251;
  and stream-valley development, 349;
  and vegetation, 176–178, **228, 230,** 251–252, **192.**

Climate classification: the establishment of climate regions based on types of vegetation, amounts of precipitation, rates of evaporation, temperature, winds, and topography:
  by Aristotle, 6, 90, **189,** 190;
  by Geiger, 190, 194;
  by Koppen, 190–194, **191, 193, 213, 214;**
  by Trewartha, 190, 194.

Climates of the world, 189–221:
  generalized diagram of, **197;**
  highland (H), **191, 192, 193,** 217;
  high latitude, **192,** 213–217, **214, 215, 216;**
    continental subarctic (boreal) (*Dwc*), 193, 213–217, **215, 216;**
    ice cap (EF), **191, 193,** 217, **219;**
    tundra (ET), **191, 192, 193, 215,** 217, **218;**
  midlatitude, **196, 197,** 206–213, **207, 208, 209;**
    dry (steppe, prairie, desert) (BSk), 210–213, **212;**
    humid continental (Dfb), **208,** 210, **211;**
    marine west coast (Cfb), 206–210, **208, 209, 211;**
    subtropical, 201–206, **204, 205;**
      east coast humid (Cfa), 206, **207;**
      west coast dry (Csa) (Mediterranean) (Csa), 201, **204, 205–**206;
  tropical, 190, **191, 192, 193,** 194–203;
    dry (BW), **195, 197, 198,** 201, **203;**
    monsoon (Am), 199;
    rain forest (Af), **195, 197, 198,** 199, **200;**
    wet and dry (Aw), **195, 197, 198,** 199–201, **202.**

Climatology, 2, **8,** 90, the earth science dealing with the study of earth's climates.

Climax vegetation, 226–227, 227n, 245–246, the plant community (vegetation) that becomes established in a specific region and is in a state of *equilibrium,* or harmony, with the region's climate. Also termed: climatic climax vegetation.

Climograph, **200**ff, a graph detailing the average monthly temperatures and precipitation for a specific geographic location.

Clouds, **128,** 129, visible masses of concentrated water droplets which form when moisture-carrying air is lifted to an elevation wherein the temperature of the air mass cools to the point of *condensation.*

Coalinga earthquake, 307.

Col, **384,** 385, a high mountain pass carved by glacial activity.

Cold front, **131, 132–137, 133,** a moving mass of cold air that forces its leading edge under a stationary, warm *air mass,* causing the warm air to be lifted, cooled, reach *dew point,* with the resultant *condensation* and, normally, *precipitation.*

Colorado Plateau, 314, **317.**

Colorado River, **164,** 338, **339, 343.**

Columbus, Christopher, 17.

Compass bearings, 39, 433–440, **439,** directions gained by using a compass based on true north at 000°, east at 090°, south at 180°, and West at 270°.
  magnetic, 30, **39–41, 40, 439,** a compass with a magnetized needle that points toward the focus of the earth's magnetic field near the Canadian island of Bathurst in the northern hemisphere, rather than toward True North.

Composite volcano, **311,** 312, a *volcano* composed of layers, which may alternate, of volcanic ash and other forms of *ejecta,* and extruded *lava flows.*

Compression, heating by, **75–76,** 75n, the warming of a moving air mass as it descends to lower altitudes where the air piles up and the air mass is reduced in volume. See: *Adiabatic heating.*

Condensation, 76, **77,** 124, the process by which water vapor is changed from its gaseous state to liquid form.
  forms of, 76, 127–129, **128;**
  heating by, 76, See: *Heat, latent.*

Conduction, 74, the movement of heat from one body to another.

Conformal map projection, **32,** 42, **43,** a *map projection* that portrays accurately the shapes of continents, islands, or coastlines, as they appear on a globe.

Conglomerate rocks, 277, **279,** a compacted mixture of various sizes of rock particles, from silt, to sand, to pebbles, to small boulders.

Conical map projection, 42, **43.**

Conifer, 227, **238,** 240, a needle-leaf tree that bears cones (from the Latin *conus,* "cone" and *ferre,* "to bear").

Contact surface, **153,** 154, see: *Interface.*

Continental drift, **269,** 291–296, **292, 294, 295,** the theory that the earth's continental landmasses were once united into one "supercontinent," which *Alfred Wegener* (in 1922) termed *Pangaea* ("all earth"). With the passage of millions of years, the continent began to break up and slide across the earth, as the seafloor spread, thus "giving birth" to the six continents: North America, South America, Eurasia, Africa, Australia, and Antarctica.

Continental glaciation, 314, **316,** 359, 370, **371,** 390–399, **392, 393, 395, 397,** sculpturing of the polar regions of the earth accomplished by huge and very thick sheets of ice that move out from a central point of snow accumulation to cover vast areas.

Continental glaciers, 167, **168,** 370, **371,** 390–399, **389, 392, 393, 397.**

Continentality, 99, 213, the presence of large landmasses prevents the oceans from asserting any moderating influence on regions away from the sea; thus the annual temperature ranges are at the extremes.

Continental shelf, 314, **315,** 356–360, **357, 358,** the edge of a landmass that extends out and under the edge of the sea, ending at the continental slope.
  development and structure of, **357, 358,** 359–360;
  origin of, 359.

Continental shields, 314, **316, 317,** massive areas of the *lithosphere's* upper surface of *bedrock* that have been exposed by the scraping, scouring, and pressure of the vast *continental glaciers* during the *Pleistocene ice ages.*

Continental slope, **357,** 360, the outer edge of the continental landmasses that bends sharply down to the sea floor.

Continental subarctic climate (boreal), 213, **215, 216.** See: *Climates.*

Contour lines, 429–**433,** 434ff, lines drawn on a *topographic map* that connect points of equal elevation above sea level.

Contraction, of the earth, 290–291, the theory that the earth's wrinkled surface is the result of the hot, liquid material cooling and shrinking, much like a plum becomes a prune.

Convection: air motion in which the movement is upward (vertical).
  within the atmosphere, 77–78, as the air rises, it expands, and cools;
  within the earth, 291, as heat rises from the *earth's core,* it causes movement of portions of the earth's *lithosphere.*

Convectional precipitation, 129–137, **131,** precipitation caused by the rapid upward movement of a mass of warm, moist air over a warm water body or a tropical desert when the air is chilled to *dew point.*

Cooling, adiabatic, see: *Adiabatic cooling.*

Cooling processes, 76–78, **77:**
  advection, 78–80, horizontal movement;
  convection, 77, vertical movement;
  evaporation, 76–77, water moving from liquid to a gaseous state;
  expansion, 77–78, convection, upward movement and the air mass expands, rather than contracts.

Coquina, 277, **279,** a rock composed of shell fragments.

Coral, skeletal fragments (mostly calcium) of once-living organisms;:
  atolls, 371–**372,** a coral reef that circles (above sea level) around the top of a submerged volcano (*sea mount*), thus creating a *lagoon;*
  barrier reef, 371–**372,** a coral ridge fronting a shoreline which may be below or above sea level;
  fringing reef, 371, a coral reef that surrounds a *lagoon;*
  shorelines, 370–**372.**

Core (earth's), 271–**272,** the center of the earth, composed of two parts: the inner core of solid iron and nickel; the outer core of liquid iron and nickel, from which heat moves outward towards the *lithosphere.*

Coriolis effect, 112–114, **113,** 145, 176, the apparent change in the direction of the forward movement of winds, air masses, and ocean currents, caused by the *rotation* of the earth. Moving objects tend to veer to their right in the Northern Hemisphere and to the left in the Southern.

Corpuscular energy, 65, energy in the form of clouds of electrified particles emitted by the sun.

Corrasion, 337, 364, 383, **388,** 406, the wearing away of rock material by moving forces, such as glaciers, winds, running water, and ocean waves. This *abrasion* is a physical action similar to scraping and sandpapering.

Corrosion, 323–326, 337, 360, the *decomposition* of rock materials by chemical action, such as the infiltration of *carbonic acid* into beds of *limestone,* which causes the rock minerals to be dissolved and car-

Corrosion (*continued*)
ried away by the moving groundwater. See: *Carbonation; Chemical weathering.*

Countercurrents, 175–176, **178,** ocean currents that flow in the opposite direction to the prevailing currents in the region.

Crater, **311, 313,** the depression at the top of a *volcano,* which, as the primary mouth or vent, may be active or inactive, depending on the condition of the liquid rock (*magma*) within or below the volcano.

Craters of the Moon National Monument, **278,** 312, 313.

Crates, 16, the 2nd century A.D. geographer who constructed the first model of a spherical earth—as a *globe.*

Crevasse (glacial), 382, **388,** a sharp and steep crack or fissure in a *glacier* caused by forces generated in its downward movement.

Crust, earth's, **272, 273,** 290ff, the outer layer of the earth, composed of two silicate materials (*sial* and *sima*), termed the *lithosphere.*

Crustal plates: see: *Continental drift; Tectonics; Plate tectonics, Seafloor spreading.*

Crystal growth, 323, the formation of crystals (ice or salt) within a rock that cause the rock to expand and break apart as it disintegrates. See: *Disintegration, Physical weathering.*

Cumulonimbus clouds, **128, 129,** a large mass of clouds, rather darkish at their base, that produce rain.

Cumulus clouds, **128, 129,** large white masses of clouds that tower up from a low-lying base to very high elevation.

Currents, ocean, **175,** 176–179, **178, 197.**

Cuspate delta, 342, a river delta that has a rather pointed, toothlike shape formed by vigorous wave action. See: *Delta.*

Cycle: 321, a series of events occurring through time, usually ending up at the starting point, wherein the cycle begins again;
 of denudation, 322, the systematic succession of erosive processes that level the land, from high mountain country to the final ending—the flat plain (see: *Peneplain*);
 geographical, 322, the *cycle of denudation* as theorized by the *geomorphologist* William Morris Davis;
 hydrologic, 124–**125,** 150, 163, 245, 336, 379, the "water cycle" in which water in its various forms (liquid, gaseous, solid) is moved through space and time from water bodies, snowfields, and glaciers into the air, then onto and into or over the land, from whence it returns it to its original sources;
 seasonal, 59–64, **60, 62,** the changing of the seasons as the earth *orbits* about the sun and *rotates* about its *axis;*
 vegetative, 224–227, **225,** the life cycle of plant forms.

Cyclone, 108, 130–132, **133, 134,** the upward and inward spiraling center of a low-pressure area, where the winds move in a counterclockwise direction in the Northern Hemisphere and the opposite in the Southern.

Cyclonic storms, **133, 134,** 137, 141, **144,** storms, such as *hurricanes, tornadoes,* and *thunderstorms,* which are created by the violence of the whirling wind about a cyclonic depression (*low*).

Cylindrical map projection, 41–**42,** see: *Map projections.*

Darwin, Charles, 6, 356, 371.

Dating of rocks, 268, 293, see: *Radioactive dating.*

Davis, William Morris, 347, see: *Cycle of denudation; Geographical cycle.*

Day, length of, 25–27, 54–55, 57, the time it takes the earth to rotate once around its *axis.* See: *Time, length of day.*

Death Valley, California, 302, **407.**

Deciduous vegetation, 235, plant forms that shed their leaves periodically, usually during a specific time of the year.

Declination, magnetic, 39–41, **40,** 433–440, **439,** the deviation of a *magnetic compass* needle away from *true north.* Example: a magnetic compass needle in Los Angeles would point 15° east of the geographic, or, *true, north.* See: *Agonic line; Isogons; Compass bearing; Compass, Magnetic.*

Decomposition, 323–326, the rotting, decay, or breaking apart of a rock and its materials due to chemical action. See: *Carbonation; Chemical weathering; Corrosion; Weathering.*

Deeps: see *Trenches.*

Deflation (by winds), 406–407, the moving of loose rock materials or particles away from their point of origin by the wind.

Deflation basin, 406, **407,** a depression in a sandy region created by the removal of silt or sand particles by the wind.

Degradation, 336, the leveling (lowering) process carried out by erosional forces (wind, running water, glaciers), in which the rock materials are broken loose and transported away. Most often used to explain the vertical downcutting of a stream valley.

Degree: a unit of measurement.
 of distance, 25–27, **28,** a unit of distance or angular measurement expressed as: one degree equals 1/360 of a circle or a sphere. One degree is subdivided into minutes (1/60 of a degree), and

one minute is subdivided into seconds (1/60 of a minute). Along the *equator,* one degree equals 69 statue (or land) miles of the earth's surface.

of temperature, 52n, see: *Celsius; Fahrenheit.*

Delta, 340–344, **342, 343, 344,** the *alluvial sediments* deposited by a stream or river at the point where it enters a standing body of water (lake or ocean), and its shape will be determined by the power and direction of incoming waves. See: *Aggradation; Estuarine delta.*

Denudation, cycle of, 322, see: *Cycle of denudation.*

Deposition: the laying down or depositing of sediments and other materials by a moving body, such as:

glaciers, 379, 389, 394, **395, 396;**

groundwater, 352;

oceans, 359, 366–370, **367, 369;**

streams, 340–345, **341, 342, 343, 344;**

winds, 408–415, **410, 412, 413.**

Desert climates, 201, **203, 208,** 210, regions where the rate of *evaporation* greatly exceeds *precipitation,* vegetation is sparse and daily temperatures throughout the year are high, averaging around 73°F (22°C).

Desertification, 252, 418, the encroachment of desert conditions on once-productive agricultural lands, such as what is happening to the Sahel region of Africa's Sahara Desert.

Desert pavement, 407, **407,** 408, the surface of a desert region which has had all of its fine-grained particles (silt, sand) carried away by the wind, leaving behind a cover of pebbles and gravels. See: *Deflation; Deflation basin.*

Deserts, 103, 201, **203, 208,** 243, **244,** 251, **262,** 407, 408–413, **409, 412:**

pebbly (*reg*), **407,** rocky, sandless desert;

sandy (*erg*), 408–413, **409,** sandy deserts, with numerous *sand dunes.*

Desert soils, 250, 251, 261, **262,** 263.

Desert vegetation, **228,** 243.

Dew, 76, moisture that collects on surfaces (soil, plants, car windows, and in the *atmosphere* as *fog*), especially at night, when the air temperature drops and the air reaches the point of saturation.

Dew point, 76, 127, the temperature at which moisture-laden air reaches saturation and can no longer retain the moisture as a gaseous water vapor. See: *Dew; Moisture; Saturation point.*

D horizon, 252–257, **255,** the unaltered parent material (bedrock) from which the layers of soil above have evolved.

Diastrophism, 299–305, **300, 301,** movements within the *lithosphere* due to *plate tectonic activity.* See: *Earthquakes; Seafloor spreading.*

landforms of, 299–302, **300, 301.**

Dike (volcanic), 307, **310, 311,** a slablike layer of *intrusive igneous rock* that has wedged its way upward through fractures in rocks overlying a chamber of *magma.*

Dip, **301,** the angle formed between an imaginary horizontal plane and the actual angle or slope of a *stratum* of rock or *fault plane.*

Discontinuity: a sudden and distinct change in the order of succession from one point, or place, to another.

Mohorovičić discontinuity, 272, **273,** the boundary between the upper region of the earth's *mantle* and the *lithosphere.*

Disintegration of rocks, 322–323, **326, 327,** the physical (mechanical) breakup of rock materials without any chemical change, caused by:

biotic action, 323, pressure applied by burrowing animals or the expansion of plant roots;

crystal growth, 323, expansion caused by the formation of ice or salt crystals;

frost action, 323, as water enters the rock and freezes, it expands, thus applying extreme pressure against both sides of the crack or fracture;

thermal (heat extremes), 323, the effect of extremes in temperature (hot during the day, very cold at night) creates stress which forces the rock apart;

unloading, 323, when a heavy overlying mass of rock material is carried away by one of the erosive processes, or *mass wasting,* the underlying rock expands, thus breaking apart;

See: *Weathering, mechanical.*

Distance, determining, 25–27, 429–440, **431.**

Diurnal range, 201, the daily cycle of changes, such as the variation between the high and low temperatures during a 24-hour period.

Doldrums, **106,** 108, the region of light winds and calms where the *northeast* and *southeast trade winds* converge and much of the wind movement is *convectional.* Also termed: *Intertropical Convergence Zone, Equatorial Belt of Low Pressures,* and the *Equatorial Trough.*

Doline, 350, **351,** a large depression or cavity in a region of *limestone* created by the entry of surface and groundwater and the resultant dissolving of much of the limestone. See: *Sinkholes; Karst topography; Carbonation; Decomposition; Weathering, chemical.*

Dome volcano, **311,** 312, a rounded *volcano,* formed by an accumulation of a succession of *lava flows.*

Drainage, air, **75,** 76, the movement of cold air down a valley slope to the floor below.

Drainage area, 347–349, a stream valley between two ridges (*interfluves*) that catches the various forms of *precipitation* that supplies water to the stream.

Drumlin, **395,** 396, a smoothly rounded mound of glacially deposited *till.*
Dunes: see *Sand dunes.*
Dust Bowl, 231, 407–408.
Dynamic inversion: see: *Inversion, static and dynamic.*

Earth, age of, 268–271, **269, 270,** see: *Geologic Time.*
  ancient theories of, 5, 15–17;
  axis of, **20, 60,** 61;
  circumference of, 16–21;
  composition of, 271–283;
  crust of, 273, **273;**
  curvature of, 15–21, **16, 17;**
  diameter of, 19, **20,** 271;
  heat balance of, 66–67, **68, 69;**
  interior of, 271–**272;**
  materials, 274–283;
    minerals, **274–275;**
    rocks, 275–283, **276, 278, 279, 281;**
  movements, 54–64, **55, 56, 60,** 62;
    revolution (orbit), 54–57, **55;**
    rotation, 57–59, **58;**
  origin and history of, 268–271, **269, 270;**
  outer margin of, **272–273, 273;**
  plates, 291ff, **292, 294;**
  sciences, 2, 3, **7, 10;**
  shape of, 14–21, **20;**
  soils, 249ff;
  and the solar system, 52–54, **53;**
  structure of, 271–273, **272, 273;**
  surface of, **273,** 290ff, **297;**
  system, 2–6, 418;
  systems, interrelationship of, 2–6, 224, 418;
  vegetation, 223, 246;
  water balance of, 124–125, **125,** 150;
  waters, 124, 150–185, **151.**
Earthflow, 328, **330,** the flowing movement of water-saturated soils down a slope.
Earthquakes, 299–305, **300, 301,** shaking of the ground by *seismic waves:*
  causes of, 291, 299–305; see: *Diastrophism; Tectonics.*
  measurement of:
    by physical damage, 304; see: *Mercalli Scale.*
    by intensity, 304; see: *Richter Scale.*
  prediction of, 304–305.
Earth-sun relationship, 52ff, **53, 55, 60,** 64ff, **65, 67, 68, 69,** 92–96.
Easterlies, **106,** 112; see: *Winds.*
Eckert IV Map Projection, 44-46, **45.**

Eclipse: the obscuring of one celestial body by another, as when the moon gets between the earth and the sun, and hides the sun.
Ecliptic, plane of, 59–64, **60,** the path made by the earth as it orbits about the sun. The shape of the earth's rotational path is *elliptical.* See: *Orbit.*
Ecology, 183–185, 268, 418–420, the science of the interrelationship of the various life forms (plant and animal) with the physical environment.
Ecosystem, 224, 268, the interaction between a group of life forms and their physical environment.
Egyptian theories of the earth, **14,** 15, 163–165.
Ejecta (volcanic), 277, **278,** 309, the volcanic material (ash, lava, volcanic bombs) thrown out (ejected) from an exploding *volcano.* See: *Pyroclastic.*
Elastic-rebound theory, 304, the sudden movement of portions of the earth's crust as they are subject to intense pressure, causing them to bend and break, and, finally, try to return to their original position. As this occurs, *seismic waves* are sent rolling through the *lithosphere,* causing *earthquakes.*
Electromagnetic radiation, 64–66, **65,** the energy emitted by the sun in varying wavelengths, covering a broad spectrum—from visible light to ultraviolet, X rays, gamma, and infrared rays.
Elements, 274, **274,** a substance, or matter, which cannot be broken apart or subdivided by natural means, such as: gold, oxygen, sodium:
  mineral-forming, 274, **274,** some minerals are composed of only one element (gold or silver), and other minerals may be composed of a combination of elements (quartz—silicon and oxygen; limestone—calcium, carbon, oxygen; table salt—sodium, carbon, oxygen).
Ellipse: oval-shaped.
Elipsoid (oblate), 20, **20,** an oval-shaped body (such as the earth), that has a greater dimension measured through the equatorial region than through the poles.
Elliptical: like an *ellipse.*
El Niño, 176, **178,** the spin-off of the Pacific equatorial countercurrent that creates dramatic changes in the climates of Peru, Ecuador, and Colombia.
Eluviation, zone of, 255, **255,** the topmost *soil* layer, where nutrients in the form of organic matter (*humus*) accumulate and, in rainy regions, are lost by the process termed *leaching,* where water pulled down by gravity leaches (robs) the *A horizon* of its nutrients.
Energy: the power or capacity to do work, to move into vigorous action:
  balance (budget), 66–67, **68,** a condition wherein a state of *equilibrium* is attained between the incoming energy from the sun and the outgoing energy

INDEX AND GLOSSARY

reflected back into space by the earth and its atmosphere;
biofuel, 246, **246,** see: *Case Study: Biofuel—Energy from Vegetation.*
corpuscular, 65;
geothermal, 158–160, **158, 160,** energy created by the harnessing of the power of rising and expanding steam from the heated rocks of the lower region of the *lithosphere;*
nuclear, 283–285, **284,** see: *Case Study: Nuclear Energy—Fission or Fusion?*
oceans, 184–185, see: *Case Study: Energy from the Oceans.*
solar, 64–67, **65,** 81–85, **82, 83, 84, 85,** the electromagnetic waves of energy emitted by the sun which supplies planet Earth with all the forms of energy we now use. See: *Case Study: Solar Energy—Friend or Foe?*

Environmental determinism, 6–9, 220, the concept that the physical environment determines (controls) the cultural destiny of any group of humans.

Epeirogeny, 296, the rising or sinking of a very large mass of the *earth's crust* due to the *tectonic* movement of the *crustal plates.*

Epiphytes, **233,** 234, plants that live detached from the soil, with their roots attached to another plant. These parasites live above ground level.

Epochs of Geologic Time, **269, 270,** 271, a subdivision of the major divisions of *Geologic Time* called *epochs.*

Equal-area map projection, 32–34, **33,** a projection on which all portions of the earth represented on the map are equal in the amount of space occupied.

Equal-distance map projection, 33–34, an azimuthal or zenithal *map projection* on which distances from the central (focal) point can be measured with accuracy. Also termed *equidistant.*

Equator, 20, 22–27, **23, 24, 28,** the *parallel of latitude* drawn about the earth at its greatest circumference, halfway between the poles, and runs in an east-west direction. Its position is 0° latitude and is the only parallel which is a *great circle.*

Equatorial belt of low pressures, see: *Doldrums.*

Equilibrium: a state of rest or action, when a system is at a stage of balance.

Equinox, **60,** 61, **62,** the time of the year when day and night are of equal length because the *earth's axis* is perpendicular to the *plane of the ecliptic,* and the *circle of illumination* passes through both *poles.* Spring (vernal) equinox occurs about March 21 and the fall (autumnal) equinox occurs about September 23.

Eras, of Geologic Time, 268–271, **269, 270,** the major divisions of the span of earth's history, as represented in *Geologic Time:* archaeozoic, proterozoic, paleozoic, mesozoic, cenozoic.

Eratosthenes, 16, the first human known to have computed the circumference of the earth. In the 3rd century B.C., he came within 100 miles (160 kilometers) of today's accurately measured distance.

Erg, 408, **409,** the Arabic word for *sandy desert.*

Erosion, 322, **325,** 328, 336, the processes responsible for the breaking up and removal of earth materials.
glacial, 383ff, **383, 384, 385, 387, 389;**
gravity, 328ff, **329, 330, 331;**
groundwater, 349ff, **351;**
mass wasting, 328ff, **329–331;**
ocean waves and currents, 359ff, **362, 363, 365, 366, 367;**
splash, 336;
stream, 336ff, **339, 347, 348;**
wind, 404ff, **407.**

Erratics (glacial), 389, **395, 396,** the debris, composed of silt particles, sand grains, pebbles, and boulders, deposited by a *glacier* that is unsorted and unstratified. See: *Moraines.*

Esker, 395, **395, 396,** a long, winding, almost snakelike mound of *alluvium* deposited in a stream channel beneath a *glacier.* These deposits are composed of stratified sand and gravels and may stretch as far as 50 miles (80 kilometers).

Estuarine delta, 340, **343,** a delta formed where a river flows into a large body of water (lake or ocean). These deltas are usually V-shaped, due to the action of the changing *tides.*

Estuary: the mouth of a stream valley where the fresh water of the stream mixes with the salty waters of the ocean due to the influence of the tides.

Evaporation, 76–77, **125**–126, 227–231, **228,** the changing of water from a liquid to a gaseous state, as water vapor.
rate of, 125–127.

Evaporites, 280, chemical rocks or sediments derived from the evaporation of the water that has held them in solution.

Evapotranspiration, 150n, **153,** the combined loss of moisture from the land and water bodies (*evaporation*) and from vegetation (*transpiration*).

Evergreen vegetation, 235, plant forms, shrubs or trees, that maintain their leaf cover throughout the year.

Exfoliation, **327,** 328, the breaking off, or peeling, of curved pieces or slabs of a solid rock mass that is uniform in structure. See: *Spalling; Weathering.*

Exosphere, 90–92, **91,** the outermost layer, or portion, of the earth's *atmosphere.* Within this layer are the *Van Allen Radiation Belts.*

Extrusive igneous rock, 276–277, **278,** 309–313, **311,**

313, rocks formed from once-liquid *magma* that was ejected from the *mantle* through the *lithosphere* by volcanic action either as *pyroclastic ejecta* or in the liquid form called *lava*.

Extrusive vulcanism:
causes of, 275–277, 305–313;
landforms of, **278, 305, 311, 313.**

Fahrenheit scale, 52n, the temperature scale developed by Gabriel Fahrenheit in the 17th century A.D., in which 32° is the freezing point of water and 212° is the boiling point. See: *Celsius;* Appendix D; and *Case Study: Thinking Metric.*

Fans, alluvial, 340, **341,** a fan-shaped collection of *alluvial sediments* deposited by a stream at the mouth of its valley.

Fault, **295,** 299ff, **300, 301,** a fracture, crack, or break, in a portion of the earth's crust.

Fault-block mountain, 299–302, **301,** a mountain or range created by the vertical uplift of the land on one side of a fault. See: *Horst; Graben; Sierra Nevada.*

Faulting, 299ff, **300, 301,** the movement of one portion of the crust past another, either vertically or horizontally. See: *Diastrophism; Tectonics; Graben; Horst; San Andreas Fault.*

Fault plane, 299, **301,** the angle of the surface of a fault relative to the perpendicular. See: *Dip.*

Fertility (soil), 231, 250ff, 415.

Fetch, 181, the distance over the ocean that a wind blows and the waves travel.

Finger lake, **384,** 387, a narrow lake occupying the floor of a *glacial trough*.

Fiord (fjord), 370, **371,** 386, **387,** a deep glacially carved valley along a coastline that has been invaded by the ocean as the glacier retreated.

Fire Island, New York, 367.

Firn, 380, **380, 381,** the accumulation of snow in a high mountain valley above the *snow line*, where the building up of the pressure from the heavy snowpack causes the formation of ice crystals—the first stage in the life of a *glacier*. See: *Accumulation, zone of, in glaciers.*

Fjord, see: *Fiord.*

Floodplain, 314, 344–345, **345, 346,** the flat area of a wide stream valley that is usually covered with water during *flood stage.*

Flood stage, 349, that point when the volume of water in the stream channel cannot be contained by the *levees*, and overflows its banks.

Fluvial, 340, from the Latin, *fluvialis*, "river," thus relating to stream activity.

Fluvial hydrologists, 340, the earth scientists who specialize in the study of earth's fresh waters in streams, rivers, and lakes.

Fluvial landforms:
depositional, 340–345, **341, 342, 343, 344,** landforms created by the *aggradational* activity of streams. See: *Alluvial fans; Alluvial plains; Deltas; Floodplain.*
erosional, 336–340, **338, 339, 347,** landforms created by the *degradational* activity of streams. See: *Incised meanders; Stream-cut terraces; Stream valleys; Waterfalls; Interfluves.*

Foehn wind, 75n, see: *Adiabatic heating; Compression; Winds.*

Fog, 127–129, **141,** 194, the forming of tiny water droplets as water vapor condenses close to the earth's surface:
advection fog, 127, the long-lasting layer of fog created as a moving warm and moist *air mass* comes in contact with cold waters or a land surface that is cold, and is chilled to the point of *condensation*.
ground-inversion fog, 140, the fog, normally short-lived, that develops as a warm air mass settles over a cold ground surface and is chilled to the point of *condensation*.
radiation fog, 140, see: *Ground-inversion fog.*

Folding of strata, 299, **300,** the bending or warping of layers (*strata*) of rock due to *tectonic* forces within the *lithosphere*.

Forests, 237–240, **237, 238, 240.**
classification of:
high latitude, 213, **238, 240;**
midlatitude, **238,** 240;
subtropical, 239–240;
tropical, 237–239, **238;**
distribution of, **238;**
soils of, 257, 259–261.

Fossil fuels, 246, 283–285, the energy-producing minerals derived from once-living organisms taken from the earth, such as coal, petroleum, oil, shale, and natural gas.

Fossils, 277, **282,** the remains of once-living organisms that have been preserved over long periods of *Geologic Time* by natural causes and are usually found in *sedimentary rocks*.

Foucault's pendulum, 57–58, **58.**

Franklin, Benjamin, 356, charted the course of the North Atlantic's *Gulf Stream.*

Frigid zone, 6, 189, 190, one of the three climatic regions devised by the ancient Greek philosopher *Aristotle*. It included the higher latitudes, the land of ice and snow.

Fringing reefs, 371, **372,** coral reefs that surround a submerged *volcano*, or guyot (sea mount). See: *Coral.*

INDEX AND GLOSSARY 461

Fronts, weather, 130–137, **130, 131, 132, 133:**
  cold, 132, **132,** see: *Cold front.*
  occluded, **133,** 135, a frontal system that develops as a portion of a warm air mass is forced aloft by the advancing *cold front* and is cut off from its ground-level source;
  polar, 103, 137, 210, 213, 379, a mass of cold fronts that emerges from the polar regions. See: *Bjerknes; Polar front theory.*
  precipitation during, 129–137;
  warm, 132, **132,** these develop when a moving mass of warm air intrudes upon an area of cold air, and the warm air is forced to rise, thus undergoing *adiabatic cooling,* with the resultant *condensation* and, usually, *precipitation.*
Frost, 79, 378, the point during the *condensation* process when the temperature of the air is at or below freezing and ice crystals will appear on earthly objects.
Frost action, 323, the mechanical (physical) breaking up (*weathering*) of solid rock as water enters the rock, and when the temperature falls to freezing, the rock endures rapid expansion, thus forcing the rock to break apart.
Fundy, Bay of, 184, 361–364, **363.**

Gabbro, 276, **276,** an *intrusive (plutonic)* rock that is dark in color and coarse-grained. (It is composed of plagioclase, pyroxene, feldspar).
Gale: a wind that moves at a velocity of over 32 miles (51 km) per hour.
Galeria, **237,** 239, the heavy, impenetrable undergrowth that evolves in *tropical rain forest* areas, usually along the edges of streams, highways, and meadows, where *insolation* can touch the ground between the trees and give life to plants that otherwise would never grow.
Galveston, Texas, 369.
Gamma rays, 64–66, **65,** a portion of the sun's *electromagnetic radiation* similar in wavelength to *ultraviolet* and *X rays,* but much shorter than *visible light.*
Gasohol, 246, a mixture of gasoline and alcohol as fuel for internal combustion engines. See *Case Study: Biofuel—Energy from Vegetation.*
Geiger, R., 190, 194, a German *climatologist* who modified *Koppen's climatic classification* system.
Geiger counter, **67,** an instrument used to detect emissions of *gamma rays* from *radioactive* substances.
Geode, 280, a round cavity within a rock that is lined with crystals.
Geodesist, 21, scientists who try to determine the earth's exact shape.

Geodesy, 2, **7,** 19–21, **20,** the earth science concerned with the study of the shape of the earth and its dimensions.
Geodetic control point, 427.
Geographic grid, 21–25, **24, 27,** 442, a place's location on the earth as fixed by the point of intersection of a specific *parallel of latitude* and a *meridian of longitude.*
Geographical cycle, 321, 322, 347, the passage of earth's landforms through stages of change similar to those endured by living organisms: a mountain is uplifted (infancy), undergoes *erosional* processes by streams, winds, or glaciers, and is sculptured into sharply defined ridges and valleys (youth); then more erosion occurs and the ridges become rounded and the valleys widened (maturity). Eventually, the hills are worn away and the final ending is a flat plain (old age), which *William Morris Davis* called the *peneplain.* See: *Cycle of denudation.*
Geographic north, **33,** 39–41, **39, 40, 439.**
Geography as a career, 418–420.
Geoid, 21, literally, this term means "like earth." If the planet Mars has a shape exactly like earth's, it could be called a "geoid," but, if not, then Mars' shape would be a "marzoid."
Geologic Time, 268–271, **269, 270,** the expanse of time from the earth's beginning to the present.
Geologic time scale, 268–271, **269, 270,** the evolutionary stages of the development of the earth from its beginning to the present.
Geology, 3, 8, 268, 299, the scientific study of the earth's structure, the materials of which it is composed, and the changes it has undergone—and is still undergoing.
Geomorphic cycle, 322, see: *Cycle of denudation; Geographical cycle.*
Geomorphology, 3, 8, 268, the scientific study of the face of the earth, primarily its *landforms.*
Geothermal energy, 158–162, **158, 160, 161,** see *Case Study: Geothermal Energy.* See also: *Energy, geothermal.*
Geysers and hot springs, 158–174, 313, **159,** when water infiltrates down to close vicinity to the extremely hot rocks beneath the earth's surface, it becomes overheated, boils, and sends roaring steam and water upward through fissures, and erupts like a spouting fountain.
Gibber Plain, 407, a rocky, sandless desert in Australia. The term is a corruption of the Gibson Desert which is south of the *Great Sandy Desert.*
Glacial deposition, 378, 389–390, 394–397, **395, 396,** deposits left behind by a receding glacier, such as *drumlins, kames, eskers, moraines, till.*

Glacial drift, 389, **395, 396,** sediments deposited by *glaciers*, ranging from the fine silt to large boulders (erratics).
Glacial erosion, 359, 360, **371,** 383–389, **384, 386, 388,** 394, **395;**
  by abrasion, 359, 370, **371,** 383, the grinding and scraping process performed by a moving glacier against the sides and floor of the valley;
  by plucking, 383, pieces of rock pulled loose from the sides or floor of the valley by the moving glacier contribute to the abrasion;
  polish, 389, when a mass of granite is polished to a shining surface by glacial abrasion;
  striations, 388, **389,** long, parallel lines (grooves) scratched into the sides or floor of the glacial valley by the rock fragments embedded in the glacier.
Glacial lakes, **384,** 385, 387, 396n.
Glacial landforms:
  alpine, depositional, 389–390;
    erosional, 383–389, **384, 386, 387, 388;**
  continental, depositional, 394–397, **395, 396;**
    erosional, **389,** 394.
Glacial troughs, 370, **384,** 385–388, a steep-sided U-shaped valley carved by an *alpine* (valley, or mountain) *glacier*.
Glaciation, 314, 377–399:
  alpine (valley, mountain), 379–390, **380, 381, 383, 384, 385, 386, 388;**
  continental, 390–397, **392, 393, 395, 396;**
  Pleistocene Ice Age, 74, 390–394, **392, 393;**
  advances, 390–394, **393;**
  interglacial stages, 74, 391, **393.**
Glaciers: masses of accumulated snow that have been compacted into ice and move down high mountain valleys by the pull of *gravity*:
  classification of, 167–169, **168,** 382–399;
  formation of, 378–382, **380, 381;**
  influence of on world climate, 167, 217, 390–394, **392;**
  movements of, 382–390, **381, 384;**
  structure of, 379–381, **381.**
Glaciers and ice caps, 167–169, **168, 169,** 217, 378–399, **380, 383, 391, 392, 397.**
Glaciology, 378, 390, the scientific study of *glaciers* and *glaciation*.
Gleization, 258, the evolution of boggy *soils* in areas of poor drainage.
Globe, 16, 27–34, **31,** a spherical (round) body on which a map of the earth's surface is printed.
Gneiss, 230, **281,** a coarse-grained *metamorphic* rock that was changed from its original form, such as the "salt and pepper" appearance of *granite*, into a darker rock striped with parallel bands of hornblende, mica, and feldspar.
Gobi Desert, Mongolia, **244,** 263.
Gondwana, 293–295, **294,** according to *Alfred Wegener's theory of continental drift*, when the super continent *Pangaea* began to split into two pieces, the northern half he called *Laurasia*, and the southern half, Gondwana. Gondwana later split and separated into the continental land masses we know as South America, Africa, India, Australia, and Antarctica.
Goode's interrupted homolosine map projection, 44, **45.**
Graben, **301,** 302, a long narrow *fault block* that has been forced to sink or drop between two faults, thus forming a deep valley, such as California's *Death Valley*.
Gradation, 322, a term meaning "leveling the land," which includes both the *degradational* (cutting down) forces and the *aggradational* (building up) forces.
Gradational forces, 322:
  glaciers, 378ff;
  groundwater, 349ff;
  ocean waves, 356ff;
  streams, 336ff;
  winds, 404ff.
Graded stream, 337–338, a stream that has attained a stage of *equilibrium*, wherein the *degradational* and *aggradational* forces have reached a balance in which there is little downcutting or deposition.
Gradient: the degree, or angle, of a *slope*, which can be gentle or steep:
  hydraulic, 155, **156,** 337, the slope of a *water table* from its highest point of elevation to its lowest point. The steeper the gradient, the faster the water will move;
  pressure, 108–112, **109,** the amount of change from one pressure region to another. The greater the change (the steeper the gradient), the stronger and faster will the wind blow;
  temperature, 99, the rate of change from one temperature area to another will determine the pattern of *isotherms*. If the isotherms are close together, it means the gradient is steep and rapid.
Grand Canyon of the Colorado, 338, **339, 434–437.** See: *Bright Angel Quadrangle*.
Granite, 273, **273,** 276, **276,** 280, **281,** 359, an *intrusive igneous rock* that is coarse-grained, composed of quartz, feldspar, and mica or hornblende. It is often used for building materials due to its resistance to *weathering*.
Granular disintegration, **326,** 328, the breakup of a rock grain by grain. See: *Disintegration, Weather, mechanical*.

Graphic map scale, **34,** 429, **434–438,** a line marked off in distance or length units. Also called linear, line, or bar scale.

Grass, **228,** 231, 234–235, **241,** a family of plants that range in size from one inch (2.5 cm) to over 100 feet (30 m) in height. Included are: cereals, such as corn, wheat, barley, rye, and oats; alfalfa; bluegrass; wild hay; and bamboo.

Grasslands, 240–243, **241:**
  distribution of, **241;**
  and soils, 199–201, 242, 251–252, 259, 261–263.

Gravitation, universal law of, 56, the attraction that one object has on another is called gravity. Basically, the law states that the larger the object, the greater the pull, and the closer two objects are to each other, the greater the pull.

Gravitational movement of water, 152–155, **153,** 250, **255.**

Gravity: the force of attraction that operates under the *Universal Law of Gravitation* between the earth and any object on or in it.
  and glaciers, 379–382;
  and mass wasting, 328–332, **329, 330, 331;**
  and running water, **125, 153,** 250–251;
  and soil moisture, 150–152, **153,** 250.

Great Barrier Reef, Australia, 371

Great circle, 21–22, **21,** 58, a circle on the earth's surface whose central point is the earth's center; thus the circle divides the spherical body of the earth into two equal halves. *Meridians* are halves of great circles that run through the poles. The *equator* is the only *parallel of latitude* that is also a great circle.

Great Lakes, **164, 166,** 394, 396n.

Great Plains, 132, 213, **317,** 251, 259, 263, 415.

Great Salt Lake, Utah, 167, 396.

Great Sandy Desert, Australia, 101, 103, 155–158, **157, 198,** 201, **244,** 250, **260, 262, 409.**

Great Soil Groups, 259ff:
  Comprehensive Soil Classification System, 259–261, **260,** also called the *7th Approximation,* this is the newest system of soil classification, devised by the U.S. Soil Conservation Service;
  1938 USDA Soil Classification System, 261–264, **262,** the classic system of soil classification, devised by cooperation between Russian and American *pedologists* around the beginning of the 20th century.

Greenhouse effect, 72–74, **74,** when incoming *shortwave* radiant energy from the sun is absorbed by the earth and its atmosphere, it is reradiated as *longwave* radiation, which cannot escape through the cloud layer. Thus the temperature within the *troposphere* rises. See: *Albedo; Atmosphere; Clouds; Electromagnetic radiation; Energy balance.*

Greenland ice cap, 167, **168,** 391–394, **393,** 397–399.

Greenwich, England, 22–27, the location of the Royal Astronomical Observatory, through which runs the *Prime (zero) Meridian.*

Greenwich meridian, 22, **23, 24,** 422–425, **422.**

Greenwich time, 22–27, 422–425.

Grid north, 39–41, a line on a map or the earth that points toward the geographical location of the North Pole. See: *Geographic north.*

Ground moraine, 389–390, **391, 395,** deposits and debris left behind by a retreating (receding) *glacier* or *ice sheet.* See: *Till plain.*

Ground radiation, **69,** 72, the *longwave* reradiation emitted by the earth's surface. See: *Terrestrial radiation.*

Ground referencing systems, 41, 442–444, **443, 444:**
  metes and bounds, 442.

Groundwater: water beneath the earth's surface which occupies the *zone of saturation* (fully saturated layer of *porous* and *permeable* rocks) beneath the *water table.*
  deposition, 352, **352;**
  erosion, 349–352, **350, 351;**
  interface, **153,** 154;
  movements of, 152–155, **153, 156;**
  supply of, **151,** 152;
  zones of, 153–155, **153.**

Gulf of Mexico, 172, 206, 213.

Gulf Stream, 177, **178,** 206, 356, 360:
  and Benjamin Franklin, 356.

Gully, 229, **325,** 337, a steep-sided valley carved out of loose earth materials by the sudden rush of water supplied by *thunderstorms.* In Mexico these gullies are called *barrancas;* in Arab lands, *wadis.*

Guyot, see: *Sea mount.*

Gypsum, 277–280, a *sedimentary rock* composed of calcium sulfate and water. It is precipitated from evaporating sea water during dry conditions.

Gyre (gyral), **175,** 177, the large, circular-flowing *ocean currents* usually found beneath *subtropical high-pressure cells.* The direction is the product of a combination of factors—the *Coriolis effect,* winds, and gravity.

Hail, 378–379, a frozen water droplet with concentric layers of ice that forms when the temperature at the point of *precipitation* is at or below freezing.

Half Dome, **266,** an exposed *intrusive granite* dome that has undergone *exfoliation (spalling),* which has given it the rounded shape.

Halomorphic soils, 264, *arid climate soils,* where *evaporation* exceeds *precipitation.*

Hanging trough, **384,** 385–386, a glacially carved tributary valley that is located high above the *trough* of the main glacier.

Hardpan, 250, a hard, *impermeable* layer of *soil* that is heavily compacted with fine particles or chemicals, thus preventing infiltration of water.

Hawaii, 150, **154,** 226, **253,** 277, 290, 304, 312, 313.

Hawaiian High, 206, the ever-present *high-pressure cell* situated north of the Hawaiian Islands (see maps, **104, 105**).

Hawaii Volcanoes National Park, 305.

"Haystacks," **351,** 352, see: *Magotes; Karst topography.*

Heat budget, 66–67, **68, 69,** the state of *balance* or *equilibrium* that exists between incoming *solar radiation* and outgoing longwave *reradiation* from the earth and its atmosphere.

Heat, latent, 76, 137, heat held within an air mass that is released when an *air mass* rises and reaches an area of cooler temperature; and, as the process of *condensation* begins, the cooling process is slowed. See: *Adiabatic cooling.*

Heating, adiabatic, see: *Adiabatic heating.*

Heating and cooling, of the atmosphere:
  controls over, 67–71, **70, 72, 73;**
    albedo, 68–70, the percentage of *insolation* reflected from the earth's surface or a cloud layer;
    land and water differences, 70–71, **78,** 99–101;
    latitude, 71, **72;**
    topography, 110–112, **110, 111.**

Heating processes, 71–76:
  compression, 75;
  condensation, 76;
  conduction, 74;
  greenhouse effect, 73, **74;**
  radiation and reradiation, **69,** 72–74;

Heimaey, 305, **305.**

Helgafel, **305,** volcano on the island of Heimaey.

Hemispheres, 58, 61, halves of a sphere, such as Northern and Southern hemispheres separated by the *equator.*

Herodotus, 16, 16n.

Heterosphere, 92–95, **93, 95,** the region above the *homosphere* where the gaseous composition changes with increase in altitude.

Highlands: mountainous areas where changes in climate, vegetation, and soils occur too rapidly because of the rapid changes in elevations.

Highland climates, see *Climates of the world, highland.*

Highland vegetation, see *Vegetation, highland.*

High latitudes: the region normally considered as situated above the 50° northern parallel, in the subarctic and polar regions.

High-latitude climates: see *Climates of the world—High latitude, boreal, continental subarctic, ice cap.*

High-latitude vegetation: see *Vegetation, high-latitude.*

High pressure, 101–108, **106,** see: *Air pressure; Horse latitudes; Subtropical belt of high pressure; Isobars; Isobaric chart.*

Hills, 314, **315.**

Himalayas, 101, **111,** 116, 167, 298, **305.**

Hindus, ancient theories of, 7, 14–15.

Homer (Greek philosopher, 6th century B.C.), 5, 15.

Homolographic map projection, 33, 44, the *equal-area projection* devised by Mollweide, on which the *parallels* are straight lines, *meridians* are ellipses.

Homolosine map projection, 44, **45,** an *equal-area projection,* similar to the *homolographic,* but, with *Goode's* version, each continent has a vertical meridian in its center and the other meridians curve toward it.

Homosphere, 92–95, **93, 95,** the region of the atmosphere surrounding the earth where the gaseous composition is uniform. It extends outward to about 72 miles (116 km), and underlies the *heterosphere.*

Horizons, soil, see: *Soil horizons.*

Horn, **384,** 385, **386,** a sharply chiseled mountain peak created at the point of intersection of three or more *cirques.*

Horse latitudes, **106,** 108, the *subtropical high-pressure belt* of the Atlantic Ocean in the vicinity of the Azores high, where periods of calm with settling air were interspersed with the powerful westerly winds, thus causing early sailing ships difficulties in attempting rapid east-to-west crossing of the ocean.

Horst, **301,** 302, an uplifted fault-block between two fracture, or fault, zones. See: *Fault; Faulting; Fault-block mountain.*

Humid continental climate regions, see: *Climates of the world.*

Humidity: moisture in the atmosphere:
  absolute, 126–127, **127,** the actual weight of the water present in a specific volume of air, normally measured in grains per cubic foot (or grams per cubic meter);
  relative, 126, the amount of water in a given mass of air as compared to the amount the air could hold if it were to be fully saturated, and is expressed as a percentage;
  specific, 126, the weight of water vapor in the air as compared to the total weight of the air mass including the water vapor.

Humid subtropical climate regions, see: *Climates of the world.*

Humus, 152, 231, 251–252, 257, the decaying organic (once-living) matter (plant or animal) in a *soil.* The greater the amount, the darker will be the soil. See: *Chernozem.*

INDEX AND GLOSSARY

Huntington, Ellsworth, 6.
Hurricanes: tropical cyclones that usually form in the Atlantic Ocean southeast of the Caribbean, and move at a speed of around 72 miles (115 km) per hour.
  distribution of, **134,** 141–145, **144;**
  formation of, 141, **144,** 206.
Hydration, 326, a process of *chemical weathering,* when a rock breaks up due to the addition of water to the chemicals within the rock.
Hydraulic action: the breakup of rock materials by the powerful impact of ocean waves striking beaches or sea cliffs, or streams pounding against valley floors or banks.
  by ocean waves, 364;
  by streams, 337.
Hydraulic gradient, 155, **156,** 337, see: *Gradient, hydraulic.*
Hydroelectric power, 165, the production of electrical energy by harnessing the powerful surge of a river as it moves through the funnels of a dam and causes electricity-generating turbines to spin. See *Case Study: Energy from the Oceans.*
Hydrologic cycle, 124, **125,** 150, 163, 245, 336, 379, the movement of water (as gaseous vapor) from water bodies (lakes, streams, oceans) to the land, where it flows back to the sea, a continuous cycle.
Hydrology, 2, **8,** 124, 340, the scientific study of the earth's waters and the *hydrologic cycle.*
Hydrolysis, 326, the addition of water to a rock wherein it unites with salts in the rock, forming an acid that then causes the rock to break up.
Hydromorphic soils, **251,** 264, water-logged soils in poor drainage areas such as swamps.
Hydrophytes, 227, **251,** plants that can thrive in or under the water surface because they can utilize the incoming sunlight to proceed with the process of *photosynthesis.*
Hydrosphere, 150, the waters of the earth and the atmosphere.
Hygroscopic attraction, 153, the retention of water molecules by soil particles beneath the earth's surface.

Ice: water in a solid state.
Ice ages, 74, 167, 314, 359, 390–394, **392, 393,** periods of earth history when great *ice sheets* formed and moved across massive areas of land.
  Pleistocene, **269, 270,** 390–394, **392, 393,** the most recent of the four great ice ages that began about 2 million years ago and came to an end about 10,000 years ago.
Ice caps and glaciers, 167–169, **168, 169,** 217, 377ff.
Icebergs, 380, 399, **399,** sections of a *glacier* that have broken off from its *terminus* where it extends out onto a body of water; this action is called *calving.*

Ice floes, 380, 398, **398,** large sections of an *ice sheet* that have broken off into the bordering ocean and appear much like floating islands.
Ice forms, 167, 378–381, **380, 381,** see: *Glaciation; Glaciers.*
Ice sheet, 167, **168, 169,** a glacier that has formed over a large expanse of a land mass and moves out in many directions. See: *Continental glaciation.*
Ice cap climates, 167, **215,** 217, 219, see: *Climates of the world, high latitude.*
Iceland, 305, **305.**
Igneous rocks: rocks that have solidified from once-liquid rock called *magma* as the result of cooling:
  extrusive, 276, **278,** 307, 313, *magma* that is pushed or ejected out of the neck of a volcano (as *ejecta* or *lava*). Because of the rapidity of the cooling, the rocks will be fine-grained, often glassy (such as *obsidian*). The most common extrusive igneous rock is *basalt;*
  intrusive, 276, **276,** 312, *magma* that moves beneath the earth's surface in a fiery, liquid state. As it cools very slowly, large-grained *granitic* rocks are formed. These rocks, often called *plutonic,* are less dense than *extrusive* rocks.
Illinoian Period, 393, **393,** the third advance of North America's *continental glacier* during the *Pleistocene Ice Age.*
Illuviation, zone of, **255,** 256, the soil region beneath the *topsoil (A horizon)* that collects nutrients and minerals carried down by *gravitational movement* of infiltrating water. This is the *B horizon,* also referred to as the *subsoil* and the *zone of accumulation.*
Incised meander, 339, **339,** a very deeply entrenched *stream valley* whose wide, meandering course has been rejuvenated by crustal uplift (*epeirogeny*), such has occurred in the Colorado Plateau, creating the *Grand Canyon of the Colorado.*
Inclination of the earth's axis, **60,** 61–64, **64,** earth's *axis* is tilted 23°30′ from the perpendicular to the *plane of the ecliptic.*
India, **111,** 298, see: *Monsoons.*
Indian Ocean, **111, 170,** 171, 172, 177, 298.
Inertia, **56,** 57, an object at rest will remain inactive unless something causes it to move, and an object moving will continue to do so, and in a straight line, unless something interferes with its motion. See: *Newton's First Law of Motion.*
Insolation, 66, **68, 69,** 73, an abbreviation of *incoming solar radiation.* See: *Solar energy; Solar radiation.*
Instability (air mass), 126, when there is rapid *convectional* (uplift) movement within an *air mass* and the *lapse rate* increases, the *weather* is subject to accelerated changes. Such unstable conditions are more frequent in *low-pressure* belts.

Interface, **153**, 154–155, the contact surface between the region of *soil moisture* (the *A* and *B* horizons) and the top of the *zone of saturation*—the *water table*.

Interfluves, 347, the ridges of land between two *stream valleys*. See: *Fluvial*.

Interglacial period, 74, 167, 391, **393**, the span of time that elapses between two *ice ages*.

Interlobate moraine, 395, **395, 396,** the *moraine* deposits that collect between two lobes of a *continental glacier*.

International Date Line (IDL), 22, 422–425, **423,** the 180° *meridian* that is 12 hours earlier or later in the day than the time at the *Prime (Zero) Meridian*, which is called *Greenwich* Mean Time. If you are west of Greenwich, you are earlier in the day; if you are east, then your time is later in the day.

International geophysical year, 171, 379.

International Hydrographic Bureau, 171.

Intertropical Convergence Zone (ITC), **106**, 107–108, the region along the *equator* wherein the *trade winds* (northeast and southeast) converge, or come together, creating a belt of *low-pressure* cells, with generally rising winds. See: *Doldrums; Equatorial belt of low pressure*.

Intrazonal soils, 259, 264, intermediate soils that show signs of immaturity, with poorly developed *soil horizons* resulting from local conditions, such as *parent material, terrain, moisture, temperatures,* and *time*. See: *Soil classification: 1938 USDA Soil Classification*.

Intrusive igneous rocks, 275–277, **276,** see: *Igneous rocks. intrusive*.

Intrusive vulcanism:
causes, 275–276, 290, 305–309, **308, 309, 310;**
landforms of, **308, 309, 310, 311.**

Inversion, temperature, 79–80, **80,** the temperature of an air mass is normally cooler in higher elevations (*normal lapse rate*), but when a warm air mass overlies a cool one, the temperature will increase rather than decrease, thus preventing rising warm air from escaping into higher altitudes. Air pollution is one of the serious consequences of such inversions in heavily populated areas.
ground, 78–79;
static and dynamic, 78–79;
and the urban environment, 79–80, **80.**

Ion, 90–92, **91,** an *atom* that is electrically charged by the addition or loss of one or more electrons.

Ionosphere, 90–92, **91,** the layer of the *atmosphere* above the *stratosphere* that contains a great number of gaseous particles that are under constant bombardment by incoming cosmic rays, whereupon they become electrically charged, or ionized.

Irkutsk, Siberia, 213, **216.**

Isorithms, 37, lines on a map connecting points of equal value which also separates regions of different values. Similar nonspecific terms are *isograms* and *isopleths*. If a map is to show temperatures over a wide area, the cartographer would use *isotherms*—lines that connect points of equal air temperature. See: *Isobars; Isogons; Isohyets; Contours; Topographic map*.

Isobar, 103, **104, 105, 109,** a line connecting points of equal *atmospheric (barometric) pressure*.

Isobaric chart, 103–108, **104, 105, 109,** a map that shows the areas of *high* and *low pressure* on a region, from which one can determine the *pressure gradient* and an estimate as to the intensity and direction of *winds*.

Isogon, **40,** 41, a line connecting points of equal *magnetic variation (declination)*.

Isohyet, 135, **134, 136,** a line connecting points of equal *precipitation*.

Isostasy, 299, from the Greek *iso*, "equal," and *stasis*, "standing," which means that a state of *equilibrium*, or balance, exists between two objects. Example: if equilibrium is maintained between two landmasses (composed of *sial*) and the floor of the ocean (*sima*), there will be no movement. If, however, one becomes heavier through acquisition of a heavier load, it will press down, causing the lighter mass to uplift until equilibrium is attained.

Isotherms, 37, 96–101, **87, 98, 100,** lines connecting points of equal *air temperature*.

Japanese Current (kuroshio), 177, **178,** a warm *ocean current* that is a spin-off from the Pacific *north equatorial current* and is darkened through the acquisition of algae and other sealife forms as it moves through the warm waters near the Philippines and heads northward, sweeping past China and Japan.

Jet streams, 94, 116–119, **117, 118,** rapidly moving air currents near the *tropopause*, where winds from the *subtropical belt of high pressure (westerlies)* collide with winds (*northeasterlies*), from the *subpolar lows*. They are narrow in width, but move at speeds up to 300 miles (480 km) per hour.

Joint-block separation, **326,** 328, the breakup of a rock mass along previously induced joints or fractures into large rectangular-shaped blocks. See: *Weathering*.

Jungle, **198,** 199, **200,** 237–239, **237, 238,** the *tropical rain forest vegetation* which develops in a climatic region of heavy annual *precipitation* ranging from 80 to 100 inches (203–500 cm), where the crowded canopy of leaves blocks the penetration of *insolation* to the ground level. Thus, there are little, if any, shrubs, grasses, or other small plants between the trunks of the tress. A commonly used term for this type of forest is a Brazilian word *Selva*.

Kalihari Desert, 103, **244:**
   Steppe, **241.**
Kame, **395**, 396, a conical hill or mound on a glacial *till plain* which is composed of sorted and stratified debris deposited by streams running beneath a glacier.
Kansan (glacial) Period, 393, **393**, the second advance of North America's *continental glacier* during the *Pleistocene Ice Age*.
Karst, 350, **350, 351,** the dissolving of *limestone* by rainfall, flowing streams, or the infiltration of groundwater, which produces *sinkholes* on the ground surface and *caverns* beneath. The name comes from the limestone region in the Adriatic Coast of Yugoslavia, where the work of water on beds of limestone was first scientifically studied.
Karst topography, 350, **350, 351,** the natural landscape created by the work of water on limestone regions. See: *Cenotes; Dolines; Ponors; Sinkholes; Uvalas.*
Katabatic winds, 79n, cool air that slides down a slope due to the pull of gravity. The temperature on the valley floor will be very cold, but agricultural crops, such as fruits, will thrive on the downslopes.
Kettle, 394, **395**, a depression left behind as a block of glacial ice that has been deposited on the *outwash plain* has melted.
Kilimanjaro, Mt., 379, a 19,340-ft (5900-m) mountain in Tanzania.
Kirghiz Steppes, 213, **214, 241.**
Klimata, 6, 90, 189, 190, the climate regions (torrid, temperature, and frigid zones) as theorized by *Aristotle*.
Köppen, 190, 245.
Köppen's climatic classification, 190–194, **191, 192, 193**.
Krakatoa, 305, a volcano west of Java.
Kuroshio Current, 177, **178**, see: *Japanese Current*.

Laccolith, 307, **310**, an *intrusive* mass of *magma* that pushes the overlying *strata* upward, thus creating a structural *dome*.
Ladurie, Emmanuel, 220.
Lagoon, **367**, 369, **369**, the body of water located between an *offshore bar* or *fringing reef* and the mainland.
Lake Agassiz, 396n.

Lake Baikal, **164,** 166, **166,** 213, the deepest lake in the world.
Lake Bonneville, 396n.
Lakes:
   freshwater, **164,** 166, 166–167;
   glacial, **384,** 385, 387, 396n;
   oxbow, 345–349, **346, 348,** a curved meander loop bypassed and cut off from the main stream channel of a *meandering* river. Called *bayou* in Louisiana, from the original Choctaw word, *bayuk;*
   saltwater, 167;
   as water supply, **151,** 166–167.
Lambert's conic conformal map projection, 43, **43.**
Land breeze, 108–110, **110,** gentle wind blowing from the land to the sea at night as the cooler air over the land moves out to replace the rising uplift (*convectional*) of the warmer air over the ocean. It is also called *offshore breeze*. The opposite movement, from sea to land during the day, is called a *sea* (or *onshore*) *breeze*.
Land bridge, 390–394.
Land ethic, 9, see: *Ecology; Ecosystems.*
Landforms, the surface features of land and sea:
   classification of, 314–318, **315, 316, 317;**
   of land surface and sea bottom, 289ff, **297, 315;**
   of the world, **297;**
   See: *Diastrophism; Glaciers; Groundwater; Oceans; Wind waves; Streams; Tetonism; Vulcanism; Waves.*
Landslides, 329, **329,** the downslope slide of large masses of rock materials due to the pull of gravity.
Land survey systems, 41, 442–444, **442, 443, 444:**
   of early California, 442, **442;**
   metes and bounds, 442;
   rectangular, 41, 443, **443;**
   township and range, 443, **444.**
Lapse rates: see *Adiabatic processes.*
Large-scale map, **36,** 36, 429, **430,** a map that shows a large amount of detail on a small amount of earth space represented on the map, such as a map of a neighborhood or a city. See: *Map, scales of.*
Latent heat, 76, 137, see: *Heat, latent.*
Lateral moraine, 390, **391,** glacial deposits that build up along the edges of a *glacier* as it moves along its valley. See: *Alpine glaciers; Glacial deposition.*
Laterite soils, 257, 259, **262,** see: *Latosols.*
Latitude, 16n, 22–25, **23, 24, 27,** the distance (measured in *degrees*) north or south of the *equator*, shown on a map or a globe as lines circling the earth that are parallel to each other. These *parallels of latitude* intersect the *meridians of longitude* at right angles.

Latitude and longitude, measurements along, 22–29, **27, 28,** see: *Geographic Grid.*

Latosols, **195,** 259, **262,** soils composed of iron and aluminum that develop in rainy climate regions where there is poor drainge. See: *Soils, classification.*

Laurasia, **294,** 295, the northern half of *Alfred Wegener's* original supercontinent, *Pangaea;* the southern half is called *Gondwana.*

Lava: extremely hot, liquid rock (*magma*) extruded onto the earth's surface.

Lava flows, 275–277, **278,** 309, **311:**

Aa, 277, **278,** a jumbled lava flow composed of rough, blocky, and very coarse pieces of basaltic rock;

Pahoehoe, **253,** 277, **278,** a smooth and billowy lava flow, sometimes having the appearance of coils of rope.

Lava plateaus, 309–312, **311.**

Leaching, 250–255, **255,** 261, the removal of minerals and other nutrients by the downward *gravitational movement* of water as it infiltrates the upper layers of *soil.*

Leaf types, 235, **236.**

Lee side, **135,** the opposite side. Example: the sheltered downwind side of a mountain as opposed to the other, *windward* side, which faces the wind. Also termed *leeward.*

Leshy (Slavic forest god), 224, **225.**

Levees:

artificial, 345, piles of boulders, sandbags, or heaps of soil dumped along the banks of streams or rivers to prevent their overflowing during *flood stage;*

natural, 344–345, **345,** ridges of *alluvium* deposited on the banks of a stream channel during a *flood stage.*

Liana, **233,** 234, climbing, woody vine of *tropical rain forests* that hang suspended from the tree branches while their roots are in the ground.

Lichens, 234, a plant combination of algae and fungus which have no root systems or stems. See: *Thallophytes.*

Life forms of vegetation, 234–235, **232, 233,** the physical characteristics of plants: the structure, shape, and size of leaves, roots, stems, trunks, as well as the variation in heights, the amount of coverage, and *periodicity.*

Lightning, 137, the flash produced by an electrical discharge within a cloud, between two clouds, or between a cloud and the earth.

Limestone: a non *clastic sedimentary rock* composed primarily of calcium carbonate (CaCO$_3$):

chemical, 280, calcite or *travertine* formed when carbon dioxide enters the earth, carried by water, and combines with calcium;

metamorphosed, **281,** when the crystalline structure of limestone is changed (metamorphosed) by tremendous heat or pressure, *marble* is created. See: *Metamorphism, contact; Metamorphism, dynamic.*

Limestone caverns, 350–352, **352,** large underground caves formed as the limestone is dissolved by the process of *carbonation* and is carried away by *groundwater movement.* See: *Carlsbad Caverns; Karst topography.*

Linear map scale: see: *Graphic map scale.*

Line of tangency, 42, **43;** see: *Tangency, line of.*

Lithosols, 264, *azonal soils* normally found on steep slopes.

Lithosphere, 2, 273, **273,** the "rocky sphere," the rigid portion of the outer layer of the earth above the more pliable *mantle.* See: *Asthenosphere; Sial; Crust, earth's.*

Llanos, 201, **241,** 242, the *savanna* grassland of southern Venezuela and eastern Colombia.

Load (of streams), 340, the material being transported by streams.

Lobeck, A. K., 46.

Local relief, 314, **315,** the difference between the highest elevation and the lowest in any given area.

Local time, 25, 422–425, the time at any point on the earth's surface relative to the position of the sun. Local noon time is the time when the sun is at the highest point (zenith) in the sky.

Loess: wind-transported and wind-deposited fine-grained soils, usually silt:

distribution of, 397, **414,** 415;

origin of, 397, 415.

Longitude, 16n, 22–25, **23, 26, 27, 28,** a point on the earth's surface measured in an arc from the center of the earth, east or west of the *Prime (Zero) Meridian.*

meridians of, 22, **26, 28;**

and time, 25, **26,** 422–425, **422, 423, 424.**

Longitudinal sand dunes, 412, mounds of *sand dunes* where the ridges are parallel to the wind direction, thus they may extend for many miles.

Longshore current, 360–361, **362,** the movement along a *shoreline* caused by the incoming oblique waves and the backwash of returning waves at right angle to the beach, which creates a down-the-coast current.

Longshore transport: the movement of beach sand down a shoreline by the *longshore current.*

Los Angeles, climate of, 206.

Low: the center of a low-pressure area.

Low-pressure cells, 101–108, **104, 105, 106, 109,** 177, areas of low *atmospheric (barometric) pressure,* where the heated air is rising *convectionally* in a counterclockwise direction (in the Northern Hemisphere). See: *Equatorial belt of low pressure; Doldrums; Convection.*

Low-latitude climates: see: *Climates of the world.*

Loxodrome, 42n, the line drawn on a map that intersects all *meridians* at the same angle. See: *Rhumb line.*

Magma, 274, 275–277, 309–313, **311,** rock in a liquid, molten state, and with a very high temperature.

Magnetic:
compass, 30, 39–41, **39, 40, 439,** 439–440, a direction-finding instrument that has a magnetized needle that points toward the magnetic poles.
declination (variation), 39–41, **39, 40, 434, 436, 439,** the deviation of the needle of a magnetic compass away from *true north.*
poles, 39–41, 42–46, **39, 40, 44,** 301, **434, 436, 439, 478, 480, 484,** the focal points of the earth's powerful magnetic field that is the result of magnetism created by the mass of iron and nickel at the earth's core.

Magnetosphere, 65, the magnetic field that surrounds the earth and extends outward to a distance of about 50,000 miles (80,000 km). See: *Corpuscular energy.*

Magotes, **351,** 352, the honeycombed remnants of beds of *limestone* in an area of *karst topography,* which have the appearance of haystacks.

Malthus, Thomas Robert, 190.

Mammoth Caves, Kentucky, 326, 352.

Mantle, 271–273, **272, 273,** 291, **309,** the region of the earth located between the *core* and the *crust,* which is extremely hot, ranging from 6332°F (3500°C) to 3632°F (2000°C), and is somewhat pliable, almost plastic.

Map projections: symbolic representations of the global earth on a flat surface:
azimuthal, 30–34, **31, 33,** 43–44, **44;**
cartograms, 46, **48;**
conformal, 30–34, **32,** 42, **43, 45;**
conical, **31,** 42–43, **43;**
cylindrical, 41–42, **42;**
Eckert IV, 44, **45;**
equal-area, 32, **33;**
equal-distance, 33;
equidistant, 32;
equivalence, 32;
Goode's homolosine, 44, **45;**
homolographic, 44, **45;**
homolosine, **45;**
interrupted, 44, **45;**
Lambert's conic conformal, 43, **43;**
mathematical, 44;
Mercator, 31, **32, 42;**
miscellaneous, 44–46;
modified conic, 43;
Mollweide, **33;**
physiographic, 46, **297;**
polyconic, 472n, **434–437;**
relief-shaded, 433, **436;**
topographic, 427–440, **430, 431, 432, 433, 434–435, 438;**

Maps, essentials of, 34–41, 427–440:
directions, 39–41, **39;**
distortions, 30–31, **31;**
evolution of, 29–30;
legends and symbols on, 37–38, **38, 432;**
properties of, 32–34;
scales of, 34–37, **34, 35, 36;**
large, medium, and small, 34–37, **36, 430;**
weather forecast, **138, 139.**

Maquis, 239, the *climax vegetation* of the *Mediterrean climate region,* also known as the *West Coastal Dry Subtropical,* and the *vegetation* is called *Mediterranean woodland and scrub.*

Marble, 280, **281,** *limestone* that has endured *metamorphosis* as it has been recrystallized through great pressure or heat.

Mariana Trench, 21, see: *Trenches.*

Marine (sea) terraces, 294, **296,** 364, **365,** a relatively flat plain along a coastline that has been cut by wave action. Also known as *wave-cut benches,* they have been elevated above sea level.

Marine West Coast Climates, see: *Climates of the World.*

Mars, 52, **53, 54.**

Martha's Vineyard, **369.**

Mass wasting, 328–332, **329, 330, 331,** the movement downward of earth materials (soils and rocks) due to the pull of gravity.

Matterhorn, 385, **386.**

Mature stream valley, 345–359, **348,** a widened stream valley that has reached a state of *equilibrium* and has a well-developed *floodplain.*

Maury, Matthew Fontaine, 356.

Meandering stream pattern, 338–349, **339, 345, 346, 348,** the wide, side-to-side wandering of a stream in an old-age stream valley.

Meanders, stream:
cut-off, 345, **346,** 349; see: *Ox bow lakes.*
incised, 339, **339;** see: *Grand Canyon of the Colorado.*

Mechanical weathering, 322–323, **324, 325, 326, 327,** the breakup of rocks without any chemical change, through such physical actions as: *thermal expansion, frost action, crystal growth, biotic activity,* and *unloading.*

Mediterranean climate region: see *Climates of the World, Mediterranean West Coastal Dry Subtropical.*

Mediterranean Sea, 172, 174, 205–206, 295.

Mediterranean vegetation, 239, **239,** see: *Mediterranean Woodland and Scrub.*

Mediterranean Woodland and Scrub, 239, **239,** low-growing woody bushes and shrubs, as well as some trees such as olive, fig, chestnut, and liveoak.

Meltwater, 380, **381, 383,** water issuing forth from a melting *glacier.*

Mercalli scale, 304, measures the intensity of an *earthquake* based on the amount of damage done to a region.

Mercator map projection, 31, **32, 42,** a *cylindrical map projection* designed by Gerhard Krämer. It was intended to be used by seamen to lay out courses across the oceans. See: *Loxodrome; Rhumb Line.*

Mercury (planet), 52–54, **53, 54.**

Mercury barometer, 102, **107,** an instrument that measures *atmospheric (barometric) pressure* by the effect of the weight of the air on liquid mercury contained in a glass tube. See: *Barometer; Air Pressure.*

Meridian: a line from the North Pole to the South Pole that is half of a great circle.

of longitude, **21,** 22–27, **23, 24, 26, 27, 28,** a line denoting a place's location east or west of the *Prime (Zero) Meridian* in *degrees;*

prime, 22, **23, 24,** the *zero meridian* which passes through Greenwich, England;

and time, 25, **26,** 422–425, **422, 423, 424, 466, 467, 468,** see: *International Date Line.*

Mesopause, **95,** 96, the upper boundary of the *mesosphere.*

Mesosphere, **95,** 96, the layer of the *atmosphere* that is characterized by a sudden and extensive decrease in temperature with increase in altitude. It is situated above the *stratosphere.*

Mesozoic Era, **269, 270,** the fourth major subdivision of *Geologic Time;* the "middle life" era, often called the Age of the Dinosaurs.

Metamorphic rocks, 280, **281,** rocks that have undergone structural change because of intensive heat or extreme pressures.

Metamorphosis: change in form or structure.

Metamorphism:

contact, 280, a structural change in a rock's structure brought on by extreme heat changes in the vicinity of the rock's place of origination;

dynamic, 280, structural change in a rock's structure caused by tremendous pressure exerted by the weight of overlying masses of rock, or by *tectonic forces.*

Meteorology, 2, **8,** 90, the scientific study of the *weather.*

Metes and bounds, 442, an ancient way of measuring and fixing the limits and the boundaries of a place.

Metric system, 9, 52n, see *Case Study: Thinking Metric.* See: *Appendix D,* 446; Metric conversion tables.

Mid-Atlantic Ridge, **292, 297, 307, 309,** the longest mountain range on the earth's surface, created by the *magma* issuing forth from the crack in the ocean floor where the seafloor is spreading apart. See: *Continental drift; Seafloor spreading; Plate tectonics; Vulcanism; Midocean Ridges.*

Midlatitudes: the area between the polar regions and the subtropical regions:

climates of, see: *Climates of the world, midlatitude.*

steppes and prairies, **208,** 213, **214, 241,** 242;

storms, 141;

vegetation, **192, 204,** 210–213, **214, 238,** 239–242, **239, 241.**

Midocean ridges, **292, 297, 305, 309,** mountain chains on the floor of the oceans, created by *vulcanism* as the *crustal plates* move apart during the process of *seafloor spreading.*

Mile: a unit of distance:

land, or statute, 26–27, **28,** one mile equals 5280 feet, 1760 yards, 63,600 inches, 880 fathoms, 1.61 kilometers, 1609 meters;

nautical, 26, **28,** distance measurement used by ships at sea or airplanes, where one nautical mile equals one minute of *latitude* and one minute of *longitude* along the equator. One nautical mile equals 1.15 statute mile or 1.85 kilometers. There are 69 statute miles for every 60 nautical miles.

Millibar, 102n, an international unit for measurement of *air pressure,* where the pressure needed to support a column of mercury 760 millimeters, or 29.92 inches (*standard atmosphere*) amounts to approximately 1013 millibars—or 15 pounds per square inch.

Mineralogy, 3, **8,** 274–275, the scientific study of the earth's minerals.

Minerals: inorganic (nonliving) substances formed naturally of one or more chemical *elements.* Examples of single-element minerals: gold, silver, copper, carbon. Minerals composed of two or more elements: sodium chloride (table salt), quartz (oxygen and silicon).

composition of, **274,** 274–275;

elements of, **274;**

families of, **275.**

Mississippi River, **164,** 165, 349, 359.

Mixed layer, 174, a layer of relatively warm water in an ocean that lies between the warmer waters near the surface and the very cold waters below. See: *Thermocline.*

Moho: the interface, or contact surface, between the *mantle* and the *earth's crust.*

Mohorovičić discontinuity, 272, **273,** the more correct term for the *Moho,* the contact surface between the *mantle* and the *earth's crust.* It was discovered by Andres Mohorovičić, a Yugoslavian seismologist. See: *Discontinuity.*

Moisture, 124–158, 150–185, **151,** 227–229, liquid present in small quantities. See: *Atmosphere; Dew point; Humidity; Precipitation; Water; Water vapor.*

Mojave Desert, **404, 405.**

Molleweide's map projection, **33.**

Monadnock, 347, a hill that is the remnant of a mountain to be found on a flat plain (*peneplain*) that was eroded during the *cycle of denudation* (the *Geographical Cycle,* as theorized by *William Morris Davis*).

Monsoon: a wind system that changes directions of flow with the changing of the seasons. During the summer a *low-pressure cell* is formed over land in the interior, and as the land heats, the air rises *convectionally,* drawing in cooler air from the nearby oceans. *Precipitation* usually results with the "monsoon rains." The opposite occurs during the winter season as the cool air from the land sweeps out onto the sea to replace the rising warm air.

in Bangladesh and India, 99–101, 110, **111,** 116, 199;

origin of, 99–101, 110, 199.

Moon: the rocky *satellite* that revolves about the earth in its lunar orbit.

phases of, 181–182, **182,** positions of the moon relative to its location with reference to the earth and the sun:

conjunction—when the moon is situated between the earth and the sun; thus, with the combined *gravitational* pull of the sun and the moon, *tidal ranges* are at their highest (called *spring tides*);

opposition—when the moon is situated on the opposite side of the earth, away from the sun, and is in full view (full moon), adding its *gravitational* pull to that of the sun; thus, again, *tidal ranges* are at their highest (*spring tides*);

quadrature—when the position of the moon is at right angles to the earth and the sun, and its *gravitational* pull is cancelled out by the pull of the sun. Thus, *tidal ranges* are at their lowest, creating *neap tides.*

syzygy—whenever the earth, moon, and sun are in line with each other (*conjunction* and *opposition*);

positions of, **182:**

crescent moon—when the moon appears as a crescent and is positioned halfway between *quadrature* and *conjunction;*

full moon—when the moon is in *opposition;*

gibbous moon—when the moon also appears as a crescent, as it is positioned between *quadrature* and *opposition;*

new moon—when the moon is situated between the earth and the sun, and the side we see receives no light from the sun. This occurs at *conjunction;*

quarter moons—whenever the moon is positioned halfway between *opposition* and *conjunction;*

and tides, 181–182, **182,** the *gravitational* pull of the moon on the earth's waters is enhanced whenever the moon and the sun combine their gravitational attraction (*conjunction* and *opposition*); thus the highest *tidal ranges* result (*spring tides*). When they are not in line, and the moon is in a right-angle position (*quadrature*), their gravitational pull is cancelled out and the low tidal ranges result (*neap tides*).

Moraines, 389–390, **391, 395,** glacial debris deposited by either *Alpine* or *continental glaciers:*

ground, 389, **395,** deposits left behind by a retreating ice sheet or glacier;

interlobate, 395, **395,** debris deposited between the lobes of a *continental glacier;*

lateral, 390, **391,** deposits along the sides of an Alpine glacier;

medial, 390, **391, 395,** the coalescing, combining, of the *lateral moraines* of two *Alpine glaciers* that have joined together;

recessional, 390, **395,** debris deposited at the *terminus* of a retreating glacier;

terminal, 390, **395,** deposits at the *terminus* of a glacier at the point of its farthest advance.

Motion, Newton's First Law of, Inertia, **56,** 57, see: *Inertia.*

Mountains: rugged, upraised portions of the earth's crust that have a *local relief* of more than 2000 ft (600 m):

building processes:

contraction, 290–291;

convection, 291;

diastrophism, 299–305, **301;**

orogeny, 296–298;

plate tectonics, 291ff;

vulcanism, 305–313, **305, 306, 308, 309, 310, 311;**

classification of, 314–318, **315;**

distribution of, **297.**

Mount Pelee, 305.

Mount Saint Helens, **306,** 307, 313.

Mudflow, 329, **330,** the movement down a slope by soil and rocks that have become saturated with water, thus forming a thick, viscous mixture.

Muir, John, 267, 268, 289.

Namib Desert, southwestern Africa, 103, 176, **244, 409.**

Nashville, Tennessee, 206, **207.**

Natural levees, 345, **345,** 349, see: *Levees, natural.*

Natural vegetation, 224–227, see: *Climax vegetation.*

Nautical mile, 26, **28,** see: *Mile, nautical.*

Neap tides, 181, **182,** see: *Moon and tides.*

Nebraskan (glacial) Period, 391, **393,** the earliest advance of the North American *continental glaciation* during the *Pleistocene Ice Ages.*

Neck, volcanic, 309, **310,** 312, the conduit of a *volcano* filled with *lava.* When it has been exposed by erosional forces, the exposed neck is called a *pipe* or *stock.*

Needle leaf vegetation, 235, **236,** evergreen vegetation with thin, needle-shaped leaves.

New Madrid fault zone, 302n.

Newton, Isaac, 57.

Newton's First Law of Motion, Inertia, 57, see: *Inertia.*

New Zealand, **343.**

Niagara Falls, 339.

Nicaragua, 290.

Nile River, 163–165, **164,** 252, **254,** 344. See also: *Deltas.*

Nimbus, **128,** 129, rain-bearing clouds.

Nip, a small *sea cliff.*

Nitrogen, 90, 92, an integral part of the air in earth's atmosphere. See: *Atmosphere, composition of.*

Normal fault, 299, **301,** see: *Faulting.*

Normal lapse rate, 78, the rate of change in the temperature of a still, nonmoving air mass that is 3.5°F (0.65°C) for every 1000 ft (100 m) change in altitude.

North, geographic (grid), **33,** 39–41, **39, 40, 434, 436, 439,** the direction along any *meridian* pointing toward true north:

magnetic, 39–41, **39, 40, 434, 436, 439,** the direction the needle of a *magnetic compass* would point, toward the magnetic north, presently located near Canada's Bathurst Island, 76°N latitude, 102°W longitude.

North Star (Polaris), 25, a star that appears in the sky directly over the earth's North Pole. The angle between your horizon and the North Star is your latitudinal position in the Northern Hemisphere. The North Star cannot be seen south of the *equator.*

Norway, 386, **387.**

Notch, 364, **365,** a horizontal groove cut into the base of a *sea cliff* by *wave action.*

Oasis, a fertile spot in a desert, supplied by groundwater, thereby able to support vegetation:

Souf, Algeria, **413;**

Timimoun, Algeria, **203.**

Oblate ellipsoid, 20, a sphere that is shorter in its top-to-bottom dimension than in its side-to-side dimension, such as the earth, which is shorter measured through its polar *axis* than the distance measured through the *equator.* See: *Geoid.*

Obsidian, 277, **278,** a highly lustrous, glassy *volcanic ejecta* that is black or very dark brown in color.

Occluded front, **133,** 135, a mass of warm air that has been uplifted by an advancing *cold front* and is cut off from its source and the ground.

Ocean current: the horizontal movement of ocean water, caused by differences in its density (*temperature* and *salinity*) and aided by movements of the winds above.

Oceanography, physical, 2, **7,** 356, the scientific study of the world's oceans, their currents, wave action, and sea floor landforms.

Oceans, 169–182:

distribution of, 169–172, **170, 173;**

and ice ages, 359;

influence of, on weather and climate, 174–179, **178,** 190, **197, 198,** 199, 206, 213;

level of, 359;

movements of:

countercurrents, 176, **178;**

currents, 175–179, **175, 178,** 360;

density, 172;

salinity, 172–174, **173;**

temperature, 174;

tides, 181–182, **182;**

wind waves, 179–181, **179, 180;**

sea-bottom features of, **297, 315,** 317–318, **357;**

shorelines of, 370–373, **371, 372, 373;**

surf zones of, 361, 364, **365;**

deposition, 366–370, **342, 367;**

erosion, 340, 364–365, **365, 366;**

temperatures of, 174;

as water supply, **151;**

as alternative energy resource, 184–185, see *Case Study: Energy from the Oceans.*

Offshore bar, **367,** 367–370, **368,** a ridge of sand deposited by wave action off the shore.

Offshore breezes, 109–110, **110,** winds blowing from a cool high-pressure land area out onto the ocean or lake to replace the rising, warmer air during the night. Also called *land breezes*.

Old-age stream valley, 346, 347–349, **348,** a stream valley with a very wide *floodplain*, in which the stream *meanders* sluggishly from side to side in great loops. It is normally in a state of *equilibrium*, as a *graded stream*, thus doing little in the way of vertical downcutting (*degradation*), except during times of *flood stage*.

Onshore breezes, 109–110, **110,** winds blowing from the ocean onto the land to replace the rising, warmer air during daylight hours. Also called *sea breezes*.

Orbit: the path followed by an object in its *revolution* about another object, or body, in space.

earth's, 54–64, **55, 60, 62,** the pathway followed by the earth during its annual revolution about the sun. See: *Ecliptic, plane of; Sidereal year; Tropical year; Aphelion; Perihelion;*

moon's, 55, 181, **182,** the pathway followed by the moon during its monthly revolution about the earth. With reference to the sun, it takes about 29½ days; with reference to the stars, it takes about 29⅓ days.

Orogeny, 296–298, the deformation of the earth's crust by the *tectonic* forces that create mountains and ranges.

Orographic precipitation, 130, **131, 135,** 135, *precipitation* caused by the forced uplift of an advancing warm, moist air mass by elevated land, such as a high plain or plateau, or a mountain range, when the air is chilled to its *dew point*.

Orthophotoquads, 433, **438,** aerial photographs of a specific land space which are similar to *topographic maps*.

Outwash plain, 394, **395,** deposits in front of a glacier's *terminus* that were laid down by *meltwater* streams issuing from the glacier.

Oxbow lake, 345, **346, 348,** 349, see: *Lake, oxbow.*

Oxidation, 326, chemical *decomposition* of a rock as oxygen unites with metallic *elements*. See: *Chemical weathering.*

Oxides, **275,** a compound of oxygen united with another *element*, such as silicon, which produces an oxide called quartz.

Oxygen, 90–95, a colorless, odorless gas, the commonest f all *elements*, that makes up about 21 percent of the air in the earth's *atmosphere*.

Ozone, 94, a triatomic (O_3) form of highly condensed oxygen.

Ozonosphere, 94, **95,** the ozone layer that lies midway in the *stratosphere* and protects the earth by absorbing incoming *ultraviolet radiation* from the sun.

Pacific Ocean, **170, 171, 173, 178.**

Pacific ring of fire, 307, **308,** the great number of active volcanoes that encircle the Pacific Ocean's bordering landmasses.

Pack ice, 167–169, **168,** 217, a large sheet of thick, floating ice that is composed of a number of pieces wedged together, usually found in the Arctic.

Padre Island, Texas, 369.

Pahoehoe lava, 277, **278,** 253, a smooth and billowy lava flow. See: *Lava; Basalt; Vulcanism.*

Paleomagnetism, 293, the orientation of magnetic particles within a rock as to the position of the earth's magnetic field when the rock was formed.

Pampa, 101, 213, **241,** 259, 263, **414,** the fertile, expansive *grasslands* of Argentina.

Pangaea, **269,** 271, 293–296, **294,** the ancient supercontinent that rose out of the congealing earth's crust billions of years ago and, according to *Alfred Wegener*, it broke up into the continents we now know. See: *Continental drift.*

Parabolic sand dunes, 413, the *U-shaped sand dunes* that have their "horns" pointed upwind, as contrasted to the *barchan*, which has its "horns" aimed downwind.

Parallels of latitude, see: *Latitude, parallels of.*

Parent material, 252, the weathered *regolith* (*unconsolidated* rock material) that underlies the A and B *horizons* of earth's *soils*. It is the beginning source of the soil.

Paricutin volcano, 307.

Pavement, desert, see: *Desert pavement.*

Pedalfer soils, 259, *zonal soils* of warm and moist regions that are rich in iron and aluminum.

Pedology, **3, 8,** 250, 259, the scientific study of earth's soils.

Pendulum, Foucault's, 57–59, **58.**

Peneplain, 347, according to *geomorphologist William Morris Davis*, the last stage of the *geographical cycle*, when elevated and rugged topography has been reduced to a flat plain, which he called the penultimate plain, meaning, "almost to the end."

Perihelion, 54–57, **55,** 182, that point in the earth's *orbit* about the sun when the earth is at its closest approach to the sun, 91.4 million miles (146 million kilometers), about January 3rd.

Periodicity (of vegetation), 235, the response of natural vegetation to the cycle of the seasons. While some plants never seem to lose their leaves at any time (evergreen), others lose their leaves when the temperature drops too low or there is insufficient moisture (deciduous). See: *Deciduous; Evergreen; Summergreen deciduous.*

Periods (of Geologic Time), 268–271, **269, 270,** the secondary division with *Geologic Time,* subdividing the *Eras.*

Permafrost, **196,** 213–217, **215, 218,** 229, 378, 406, ground that is frozen permanently throughout the year. In some areas, such as the northern portions of Alaska, Canada, and Siberia, in the *tundra* regions, it may be frozen as much as a mile down. Spring thaws normally melt the top 12 inches (30 cm), forming a wet, slushy condition. See: *Solifluction; Tundra.*

Permeability of rocks, 152, 152n, the capacity to allow movement of water between and through the crystals and grains of a rock.

Petrified Forest, Arizona, **279.**

Petrified wood, **279,** 280, a *sedimentary rock* formed when the woody cells of a fallen tree that has been covered by soils and other deposits over a long period of time have been replaced by the mineral silica, and the wood's original shape and form is retained.

Petrology, **3, 8,** 275, the scientific study of the earth's rocks.

Phases of the moon, 181–182, **182,** see: *Moon, phases of.*

Photogrammetry, 427–429, the use of aerial photography (by airplanes or satellites) to produce stereoscopic photographs of the land that allow earth scientists the opportunity to have a three-dimensional view of a selected portion of earth space.

Photosynthesis, 227n, the process by which plants take carbon dioxide out of the air or water and convert it into the food (carbohydrates: sugar and starch) they need.

Physical geography, 2–9, the scientific study of planet earth, its atmosphere, waters, vegetation, and landforms.

Physical oceanography, **2, 7,** 293, 356, see: *Oceanography, physical.*

Physical weathering, 322–323, **324, 325, 326, 327,** see: *Mechanical weathering.*

Piedmont: at the foot of a mountain.

Piedmont alluvial plain, 314, **341,** 340, a plain at the foot of a mountain range created by the coalescing, coming together, of a number of *alluvial fans.*

Plain, 314–318, **315, 317,** a wide expanse of flat land at low elevation [less than 500 ft (150 m)] and with a *local relief* of less than 300 ft (91 m).

Plane of the ecliptic, **60,** 61, see: *Ecliptic, plane of.*

Planets, 52, **53.**

Plant geography, **2, 8,** 224ff, the scientific study of the earth's vegetation. See: *Biogeography.*

Plants, see: *Vegetation.*

Plate tectonics, 291–296, **292, 294,** the internal forces within the earth that cause movements within the earth's crust, thus moving and thrusting the pieces of the earth's crust, *plates* one against the other. See: *Tectonics.*

Plateaus, 314–318, **315, 316,** relatively flat areas with a *local relief* of less than 300 ft (91 m) and an elevation up to 10,000 ft (3000 m) above sea level.

Plates, 291ff, **292, 294, 297,** the *earth's crust* has been fractured and broken into a number of pieces by the inner *tectonic forces* operating within the earth.

Pleistocene Epoch (of Geologic Time), **269, 270,** 271, 390–394, **392, 393,** the third stage of the *Cenozoic Era* that endured for about 1½ million years, ending about 10,000 years ago.

Pleistocene Ice Age, 74, **269,** 390–394, **392, 393,** the time when great *continental glaciers* covered much of the Northern and Southern hemispheres. See: *Glaciation, continental; Greenhouse effect.*

  interglacial periods, 74, 393, **393;**

  origin and extent of, 390–394, **392, 393;**

  stages (advances) of, 390–394, **393.**

Pliocene Epoch (of Geologic Time), **269, 270.**

Plucking, 383, as moisture from a glacier penetrates the rock wall of the glacial valley and freezes, rocks will be broken loose and pulled from the wall by the moving glacier. See: *Glacial erosion, by plucking.*

Plutonic rocks, 282, **282,** *intrusive igneous rock* that solidifies slowly underground, forming large grains. The most common plutonic rock is *granite.*

Podzolization, 257, **262,** the process of soil development that occurs in cool, humid climate regions, under a *coniferous needle-leaf forest.* These soils are rich in *humus* and are not subjected to excessive *leaching.*

Podzols, 257, **262,** 261, fertile soils developed by the podzolization process.

Pt. Barrow, Alaska, 217, **218.**

Polar easterlies, **106,** 112, cold winds that blow down from the *polar regions,* forced to veer to their right in the *Northern Hemisphere* by the *Coriolis effect,* thus appearing to come from the east.

Polar-front theory, 107, **132,** 137, developed by *Vilhelm Bjerknes,* a Norwegian *meteorologist,* during World War I to explain the constant onslaught of *cyclonic storms* (polar *air masses*) that sweep onto the continental landmasses and collide with warm air from the subtropical regions.

Polar highs, **106,** high-pressure cells that develop over the poles due to the settling of cold air that originally rose from the equatorial region.

Polaris (North Star), 25, see: *North Star.*

Polar regions: areas above 75° N latitude in the Northern Hemisphere, and below 75° S latitude in the Southern Hemisphere.

  climates of, **215,** 217, **218, 219,** 398.

Polish, glacial, 388–389, shining *intrusive* rock surfaces, such as the *granite* in *Yosemite National Park*, which have been polished to a high luster by a moving *glacier*.

Poljes, **351,** 352, see: *Uvala.*

Pollution, of the atmosphere, 79–**80:**
of the earth, 217–220.

Ponors, **351,** 352, deep, steep-sided *sinkholes* in an area of *karst topography* which has reached the stage of maturity.

Population, effect on our physical environment, **4,** 7–9, 190, 217–220:
map of the world, **4.**

Porosity: the capacity to absorb or take in moisture.

Porous rocks, 152, 152n, rocks that have a great number of spaces (pores) between crystals or grains, which allow water molecules to enter. See: *Permeability.*

Port Darwin, Australia, 201, **202.**

Portraying the earth, 13–47.

Position, determination of, 21–25, **23, 24,** 427:
by latitude, **24,** 25, 427;
by longitude, **24,** 25, 427.
See: *Geographic grid.*

Prairies: *grasslands* composed of relatively tall grasses, devoid of trees, to be found in subhumid climate regions such as the regions around the Mississippi Valley.
climates of, 213, **214;**
distribution of, **208,** 241, **262;**
soils of, **241, 260,** 261–263, **262;**
vegetation of, 213, **241,** 242.

Precambrian time, **269, 270,** 271, the expanse of earth history (*Geologic Time*) that transpired before the *Cambrian Period* of the *Paleozoic Era.*

Precipitation: water in the form of solid (snowflakes, ice crystals, sleet, hail) or liquid (rain, dew, fog) that forms in the atmosphere and falls toward the earth's surface:
causes of, 124, **125,** 129–137, **131, 132, 133, 134, 135;**
convectional, 130, **131,** 199;
frontal (cyclonic), 130–137, **131, 132, 133;**
orographic, 130, **131, 135;**
forms of, 129–**131;**
global patterns of, **134;**
record amounts of, 199.

Pressure, air, 92–93, 101–112, **104, 105, 106, 109,** see: *Air pressure; Gradient, pressure.*

Pressure gradient, see: *Gradient, pressure.*

Prime meridian, 22, **23, 24, 422,** the universally accepted *zero meridian* from which one can determine the longitudinal position of the *geographic grid.* See: *Greenwich Meridian, prime.*

Principal meridian, 443, **443, 444,** a north-south *meridian* used as the reference meridian to establish land boundaries by the United States Land Survey System.

Projections, map, see: *Map projections.*

Proterozoic Era, **269, 270,** 271, an early stage of *Geologic Time,* the span of time when living organisms first began to appear on earth.

Ptolemy, 16–17, **18,** a 2nd century mapmaker whose world map was based on an equatorial circumference that was considerably shorter than the actual distance, thus misleading Columbus into thinking he could travel from Spain to far Cathay (China) in a few weeks' time. See: *Equator.*

Public land survey system, 41, 441–444, **443.**

Pumice, 277, **278,** cellular, glassy volcanic *pyroclastic ejecta* that is light and powdery in texture.

P-waves, 302n, the primary, compressional shock waves emitted by an *earthquake* that are the first to arrive at a point some distance from the epicenter (the focal point of the earthquake).

Pyroclastic ejecta, 277, **278,** 309, **311,** fragments of *lava* that have been violently ejected from the vent of a *volcano* as bits of liquid rock, ashes, or cooling pieces of hardened lava. From *pyro* ("born of fire") and *clastic* ("fragments"). See: *Extrusive igneous rocks.*

Pythagoras, 5, 15.

Qanats, 160–162, **163,** the underground aqueducts dug by Persians in the Iranian Plateau some 4500 years ago.

Quadrangle maps, 427–440, **434–435, 436–437,** a series of maps made by the U.S. Geological Survey with the four-sided area bounded by *parallels of latitude* and *meridians of longitude.*

Quakes, see: *Earthquakes.*

Quartz, 274, **275,** a *mineral* (silicon dioxide) of a clear, almost glasslike quality. Found in a variety of colors, from a pink to pale brown. The most abundant mineral in the *lithosphere.*

Quartzite, **281,** *metamorphosed sandstone* composed primarily of *quartz.*

Quaternary Period, **270,** the most recent subdivision of the *Cenozoic Era,* of *Geologic Time* which includes the *Pleistocene* and *Recent* epochs. It has endured approximately 2 million years.

Radiation, 64, **65, 68,** 72–74, **73, 74,** the emission of *energy* in the form of waves or particles:
ground, 72, longwave radiant energy emitted from the surface of the earth;

Radiation (*continued*)
- infrared, **65**, 72, invisible heat rays emitted by the sun that are beyond the red end of the color spectrum;
- longwave, **69**, 79n, 72–**74**, energy reradiated into the atmosphere by the earth, and when it is reradiated by the atmosphere back to earth, the *Greenhouse Effect* takes place;
- shortwave, 69, 79n, 72–**74**, the *electromagnetic radiation* emitted by the sun, which includes infrared, ultraviolet, and visible light rays;
- solar, 64–70, **67, 68,** shortwave electromagnetic radiation emitted by the sun in the form of visible light, infrared, heat, ultraviolet, gamma, and X rays. See *Case Study: Solar Energy—Friend or Foe?*
- terrestrial, 72, longwave radiation emitted by the earth's surface—land or waters.

Radiation belts (Van Allen), 65–66, **67, 91,** two belts of high-energy radiation particles (protons and electrons) that encircle the earth and absorb much of the dangerous incoming *corpuscular energy*.

Radiation fog, 79, 127, fog that is produced by the contact of the bottom layer of a warm, moist air mass that moves over ground that has cooled; thus, *dew point* is reached and *condensation* occurs. See: *Fog, ground inversion*.

Radioactive dating, 268, 293, methods for assessing the age of earth materials (rocks, fossils) by determining the rate of radioactive decay of the atomic particles within. Included are: radiocarbon and radiometric dating.
- radiocarbon: as all living organisms contain large amounts of carbon, when they die they begin to lose the carbon atoms at a specific rate. By measuring the amount of carbon remaining in a fossil, one can arrive at a fairly close estimate as to the time in history the creature once lived up to 50,000 years ago.
- radiometric: this dating method is based on the type of radioactive substances within the material under question. The potassium-argon method can be used to date the age of the earth beyond 50,000 years ago.

Rainfall, annual average, 135:
- of the United States, **136**;
- of the world, **134**.

Rain forests, 135–136, **195, 197,** 237–239, **238**.

Rainshadow, 135–136, the *lee* (leeward), or downwind, side of a mountain or range that receives little, if any, precipitation.

Raisz, Erwin, 46.

Recessional moraines, 390, **395**, see: *Moraines: recessional*.

Rectangular survey System, United States, 41, 443, **443**.

Reefs, see: *Coral, barrier reefs,* and *Coral, fringing reefs*.

Reflection of solar radiation, **68**, 68–69, **69**, the bending back into space the incoming solar radiation (*insolation*) by particles within the earth's *atmosphere,* as well as by *clouds* and the earth's surface. See: *Albedo*.

Refraction, wave, 360–361, **362,** the changing of the direction of a wave movement by an alteration in the depths below, or the angle of the surface over which it moves. See: *Longshore current; Longshore transport*.

Reg, 407, **407,** 408, Arabic word for a rocky, pebbly *desert* nearly devoid of any sand grains because of blowing wind and the *deflation* process. See: *Deflation basin; Desert pavement*.

Regional approach to the study of the earth, 3.

Regolith, **255,** 256, 328, loose rock fragments overlying the bedrock. See: *Parent material*.

Regosols, 264, pebbles, gravels that make up the soil of rocky *deserts* (*regs*).

Relative humidity, 126–127, see: *Humidity, relative*.

Representative fraction (RF) map scale, **34, 430, 431,** a map scale given as a fraction, which states that one unit of distance on the map represents a specified number of the same kind of units on the earth's surface. Thus: 1/63360 means that one inch (centimeter) on the map would represent 63,360 inches (centimeters) on the actual surface of the earth.

Remote sensing, 46, **428,** the study of the earth from space, using special equipment (electromagnetic) and cameras with infrared film to detect (sense) the differences in the infrared radiation emitted by soils, vegetation, human-made features, and cool or warm ocean currents.

Reradiation, 71, 72–74, **74,** after the earth and its waters have absorbed the *shortwave radiation* issued by the sun, heat in the form of *longwave radiation* is emitted back into the *atmosphere*.

Reverse fault, 299, **301**, see: *Faulting*.

Revolution: the movement of a planetary satellite in its orbital path around a focal point, such as the earth's revolution about the sun, its focal point, of the moon's revolution about the earth. See: *Ecliptic, plane of*.

Rhumb line, 42n, a line, a section of a *great circle,* drawn on a map, that intersects all *meridians* at the same angle and can be used for navigational purposes. Also called a *loxodrome*.

Richter scale, 304, the measurement of the magnitude (intensity) of an earthquake expressed in numbers of increasing strength from 1 to 10. Each step increases in magnitude at a terrific rate. Example: a magnitude 4 earthquake is not twice as rough as a magnitude 2, but is 100 times bigger!

Rift valley, **301,** 302, a section of land between two faults that has dropped, creating a long, narrow,

INDEX AND GLOSSARY 477

trenchlike valley. Examples: Death Valley, California, the Jordan River, Israel. See: *Graben.*
Rip currents, 361.
Ripple marks, 280, the imprint of wind or wave action left on a bed of sand.
Riptide, 361.
Rivers, 162–165, **164.**
Roches moutonnées, 388, **388,** glacially rounded rocks found in a glacial trough. The term means "sheep rocks."
Rock cycle, 280–283, **282,** the constantly recurring series of changes in the composition and structure of rocks as they are pressured, heated, or weathered into small pieces.
Rockfall, 332, the falling of rocks down a steep slope and collecting at the base as *talus.*
Rock : naturally occurring substances composed of one or more minerals:
  breakup of, by exfoliation (spalling), **327,** 328;
    by granular disintegration, **326,** 328;
    by joint-block separation, **326,** 328;
    by shattering, **327,** 328;
  composition of:
    igneous, 275–277, **276, 278;**
    metamorphic, 280, **281;**
    sedimentary, 277–280, **279;**
  consolidated and unconsolidated, 275n.
Rocky Mountains, 243, 307, **312,** 379.
Ross Ice Shelf, 398.
Rostov-on-Don, **214.**
Rotation of the earth, **26,** 57–59, **58, 59,** the spinning west-to-east movement of the earth about its *axis.* One complete rotation every 24 hours.
Royal Geographical Society of Great Britain, 356.
Running water, 336ff:
  sheetwash and overland flow, 336–337;
  splash erosion, 336;
  streams and rivers, 336–349;
  as water supply, **151,** 162–166.
Runoff, stream, 165:
  infiltration into groundwater zones, 152–155, **153, 156,** 250–257, **255,** 350;
  part of the *hydrologic cycle,* **125,** 336.
Ruwenzori Mountains, 245.

Sahara Desert, 103, 172, **198,** 199, 227, 242, **244,** 252, **260, 262,** 263, **409, 413.**
St. Andrews State Park, Florida, **324.**
Salinity: the amount of salt accumulation:
  of lakes, 167;
  of oceans, 172–174, **173,** 174n;
  of seas, 172.
Saltation, 408, **408,** the movement of sand grains over the ground by the wind, or within a stream.
San Andreas Fault, **295,** 296, 299, the lengthy fault zone that stretches from the Gulf of California through southwesterly California, along the Transverse mountain ranges, then northerly near the coast, exiting the continent at the edge of San Francisco. It is the boundary line between the Pacific and North American crustal *plates.* See: *Continental drift; Plate tectonics.*
Sandbar, 366–370, **367, 368, 369,** a ridge of sand located off the shore of a coast.
Sand dunes: mounts or ridges of sand transported and deposited by the wind:
  distribution of, 408–413;
  formation of, 408–413, **410, 412, 413;**
  types of:
    barchans, 411, **412;**
    longitudinal, 412;
    self, 412;
    star, 413;
    transverse, 413;
    U-shaped (parabolic), 413;
    whalebacks, 413.
Sand shadow, 413.
Sand spit, **367, 368,** 369, a narrow sand deposit that stretches fingerlike downcurrent at the mouth of a bay or from a protrusion of land on an irregular coastline.
Sandstone, 277, **279,** a *sedimentary rock* composed of sand grains that are normally derived from weathered *granite.*
Sandy deserts (ergs):
  distribution of, **198,** 201, **244,** 408, **409;**
  formation of, 411–415;
  human imprint on, **413.**
Santa Ana winds, 75–76, 75n, **75,** the descending winds that become heated by *compression* as they move from high elevations to lower ones. See: *Chinook; Adiabatic heating.*
Sargasso Sea, 177, see: *Bermuda Triangle; Gyrals.*
Satellites: a body that orbits another larger body, such as the moon orbits planet Earth.
  meteorological, 58, **59, 428,** human-made vehicles launched into space to scientifically observe our planet and to learn more concerning our weather and climate.
Saturation, zone of, **153,** 154, the lowest of the *groundwater* zones, which is fully saturated with water. The top of this zone is the *water table.*
Saturation point, 127, the point at which an *air mass,* at a given temperature and pressure can no longer

Saturation point (*continued*)
retain the *moisture* within it as a gaseous vapor; thus *condensation* occurs. See: *Dew point*.

Savanna, 240–242, **241**, the grasslands in tropical and subtropical regions that are dotted with scattered clumps of trees. These grasses are tall and tough. Most of the savannas are to be found in *tropical wet-and-dry climate* (Aw) regions.

Scales, map, see: *Maps, scales of*.

Sclerophyllus vegetation, 235, plants (shrubs or trees) with leathery, coarse, and hardened leaves or spines, typical of the semiarid *Mediterranean (west coastal dry subtropical) climate* regions.

Sea breezes, 109–110, **110**, normally gentle breezes blowing from the sea onto the land during daylight hours. Also called: *Onshore breezes*.

Sea cave, 364, **365**, a cave situated at the base of a *sea cliff*, created by erosional action of ocean waves.

Sea cliff, 364, **365, 366, 367**, a sheer, steep face of an elevated *sea (marine) terrace* created by the erosional power of ocean waves.

Seafloor spreading, 291–296, **292, 294, 297, 309**, the moving, spreading apart of the *tectonic crustal plates* along a *mid ocean ridge*. See: *Continental drift; Plate tectonics; Mid-Atlantic ridge*.

Sea mount, 371, **372**, a *volcano* whose top is below sea level, usually flat-topped and rimmed with *coral* growth. Also known as a *guyot*.

Season, cycle of, 59–64, **60, 62**, the changing of climatic conditions from summer to fall to winter to spring, and on to summer again. This is due to the inclination of the earth's *axis* and its orbital path around our sun. See: *Ecliptic, plane of*.

Sea terraces, 364, **365**, see: *Marine terraces*.

Sedge (vegetation), 217, 227, 243, grasslike plants that have stems without joints and are typical of *tundra* vegetation.

Sedimentary rocks, 277–280, **279**:
formation of, 277;
types of:
chemical, 277, **279**;
clastic, 277, **279**;
organic, 277, **279**.

Sediments, 152n, 277, fine particles, deposited by wind or water, that are composed of silt, sand, or clay, or the skeletal remains of once-living organisms. See: *Alluvium*.

Seif sand dunes, 412, long, sharply ridged mounds of sand that lie parallel to the prevailing wind direction.

Seismic sea waves, 179, 304, 360, great ocean waves created by *diastrophic (earthquake)* movement on the floor of the ocean. Also called *tsunamis*.

Seismology, 304–305, the scientific study of *earthquakes*.

Selva, 237, the *tropical rain forest* that has little, if any, undergrowth because the heavy cover of foliage does not allow much sunlight to enter the lower levels; thus there is little in the way of shrubs or grasses situated between the trunks at ground level. See: *Jungle*.

Semple, Ellen Churchill, 6, an early proponent of the concept called *environmental determinism*, which argued that the physical environment determines the cultural destiny of any human culture group.

Sensible air temperature, 101, the degree of *temperature* felt by a human body, despite what the *thermometer* says.

7th Approximation, 259–261, **260**, the most recent soil classification system, developed in 1960 by pedologists of the Soil Conservation Service of the U.S. Department of Agriculture. It is based primarily on environmental controls.

Sextant, 25, an instrument to measure angular distances and to determine one's position of *latitude*.

Shale, 277, *sedimentary rock* composed of compacted flakes of clay or silt particles.

Shattering, of rocks, **327**, 328, the breakup of rocks into small pieces due to physical action and stresses. See: *Mechanical weathering; Disintegration*.

Sheep rocks, 388, **388**, see: *Roche moutonnéees*.

Sheetwash, 336–337, the overland flow of water down a slope.

Shelves, continental, **315**, 317, 356–360, **357, 358**, 370, the extension of a continental landmass (continent) into the ocean:
development and structure of, **357, 358**, 359–360;
edges, **315**, 356, **357**;
origin of, 359.

Shield volcano, **311**, 312, a rounded, dome-shaped *volcano* composed of the accumulation of repeated *basaltic lava* flows.

Shield, continental, 314–317, **316, 317**, a large portion of a continent that had been subjected to extreme *glacial action* during the *ice ages*, when overlying *sedimentary rock strata*, as well as *soils*, were removed by abrasive scouring, which left the pre-Cambrian *granite* exposed.

Ship Rock, New Mexico, 309.

Shorelines, classification of, **357**, 370–373:
coral, 370–372, **372**;
elevated, 296, **296**;
emerged, 296, **296**, 366, 370;
glaciated, 370, **371**;
submerged, 370;
volcanic, 372, **373**.

Shrubs: woody plants, with two or more stems branching out near the ground:
  desert, **228**, 243;
  grassland, **228, 232**.
Sial, 273, **273**, 298–299, **357**, 359, an abbreviated term for the chemical elements silicon and aluminum, which make up the *granite* landmasses, the upper portion of the *lithosphere*. Sometimes referred to as the *sialic layer*.
Siberia, 213, **214, 215**, 238, 240, 245, 257, 391, **392**.
Sidereal year, 55, the number of days (365) it takes the earth to complete its *orbit* about the sun, relative to measurement from a distant star.
Sierra Club, 268.
Sierra Nevadas, California, 302, range of uplifted *fault-block mountains (horst)*.
Silicate minerals, **275**.
Silicon, **274**, 274, **275**, the second-most abundant mineral-forming *element*.
Sill, 307, **310, 311,** a layer of *intrusive igneous rock* between two layers of sedimentary rock. Its thickness may vary from a few inches (centimeters) to many feet (meters), and it may extend horizontally for many miles (kilometers). See: *Laccolith*.
Sima, 273, **273**, 298–299, **357**, an abbreviated term for the chemical *elements* silicon and magnesium, which make up the heavy, dense, *basaltic* lower portion of the earth's *lithosphere* that underlies the continents and is the ocean floor.
Sinkholes, 350, **350, 351,** concave depressions found in regions of *limestone* that are caused by the collapsing of the roof of an underground cavern due to the *decomposition* of the limestone by the infiltration of *groundwater*. See: *Carbonation; Dolines; Chemical weathering; Karst topography*.
Slate, **281**, a fine-grained rock *metamorphosed* by extreme heat from its original form as *shale*. Because of its well-developed cleavage, it splits into smooth, thin plates.
Sleet, 379, a form of *precipitation* that occurs when raindrops fall through an area where the temperature is at or below freezing and become frozen pellets.
Slope: an inclination, or slanting of a plane, whether upward or down, as opposed to vertical or horizontal position. See: *Gradient*.
Slope, continental, **357**, 360, see: *Continental slope*.
Slopes, aspect of, 230, **230**, 322, the direction or mountain slope is facing with respect to the sun.
Slump, **331**, 332, the vertical or inclined drop of earth materials down a *slope* due to the pull of *gravity*.
Small circles, **21**, 21–22, circles drawn on a sphere that do not divide the sphere into two equal halves. All *parallels of latitude* (except the *equator*) are small circles. See: *Circles, great and small; Great Circles*.
Small-scale map, 34–37, **36,** a map that shows a small amount of detail but covers a large amount of earth space.
Smog, 79–80, **80,** the pollution rising over heavily urbanized or industrialized areas, which is a mixture of smoke, carbon monoxide, and other pollutants, creating a hazy, foglike gloom. Most often occurs at times of *temperature inversions*.
Snow, 379–382, **380, 381,** a form of *precipitation* consisting of fine ice crystals, or large, feathery snowflakes, which are the result of below-freezing *temperature* during *condensation*.
Snow line, 379, the elevation on a hill or mountain above which snow can accumulate and remain throughout the year. It is the lowest elevation at which a permanent snowfield or pack can be perpetuated. See: *Timberline*.
Sodcover, 231, the upper layer of *soil* that is tightly knit because of extensive root systems of the vegetation.
Soil, 248–265, a mixture of rock materials and organic materials, as well as water, gases, and air.
  classification of:
    1938 USDA, 261–264, **262, 263;**
    7th Approximation, 259–261, **260;**
  colors, 257–258;
  creep, 328;
  formation of:
    controls on, 250–252;
      climate, 150, 199, 250–251;
      parent material, 251–252;
    terrain, 229–230, 252;
    time, 252, **253;**
    vegetation, 251–252;
  processes of, 257–259;
    calcification, 259;
    laterization, 259;
    podzolization, 257.
Soil groups, 259–264, **260, 262**.
Soil horizons, 252–257, **254, 255, 256,** the several horizontal layers of a soil, each of which has a distinctive structure, color, and texture.
Soil moisture, 150–152, **151,** 227–229, **228**.
Soil orders (of the 1938 USDA Classification), 261–264, **262, 263:**
  azonal, 261–264, **263;**
  intrazonal, 261–264;
  zonal, 261–263.
Soil profile, 252–257, **254, 255. 256**.

Soil structure, 257, **258,** the alignment of the soil particles—as grains (*granular*), thin sheets (*platy*), or irregular pieces (*blocky*).

Soil texture, 152, 257, the composition of the soil horizons, which vary from the smallest particles, clay, to the middle-sized, silt, and the largest grains—sand.

Soil water, zone of, 150–155, **153,** 250, the topmost layer of the *groundwater* zones, normally within a few inches (cm) of the earth's surface.

Solar energy: the electromagnetic waves of *energy* emitted by radiation from the sun. See: *Case Study: Solar Energy—Friend or Foe?*
and the atmosphere, 64–67, **65, 67;**
and the heat balance, 66–70, **68, 69;**
radiation, 64–66, **65,** 81–85.

Solar system, 52–54, **53,** the group of celestial bodies, consisting of the planets, satellites (like the various moons), meteors, meteorites, and the sun, which exist in our galaxy.

Solifluction, 328, **331,** the slow movement down a slope of rocks and soils in a *tundra,* or *permafrost,* region, resulting from the spring thawing of the upper layers of frozen ground.

Solstices, 59–64, **60,** 61n, **62, 63, 72,** the "time of the sun," when the sun's rays are vertical over the northernmost or southernmost distance from the equator.
summer, 59–64, **60,** 61n, **62, 72,** in the Northern Hemisphere, that point in time when the sun's vertical rays are directly hitting the earth at the *Tropic of Cancer* (23°30′N) as the North Pole is tilted towards the sun. This occurs about June 21 and is the Winter Solstice in the Southern Hemisphere;
winter, 59–64, **60,** 61n, **62, 72,** in the Northern Hemisphere, the time when the sun's vertical rays are hitting the earth directly over the *Tropic of Capricorn* (23°30′S) as the North Pole is tilted away from the sun. This occurs about September 21, and is the Summer Solstice for the Southern Hemisphere.

Solum, **255,** the "true soil," a combination of the *A* and *B horizons* of a soil.

Solution, 326, the process by which moisture enters a rock and dissolves the chemicals that bind the rock particles together, thus causing the rock to decompose and break apart. See: *Chemical weathering; Decomposition.*

Sonoran Desert, 229, 243, **244, 262.**

Sorted and unsorted rocks, 277.

Spalling, **327,** 328, see: *Exfoliation.*

Spanish land grants in California, **442.**

Spit, **367, 368,** 369, see: *Sand spits.*

Splash erosion, 336, erosion caused by the impact of falling raindrops on loose soil grains.

Springs, 155, **156.**

Spring tides, 181–182, **182,** see: *Moon, and tides.*

Specific humidity, 126, see: *Humidity, specific.*

Stability, air mass, 126, when a *temperature inversion* exists and the *lapse rate* is lower than normal, there will be little, if any *convectional* uplift of warm air, and the *weather* will be calm and stable. See: *Instability.*

Stacks, 364, **365,** columns of resistant rock left standing alone and separated from the nearby *sea cliff.*

Stalactites, 352, **352,** deposits of *calcium carbonate* hanging from a ceiling of a *limestone cavern* that have the appearance of pointing fingers.

Stalagmites, 352, **352,** deposits of *calcium carbonate* on the floor of a *limstone cavern,* made by the building up of a columnlike mound as the moisture evaporates, leaving the minerals behind.

Star-shaped dune, 413, see: *Sand dunes.*

Stated (veral) map scale, **34,** 34–37, 429, the ratio of the distance on a map corresponding to the actual earthspace distance it represents in words, such as: "One inch (2.5 cm) represents 100 miles (160 km)."

Static inversion, see: inversion, static and dynamic.

Steppes, **191, 192, 208,** 213, **214, 241,** 242, **242,** 259, **260,** 261–263, 415, the *grasslands* of the *midlatitudes,* such as the Ukraine, which are usually treeless.

Storms, 137–145:
hurricanes, 141–145, **144;**
sand, 404, 406;
thunder, 137–140, **140;**
tornadoes, 140–141, **143;**
tropical cyclones, **134,** 141–145, **144.**

Stony desert, 407, **407,** see: *Reg; Deflation basin; Desert pavement.*

Stoss side, 387–389, **388,** the smooth, somewhat polished side of a rock mound (*roches moutonnées*) abraded by a moving glacier as it flows over it. The opposite, lee, side is then plucked of loosened rock materials and is much rougher.

Strata, 280, distinctive layers or beds, of rocks. The singular term is *stratum.*

Stratosphere, **91, 93,** 94, **95**–**96,** the uppermost layer of the lower atmosphere, separated from the lowest layer, the *troposphere,* by the *tropopause* at about 4 miles (6 km) over the poles and 10–12 miles (16–32 km) over the *equator.* Temperatures are very low and there are no clouds. Its upper boundary

INDEX AND GLOSSARY **481**

is the *stratopause*, about 30 miles (47 km) above the earth.
Stratocumulus clouds, **128,** 129, a broad layer of cumulus clouds.
Stratus clouds, **128,** 129, a thin layer of grayish clouds that seem to cover the sky from horizon to horizon.
Streams, agrading by, 340–345:
 base level of, 337–338, **338;**
 capacity of, 336–338;
 degrading by, 337, 347–349, **347, 348;**
 deposition by, 340–345, **341, 342, 343, 344;**
 depositional landforms of, 340–345, **341, 342, 343, 344;**
 drainage area of, 349;
 equilibrium of, 338, 349;
 erosion by, 336–340
 erosional landforms of:
  incised meanders, 339, **339;**
  stream-cut terraces, 340;
  valleys, 345–349, **347, 348;**
  waterfalls, **334,** 338–339;
 floodplains of, 344–345, **345, 346;**
 graded, 338, **338;**
 rejuvenated, 347;
 transport by, 340;
 velocity of, 337;
 as water supply, 162–167, **166.**
Stream-valley development, 345–349:
 stages of:
  maturity, **348,** 349;
  old age, **348,** 349;
  youth, 347–349, **347.**
Striations (glacial), 388, **389,** long scratches ground into a rock surface by rock fragments carried by a moving glacier.
Strike-slip fault, **295,** 299, **301,** when the movement in a *fault* is primarily a horizontal displacement.
Structural hills, 314, hills formed by *tectonic forces*, such as *diastrophism* or *vulcanism*.
Sub: below, underneath, or "away from."
Subarctic: below or south of the Arctic region.
Sublimation, 125–126, 380, **381,** the change in form of water from a solid state, as in ice or snow, to a gaseous state (vapor), without going through the liquid state.
Submarine canyons, **357,** 360, canyons (deep valleys) on the floor of the ocean or in the *continental slope*.
Subpolar lows, 103–107, **106,** the belt of *low-pressure cells* located at the points of confrontation between the cold polar *air masses* and warmer subtropical air masses. See: *Polar front*.
Subsoil, **255,** 255–256, the layer of soil beneath the *topsoil*. The B horizon. It is the *zone of accumulation (illuviation)* of nutrients removed from the topsoil by *gravitational* movement of penetrating waters.
Subtropical climate regions, see: *Climates of the world*.
Subtropical highs, 103, **106,** 205–206, a series of *high-pressure cells* located near the 30° parallels, north and south, the result of subsidence of air that has cooled in the upper reaches of the troposphere. The winds in the Northern Hemisphere "highs" rotate clockwise and are the source of the *prevailing westerlies* in the subtropics and the *trade winds* in the tropical regions. See: *Air pressure; Horse Latitudes*.
Succession (of plant systems), 226–227, the gradual change of the various forms of plant life in a specific area or region that leads ultimately to the *climax vegetation*.
Succulents, 235, **236,** plants that have thick, spongy leaves and are able to contain a considerable amount of moisture and avoid losing it, a necessity for plant life in arid regions.
Sudan, 199–201, **241,** 242, the broad savanna grassland that stretches across Africa south of the Sahara Desert. It is now experiencing serious problems of drought and *desertification*.
Sulfur and sulfides, **275.**
Sun, 52–54, **53, 55, 60,** 64–70, **65, 68.**
Sunrise and sunset, 57, the apparent rise of the sun above the eastern horizon, and the setting of the sun beyond the western horizon, which is, actually, the rotation of the earth from west to east. Thus, as we move toward the sun, it seems to rise, and as we move away from it, it appears to set.
Surface runoff, 165, movement of water across the land, usually from high to lower elevations as it is pulled by *gravity*.
Surf zone, **362,** 364, **365,** the area along a *shoreline* that is subjected to the action of breaking *waves* and the *longshore current*.
 deposition, 366–370, **367;**
 erosion, **362,** 364–365, **365;**
  corrasion, 364;
  hydraulic action, 364.
Swamps, 230, **233,** low-lying land overly saturated with water, and with vegetation, such as *hydrophytes*, that can survive only when it can live in water.
Swash, 361, the movement of an ocean wave up a beachfront.
S-waves, 302n, the swaying, almost rolling motion that

S-waves (*continued*) occurs after the initial shock waves (*P-waves*) produced by an *earthquake* have passed through the affected region.

Symbols, of air masses, 114–116, **115, 132.**

of map legends, **35,** 37, **38, 432.**

Syncline, 299, **300,** a downfold of rock strata, where the center of the fold is at the bottom of a troughlike depression.

Systematic geography, 3, the study of the earth's physical environment system by system or topic by topic, as compared to the *regional approach.*

Systems, earth, 2–10.

Syzygy, 181, **182,** that point in time of the moon's orbit about the earth, when the earth, sun, and moon are in line. See: *Conjunction; Opposition; Moon, phases of.*

Tablelands, 314, see: *Plateaus.*

Tahitian concept of the earth, 15.

Taiga, 213, **216, 238, 240,** 245, 257, 261, the vast *coniferous forest* located in the *high latitudes* between the *tundra* on the north and the *humid continental* regions to the south. See: *Boreal; Continental subarctic climate.*

Talus, 332, the rock fragments that collect at the foot of a cliff, or on a steep slope, the result of *rockfall.*

Tangency, line of, 42, **43, 44,** the line on the surface of a globe, where a cylinder or cone of clear plastic touches the globe, in the process of *map making.* At this point, there is the least amount of *distortion.*

Tarn, **384,** 385, a small lake left behind in a *cirque* after the *glacier* has melted and disappeared.

Tectonic forces, 290ff, see: *Diastrophism; Vulcanism.*

Tectonics: the internal forces within the earth that cause crustal movements (*diastrophism* and *vulcanism*).

Tectonics, plate, 290ff, **292, 294,** the movement of the large crustal plates of the earth's *lithosphere,* brought about by the spreading of the seafloor and the drifting of the continental landmasses, as the result of heat rising from the earth's interior through *convectional* movement. See: *Continental drift; Seafloor spreading.*

Tectonism, 290ff:

causes of, 290ff;

forces of:

diastrophism, 299–305, **300, 301, 303;**

vulcanism, 305–313, **305, 306, 308, 309, 310, 311, 313;**

theories of the ancients about, 290.

Temperate zone, 6, 189, the middle latitude regions of the earth, as theorized by the ancient Greek philosopher *Aristotle,* situated between the icy Frigid Zone of the northern regions and the hot Torrid Zone of the south.

Temperature: the amount of heat expressed in degrees *Fahrenheit* or *Celsius (centigrade),* determined by a *thermometer.*

of the atmosphere, 94, 95, 101, **95;**

of the oceans, 174.

Temperature gradient, 99, see: *Gradient, temperature.*

Temperature, record extremes of, 101, 201, 398.

Temperature inversion, 78–80, **80,** see: *Inversion, temperature.*

Terminal moraine, **381,** 390, **395,** see: *Moraines, terminal.*

Terminator, **60,** 61, see: *Circle of Illumination.*

Terminus (of glacier), **381, 383,** 389, 390, the foot, or front edge, of a glacier.

Terrace: a relatively flat surface with a usually sharp, steep face, composed of alluvial sediments, that borders a stream, lake, or ocean.

alluvial, 345;

marine (sea), **296,** 364–365, **365;**

stream-cut, 340, **348;**

wave-built, 365, **365.**

Terrestrial radiation: see, *Radiation, ground.*

Tethys, Sea of, **294,** 298.

Texas, 369, **412.**

Thallophytes, **232, 233,** 234, the simplest form of plant life—single-celled organisms with little, if any, differentiation between leaves, stems, or roots.

Thar (Indian) Desert, 103, 201, **244,** 263.

Thawing: the melting (liquifying) of ice or snow due to an increase in temperature above the freezing point.

of *permafrost,* **196,** 217, **218,** 229, 378, see: *Solifluction.*

Thermal: relating to heat.

Thermocline, 174, a region in a body of water where there is a rapid change in temperature vertically.

Thermosphere, **95,** 96, the upper atmosphere where the temperature increases with the altitude at a steady rate. Includes both the *ionosphere* and the *exosphere.*

Thunderstorms, 137–140, **140,** storms formed by very strong upward (*convectional*) wind currents, creating large *cumulonimbus clouds,* from which comes heavy *precipitation* (rain and, occasionally, hail) and flashes of lightning.

Tidal bore, 361–364, **363,** see: *Bore, tidal.*

Tidal currents, 361–364, **363,** the incoming and outgoing movement of ocean water in a river *estuary* caused by the changing of the *tides* from high to low.

INDEX AND GLOSSARY

Tidal range, 181–182, 381, the difference between the level of sea water at high tide and low tide. See: *Moon, and tides*.

Tidal waves, 181, 361–364, **363**, waves created by an incoming high tide into an *estuary* which has a *baselevel* roughly equal to the level of the ocean at low tide. The term is often misused as a label for a *seismic sea wave*, or *tsunami*.

Tides, 181–182, **182**, 361–364, **363**, the rising and lowering of the level of the sea caused by the gravitational pull of the moon on the waters of the earth as it moves in orbit. See: *Moon, tides; Neap tides; Spring tides*.

Tien Shan Mountains, 382, **385**.

Till plain, 389, 389n, **395**, a flat plain scoured by a glacier and covered with glacial deposits after the glacier has receded. See: *Moraines*.

Timberline, 229, 245, 379, the line of elevation above sea level beyond which trees will not grow. The elevation of the line depends on local conditions and will be at a higher elevation in tropical regions, and close to sea level in the higher latitudes. It is sometimes called the *Tree line*.

Time: a specific moment, a measurable period of duration, involving a sequence of processes or events:

   Geologic, 268–271, **269, 270;**

   and Lord Rutherford, 268;

   length of day, 25, **26,** 55, with respect to the sun (a tropical or solar day), 24 hours; with respect to the stars (sidereal day), 23 hours, 56 minutes, 45 seconds;

   length of year, 54–55, **55**, see: *Sidereal year; Tropical (solar) year*.

   local, 25, 422–423;

   and longitude, 25–27, **26,** 422–425, 422, 423, 424;

   and the Sumerians, 55.

Time zones, 422–425, **422, 423, 424,** see: Appendix A: Canada and the United States, **424,** 425;

   world, 422–425, **422**.

Tombolo, **367,** 370, a ridge of sand connecting an *offshore sandbar* or island to the mainland, resulting from the sandbars disruption of the *longshore current*.

Topographic map, 426–440, **430, 431, 432, 433, 478–479, 480–481,** a *large-scale* map which shows a considerable amount of detail, such as differences in an area's *topography* (by using *contour lines*), intermittent streams, roads, campsites, residential areas, railways, and power lines.

Topography, 229–230, 360, the terrain, the shape of the land and its physical features.

Topsoil, 252–255, **255,** the uppermost layer of *soil*, the A *horizon*, which receives nutrients (*humus*) from plant and animal life above. It is also called the *Zone of Eluviation* ("wash out" or "robbery") because of the leaching of these nutrients by the *gravitational flow* of water down to the *subsoil*, the B *horizon*.

Tornadoes, 140–141, **142, 143,** violent, whirling storms, small in width, but extremely devastating as they move across the land.

Torrid Zone, 6, 189, the extremely hot tropical region of Africa, south of the *Mediterranean*, as theorized by the ancient Greek philosopher *Aristotle*. See: *Frigid Zone; Temperate Zone*.

Township and Range, 443, **443, 444,** the *rectangular land survey* system used in the United States to establish legal property boundaries.

Trades (trade winds), **106,** 112–114, the winds that blow towards the *Equatorial regions* from the *subtropical high-pressure belts,* and their direction across the earth's surface is influenced by the *Coriolis effect*. In the Northern Hemisphere they are called the northeast trades; in the Southern, the southeast trades. See: *Intertropical convergence zone* (ITC).

Transpiration, 125, **125,** 150, 229, the loss of moisture from vegetation through the pores in the leaves. See: *Evapotranspiration*.

Transverse dunes, 413, *sand dunes* that have a wavelike appearance and ridges that are at right angles to the prevailing wind direction.

Travertine, 280, deposits of *calcium carbonate (calcite)* that builds up in *limestone caverns*, hot springs, or *geyser* areas, that are rich in limestone.

Tree, 234ff, a large woody plant with a single trunk and many branches:

   deciduous, 235;

   evergreen, 235;

   semideciduous, 235.

Trenches (deeps), 294, **297, 309, 315,** deep, steep-sided valleys on the floor of the ocean, created by the collision of *crustal plates*, when one side is forced downward and plunges into the *asthenosphere*.

Trewartha, Glenn T., 190, 194.

Tropical air masses, 114–116, **115,** warm *air masses* generated over land and sea in the region between the *parallels of latitude* 30°N and 30°S.

Tropical climates, see: *Climates of the world*.

Tropical cyclones, **134,** 141–145, **144,** very intense *cyclonic storms*, whirling counterclockwise in the Northern Hemisphere (opposite in the Southern). They are called *hurricanes* in the Caribbean, *typhoons* in East Asia and the China Sea, *baguios* in the Philippines, and *cyclones* in the Indian Ocean.

Tropical deciduous forests, **238,** 239, forests in the

Tropical deciduous forests (*continued*)
tropical wet-and-dry (AW) *climate regions*, where leaves are lost during the dry seasons.

Tropical grasslands, **198**, 199–201, 240–242, **241**, see. *Prairies; Savannas.*

Tropical (solar) year, 54, the number of days in a year as determined by the length of time elapsed in the earth's orbit with respect to the sun: 365¼ days. See: *Time, length of year; Sidereal year.*

Tropic of Cancer, **24, 60,** 61, **64.**

Tropic of Capricorn, **24, 60,** 61, **64.**

Tropopause, **69,** 94, **95,** the dividing line between the *troposphere* and the *stratosphere.*

Troposphere, **69, 91,** 92–95, **95,** 150, the lowest layer of the earth's *atmosphere,* extending about 4 miles (6 km) above the poles, and about 12 miles (19 km) above the *equator.*

Trough, glacial, 370, **371, 384,** 385–386, **387,** the *U-shaped valley* carved by a *glacier.*

Tsunami, 179, 304, 360, the Japanese term for the powerful *seismic sea waves* generated by an *earthquake* in the ocean floor. See: *Tidal waves.*

Tuff, 277, **278,** tiny fragments of *volcanic ejecta* that look like chips of broken black glass.

Tundra, **196, 215,** 217, **218,** 242–243, **243, 262,** 263, regions with a permanently frozen *subsoil (permafrost),* and a vegetation composed of tiny plants (herbs, sedges, lichens, dwarf willow trees) that must adapt to a very short growing season.
  alpine, 245, high valleys in mountains that have a tundralike quality and appearance in soils and vegetation;
  Arctic, **215,** 217, **218, 241,** 328–329, **331,** land areas north of the 60th parallel, bordering the Arctic Ocean, that are permanently frozen (except for the top few inches, which thaw during the short summer) due to the average temperature of less than 10°F (−12°C);
  climate regions, **215,** 217;
  soils, 229, **262,** 263, dark, organic, and slimy soil during the periods of thawing;
  vegetation, **241,** 242–243, **243,** grasses, herbs, mosses, lichens, dwarf willow trees.

Turbidity current, 360, 370, the swift-moving stream of heavy, dense, silt-laden water down a *continental slope,* sometimes carving *submarine canyons* along the face of the slope.

Typhoon, **134,** 141, **144,** *tropical cyclonic storms* in the oceans off eastern Asia.

U-shaped glacial valley, 385–387, the transformation of a V-shaped stream valley by the abrasion of a moving glacier into a rounded valley that has rather vertical sides. See: *Troughts, glacial.*

U-shaped (parabolic) sand dunes, 413, see: *Parabolic sand dunes.*

Ukraine, steppes of, 213, **214, 241,** 242, **242,** 259, 263, **414,** 415.

Ultraviolet rays, 64–66, **65, 67,** 94, a portion of the sun's *electromagnetic radiation* that is beyond the visible light spectrum because its wavelength is shorter.

Unconsolidated rocks, 275n, 328, rocks composed of loose materials not permanently joined or cemented together, such as a handful of beach sand.

Universal law of gravitation, 56, **56,** see: *Gravitation, universal law of.*

Unloading, 323, the removal by the forces of *erosion* of overlying rock materials causes the expansion of the compressed *bedrock* below, and the resultant fracturing and breaking apart of the bedrock.

Unstable air mass, see: *Instability.*

United States Coast and Geodetic Survey (USCGS), 429.

United States Geological Survey (USGS), 427ff.

United States Public Land Survey System, 41, 443–444, **433, 434,** see: *Land survey systems.*

Uplands, **317,** 318, regions of hilly land at higher elevations.

U.S.S.R., 210, 213, **214, 216,** 240, 242, 245, 398.

Uvala (polje), **351,** 352, the flat area of *karst topography,* where a wide depression is created when a great number of *sinkholes* have coalesced.

Upwelling, 176, the movement of a cold mass of water upward into warmer, less dense waters, as the *Benguela Current* spins off of the cold *West Wind Drift,* sweeps past southwestern Africa, and rises near 20°S latitude.

V-shaped (stream) valley, 347–349, **347,** 385.

Vadose zone, **153,** 154, the portion of the ground beneath the earth's surface and the *water table;* also called the *Zone of Aeration.*

Valley glaciers, see: *Alpine glaciers.*

Valley train: outwash deposits composed of sand and gravels on the floor of a *glacial trough.*

Van Allen Radiation Belts, 65–66, **67, 91,** see: *Radiation belts (Van Allen).*

Vancouver, British Columbia, Canada, 206–210, **209.**

Variation (magnetic), **39,** 39–41, **40,** 433–440, **439,** see: *Declination, magnetic.*

Vegetation, natural, 223ff:
  and climate, 217–220, 223, 245–266;
  climatic climax, 226–227, 227n;
  climax, 226–227, 227n;
  coverage, 235;
  cycle of, 224–226, **225;**
  distribution of, 237–245, **238, 241;**

deserts, 243, **244**;
forests, 237–240, **238**;
grasslands, 240–242, **241**;
highland, 243–245;
energy from, 246, see: *Case Study: Biofuel—Energy from Vegetation.*
environmental controls on, 227–231, **228, 230**;
air temperature, 229;
aspect, 229–230, **230**;
biological (biotic) activity, 231;
moisture, 227–229, **228**;
soils, 231;
topography, 229–230, **230**;
and humanity, 245–246;
life forms of, **232, 233**, 234–235;
and religion, 224–226, **225**, 243;
structure of, 231–236;
succession, 226–227;
vertical zones of, 245.
Veldt, 213, **241**, 242, the *savanna grasslands* of southern Africa.
Vent, **310**, the throat (pipe) of a *volcano* through which flows the *magma*.
Ventifacts, 406, sharp-edged stones or pebbles that have been shaped by the *abrasive* action of winds.
Venus (planet), 52, **53**.
Verbal (stated) map scale, 34–37, **34, 35, 36, 429, 430**, see: *Map, scales of*.
Vernal equinox, **60**, 61, **62**.
Vertical rays, 59–64, **64, 72, 73**, sun's *radiation* which strikes the earth and its atmosphere at a 90° angle.
Visible light, 64–67, **65, 69**, that portion of the *electromagnetic shortwave radiation* by the sun which has a longer *wavelength* than the other components of the spectrum; thus it can be seen.
Volcanic ash, 276–277, 309, **311, 313**, very fine particles of *volcanic ejecta* blown from a volcano's *vent* during an eruption.
Volcanic bombs, 277, **278**, rounded pieces of *lava* blown from a *volcano* in liquid form then congeal rapidly as they move through the air. See: *Ejecta; Obsidian*.
Volcanic eruptions, 275–277, 290, 305–313, **305, 306, 308, 309, 310, 311, 312, 313**, the passive or violent ejection of gaseous, solid (*ejecta*), or liquid (*lava*) materials from a *volcano* onto the earth's surface. This form of *vulcanism* is termed *extrusive*.
Volcanic islands, 307, **308, 309, 373**.
Volcanic neck, 309, **310, 311**, the remnant of solidified *lava* that occupies the *vent* of a *volcano*, and has been exposed by *weathering* and *erosion* of the overlying layers of *extrusive igneous rock*.

Volcanic shorelines, 372, **373**.
Volcanoes:
causes of, 305–313;
distribution of, 305–307, **308**;
famous eruptions of, 305–307, **305, 306**;
types of, 307–309, **309**, 311, 312.
Vulcanism, 305–313, the forms of volcanic activity attributed to the movement of molten rock (*magma*) beneath the earth's surface (*intrusive*) and onto the surface (*extrusive*).
ancient theories of, 290;
extrusive, **305, 306, 309**, 309–313, **311, 313, 373**;
intrusive, 307–309, **310, 312**;

Wadis, 229, 408, an Arabic term *wādiy*, for stream valleys in *arid regions* that are only intermittently filled with running water. See: *Gully*.
Warm front, **132**, 132–135, **133**, see: *Fronts, warm*.
Water, 124–145, 149–185, **151**:
atmospheric, 150, **151**;
earth's supply of, **151**;
ground, 151, 152–162, 168, **156, 157**;
ice caps and glaciers, **151**, 167–169, **168, 169**;
lakes, **151**, 162–169, **164, 166**;
oceans, **151**, 169–183, **170, 171, 173**;
soil, 150–152, **151**, 250 ff;
streams and rivers, **151**, 162–167, **164**.
Water balance (budget), 124–125, **150**, 245, the ratio between the receipt and loss of *moisture* by a region of land, water, or vegetation.
Waterfalls, **334**, 338.
Waterlogged soils, 250–251, **251**, 264, soils in poor drainage areas where the *zone of saturation* has risen close to the earth's surface and the roots of plants are denied the air they need. See: *Hydromorphic*.
Water table, **153**, 154, **156**, the top, or upper boundary, of the *zone of saturation*. It serves as an *interface*, the boundary line between the *vadose zone* and the *zone of saturation*.
Water vapor, 124, **125**, 150, water in gaseous form.
Wave action, 179–181, **180**, 360–364, **362, 363, 365, 366, 367, 368, 369**.
Wave-built terrace, 365, **365**, the raised portion of a shoreline constructed by wave-deposited materials.
Wave-cut bench, 365, **365**, the sloping face of a beach scoured by the abrasive action of incoming and outgoing ocean waves. Also called an *abrasion platform* wave-cut terrace.
Wave deposition, 366–370, **367, 368, 369**.
Wave erosion, **362**, 364–365, **365**.

Wavelength: the distance from wave crest to the next wave crest in any form of wave action: ocean waves or electromagnetic shortwave and longwave radiation.

Wave motion, 179–181, **180.**

Wave refraction, 360–361, **362,** see: *Refraction, wave; Longshore current.*

Waves, seismic sea, 179, 304, 360, see: *Tsunamis.*
   tidal, 181, 361–364, **363;**
   tsunamis, 179, 304, 360;
   wind, 179–181, **180,** 360, **362.**

Weather, 90, the condition or state of the atmosphere at a given place for a short period of time. See: *Climate.*

Weathering, 322–328, **324, 325, 326, 327,** the *decomposition* or *disintegration* of rock materials in or on the earth's surface. See: *Chemical weathering; Mechanical weathering; Physical weathering.*

Weather (forecast) maps, **138, 139.**

Wegener, Alfred, 293.

Wells, 155–158:
   artesian, 155–158, **156, 157,** see: *Artesian wells.*

Westerlies, **106,** 112–114, the wind system of the mid-latitudes, where the winds blow from west to east, poleward from the *subtropical belt of high pressure.* Also called the *prevailing westerlies.*

Wet adiabatic lapse rate, **77,** 78, see: *Adiabatic processes.*

Whaleback, 413, a large rounded mound of sand often found in sandy deserts.

Wind: the horizontal movement of air within the *troposphere:*
   controls:
      Coriolis effect, 112–114, **113;**
      friction, 112;
      land and sea breezes, 109–110, **110;**
      pressure gradient, 108–112, **109;**
   deposition, 408–415, see: *Sand dunes.*
   erosion, 406–408;
      abrasion, 406;
      deflation, 406–407, **407;**
      landforms of, **407,** 410, 411, 413;
   systems, **106,** 108, 112–14;
   terminology, 112n;
   transportation, 407, **408;**
      saltation, 407, **408;**
      surface creep, 407.

Winds, 94, 108–114:
   adiabatic, 76–78, **77;**
   easterlies, **106,** 112;
   global, **106,** 108–114;
   jet streams, 94, 116–119, **117, 118;**
   local, 109, **110;**
   and the pressure gradient, 108–112, **109;**
   surface, 112;
   trade, **106,** 108, 112n;
   upper, 94;
   westerlies, **106,** 112.

Wind chill factor, 101, **102,** the lowering of the *sensible temperature* due to the blowing of the wind. This increases an individual's sensitivity to the temperature.

Windward: the side which bears the brunt of oncoming winds.
   coasts, 135–136, **209,** 210.

Winter solstice, 59–64, **60,** 61n, **62, 72.**

Wisconsin (glacial) Period, 393, **393,** the final advance of the *continental glaciers,* which occurred during the *Pleistocene Ice Age* in North America.

Xerophytes, **228,** 231, vegetation, of *arid regions* where little *moisture* is available; thus the plants must be drought-resistant.

X-rays, 64–66, **65,** a portion of the sun's *electromagnetic radiation* at the high end of the spectrum.

Year, length of, see: *Time, length of year.*

Yellow River, China, 397, **414,** 415.

Yellowstone National Park, 159, **159,** 313.

Yosemite National Park, 227, **266,** 267, 268, **327.**

Zenith: the point in the sky directly over an individual's head.

Zero meridian, 22–25, **26,** 422–425, **422.**

Zonal soils, **254,** 261, **262,** soils of the *1938 USDA Soil Classification System* which have well-developed *soil horizons.*

Zones of:
   ablation (glacier), 380, **381,** 383;
   accumulation:
      in glaciers, 379–380, **380, 381;**
      in soils, **255,** 256;
   aeration, **153–**154;
   eluviation, 255, **255;**
   illuviation, **255,** 256;
   saturation, 153, 154;
   soil water, 150–152, **151;**
   time, 423–425, **422, 423, 424.**